Student's Solutions Manual

Volume 1

to Accompany Swokowski's

Calculus: The Classic Edition

Jeffery A. Cole
Anoka-Ramsey Community College

Gary K. Rockswold
Minnesota State University, Mankato

Brooks/Cole
Thomson Learning.

Australia • Canada • Mexico • Singapore • Spain • United Kingdom • United States

Sponsoring Editor: *Gary Ostedt*
Assistant Editor: *Carol Ann Benedict*
Production Editor: *Scott Brearton*
Cover Design: *Vernon Boes*
Marketing Team: *Karin Sandberg, Beth Kroenke*
Print Buyer: *Tracy Brown*
Cover Printing: *Maple-Vail Book Mfg. Group*
Printing and Binding: *Maple-Vail Book Mfg. Group*

For more information about this or any other Brooks/Cole products, contact:
BROOKS/COLE
511 Forest Lodge Road
Pacific Grove, CA 93950 USA
www.brookscole.com
1-800-423-0563 (Thomson Learning Academic Resource Center)

For permission to use material from this work, contact us by
Web: www.thomsonrights.com
fax: 1-800-730-2215
phone: 1-800-730-2214

Printed in the United States of America

5   4

ISBN 0-534-38273-8

PREFACE

This *Student's Solutions Manual* contains selected solutions and strategies for solving typical exercises in Chapters 1 through 11 of the text, *Calculus: The Classic Edition*, by Earl W. Swokowski. No particular problem pattern has been followed in this manual, other than to include solutions to approximately two-thirds of the odd-numbered problems in each section, with extra attention paid to the applied "word" problems. We have tried to illustrate enough solutions so that the student will be able to obtain an understanding of all types of problems in each section.

This manual is *not* intended to be a substitute for regular class attendance. However, a significant number of today's students are involved in various outside activities, and find it difficult, if not impossible, to attend all class sessions. This manual should help supplement the needs of these students. In addition, it is our hope that this manual's solutions will enhance the understanding of all readers of the material and provide insights to solving other exercises.

We have included extra figures when appropriate to enhance a solution. All figures are new and have been plotted using computer software to provide a high degree of precision.

We would appreciate any feedback concerning errors, solution correctness, solution style, or manual style. These and any other comments may be sent directly to us or in care of the publisher.

We would like to thank: Editor Dave Geggis, for entrusting us with this project and continued support; Sally Lifland and Gail Magin of Lifland, et al., Bookmakers, for assembling the final manuscript; and George and Bryan Morris, for preparing the new figures. We dedicate this book to Earl Swokowski, and thank him for his confidence in us and for being a great mentor.

Jeffery A. Cole
Anoka-Ramsey Community College
11200 Mississippi Blvd. NW
Coon Rapids, MN 55433

Gary K. Rockswold
Minnesota State University, Mankato
P.O. Box 41
Mankato, MN 56002

To the Student

This is a text supplement and should be read along *with* the text. Read all exercise solutions in this manual since explanations of concepts are given and then these concepts appear in subsequent solutions. We do not usually review all concepts necessary to solve a particular problem. If you are having difficulty with a previously covered concept, look back to the section where it was covered for more complete help. The writing style we have used in this manual reflects the way we explain concepts to our own students. It is not as mathematically precise as that of the text, but our students have found that these explanations help them understand difficult concepts with ease.

The most common complaint about solutions manuals that we receive from our students is that there are not enough exercise solutions in them. We believe there is a sufficient number of solutions in this manual, with about one-third of the exercises solved in every section—all of them are odd-numbered exercises.

Lengthier explanations and more steps are given for the more difficult problems. We have included additional intuitive information that our students have found helpful—see page 32. We have followed the guidelines given in the text for some solutions—see page 98.

In the review sections, the solutions are somewhat abbreviated since more detailed solutions were given in previous sections. However, this is not true for the word problems in these sections since they are unique. In easier groups of exercises, representative solutions are shown. Occasionally, alternate solutions are also given. When possible, we tried to make each piece of art with the same scale to show a realisitic and consistent graph.

This manual was done using EXP: *The Scientific Word Processor*. There are some limitations to the number of boxes and characters per line, which sometimes causes the appearance of an "incomplete" line. We have used a variety of display formats for the mathematical equations, including centering, vertical alignment, and flushing text to the right. We hope that these make reading and comprehending the material easier for you.

Notations

The following notations are used in the manual.

Note: Notes to the student pertaining to hints on solutions, common mistakes, or conventions to follow.

{ }	{ comments to the reader are in braces }
LHS, RHS	{ Left Hand Side, Right Hand Side – used for identities }
$\Rightarrow$	{ implies, next equation, logically follows }
$\Leftrightarrow$	{ if and only if, is equivalent to }
•	{ bullet, used to separate problem statement from solution or explanation }
★	{ used to identify the answer to the problem }
§	{ section references }
$\forall$	{ For all, i.e., $\forall x$ means "for all x". }
$\mathbb{R} - \{a\}$	{ The set of all real numbers except a. }
$\therefore$	{ therefore }

The following notations are defined in the manual, but also listed here for convenience.

DNE	{ Does Not Exist }
L, I, S	{ the original limit, integral, or series }
T, S	{ the result is obtained from using the trapezoidal rule or Simpson's rule }
$\overset{A}{=}$	{ integration by parts has been applied— the parts are defined following the solution }
$\{\frac{\infty}{\infty}\}$, $\{\frac{0}{0}\}$	{ L'Hôpital's rule is applied when this symbol appears }
C, D	{ converges or convergent, diverges or divergent }
AC, CC	{ absolutely convergent, conditionally convergent }
DERIV	{ see notes in §11.8 and §11.9 for this notation }
↑, ↓	{ increasing, decreasing }
CN	{ critical number(s) }
PI	{ point(s) of inflection }
CU, CD	{ concave up, concave down }
MAX, MIN	{ absolute maximum or minimum }
LMAX, LMIN	{ local maximum or minimum }
VA, HA, OA	{ vertical, horizontal, or oblique asymptote }
QI, QII, QIII, QIV	{ quadrants I, II, III, IV }
NTH, INT, BCT, LCT, RAT, ROT, AST	{ various series tests: nth-term, integral, basic comparison, limit comparison, ratio, root, the alternating series }

Table of Contents

Chapter 1: Precalculus Review

Note: An informal definition of absolute value that may be helpful is

$$|\text{something}| = \begin{cases} \text{itself} & \text{if itself is positive or zero} \\ -(\text{itself}) & \text{if itself is negative} \end{cases}$$

This idea can be used for Exercises 1–8.

3 (a) Since $(4 - \pi)$ is positive, $|4 - \pi| = 4 - \pi$.

 (b) Since $(\pi - 4)$ is negative, $|\pi - 4| = -(\pi - 4) = 4 - \pi$.

 (c) Since $(\sqrt{2} - 1.5)$ is negative, $\left|\sqrt{2} - 1.5\right| = -(\sqrt{2} - 1.5) = 1.5 - \sqrt{2}$.

5 If $x < -3$, then $x + 3 < 0$. That is, the expression $x + 3$ (or $3 + x$) is negative. If you are unsure that this is true, let x equal a number less than -3, say -10, and examine the sign of $x + 3$. Now, since $3 + x < 0$, $|3 + x| = -(3 + x) = -x - 3$.

9

$$15x^2 - 12 = -8x \qquad \{\,\text{given equality}\,\}$$

$$15x^2 + 8x - 12 = 0 \qquad \{\,\text{move all terms to one side of the equals sign}\,\}$$

$$(5x + 6)(3x - 2) = 0 \qquad \{\,\text{factor}\,\}$$

From the last equation, we conclude that $5x + 6 = 0$ or $3x - 2 = 0$, or equivalently, $5x = -6$ or $3x = 2$. Solving for x yields the solutions $x = -\frac{6}{5}$ or $x = \frac{2}{3}$.

15 To solve $2x^2 - 3x - 4 = 0$, use the quadratic formula, $x = \dfrac{-b \pm \sqrt{b^2 - 4ac}}{2a}$,

with $a = 2$, $b = -3$, and $c = -4$. Thus, $x = \dfrac{-(-3) \pm \sqrt{(-3)^2 - 4(2)(-4)}}{2(2)} =$

$\dfrac{3 \pm \sqrt{41}}{4}$ or $\dfrac{3}{4} \pm \dfrac{1}{4}\sqrt{41}$. Note that $\dfrac{3 \pm \sqrt{41}}{4} \neq 3 \pm \dfrac{\sqrt{41}}{4}$.

Obtaining two-decimal-place approximations for these values,

$$\text{we have } \frac{3 + \sqrt{41}}{4} \approx 2.35 \text{ and } \frac{3 - \sqrt{41}}{4} \approx -0.85.$$

17

$$2x + 5 < 3x - 7 \qquad \{\,\text{given inequality}\,\}$$

$$-x < -12 \qquad \{\,\text{subtract } 3x \text{ and } 5 \text{ from both sides}\,\}$$

$$x > 12 \qquad \{\,\text{multiply by } -1 \text{ and change the inequality direction}\,\}$$

Remember to change the direction of the inequality when multiplying or dividing by a negative value. The last inequality is equivalent to the interval $(12, \infty)$.

$\boxed{19}$

$$3 \leq \frac{2x-3}{5} < 7 \qquad \{\text{given inequality}\}$$

$$15 \leq 2x - 3 < 35 \qquad \{\text{multiply all three expressions by 5}\}$$

$$18 \leq 2x < 38 \qquad \{\text{add 3 to each part}\}$$

$$9 \leq x < 19 \qquad \{\text{divide each part by 2}\}$$

Hence, the solutions are the numbers in the half-open interval [9, 19).

Note: For the problems using sign charts the following procedure is used.

   (1) Factor the expression into linear and/or quadratic factors.

   (2) Construct the sign chart. {The sign of a *linear* factor changes from negative to positive as x goes through values from $-\infty$ to ∞ (left to right) if the coefficient of x is positive, as in $x + 3$, $-4 + x$, $2x - 7$, etc. The sign changes from positive to negative as x goes through values from $-\infty$ to ∞ if the coefficient of x is negative, as in $-x + 3$, $4 - x$, $-2x + 5$, etc.}

   (3) Determine the regions containing the desired sign. {The sign of the region is negative if the region contains an odd number of negative signs, positive if the region contains an even number of negative signs.}

$\boxed{21}$ $x^2 - x - 6 < 0$ {factor} $\Rightarrow (x - 3)(x + 2) < 0$. The factor $x - 3$ is 0 at $x = 3$ and its sign goes from $-$ to $+$ since the coefficient of x, 1, is positive. See *Chart 21*. The factor $x + 2$ is 0 at $x = -2$ and its sign also goes from $-$ to $+$. The region from $-\infty$ to -2 has two negative signs (an even number) and hence any value in that region will make the original product, $(x - 3)(x + 2)$, greater than 0. Similarly, the region from 3 to ∞ is a "positive" region. However, we wanted the values that would make the product, $(x - 3)(x + 2)$, less than 0. These will be the values from -2 to 3, i.e., $-2 < x < 3$.

Value of x:	-2		3	
Sign of $x - 3$:	$-$	$-$	$+$
Sign of $x + 2$:	$-$	$+$	$+$

$$\boxed{\textit{Chart 21}}$$

$\boxed{25}$ Note that the $\geq$ sign indicates that we want to include the values that make the LHS (left-hand side) and RHS (right-hand side) equal, in addition to the values that satisfy the inequality.

$$x(2x + 3) \geq 5 \qquad \{\text{given inequality}\}$$

$$2x^2 + 3x - 5 \geq 0 \qquad \{\text{multiply terms and subtract 5}\}$$

$$(2x + 5)(x - 1) \geq 0 \qquad \{\text{factor and set up sign chart}\}$$

Value of x:		$-5/2$		1	
Sign of $x - 1$:		$-$		$-$	$+$
Sign of $2x + 5$:		$-$		$+$	$+$

Chart 25

From the sign chart, we see that the positive regions occur when $x < -\frac{5}{2}$ or $x > 1$. Combining these intervals with the values that make the inequality equal, $-\frac{5}{2}$ and 1, we have the union of intervals $(-\infty, -\frac{5}{2}] \cup [1, \infty)$ for our solution.

27 Note that you should <u>not</u> multiply by the factor $2x - 3$, as you may have done in the past with rational equations. This is because $2x - 3$ may be positive or negative, and multiplying by it would require solving two inequalities. This tends to be more difficult than the method presented.

$$\frac{x + 1}{2x - 3} > 2 \qquad \{\text{given inequality}\}$$

$$\frac{x + 1}{2x - 3} - 2 > 0 \qquad \{\text{set one side equal to } 0\}$$

$$\frac{x + 1 - 2(2x - 3)}{2x - 3} > 0 \qquad \{\text{combine expressions with lcd } 2x - 3\}$$

$$\frac{-3x + 7}{2x - 3} > 0 \qquad \{\text{simplify numerator}\}$$

Value of x:		$3/2$	$7/3$	
Sign of $-3x + 7$:	$+$	$+$	$-$
Sign of $2x - 3$:	$-$	$+$	$+$

Chart 27

The inequality holds true if $\frac{3}{2} < x < \frac{7}{3}$, which is the interval $(\frac{3}{2}, \frac{7}{3})$.

29 $\dfrac{1}{x - 2} \ge \dfrac{3}{x + 1} \Rightarrow \dfrac{1}{x - 2} - \dfrac{3}{x + 1} \ge 0 \Rightarrow \dfrac{1(x + 1) - 3(x - 2)}{(x - 2)(x + 1)} \ge 0 \Rightarrow$

$$\dfrac{-2x + 7}{(x - 2)(x + 1)} \ge 0 \Leftrightarrow (-\infty, -1) \cup (2, \tfrac{7}{2}]$$

Value of x:		-1	2	$7/2$	
Sign of $-2x+7$:	$+$	$+$	$+$	$-$
Sign of $x - 2$:	$-$	$-$	$+$	$+$
Sign of $x + 1$:	$-$	$+$	$+$	$+$

Chart 29

Note: It is extremely important that you understand the properties of absolute values (1.2) in the text. Another way to think of (i), (ii), and (iii) is $|a| < (>, =) b$ means that the distance that a lies from the origin is less than (greater than, equal to) b units.

$\boxed{31}$

$$|x + 3| < 0.01 \qquad \{\text{given inequality}\}$$

$$-0.01 < x + 3 < 0.01 \qquad \{\text{applying (1.2)(i)}\}$$

$$-3.01 < x < -2.99 \qquad \{\text{subtract 3 from all three expressions}\}$$

The equivalent interval notation is $(-3.01, -2.99)$.

$\boxed{33}$

$$|x + 2| \geq 0.001 \qquad \{\text{given inequality}\}$$

$$x + 2 \geq 0.001 \text{ or } x + 2 \leq -0.001 \qquad \{\text{applying (1.2)(ii)}\}$$

$$x \geq -1.999 \text{ or } x \leq -2.001 \qquad \{\text{subtract 2}\}$$

The equivalent interval notation is $(-\infty, -2.001] \cup [-1.999, \infty)$.

$\boxed{37}$

$$|6 - 5x| \leq 3 \qquad \{\text{given inequality}\}$$

$$-3 \leq 6 - 5x \leq 3 \qquad \{\text{applying (1.2)(i)}\}$$

$$-9 \leq -5x \leq -3 \qquad \{\text{subtract 6 from all three expressions}\}$$

$$\tfrac{9}{5} \geq x \geq \tfrac{3}{5} \qquad \{\text{divide by } -5, \text{ change the direction of } \textit{both} \text{ inequalities}\}$$

$$\tfrac{3}{5} \leq x \leq \tfrac{9}{5} \qquad \{\text{equivalent inequality}\}$$

The equivalent interval notation is $[\tfrac{3}{5}, \tfrac{9}{5}]$.

$\boxed{39}$ (a) $x = -2$ is the set of points that has x-coordinate -2 and any real number for the y-coordinate. It is the line parallel to the y-axis that intersects the x-axis at $(-2, 0)$.

(b) $y = 3$ is the set of points that has any real number for the x-coordinate and 3 for the y-coordinate. It is the line parallel to the x-axis that intersects the y-axis at $(0, 3)$.

(c) $x \geq 0$ is the set of all points whose x-coordinate is zero or positive, and has any real number for the y-coordinate. This is the set of all points to the right of and on the y-axis.

(d) For the product xy to be positive, x and y must both be positive or both be negative. Hence, $xy > 0$ is the set of all points in quadrants I and III.

(e) $y < 0$ is the set of all points that has negative values for the y-coordinate and any real number for the x-coordinate. This is the set of all points below the x-axis.

(f) The points in this set must satisfy *both* $|x| \leq 2$ and $|y| \leq 1$. Hence, it is the set of all points within the rectangle such that $-2 \leq x \leq 2$ and $-1 \leq y \leq 1$.

$\boxed{41}$ (a) Applying the distance formula (1.4) with $A(4, -3)$ and $B(6, 2)$, we have

$$d(A, B) = \sqrt{(6 - 4)^2 + [2 - (-3)]^2} = \sqrt{4 + 25} = \sqrt{29}.$$

(b) Applying the midpoint formula (1.5) with $A(4, -3)$ and $B(6, 2)$, we have

$$M(A, B) = \left(\frac{4 + 6}{2}, \frac{-3 + 2}{2}\right) = \left(5, -\tfrac{1}{2}\right).$$

43 We need to show that the sides satisfy the Pythagorean theorem. Finding the distances, we have $d(A, B) = \sqrt{98}$, $d(B, C) = \sqrt{32}$, and $d(A, C) = \sqrt{130}$. Since $d(A, C)$ is the largest of the three values, it must be the hypotenuse, hence, we need to check if $d(A, C)^2 = d(A, B)^2 + d(B, C)^2$. Since $(\sqrt{130})^2 = (\sqrt{98})^2 + (\sqrt{32})^2$, we know $\triangle ABC$ is a right triangle. The area of a triangle is given by area $= \frac{1}{2}$(base)(height). We can use $d(B, C)$ for the base b and $d(A, B)$ for the height h. Hence, area $= \frac{1}{2}bh = \frac{1}{2}(\sqrt{32})(\sqrt{98}) = \frac{1}{2}(4\sqrt{2})(7\sqrt{2}) = \frac{1}{2}(28)(2) = 28$.

45 $y = 2x^2 - 1$ • Since we can substitute $-x$ for x in the equation and obtain an equivalent equation, we know the graph is symmetric with respect to the y-axis. We will make use of this fact when constructing our table. As in Example 5(a), we obtain a parabola.

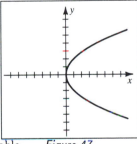

Figure 45

x	± 3	± 2	± 1	$\pm \frac{1}{\sqrt{2}}$	0
y	17	7	1	0	-1

47 From Example 5(b), the graph of $x = y^2$ is a parabola that opens to the right. The graph of $x = \frac{1}{4}y^2$ is similar to that graph, but the effect of the $\frac{1}{4}$ is to widen the parabola $x = y^2$. Since we can substitute $-y$ for y in the equation and obtain an equivalent equation, we know the graph is symmetric with respect to the x-axis. Again, we can make use of this fact to help us construct a table.

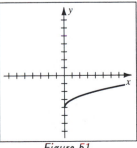

Figure 47

y	± 8	± 4	± 2	± 1	0
x	16	4	1	$\frac{1}{4}$	0

51 $y = \sqrt{x} - 4$ • Since we can take the square root of a number only if it is nonnegative, we see that x must be greater than or equal to 0. If x is less than 0, y is undefined and there is no point on the graph. The -4 has the effect of lowering the graph of $y = \sqrt{x}$ four units.

Figure 51

x	0	1	4	9	16
y	-4	-3	-2	-1	0

$\boxed{55\text{--}56}$ Solving the circle equation $x^2 + y^2 = r^2$ for y
yields $y^2 = r^2 - x^2 \Rightarrow y = \pm \sqrt{r^2 - x^2}$. The positive
radical corresponds to the upper semicircle of radius r
and the negative radical corresponds to the lower
semicircle.   Solving $x^2 + y^2 = r^2$ for x yields
$x = \pm \sqrt{r^2 - y^2}$. The positive radical corresponds to
the right semicircle of radius r and the negative
radical corresponds to the left semicircle. Hence,

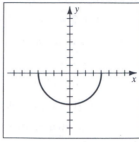

Figure 55

$y = -\sqrt{16 - x^2}$ is the lower semicircle of the circle $x^2 + y^2 = 16$. The radius is 4.

$\boxed{59}$ Tangent to both axes; center in the second quadrant; radius 4 • "Tangent to the
y-axis" means that the circle will intersect the y-axis at exactly one point. A similar
statement holds for the x-axis. The distance from the center to this point of
tangency, 4, is the length of the radius. Since the center $C(h,\ k)$ is in QII, $h = -4$
and $k = 4$.   The equation is $[x - (-4)]^2 + [y - 4]^2 = 4^2$, or equivalently,
$(x + 4)^2 + (y - 4)^2 = 16$.

$\boxed{63}$ x-intercept 4; y-intercept -3 • These are the points $A(4, 0)$ and $B(0, -3)$.

The slope between A and B is $m_{AB} = \dfrac{-3 - 0}{0 - 4} = \dfrac{-3}{-4} = \dfrac{3}{4}$.

Using the slope-intercept form, we have $m_{AB} = \frac{3}{4}$ and $b = -3 \Rightarrow y = \frac{3}{4}x - 3$.

Equivalent equations are $4y = 3x - 12$ and $3x - 4y = 12$.

$\boxed{65}$ Through $A(2, -4)$; parallel to the line $5x - 2y = 4$ •

$5x - 2y = 4 \Leftrightarrow y = \frac{5}{2}x - 2$. The slope of the given line is $\frac{5}{2}$. Using the same slope
and the point-slope form, $y - (-4) = \frac{5}{2}(x - 2) \Rightarrow 2(y + 4) = 5(x - 2) \Rightarrow$
$2y + 8 = 5x - 10 \Rightarrow 5x - 2y = 18$.

$\boxed{67}$ The perpendicular bisector of AB is the line that passes through the midpoint of
segment AB and intersects segment AB at a right angle. $M(A,\ B) = (\frac{1}{2},\ \frac{5}{2})$ and
$m_{AB} = -\frac{7}{5}$. We use the negative reciprocal of $-\frac{7}{5}$, namely $\frac{5}{7}$, for the slope of the
required line. Using the point-slope form with midpoint M yields $y - \frac{5}{2} = \frac{5}{7}(x - \frac{1}{2})$
$\Rightarrow 7(y - \frac{5}{2}) = 5(x - \frac{1}{2}) \Rightarrow 7y - \frac{35}{2} = 5x - \frac{5}{2} \Rightarrow 5x - 7y = -15$.

71 We want to choose multipliers for the equations so that one of the variables will be eliminated when we add the equations together. If we multiply the first equation by 7 and the second equation by 5, the coefficients of y will be additive inverses of each other.

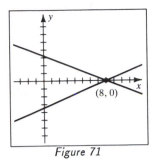

Figure 71

Hence, $\begin{cases} 7(2x + 5y = 16) \\ 5(3x - 7y = 24) \end{cases}$ is equivalent to $\begin{cases} 14x + 35y = 112 \\ 15x - 35y = 120 \end{cases}$

Adding the two equations gives us $29x = 232$, or $x = 8$.

Substituting $x = 8$ into $2x + 5y = 16$ yields $2(8) + 5y = 16 \Rightarrow 5y = 0 \Rightarrow y = 0$.

Thus, the lines intersect at the point $(8, 0)$.

75 (a) Surface area of the capsule

= surface area of a sphere (two hemispheres) + surface area of a cylinder

$= 4\pi r^2 + 2\pi rh$.

The radius of the hemisphere is $\frac{1}{4}$ cm. The height of the cylinder is the total length, 2, minus the radius on *both* ends, $2(\frac{1}{4}) = \frac{1}{2}$ cm. Thus, the surface area of the capsule is $4\pi(\frac{1}{4})^2 + 2\pi(\frac{1}{4})(2 - \frac{1}{2}) = \frac{\pi}{4} + \frac{3\pi}{4} = \pi$ cm^2.

Surface area of the tablet

= surface area of top and bottom + surface area of cylinder

$= 2(\pi r^2) + 2\pi rh = 2\pi r^2 + 2\pi r(\frac{1}{2}) = 2\pi r^2 + \pi r$.

Equating the two surface areas yields

$\qquad 2\pi r^2 + \pi r = \pi \qquad$ { area of tablet = area of capsule }

$\qquad\quad 2r^2 + r = 1 \qquad$ { divide by π }

$\qquad 2r^2 + r - 1 = 0 \qquad$ { set in standard form }

$\quad (2r - 1)(r + 1) = 0 \qquad$ { factor }

Hence, $r = \frac{1}{2}$ or $r = -1$. Since r cannot be negative, $r = \frac{1}{2}$ or $d = 1$ cm.

(b) Volume of the capsule = Volume$_{sphere}$ + Volume$_{cylinder}$

$$= \tfrac{4}{3}\pi r^3 + \pi r^2 h$$

$$= \tfrac{4}{3}\pi(\tfrac{1}{4})^3 + \pi(\tfrac{1}{4})^2(\tfrac{3}{2})$$

$$= \tfrac{\pi}{48} + \tfrac{3\pi}{32} = \tfrac{11\pi}{96} \approx 0.360 \text{ cm}^3. \qquad\qquad \text{(cont.)}$$

$$\text{Volume of the tablet} = \text{Volume}_{cylinder}$$
$$= \pi r^2 h$$
$$= \pi(\tfrac{1}{2})^2 (\tfrac{1}{2})$$
$$= \tfrac{\pi}{8} \approx 0.393 \text{ cm}^3.$$

[77] We want to know what condition will assure us that an object's image is at least 3 times as large as the object, or equivalently, when $M \ge 3$. $M \ge 3 \{f = 6\} \Rightarrow$ $\dfrac{6}{6-p} \ge 3 \Rightarrow 6 \ge 18 - 3p$ {Since $6 - p > 0$, we can multiply by $6 - p$ and not change the direction of the inequality.} $\Rightarrow 3p \ge 12 \Rightarrow p \ge 4$. However, $p < 6$ since $p < f$. Thus, $4 \le p < 6$.

[81] (a) $R = R_0 \Rightarrow R_0 = R_0(1 + aT) \Rightarrow 1 = 1 + aT \Rightarrow aT = 0.$

       Since $a > 0$, T must be $0\,^{\circ}$C. Thus, R_0 is the resistance when $T = 0\,^{\circ}$C.

  (b) $R = 0$ and $T = -273 \Rightarrow 0 = R_0(1 - 273a) \Rightarrow$

                                 {since $R_0 > 0$} $1 - 273a = 0 \Rightarrow a = \frac{1}{273}.$

  (c) $R = 2$, $R_0 = 1.25 = \frac{5}{4}$, and $a = \frac{1}{273} \Rightarrow 2 = \frac{5}{4}(1 + \frac{1}{273}T) \Rightarrow$

               $\frac{8}{5} = 1 + \frac{1}{273}T \Rightarrow \frac{3}{5} = \frac{1}{273}T \Rightarrow T = \frac{819}{5} \Rightarrow T = 163.8\,^{\circ}$C.

Exercises 1.2

[3] $f(x) = 5x - 2$ •

  (a) Substituting a for x, $f(a) = 5(a) - 2 = 5a - 2.$

  (b) Substituting $-a$ for x, $f(-a) = 5(-a) - 2 = -5a - 2.$

  (c) This is the negative of the result in part (a).
$$-f(a) = -1 \cdot f(a) = -1 \cdot (5a - 2) = -5a + 2$$

  (d) Substituting $(a + h)$ for x, $f(a + h) = 5(a + h) - 2 = 5a + 5h - 2.$

  (e) Substituting a for x, and then h for x, $f(a) + f(h) = (5a - 2) + (5h - 2) =$

              $5a + 5h - 4$. Note that parts (d) and (e) are not equal.

  (f) Using part (d) for $f(a + h)$ and part (a) for $f(a)$, we have
$$\frac{f(a + h) - f(a)}{h} = \frac{(5a + 5h - 2) - (5a - 2)}{h} = \frac{5h}{h} = 5.$$

[5] $f(x) = x^2 - x + 3$ •

  (a) Substituting a for x, $f(a) = (a)^2 - (a) + 3 = a^2 - a + 3.$

  (b) Substituting $-a$ for x, $f(-a) = (-a)^2 - (-a) + 3 = a^2 + a + 3.$

  (c) Negating the result in part (a),
$$-f(a) = -1 \cdot f(a) = -1 \cdot (a^2 - a + 3) = -a^2 + a - 3.$$

  (d) Substituting $(a + h)$ for x,
$$f(a + h) = (a + h)^2 - (a + h) + 3 = a^2 + 2ah + h^2 - a - h + 3.$$

(e) Substituting a for x, and then h for x,

$$f(a) + f(h) = (a^2 - a + 3) + (h^2 - h + 3) = a^2 + h^2 - a - h + 6.$$

(f) Using part (d) for $f(a + h)$ and part (a) for $f(a)$, we have

$$\frac{f(a + h) - f(a)}{h} = \frac{(a^2 + 2ah + h^2 - a - h + 3) - (a^2 - a + 3)}{h} =$$

$$\frac{2ah + h^2 - h}{h} = \frac{h(2a + h - 1)}{h} = 2a + h - 1.$$

$\boxed{7}$ $f(x) = \dfrac{x + 1}{x^3 - 4x}$ •   This function is defined for all values of x except those that make the denominator equal to zero. Setting the denominator equal to 0 and solving for x yields $x^3 - 4x = 0 \Rightarrow x(x^2 - 4) = 0 \Rightarrow x(x + 2)(x - 2) = 0$. Hence, the domain of f is all real numbers except -2, 0, and 2. We may denote this by $\mathbb{R} - \{0, \pm 2\}$.

$\boxed{9}$ $f(x) = \dfrac{\sqrt{2x - 3}}{x^2 - 5x + 4}$ •   For this function, we must have a nonnegative (zero or positive) radicand (the expression under the radical sign) *and* a nonzero denominator. The denominator is zero if $x^2 - 5x + 4 = 0 \Rightarrow (x - 1)(x - 4) = 0 \Rightarrow x = 1$ or 4. The radicand is nonnegative if $2x - 3 \geq 0$ or $x \geq \frac{3}{2}$. Hence, the domain of f is all real numbers greater than or equal to $\frac{3}{2}$, except 1 and 4. Excluding 4 $\{1 \notin [\frac{3}{2}, \infty)\}$, we have $[\frac{3}{2}, 4) \cup (4, \infty)$. Note that if the radical was in the denominator, the radicand would have to be positive instead of nonnegative.

$\boxed{11}$ f is even if $f(-x) = f(x)$ and f is odd if $f(-x) = -f(x)$.

(a) $f(-x) = 5(-x)^3 + 2(-x) = -5x^3 - 2x = -(5x^3 + 2x) = -f(x)$.

Thus, $f(x) = 5x^3 + 2x$ is an odd function.

(b) $f(-x) = |-x| - 3 = |-1 \cdot x| - 3 = |-1||x| - 3 = 1|x| - 3 = f(x)$.

Thus, $f(x) = |x| - 3$ is an even function.

(c) $f(-x) = \left[8(-x)^3 - 3(-x)^2\right]^3 = (-8x^3 - 3x^2)^3 = \left[-1(8x^3 + 3x^2)\right]^3$

$= (-1)^3(8x^3 + 3x^2)^3 = -(8x^3 + 3x^2)^3$. This is not equal to $f(x)$ or

$-f(x)$, and hence, $f(x) = (8x^3 - 3x^2)^3$ is neither even nor odd.

$\boxed{15}$ $f(x) = 2\sqrt{x} + c;$ $c = 0,$ 3, -2 • $f(x) = \sqrt{x}$ is pictured on page 16 of the text. The 2 in front of the radical has the effect of vertically stretching the graph of $f(x) = \sqrt{x}$ by a factor of 2. The value c is the vertical shift factor. The value 3 will shift $f(x) = 2\sqrt{x}$ up 3 units and -2 will shift $f(x) = 2\sqrt{x}$ down 2 units. You should try to think of these graphs in general terms with minimal point plotting.

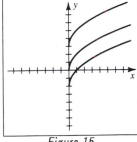

Figure 15

19 $f(x) = c\sqrt{4 - x^2}$;　$c = 1,\ 3,\ -2$　●　The graph of $f(x) = \sqrt{4 - x^2}$ is the upper semicircle of the circle $x^2 + y^2 = 4$.　See §1.1.55–56 for a discussion on semicircle equations.　The value c is a vertical stretching factor.　$c = 3$ will vertically stretch $f(x) = \sqrt{4 - x^2}$ by a factor of 3.　$c = -2$ will vertically stretch $f(x) = \sqrt{4 - x^2}$ by a factor of 2 and reflect the graph through the x-axis.

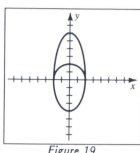

Figure 19

21 $f(x) = (x - c)^{2/3} + 2$; $c = 0,\ 4,\ -3$　●　The graph of $f(x) = x^{2/3}$ is shown on page 17 of the text.　The value c will have the effect of horizontally shifting that graph.　$c = 4$ shifts $f(x) = x^{2/3}$ to the right 4 units and $c = -3$ shifts $f(x) = x^{2/3}$ to the left 3 units.　The " $+2$ " will vertically shift all three of these graphs up 2 units.

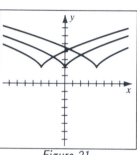

Figure 21

23 (a) $y = f(x + 3)$　●　shift f left 3 units

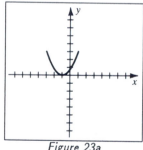

Figure 23a

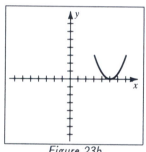

Figure 23b

(b) $y = f(x - 3)$　●　shift f right 3 units
(c) $y = f(x) + 3$　●　shift f up 3 units
(d) $y = f(x) - 3$　●　shift f down 3 units
(e) $y = -3f(x)$　●　stretch f by a factor of 3 and reflect through the x-axis

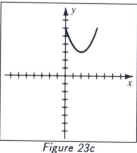

Figure 23c

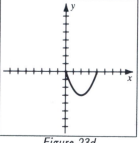

Figure 23d

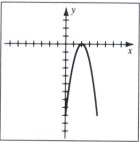

Figure 23e

(f) $y = -\frac{1}{3}f(x)$ •   stretch f by a factor of $\frac{1}{3}$ and reflect through the x-axis

(g) $y = -f(x + 2) - 3$ •   reflect through the x-axis, shift left 2 units and down 3

(h) $y = f(x - 2) + 3$ •   shift f right 2 units and up 3

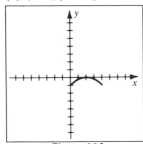

Figure 23f

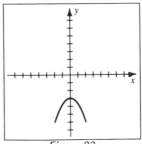

Figure 23g

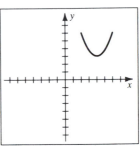

Figure 23h

$\boxed{27}$ $f(x) = \begin{cases} \dfrac{x^2 - 1}{x + 1} & \text{if } x \neq -1 \\ 2 & \text{if } x = -1 \end{cases}$ •

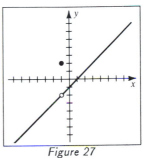

Figure 27

Simplifying f we have $\dfrac{x^2 - 1}{x + 1} = \dfrac{(x + 1)(x - 1)}{x + 1} = $

$x - 1$, if $x \neq -1$. Thus, if $x \neq -1$, $f(x) = x - 1$, a

line. If $x = -1$, $f(x) = 2$, which is just the point

$(-1, 2)$. Hence, f is a line with a hole at $(-1, -2)$

and the point $(-1, 2)$.

$\boxed{29}$ The graph of $f(x) = [\![x]\!]$ is shown in the text on page 18.

(a) shift $f(x) = [\![x]\!]$ right 3 units

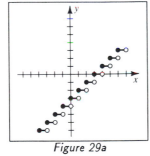

Figure 29a

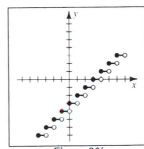

Figure 29b

(b) shift $f(x) = [\![x]\!]$ down 3 units — this is the same as the graph in part (a).

(c) The normal range for this function is the set of integers. The 2 in front of $[\![x]\!]$

has the effect of making the range consist of the even integers.

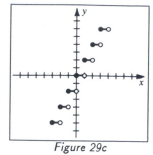

Figure 29c

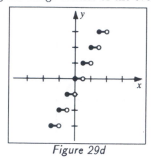

Figure 29d

(d) To determine the pattern of "steps" for this function, find the values of x that make $f(x)$ change from 0 to 1, then from 1 to 2, etc. If $2x = 0$, then $x = 0$, and if $2x = 1$, then $x = \frac{1}{2}$. Thus, the function will equal 0 from $x = 0$ to $x = \frac{1}{2}$ and then jump to 1 at $x = \frac{1}{2}$. If $2x = 2$, then $x = 1$. The pattern is established: each step will be $\frac{1}{2}$ unit long.

33 (a) $(f + g)(x) = f(x) + g(x)$

$$= \frac{2x}{x - 4} + \frac{x}{x + 5} = \frac{2x(x + 5) + x(x - 4)}{(x - 4)(x + 5)} = \frac{3x^2 + 6x}{(x - 4)(x + 5)}$$

$(f - g)(x) = f(x) - g(x)$

$$= \frac{2x}{x - 4} - \frac{x}{x + 5} = \frac{2x(x + 5) - x(x - 4)}{(x - 4)(x + 5)} = \frac{x^2 + 14x}{(x - 4)(x + 5)}$$

$(fg)(x) = f(x) \cdot g(x) = \frac{2x}{x - 4} \cdot \frac{x}{x + 5} = \frac{2x^2}{(x - 4)(x + 5)}$

$\left(\frac{f}{g}\right)(x) = \frac{f(x)}{g(x)} = \frac{2x/(x - 4)}{x/(x + 5)} = \frac{2x(x + 5)}{x(x - 4)} = \frac{2(x + 5)}{x - 4} \{\text{if } x \neq 0\} = \frac{2x + 10}{x - 4}$

(b) The domain of f is all real numbers except 4. The domain of g is all real numbers except -5. The intersection of these two domains, all reals except -5 and 4, is the domain of the first three functions. To determine the quotient's domain, we also exclude any values that make the denominator equal to zero. Hence, we exclude $x = 0$ and the domain of the quotient is all real numbers except -5, 0, and 4.

35 $f(x) = x^2 - 3x;\ g(x) = \sqrt{x + 2}$ •

(a) $(f \circ g)(x) = f(g(x)) = (\sqrt{x + 2})^2 - 3(\sqrt{x + 2}) = x + 2 - 3\sqrt{x + 2}$.   The domain of $(f \circ g)(x)$ is the set of all x in the domain of g, which is $x \geq -2$, such that $g(x)$ is in the domain of f. Since the domain of f is $\mathbb{R}$, any value of $g(x)$ is in f's domain. Thus, the domain is all x such that $x \geq -2$.

(b) $(g \circ f)(x) = g(f(x)) = \sqrt{(x^2 - 3x) + 2} = \sqrt{x^2 - 3x + 2}$.   The   domain   of

$(g \circ f)(x)$ is the set of all x in the domain of f, which is $\mathbb{R}$, such that $f(x)$ is in

the domain of g. Since the domain of g is $x \geq -2$, we must solve $f(x) \geq -2$.

$x^2 - 3x \geq -2 \Rightarrow x^2 - 3x + 2 \geq 0 \Rightarrow (x - 1)(x - 2) \geq 0 \Rightarrow x \in (-\infty, 1] \cup$

$[2, \infty)$ { use a sign chart as before to solve the quadratic inequality }. Thus, the

domain is all x such that $x \in (-\infty, 1] \cup [2, \infty)$.

$\boxed{39}$ $f(x) = \sqrt{25 - x^2}$; $g(x) = \sqrt{x - 3}$ •

(a) $(f \circ g)(x) = f(g(x)) = \sqrt{25 - (\sqrt{x - 3})^2} = \sqrt{25 - (x - 3)} = \sqrt{28 - x}$.   The

domain of $(f \circ g)(x)$ is the set of all x in the domain of g, which is $x \geq 3$, such

that $g(x)$ is in the domain of f. Since the domain of f is $-5 \leq x \leq 5$ and $g(x)$

cannot be less than 0, we must solve $g(x) \leq 5$. $\sqrt{x - 3} \leq 5 \Rightarrow x - 3 \leq 25$

{ squaring both sides } $\Rightarrow x \leq 28$. Remembering that $x \geq 3$, we have $[3, \infty) \cap$

$(-\infty, 28] = [3, 28]$.

(b) $(g \circ f)(x) = g(f(x)) = \sqrt{\sqrt{25 - x^2} - 3}$. See the discussion for Exercise 33(b) to

determine the inequality. $f(x) \geq 3 \Rightarrow \sqrt{25 - x^2} \geq 3 \Rightarrow 25 - x^2 \geq 9 \Rightarrow$

$x^2 \leq 16$. Taking the square root of both sides yields $|x| \leq 4$ since $\sqrt{x^2} = |x|$.

Thus, the solution is $-4 \leq x \leq 4$, or $x \in [-4, 4]$.

$\boxed{41}$ $f(x) = \frac{x}{3x + 2}$; $g(x) = \frac{2}{x}$ •

(a) $(f \circ g)(x) = f(g(x)) = \frac{2/x}{3(2/x) + 2} = \frac{2}{6 + 2x} = \frac{1}{x + 3}$.

Domain of $g = \mathbb{R} - \{0\}$. Domain of $f = \mathbb{R} - \{-\frac{2}{3}\}$. We must have $g(x)$ in the

domain of f. Since f cannot have $-\frac{2}{3}$ in its domain, we must solve $g(x) \neq -\frac{2}{3}$.

Thus, $\frac{2}{x} \neq -\frac{2}{3} \Rightarrow x \neq -3$. Hence, the domain of $f \circ g$ is $\mathbb{R} - \{-3, 0\}$.

(b) $(g \circ f)(x) = g(f(x)) = \frac{2}{x/(3x + 2)} = \frac{6x + 4}{x}$.   Domain   of $f = \mathbb{R} - \{-\frac{2}{3}\}$.

Domain of $g = \mathbb{R} - \{0\}$. We must have $f(x)$ in the domain of g. Since g cannot

have 0 in its domain, we must solve $f(x) \neq 0$. Thus, $\frac{x}{3x + 2} \neq 0 \Rightarrow x \neq 0$.

Hence, the domain of $g \circ f$ is $\mathbb{R} - \{-\frac{2}{3}, 0\}$.

$\boxed{43}$ $y = (x^2 + 3x)^{1/3}$ • Suppose you were to find the value of y if x was equal to 3.

Using a calculator, you might compute the value of $x^2 + 3x$ first, and then raise that

result to the $\frac{1}{3}$ power. The last calculator operation that you would perform to make

such an evaluation is often a good choice for y. Thus, we would choose $y = u^{1/3}$ and

$u = x^2 + 3x$.

$\boxed{49}$ $y = \dfrac{\sqrt{x+4} - 2}{\sqrt{x+4} + 2}$ •   There is not a "simple" choice for y here as there was in

Exercise 43. One choice for u would be $u = x + 4$. Then y would be $\dfrac{\sqrt{u} - 2}{\sqrt{u} + 2}$.

Another choice for u would be $u = \sqrt{x+4}$. Then y would be $\dfrac{u - 2}{u + 2}$.

$\boxed{53}$ The formula for the volume of a box is $V = lwh$.

{ Volume = length × width × height }. In this case, the length is $30 - 2x$ { since x

is taken off each end }, the width is $20 - 2x$, and the height is x. Thus,

$V = lwh = (30 - 2x)(20 - 2x)(x) = 4x^3 - 100x^2 + 600x$.

$\boxed{57}$ (a) CTP forms a right angle, and the Pythagorean theorem may be applied.

$$r^2 + y^2 = (h + r)^2 \Rightarrow y^2 = h^2 + 2rh \Rightarrow y = \pm\sqrt{h^2 + 2hr} \ \{y > 0\} \Rightarrow$$

$$y = \sqrt{h^2 + 2hr}.$$

(b) $y = \sqrt{(200)^2 + 2(200)(4000)} = \sqrt{(200)^2(1 + 40)} = 200\sqrt{41} \approx 1280.6$ mi.

$\boxed{59}$ Form a right triangle with the control booth and the beginning of the runway. Let y

denote the distance from the control booth to the beginning of the runway and apply

the Pythagorean theorem. $y^2 = 300^2 + 20^2 \Rightarrow y^2 = 90{,}400$. Now form a right

triangle with sides y and x and hypotenuse d. Then $d^2 = y^2 + x^2 \Rightarrow$

$d^2 = 90{,}400 + x^2 \Rightarrow d = \sqrt{90{,}400 + x^2}$.

$\boxed{61}$ (a) For similar triangles, use a little triangle at the top of the figure and a large

triangle. We conclude that "y is to b" as "$y + h$ is to a". Thus, $\dfrac{y}{b} = \dfrac{y + h}{a} \Rightarrow$

$ay = by + bh \Rightarrow y(a - b) = bh \Rightarrow y = \dfrac{bh}{a - b}$.

(b) The volume of the frustum can be calculated by finding the volume of the large

cone minus the volume of the small cone. Using $V = \frac{1}{3}\pi r^2 h$, we have

$$V = \tfrac{1}{3}\pi a^2(y + h) - \tfrac{1}{3}\pi b^2 y \qquad \{ \text{Volume}_{large\ cone} - \text{Volume}_{small\ cone} \}$$

$$= \tfrac{\pi}{3}\left[(a^2 - b^2)y + a^2 h\right] \qquad \{ \text{factor out } \tfrac{\pi}{3} \text{ and combine } y \text{ terms} \}$$

$$= \tfrac{\pi}{3}\left[(a^2 - b^2)\dfrac{bh}{a - b} + a^2 h\right] \qquad \{ \text{substitute for } y \text{ from part (a)} \}$$

$$= \tfrac{\pi}{3}h\left[(a + b)b + a^2\right] \qquad \{ \text{factor out } h \text{ and cancel } a - b \}$$

$$= \tfrac{\pi}{3}h(a^2 + ab + b^2) \qquad \{ \text{equivalent expression} \}$$

(c) Letting $a = 6$, $b = 3$, and $V = 600$ in the formula found in part (b) yields

$600 = \frac{\pi}{3}h(6^2 + 6\cdot 3 + 3^2) \Rightarrow 1800 = \pi h(63) \Rightarrow h = \frac{1800}{63\pi} = \frac{200}{7\pi} \approx 9.1$ ft.

Exercises 1.3

Note: Multiply each degree measure by $\frac{\pi}{180}$ to obtain the listed radian measure.

1 (a) $150° \cdot \frac{\pi}{180} = \frac{5 \cdot 30\pi}{6 \cdot 30} = \frac{5\pi}{6}$ (b) $120° \cdot \frac{\pi}{180} = \frac{2 \cdot 60\pi}{3 \cdot 60} = \frac{2\pi}{3}$

(c) $450° \cdot \frac{\pi}{180} = \frac{5 \cdot 90\pi}{2 \cdot 90} = \frac{5\pi}{2}$ (d) $-60° \cdot \frac{\pi}{180} = -\frac{60\pi}{3 \cdot 60} = -\frac{\pi}{3}$

Note: Multiply each radian measure by $\frac{180}{\pi}$ to obtain the listed degree measure.

3 (a) $\frac{2\pi}{3} \cdot \left(\frac{180}{\pi}\right)° = \left(\frac{2 \cdot 60 \cdot 3\pi}{3\pi}\right)° = 120°$ (b) $\frac{5\pi}{6} \cdot \left(\frac{180}{\pi}\right)° = \left(\frac{5 \cdot 30 \cdot 6\pi}{6\pi}\right)° = 150°$

(c) $\frac{3\pi}{4} \cdot \left(\frac{180}{\pi}\right)° = \left(\frac{3 \cdot 45 \cdot 4\pi}{4\pi}\right)° = 135°$ (d) $-\frac{7\pi}{2} \cdot \left(\frac{180}{\pi}\right)° = -\left(\frac{7 \cdot 90 \cdot 2\pi}{2\pi}\right)° = -630°$

5 A measure of $50°$ is equivalent to $\left(50 \cdot \frac{\pi}{180}\right)$ radians. The radius is one-half of the diameter. Thus, $s = r\theta = \left(\frac{1}{2} \cdot 16\right)\left(50 \cdot \frac{\pi}{180}\right) = \frac{20\pi}{9} \approx 6.98.$

Note: For Exercises 7-8, the method of solution is to first find a trigonometric function involving: the angle, the side whose value is given, and the unknown side (x or y). Next use the special values of the trigonometric functions and solve for the variable.

7 $\sin 30° = \frac{4}{x} \Rightarrow \frac{1}{2} = \frac{4}{x} \Rightarrow x = 8;\ \tan 30° = \frac{4}{y} \Rightarrow \frac{\sqrt{3}}{3} = \frac{4}{y} \Rightarrow y = 4\sqrt{3}$

Note: Answers are in the order **sin, cos, tan, cot, sec, csc** for any problems requiring the trigonometric functions. For Exercises 9–12, sketch a triangle as in Exercise 9. Use the Pythagorean theorem to find the remaining side.

9 $\sin \theta = \frac{3}{5}$ •

(adj)^2 + (opp)^2 = (hyp)^2 $\Rightarrow$

(adj)^2 + $3^2 = 5^2 \Rightarrow$ adj = 4.

★ $\frac{3}{5}, \frac{4}{5}, \frac{3}{4}, \frac{4}{3}, \frac{5}{4}, \frac{5}{3}$

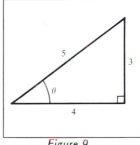

Figure 9

13 For the point $Q(x,\ y)$, we let r denote the distance from the origin to Q.

$x = 4$ and $y = -3 \Rightarrow r = \sqrt{x^2 + y^2} = \sqrt{4^2 + (-3)^2} = \sqrt{25} = 5.$

$\sin \theta = \frac{y}{r} = \frac{-3}{5} = -\frac{3}{5}.$ $\cos \theta = \frac{x}{r} = \frac{4}{5}.$

$\tan \theta = \frac{y}{x} = \frac{-3}{4} = -\frac{3}{4}.$ $\cot \theta = \frac{x}{y} = \frac{4}{-3} = -\frac{4}{3}.$

$\sec \theta = \frac{r}{x} = \frac{5}{4}.$ $\csc \theta = \frac{r}{y} = \frac{5}{-3} = -\frac{5}{3}.$

15 $2y - 7x + 2 = 0 \Leftrightarrow y = \frac{7}{2}x - 1$. Thus, the slope of the given line is $\frac{7}{2}$. The line through the origin with that slope is $y = \frac{7}{2}x$. If $x = -2$, then $y = \frac{7}{2}(-2) = -7$ and $(-2, -7)$ is a point on the terminal side of θ. $x = -2$ and $y = -7 \Rightarrow$ $r = \sqrt{(-2)^2 + (-7)^2} = \sqrt{53}$.   ★ $-\frac{7}{\sqrt{53}}, -\frac{2}{\sqrt{53}}, \frac{7}{2}, \frac{2}{7}, -\frac{\sqrt{53}}{2}, -\frac{\sqrt{53}}{7}$

19 (a) The tangent and secant functions are related by the identity $1 + \tan^2\theta = \sec^2\theta$. Solving for $\tan\theta$, we have $\tan^2\theta = \sec^2\theta - 1$, or $\tan\theta = \pm\sqrt{\sec^2\theta - 1}$. The "$-$" is not used since θ is acute and all six trigonometric functions are positive. Thus, $\tan\theta = \sqrt{\sec^2\theta - 1}$.

(b) The sine isn't directly related to the secant, but it is related to the reciprocal of the secant, namely, the cosine. Hence, $\sin^2\theta + \cos^2\theta = 1 \Rightarrow \sin^2\theta = 1 - \cos^2\theta$

$\Rightarrow \sin\theta = \sqrt{1 - \cos^2\theta} = \sqrt{1 - \dfrac{1}{\sec^2\theta}} = \sqrt{\dfrac{\sec^2\theta - 1}{\sec^2\theta}} = \dfrac{\sqrt{\sec^2\theta - 1}}{|\sec\theta|} =$

$\dfrac{\sqrt{\sec^2\theta - 1}}{\sec\theta}$, since $|\sec\theta| = \sec\theta$ for acute θ.

23 Let $x = 5\tan\theta$. $\dfrac{x}{\sqrt{25 + x^2}} = \dfrac{5\tan\theta}{\sqrt{25 + (5\tan\theta)^2}} = \dfrac{5\tan\theta}{\sqrt{25 + 25\tan^2\theta}} = \dfrac{5\tan\theta}{\sqrt{25(1 + \tan^2\theta)}}$

$= \dfrac{5\tan\theta}{5\sqrt{\sec^2\theta}} = \dfrac{\tan\theta}{|\sec\theta|} = \dfrac{\tan\theta}{\sec\theta}$ { since $\sec\theta > 0$ if $-\frac{\pi}{2} < \theta < \frac{\pi}{2}$ } $= \dfrac{\sin\theta/\cos\theta}{1/\cos\theta} = \sin\theta$.

25 Let $x = 3\sec\theta$. $\dfrac{\sqrt{x^2 - 9}}{x} = \dfrac{\sqrt{(3\sec\theta)^2 - 9}}{3\sec\theta} = \dfrac{\sqrt{9\sec^2\theta - 9}}{3\sec\theta} = \dfrac{\sqrt{9(\sec^2\theta - 1)}}{3\sec\theta} =$

$\dfrac{3\sqrt{\tan^2\theta}}{3\sec\theta} = \dfrac{|\tan\theta|}{\sec\theta} = \dfrac{\tan\theta}{\sec\theta}$ { since $\tan\theta > 0$ if $0 < \theta < \frac{\pi}{2}$ } $= \dfrac{\sin\theta/\cos\theta}{1/\cos\theta} = \sin\theta$.

27 (a) Since $\theta = \frac{2\pi}{3}$ is in the second quadrant, the reference angle θ_R is $\pi - \theta = \pi - \frac{2\pi}{3} = \frac{\pi}{3}$. The sine is positive in both the first and second quadrants, so $\sin\frac{2\pi}{3} = \sin\frac{\pi}{3} = \frac{\sqrt{3}}{2}$.

(b) Since the sine is an odd function, $\sin(-\frac{5\pi}{4}) = -\sin\frac{5\pi}{4}$ (see *formulas for negatives* on page 36 of the text). $\theta = \frac{5\pi}{4}$ is in the third quadrant and its reference angle θ_R is $\theta - \pi = \frac{5\pi}{4} - \pi = \frac{\pi}{4}$. Because the sine is negative in the third quadrant and positive in the first quadrant, we must change the sign as we write the expression in terms of θ_R. Thus, $-\sin\frac{5\pi}{4} = -(-\sin\frac{\pi}{4}) = \sin\frac{\pi}{4} = \frac{\sqrt{2}}{2}$.

31 (a) Since $\frac{2\pi}{3}$ is in QII, $\theta_R = \pi - \theta = \pi - \frac{2\pi}{3} = \frac{\pi}{3}$. The secant is negative in QII and positive in QI. Thus, $\sec\frac{2\pi}{3} = -\sec\frac{\pi}{3} = -\dfrac{1}{\cos\frac{\pi}{3}} = -\dfrac{1}{\frac{1}{2}} = -2$.

(b) Because the secant is an even function, $\sec(-\frac{\pi}{6}) = \sec\frac{\pi}{6}$.

Thus, $\sec(-\frac{\pi}{6}) = \sec\frac{\pi}{6} = \dfrac{1}{\cos\frac{\pi}{6}} = \dfrac{1}{\sqrt{3}/2} = \dfrac{2}{\sqrt{3}}$.

$\boxed{35}$ (a) $f(x) = 2\cos(x + \pi)$ • Shift $f(x) = \cos x$ to the left π units and then vertically stretch it by a factor of 2.

(b) $f(x) = 2\cos x + \pi$ •

Vertically stretch $f(x) = \cos x$ by a factor of 2 and then shift it up π units.

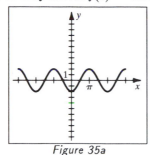

Figure 35a

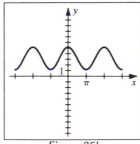

Figure 35b

$\boxed{39}$ One choice would be $u = \tan^2 x + 4$ and $y = \sqrt{u}$.

Another choice is $u = \tan^2 x$ and $y = \sqrt{u + 4}$.

$\boxed{43}$
$$\frac{f(x + h) - f(x)}{h} = \frac{\cos(x + h) - \cos x}{h} \qquad \{\text{definition of } f\}$$

$$= \frac{\cos x \cos h - \sin x \sin h - \cos x}{h} \qquad \{\text{definition of } \cos(x + h)\}$$

$$= \frac{\cos x \cos h - \cos x}{h} - \frac{\sin x \sin h}{h} \qquad \{\text{break up terms}\}$$

$$= \cos x \left(\frac{\cos h - 1}{h}\right) - \sin x \left(\frac{\sin h}{h}\right) \qquad \{\text{factor out } \cos x\}$$

$\boxed{45}$ $(1 - \sin^2 t)(1 + \tan^2 t) = (\cos^2 t)(\sec^2 t) = (\cos^2 t)(1/\cos^2 t) = 1$

$\boxed{49}$ $\dfrac{1 + \csc \beta}{\sec \beta} - \cot \beta = \dfrac{1}{\sec \beta} + \dfrac{\csc \beta}{\sec \beta} - \cot \beta = \cos \beta + \dfrac{1/\sin \beta}{1/\cos \beta} - \cot \beta =$

$$\cos \beta + \frac{\cos \beta}{\sin \beta} - \cot \beta = \cos \beta + \cot \beta - \cot \beta = \cos \beta$$

$\boxed{51}$ $\sin 3u = \sin(2u + u)$ $\{\text{break up } 3u\}$

     $= \sin 2u \cos u + \cos 2u \sin u$ $\{\text{addition formula for sine}\}$

     $= (2\sin u \cos u)\cos u + (1 - 2\sin^2 u)\sin u$ $\{\text{double-angle formulas}\}$

         Note: We chose $\cos 2u = 1 - 2\sin^2 u$ because we are working toward an

            expression that only has $\sin u$ in it.

     $= 2\sin u \cos^2 u + \sin u - 2\sin^3 u$ $\{\text{multiply terms}\}$

     $= 2\sin u(1 - \sin^2 u) + \sin u - 2\sin^3 u$ $\{\cos^2 u + \sin^2 u = 1\}$

     $= 2\sin u - 2\sin^3 u + \sin u - 2\sin^3 u$ $\{\text{multiply terms}\}$

     $= 3\sin u - 4\sin^3 u$ $\{\text{combine terms}\}$

     $= \sin u(3 - 4\sin^2 u)$ $\{\text{factor out } \sin u\}$

$\boxed{53}$ $\cos^4 \frac{\theta}{2} = \left(\cos^2 \frac{\theta}{2}\right)^2 = \left(\frac{1 + \cos\theta}{2}\right)^2$ {half-angle formula} $= \frac{1 + 2\cos\theta + \cos^2\theta}{4} =$

$\frac{1}{4} + \frac{1}{2}\cos\theta + \frac{1}{4}\left(\frac{1 + \cos 2\theta}{2}\right)$ {half-angle formula} $= \frac{1}{4} + \frac{1}{2}\cos\theta + \frac{1}{8} + \frac{1}{8}\cos 2\theta =$

$$\frac{3}{8} + \frac{1}{2}\cos\theta + \frac{1}{8}\cos 2\theta.$$

$\boxed{55}$ $2\cos 2\theta - \sqrt{3} = 0 \Rightarrow \cos 2\theta = \frac{\sqrt{3}}{2}$ {2θ is just an angle—so we solve this equation for

2θ and then divide those solutions by 2} $\Rightarrow 2\theta = \frac{\pi}{6} + 2\pi n, \frac{11\pi}{6} + 2\pi n \Rightarrow$

$$\theta = \frac{\pi}{12} + \pi n, \frac{11\pi}{12} + \pi n, \text{ where } n \text{ denotes any integer.}$$

$\boxed{57}$ $2\sin^2 u = 1 - \sin u \Rightarrow 2\sin^2 u + \sin u - 1 = 0 \Rightarrow (2\sin u - 1)(\sin u + 1) = 0 \Rightarrow$

$$\sin u = \frac{1}{2}, -1 \Rightarrow u = \frac{\pi}{6}, \frac{5\pi}{6}, \frac{3\pi}{2}$$

$\boxed{61}$ $\sin 2t + \sin t = 0 \Rightarrow 2\sin t\cos t + \sin t = 0 \Rightarrow \sin t(2\cos t + 1) = 0 \Rightarrow$

$$\sin t = 0 \text{ or } \cos t = -\frac{1}{2} \Rightarrow t = 0, \pi \text{ or } \frac{2\pi}{3}, \frac{4\pi}{3}$$

$\boxed{63}$ A first approach uses the concept that if $\tan\alpha = \tan\beta$, then $\alpha = \beta + \pi n$.

$$\tan 2x = \tan x \Rightarrow 2x = x + \pi n \Rightarrow x = \pi n \Rightarrow x = 0, \pi.$$

Another approach is: $\tan 2x = \tan x \Rightarrow \frac{\sin 2x}{\cos 2x} = \frac{\sin x}{\cos x} \Rightarrow$

$\sin 2x \cos x = \sin x \cos 2x \Rightarrow \sin 2x \cos x - \sin x \cos 2x = 0 \Rightarrow$

$\sin(2x - x)$ {subtraction formula for the sine} $= 0 \Rightarrow \sin x = 0 \Rightarrow x = 0, \pi.$

$\boxed{65}$ In degree mode, enter -0.5640 and press $\boxed{\text{INV}}$ $\boxed{\text{SIN}}$. This is approximately

$-34.3328\ldots$ degrees. Multiplying $0.3328\ldots$ by 60 gives us approximately 19.97

minutes. Rounding to the nearest 10 minutes, we obtain $-34°20'$. Since the sine is

negative in QIII and QIV, we want the reference angles for $34°20'$ in those quadrants.

$180° + 34°20' = \underline{214°20'}$ and $360° - 34°20' = \underline{325°40'}$.

$\boxed{69}$ After entering -1.116, use $\boxed{1/x}$ and then $\boxed{\text{INV}}$ $\boxed{\text{COS}}$ to obtain $153°40'$ after

rounding to the nearest 10 minutes. The reference angle is $180° - 153°40' = 26°20'$.

The secant is negative in QII and QIII. QII: $153°40'$, QIII: $206°20'$

Chapter 2: Limits of Functions

Note: DNE denotes Does Not Exist.

1 To evaluate this limit, we only need to substitute -2 for x.

$$\lim_{x \to -2} (3x - 1) = 3(-2) - 1 = -7.$$

5 As x approaches *any* value, the limit of a constant function is the constant.

That is, the value 100 has no bearing on the answer.

Hence, $\lim_{x \to 100} 7 = 7$. Remember, the graph of $f(x) = 7$ is a horizontal line.

9 Since $\lim_{x \to -1} (2x + 1) = -1 \neq 0$, we only need to substitute -1 for x into the

expression. $\lim_{x \to -1} \dfrac{x + 4}{2x + 1} = \dfrac{-1 + 4}{2(-1) + 1} = \dfrac{3}{-1} = -3.$

11 If we substitute -3 for x, the expression evaluates to $\frac{0}{0}$, which is undefined.

Since $x \neq -3$, we can cancel like terms and evaluate the resulting limit.

$$\lim_{x \to -3} \frac{(x + 3)(x - 4)}{(x + 3)(x + 1)} = \lim_{x \to -3} \frac{x - 4}{x + 1} = \frac{-7}{-2} = \frac{7}{2}.$$

13 This exercise is similar to Exercise 11 except that we must factor the numerator first.

$$\lim_{x \to 2} \frac{x^2 - 4}{x - 2} = \lim_{x \to 2} \frac{(x + 2)(x - 2)}{x - 2} = \lim_{x \to 2} (x + 2) = 4.$$

17 A substitution of 4 for k produces a zero denominator and hence, an undefined expression. Thus, we must change the form of the expression. We factor the numerator using the difference of two squares twice.

$$k^2 - 16 = (k + 4)(k - 4) = (k + 4)(\sqrt{k} + 2)(\sqrt{k} - 2)$$

This last factorization is valid because k is positive and so $\sqrt{k} \cdot \sqrt{k} = k$. Hence,

$$\lim_{k \to 4} \frac{k^2 - 16}{\sqrt{k} - 2} = \lim_{k \to 4} \frac{(k + 4)(\sqrt{k} + 2)(\sqrt{k} - 2)}{\sqrt{k} - 2} = \lim_{k \to 4} (k + 4)(\sqrt{k} + 2) = 32.$$

Alternatively, we could have rationalized the denominator by multiplying the numerator and denominator by $\sqrt{k} + 2$ and then canceling the factor $k - 4$.

19
$$\begin{aligned}
\lim_{h \to 0} \frac{(x + h)^2 - x^2}{h} &= \lim_{h \to 0} \frac{(x^2 + 2xh + h^2) - x^2}{h} && \{\text{expand } (x + h)^2\} \\
&= \lim_{h \to 0} \frac{2xh + h^2}{h} && \{\text{combine like terms}\} \\
&= \lim_{h \to 0} \frac{h(2x + h)}{h} && \{\text{factor out } h\} \\
&= \lim_{h \to 0} (2x + h) && \{\text{cancel } h \text{ since } h \neq 0\} \\
&= 2x && \{\text{let } h = 0\}
\end{aligned}$$

21 In order to find this limit, we must factor the sum of two cubes, that is,

$$h^3 + 8 = h^3 + 2^3 = (h + 2)(h^2 - 2h + 4).$$

Hence, $\lim\limits_{h \to -2} \dfrac{h^3 + 8}{h + 2} = \lim\limits_{h \to -2} \dfrac{(h + 2)(h^2 - 2h + 4)}{h + 2} = \lim\limits_{h \to -2} (h^2 - 2h + 4) = 12.$

23 $\lim\limits_{z \to -2} \dfrac{z - 4}{z^2 - 2z - 8} = \lim\limits_{z \to -2} \dfrac{z - 4}{(z - 4)(z + 2)} = \lim\limits_{z \to -2} \dfrac{1}{z + 2}.$

Since $\lim\limits_{z \to -2} (z + 2) = 0$, the expression $\dfrac{1}{z + 2}$ does not approach some real number L

as z approaches -2, and the limit DNE.

25 (a) If $x \to 4^-$, then $x < 4$, $x - 4 < 0$, and $|x - 4| = -(x - 4)$.

Hence, $\lim\limits_{x \to 4^-} \dfrac{|x - 4|}{x - 4} = \lim\limits_{x \to 4^-} \dfrac{-(x - 4)}{x - 4} = \lim\limits_{x \to 4^-} (-1) = -1.$

(b) If $x \to 4^+$, then $x > 4$, $x - 4 > 0$, and $|x - 4| = x - 4$.

Hence, $\lim\limits_{x \to 4^+} \dfrac{|x - 4|}{x - 4} = \lim\limits_{x \to 4^+} \dfrac{x - 4}{x - 4} = \lim\limits_{x \to 4^+} (1) = 1.$

(c) Since the right-hand and left-hand limits are not equal, the limit DNE.

27 (a) If $x \to -6^-$, then $x < -6$, and $x + 6 < 0$.

Thus, $\lim\limits_{x \to -6^-} (\sqrt{x + 6} + x)$ DNE, since $\sqrt{x + 6}$ is undefined for $x + 6 < 0$.

(b) If $x \to -6^+$, then $x > -6$, and $x + 6 > 0$. Thus,

$\lim\limits_{x \to -6^+} (\sqrt{x + 6} + x) = \sqrt{0} + (-6) = -6$, since $\sqrt{x + 6}$ is defined for $x > -6$.

(c) The limit DNE,

since $\sqrt{x + 6}$ is not defined throughout an open interval containing -6.

29 (a) If $x \to 0^-$, then $x^3 \to 0^-$. If x^3 is a small negative number, then $1/x^3$ is a large

negative number. Thus, $\lim\limits_{x \to 0^-} (1/x^3)$ DNE, since the function becomes

unbounded in the negative sense.

(b) If $x \to 0^+$, then $x^3 \to 0^+$. If x^3 is a small positive number, then $1/x^3$ is a large

positive number. Thus, $\lim\limits_{x \to 0^+} (1/x^3)$ DNE, since the function becomes

unbounded in the positive sense.

(c) Since $1/x^3$ does not approach some real number L as x approaches 0, the limit

DNE.

33 (a) As $x \to 2^-$, f approaches the hole in the graph. The hole has a y-coordinate of 1.

Thus, $\lim\limits_{x \to 2^-} f(x) = 1.$

(b) As $x \to 2^+$, f approaches the hole in the graph. Thus, $\lim\limits_{x \to 2^+} f(x) = 1.$

(c) Since the right-hand and left-hand limits are equal, we have $\lim\limits_{x \to 2} f(x) = 1$.

Note that even though $f(2) = 4$, this does not affect the limit.

Remember, the notation $\lim\limits_{x \to 2} f(x) = 1$ is just shorthand for saying

"as x gets close to 2, y gets close to 1."

(d) As $x \to 0^-$, f approaches 3. Thus, $\lim\limits_{x \to 0^-} f(x) = 3$.

(e) As $x \to 0^+$, f approaches 3. Thus, $\lim\limits_{x \to 0^+} f(x) = 3$.

(f) Since the left-hand and right-hand limits are equal, we have $\lim\limits_{x \to 0} f(x) = 3$.

$\boxed{35}$ (a) As $x \to 2^-$ { typical values are 1.99 and 1.999 }, f approaches a dot with a

y-coordinate of 1. Thus, $\lim\limits_{x \to 2^-} f(x) = 1$.

(b) As $x \to 2^+$ { typical values are 2.01 and 2.001 }, f approaches a hole with a

y-coordinate of 0. Thus, $\lim\limits_{x \to 2^+} f(x) = 0$.

(c) Since the right-hand and left-hand limits are not equal, $\lim\limits_{x \to 2}$ DNE.

(d) The graph shown has domain $-3 < x \le 4$. As $x \to 0$, the values of y are exactly

the same as if $x \to 2$, or 1, or 3, etc. As in part (a), $\lim\limits_{x \to 0^-} f(x) = 1$.

(e) As in part (b), $\lim\limits_{x \to 0^+} f(x) = 0$.

(f) As in part (c), $\lim\limits_{x \to 0} f(x)$ DNE.

$\boxed{41}$ (a) If $x \to 1^-$, then $x < 1$ and we use $x^2 - 1$ for the value of $f(x)$. Thus, $\lim\limits_{x \to 1^-} f(x) = \lim\limits_{x \to 1^-} (x^2 - 1) = 0$.

(b) If $x \to 1^+$, then $x > 1$ and we use $4 - x$ for the value of $f(x)$. Thus, $\lim\limits_{x \to 1^+} f(x) = \lim\limits_{x \to 1^+} (4 - x) = 3$.

(c) Since the left-hand and right-hand limits are not

equal, $\lim\limits_{x \to 1} f(x)$ DNE.

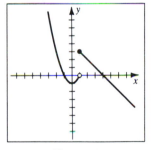

Figure 41

$\boxed{47}$ (a) If $x \le 20{,}000$, then the tax is 15% of x, or, equivalently, $0.15x$. If $x > 20{,}000$,

then the tax on the first \$20,000 is 15% of \$20,000, or \$3000. In addition,

$x - 20{,}000$ is the amount over \$20,000, which is taxed at 20%. Thus, the tax in

this case is the sum of the two amounts, namely, $3000 + 0.20(x - 20{,}000)$.

$$T(x) = \begin{cases} 0.15x & \text{if } x \le 20{,}000 \\ 3000 + 0.20(x - 20{,}000) & \text{if } x > 20{,}000 \end{cases}$$

$$= \begin{cases} 0.15x & \text{if } x \le 20{,}000 \\ 0.20x - 1000 & \text{if } x > 20{,}000 \end{cases}$$

(b) If $x \to 20{,}000^-$, then we use $0.15x$ for $T(x)$ and

$$\lim_{x \to 20{,}000^-} T(x) = \lim_{x \to 20{,}000^-} (0.15x) = 3000.$$

If $x \to 20{,}000^+$, then we use $(0.20x - 1000)$ for $T(x)$ and

$$\lim_{x \to 20{,}000^+} T(x) = \lim_{x \to 20{,}000^+} (0.20x - 1000) = 3000.$$

$\boxed{49}$ (a) As $t \to 0^+$, $F(t)$ approaches the hole at $(0, 2)$.

Thus, $\lim_{t \to 0^+} F(t) = 2$; at liftoff there is a force of 2 g's.

(b) As $t \to 3.5^-$, $F(t)$ approaches the dot at $(3.5, 8)$. Thus,

$\lim_{t \to 3.5^-} F(t) = 8$; just before the second booster is released, the force is 8 g's.

As $t \to 3.5^+$, $F(t)$ approaches the hole at $(3.5, 1)$.

Thus, $\lim_{t \to 3.5^+} F(t) = 1$; just after the second booster is released, the force is 1 g.

(c) As $t \to 5^-$, $F(t)$ approaches the dot at $(5, 3)$. Thus,

$\lim_{x \to 5^-} F(t) = 3$; just before the spacecraft's engines shut off, the force is 3 g's.

As $t \to 5^+$, $F(t)$ approaches the hole at $(5, 0)$ along the horizontal line $F(t) = 0$.

Thus, $\lim_{x \to 5^+} F(t) = 0$; just after the spacecraft's engines shut off,

there is no force.

Note: In Exercises 51–56, answers may vary depending on the type of calculator used. Round-off will affect answers. The values in the tables were found using double precision. Since we cannot enter arbitrarily small values on a calculator, we cannot even begin to use a calculator to prove that a limit exists.

$\boxed{51}$

x	$1 + x$	$1/x$	$(1 + x)^{1/x}$
0.1	1.10	10	2.5937
-0.1	0.90	-10	2.8680
0.01	1.01	100	2.7048
-0.01	0.99	-100	2.7320
0.001	1.001	1000	2.7169
-0.001	0.999	-1000	2.7196

As x becomes smaller, $(1 + x)^{1/x}$ approaches 2.72 (approximately).

$\boxed{55}$

$x*$	$\dfrac{1}{|x|}$	$\dfrac{4^{|x|} + 9^{|x|}}{2}$	$\left(\dfrac{4^{|x|} + 9^{|x|}}{2}\right)^{1/|x|}$
0.1	10	1.1972146	6.0495
0.01	100	1.0180874	6.0049
0.001	1000	1.0017934	6.0005

$*$ Only positive values for x were used since the sign of x does not affect the value of any of the expressions.

Exercises 2.2

$\boxed{1}$ Substitute v for f, t for x, c for a, and K for L in (2.4) and (2.5).

(a) $\lim\limits_{t \to c} v(t) = K$ means that for every $\epsilon > 0$,

there is a $\delta > 0$ such that if $0 < |t - c| < \delta$, then $\left| v(t) - K \right| < \epsilon$.

(b) $\lim\limits_{t \to c} v(t) = K$ means that for every $\epsilon > 0$,

there is a $\delta > 0$ such that if t is in the open interval $(c - \delta, c + \delta)$

and $t \neq c$, then $v(t)$ is in the open interval $(K - \epsilon, K + \epsilon)$.

$\boxed{3}$ Substitute g for f, p for a, and C for L in (2.4) and (2.5). Since $x \to p^-$, $x < p$. Replace $0 < |x - a| < \delta$ by $p - \delta < x < p$ in (2.4). Replace $(a - \delta, a + \delta)$ by $(p - \delta, p)$ in (2.5).

(a) $\lim\limits_{x \to p^-} g(x) = C$ means that for every $\epsilon > 0$,

there is a $\delta > 0$ such that if $p - \delta < x < p$, then $\left| g(x) - C \right| < \epsilon$.

(b) $\lim\limits_{x \to p^-} g(x) = C$ means that for every $\epsilon > 0$,

there is a $\delta > 0$ such that if x is in the open interval $(p - \delta, p)$,

then $g(x)$ is in the open interval $(C - \epsilon, C + \epsilon)$.

$\boxed{7}$ Since $\dfrac{4x^2 - 9}{2x - 3} = \dfrac{(2x - 3)(2x + 3)}{2x - 3} = 2x + 3$ if $x \neq \frac{3}{2}$, the graph of $y = \dfrac{4x^2 - 9}{2x - 3}$ is

the line $y = 2x + 3$ with a hole at the point $(\frac{3}{2}, 6)$. To obtain the 6, substitute $\frac{3}{2}$ for

x in the reduced expression, $y = 2x + 3$. For $\epsilon = 0.01$, use the lines $y = L - \epsilon =$

$6 - 0.01$ and $y = L + \epsilon = 6 + 0.01$. Hence, $L - \epsilon < \dfrac{4x^2 - 9}{2x - 3} < L + \epsilon \Leftrightarrow$

$5.99 < 2x + 3 < 6.01 \Leftrightarrow 2.99 < 2x < 3.01$, or, equivalently, $1.495 < x < 1.505$.

Thus, δ must be within $1.505 - \frac{3}{2} = \frac{3}{2} - 1.495 = \underline{0.005}$ units of $\frac{3}{2}$.

$\boxed{9}$ Since $\epsilon = 0.1$, $L - \epsilon = 16 - 0.1 = 15.9$ and $L + \epsilon = 16 + 0.1 = 16.1$.

Thus, the graph of $y = x^2$ must lie between the lines $y = 15.9$ and $y = 16.1$, or

$15.9 < x^2 < 16.1$. Solving for x, we see that $\sqrt{15.9} < x < \sqrt{16.1}$.

Now, $4 - \delta < x < 4 + \delta \Rightarrow$ that δ is the *minimum* of $4 - \sqrt{15.9}\ \{\approx 0.01252\}$ and

$\sqrt{16.1} - 4\ \{\approx 0.01248\}$. Thus, $\delta = \sqrt{16.1} - 4$.

Note: If $\delta = 4 - \sqrt{15.9}$ and $x = 4 + \delta = 4 + 4 - \sqrt{15.9} \approx 4.01252$,

then $x^2 \approx 16.1003 > 16.1$. This is why we must pick the minimum.

13 *Note:* When proving that a limit exists, remember that someone else picks the $\epsilon > 0$ that they want. Then, using that ϵ, we must choose a δ that satisfies (2.4) or (2.5).

Here, $f(x) = 5x$, $L = 15$, and $a = 3$. Then $\left|f(x) - L\right| = \left|(5x) - 15\right| = 5\left|x - 3\right|$.

Thus, $\left|f(x) - L\right| < \epsilon \Leftrightarrow 5\left|x - 3\right| < \epsilon \Leftrightarrow \left|x - 3\right| < \frac{\epsilon}{5}$.

Hence, we may choose $\delta = \frac{\epsilon}{5}$. Now if $0 < \left|x - 3\right| < \delta$,

$$\text{then } \left|f(x) - L\right| = 5\left|x - 3\right| < 5\delta = 5\left(\tfrac{\epsilon}{5}\right) = \epsilon, \text{ as desired.}$$

Note: We could have selected *any* positive $\delta < \frac{\epsilon}{5}$.

It is only required that we satisfy the definition with one particular value of δ.

Note: This is often a confusing topic for calculus students. It is sometimes helpful to substitute specific values for ϵ and δ to help illustrate and solidify the concept. For example, if ϵ is chosen to be 0.05, then δ must be less than 0.01 to assure that $5x$ is within ϵ of 15.

Note: As a physical example, suppose that four 1-foot boards are to be stacked end-to-end, and that their total length must be within $\frac{1}{8}$ inch of 4 feet. We can guarantee this final condition if each board is cut within $\frac{1}{4}$ of $\frac{1}{8}$ inch, or $\frac{1}{32}$ inch, of its desired 1-foot length. Identifying the notations of Definition (2.4) in this example, we have x denoting the length of an individual board, $f(x) = 4x$ {the sum of the boards}, $L = 4$ ft {the desired total length}, $a = 1$ ft {the desired length of each board}, $\epsilon = \frac{1}{8}$ inch {the total error tolerance}, and $\delta = \frac{1}{32}$ inch {the individual error tolerance}.

15 Here, $f(x) = 2x + 1$, $L = -5$, and $a = -3$. Then $\left|f(x) - L\right| = \left|(2x + 1) + 5\right| = \left|2x + 6\right| = 2\left|x + 3\right|$. Thus, $\left|f(x) - L\right| < \epsilon \Leftrightarrow 2\left|x + 3\right| < \epsilon \Leftrightarrow \left|x + 3\right| < \frac{\epsilon}{2}$.

Hence, we may choose $\delta = \frac{\epsilon}{2}$. Now if $0 < \left|x + 3\right| < \delta$,

$$\text{then } \left|f(x) - L\right| = 2\left|x + 3\right| < 2\delta = 2\left(\tfrac{\epsilon}{2}\right) = \epsilon, \text{ as desired.}$$

19 Here, $f(x) = 3 - \frac{1}{2}x$, $L = 0$, and $a = 6$. Then $\left|f(x) - L\right| = \left|(3 - \frac{1}{2}x) - 0\right| = \left|-\frac{1}{2}\right|\left|x - 6\right| = \frac{1}{2}\left|x - 6\right|$. Thus, $\left|f(x) - L\right| < \epsilon \Leftrightarrow \frac{1}{2}\left|x - 6\right| < \epsilon \Leftrightarrow \left|x - 6\right| < 2\epsilon$.

Hence, we may choose $\delta = 2\epsilon$. Now if $0 < \left|x - 6\right| < \delta$,

$$\text{then } \left|f(x) - L\right| = \tfrac{1}{2}\left|x - 6\right| < \tfrac{1}{2}\delta = \tfrac{1}{2}(2\epsilon) = \epsilon, \text{ as desired.}$$

23 *Note:* The graph of $f(x) = c$ is a horizontal line.

Thus, $\left|f(x) - c\right| = 0$ is *always* less than $\epsilon > 0$, regardless of the value of x.

Here, $f(x) = c$, $L = c$, and a is arbitrary. Then $\left|f(x) - L\right| = \left|c - c\right| = 0 < \epsilon \; \forall x$ and any $\epsilon > 0$. So any $\delta > 0$ will satisfy (2.4), that is, δ can be chosen arbitrarily.

25 Let $f(x) = x^2$. For any small positive ϵ consider the lines $y = L + \epsilon = a^2 + \epsilon$ and $y = L - \epsilon = a^2 - \epsilon$ in *Figure 25*. These lines intersect the graph of f at points with x-coordinates $-\sqrt{a^2 + \epsilon}$ and $-\sqrt{a^2 - \epsilon}$. $(y = x^2 = a^2 - \epsilon \Rightarrow x = -\sqrt{a^2 - \epsilon}$ and $y = x^2 = a^2 + \epsilon \Rightarrow x = -\sqrt{a^2 + \epsilon}, x < 0.)$ If $x \in (-\sqrt{a^2 + \epsilon}, -\sqrt{a^2 - \epsilon})$, then $f(x) \in (a^2 - \epsilon, a^2 + \epsilon)$. Thus, if we choose δ less than or equal to the minimum of $\left[-\sqrt{a^2 - \epsilon} - (-a) \right] = a - \sqrt{a^2 - \epsilon}$ and $\left[(-a) - (-\sqrt{a^2 + \epsilon}) \right] = \sqrt{a^2 + \epsilon} - a$, it follows that $x \in (a - \delta, a + \delta) \Rightarrow x \in (-\sqrt{a^2 + \epsilon}, -\sqrt{a^2 - \epsilon}) \Rightarrow$

$$f(x) \in (a^2 - \epsilon, a^2 + \epsilon). \text{ By (2.5), } \lim_{x \to a} x^2 = a^2.$$

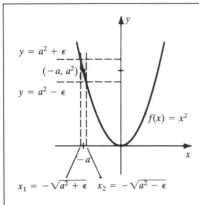

Figure 25

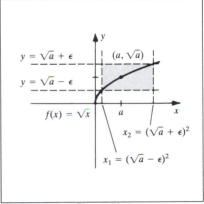

Figure 29

29 Let $f(x) = \sqrt{x}$. For any small positive ϵ consider the lines $y = L + \epsilon = \sqrt{a} + \epsilon$ and $y = L - \epsilon = \sqrt{a} - \epsilon$ in *Figure 29*. These lines intersect the graph of f at points with x-coordinates $(\sqrt{a} + \epsilon)^2$ and $(\sqrt{a} - \epsilon)^2$. $(y = \sqrt{x} = \sqrt{a} - \epsilon \Rightarrow x = (\sqrt{a} - \epsilon)^2$ and $y = \sqrt{x} = \sqrt{a} + \epsilon \Rightarrow x = (\sqrt{a} + \epsilon)^2.)$ If $x \in ((\sqrt{a} - \epsilon)^2, (\sqrt{a} + \epsilon)^2)$, then $f(x) \in (\sqrt{a} - \epsilon, \sqrt{a} + \epsilon)$. Thus, if we choose δ less than or equal to the minimum of $\left[(\sqrt{a} + \epsilon)^2 - a \right]$ and $\left[a - (\sqrt{a} - \epsilon)^2 \right]$, it follows that $x \in (a - \delta, a + \delta) \Rightarrow x \in ((\sqrt{a} - \epsilon)^2, (\sqrt{a} + \epsilon)^2) \Rightarrow f(x) \in (\sqrt{a} - \epsilon, \sqrt{a} + \epsilon)$. By (2.5), $\lim_{x \to a} \sqrt{x} = \sqrt{a}$.

33 Since $f(x) = \dfrac{3x + 3}{|x + 1|} = \dfrac{3(x + 1)}{|x + 1|}$, we have $f(x) = 3$ when $x + 1 > 0$ $\{x > -1\}$ and $f(x) = -3$ when $x + 1 < 0$ $\{x < -1\}$. We will use proof by contradiction. Assume the limit L exists. For each $\epsilon \le 1$ there is a δ that satisfies (2.4). Let $x_1 \in (-1 - \delta, -1)$ and $x_2 \in (-1, -1 + \delta)$. Then $x_1 < -1 \Rightarrow f(x_1) = -3$ and $x_2 > -1 \Rightarrow f(x_2) = 3$. So $6 = |f(x_2) - f(x_1)| = \left| \left[f(x_2) - L \right] - \left[f(x_1) - L \right] \right| \le |f(x_2) - L| + |f(x_1) - L| < \epsilon + \epsilon \le 2$.

This is a contradiction and hence the limit does not exist.

$\boxed{35}$ Note that as x becomes small {close to 0}, $1/x^2$ becomes a very large number. By making x small enough, $1/x^2$ can be made larger than any $L + \epsilon$. Formally, we will use proof by contradiction. Assume the limit L exists. Then by (2.4), there is a $\delta > 0$ such that $x \in (-\delta, \delta)$, $x \neq 0 \Rightarrow 1/x^2 \in (L - \epsilon, L + \epsilon)$. But this is impossible since $1/x^2$ can be made arbitrarily large by picking $|x|$ small enough — that is, $0 < |x| < \dfrac{1}{\sqrt{|L + \epsilon|}} \Rightarrow 1/x^2 > L + \epsilon$. This is a contradiction and the limit does not exist.

$\boxed{37}$ If $\lim\limits_{x \to -5} \dfrac{1}{x + 5} = L$ exists, then given any $\epsilon > 0$, we could find $\delta > 0$ such that if $-5 - \delta < x < -5 + \delta$, $x \neq -5$, then $L - \epsilon < \dfrac{1}{x + 5} < L + \epsilon$ by (2.5).

But this is impossible since $\dfrac{1}{x + 5}$ can be made larger than $L + \epsilon$ by

making $x + 5 > 0$ small enough.

$\boxed{41}$ Any interval containing a contains both rational and irrational numbers. As a result, $f(x) = 0$ and $f(x) = 1$ near a. Regardless of the value of L, $|f(x) - L| \geq \frac{1}{2}$ because x can be either rational or irrational. Formally, we use proof by contradiction. Assume L exists. Let $\epsilon \leq \frac{1}{4}$ and δ satisfy (2.4). The interval $(a - \delta, a + \delta)$ contains both rational and irrational numbers. Let x_1 and x_2 be rational and irrational numbers, respectively, in this interval. Then, $1 = |f(x_2) - f(x_1)| = \left| \left[f(x_2) - L \right] - \left[f(x_1) - L \right] \right| \leq |f(x_2) - L| + |f(x_1) - L| < \epsilon + \epsilon \leq \frac{1}{2}$.

This is a contradiction and so L does not exist.

Exercises 2.3

Note: L denotes the original limit for the exercise.

$\boxed{5}$ By (2.9), we can substitute 4 for x into the expression.

Hence, $\lim\limits_{x \to 4} (3x - 4) = 3(4) - 4 = 8$. Note that (2.9) is just one particular case that is handled by the more general theorem in (2.11), which says we can simply

substitute a value a for the variable x to find $\lim\limits_{x \to a} f(x)$.

$\boxed{7}$ By (2.8)(iii) with $f(x) = x - 5$ and $g(x) = 4x + 3$, we can substitute -2 for x since

$$g(-2) = -5 \neq 0. \text{ Hence, } \lim\limits_{x \to -2} \frac{x - 5}{4x + 3} = \frac{-7}{-5} = \frac{7}{5}.$$

$\boxed{9}$ By (2.10)(ii) with $f(x) = -2x + 5$ and $n = 4$,

$$\lim\limits_{x \to 1} (-2x + 5)^4 = \left[\lim\limits_{x \to 1} (-2x + 5) \right]^4 = \left[-2(1) + 5 \right]^4 = 3^4 = 81.$$

$\boxed{13}$ By (2.11) with $f(x) = 3x^3 - 2x + 7$,

$$\lim\limits_{x \to -2} (3x^3 - 2x + 7) = (-24 + 4 + 7) = -13.$$

$\boxed{15}$ By (2.8)(ii) with $f(x) = x^2 + 3$ and $g(x) = x - 4$, we have

$$\lim_{x \to \sqrt{2}} (x^2 + 3)(x - 4) = \lim_{x \to \sqrt{2}} (x^2 + 3) \cdot \lim_{x \to \sqrt{2}} (x - 4) = 5(\sqrt{2} - 4) = 5\sqrt{2} - 20.$$

$\boxed{21}$ If we substitute $\frac{1}{2}$ for x into the given expression, we obtain $\frac{0}{0}$, which is undefined.

If we factor the numerator and the denominator, there will be a common factor,

$2x - 1$, to cancel. The limit may then be evaluated by substituting $\frac{1}{2}$ for x since the

denominator is not zero. $\displaystyle\lim_{x \to 1/2} \frac{2x^2 + 5x - 3}{6x^2 - 7x + 2} = \lim_{x \to 1/2} \frac{(2x - 1)(x + 3)}{(2x - 1)(3x - 2)} =$

$$\lim_{x \to 1/2} \frac{x + 3}{3x - 2} = \frac{\frac{1}{2} + 3}{3(\frac{1}{2}) - 2} = -7.$$

$\boxed{23}$ $\displaystyle\lim_{x \to 2} \frac{x^2 - x - 2}{(x - 2)^2} = \lim_{x \to 2} \frac{(x - 2)(x + 1)}{(x - 2)(x - 2)} = \lim_{x \to 2} \frac{x + 1}{x - 2}.$

Now $\displaystyle\lim_{x \to 2} \frac{x + 1}{x - 2}$ DNE because the numerator approaches 3 while the denominator

approaches 0. The absolute value of this ratio will become arbitrarily large.

$\boxed{25}$ Substitution of -2 for x leads to the undefined expression $\frac{0}{0}$.

Thus, we must change the form of the given expression by factoring and reducing.

$$\lim_{x \to -2} \frac{x^3 + 8}{x^4 - 16} = \lim_{x \to -2} \frac{(x + 2)(x^2 - 2x + 4)}{(x + 2)(x - 2)(x^2 + 4)} = \lim_{x \to -2} \frac{x^2 - 2x + 4}{(x - 2)(x^2 + 4)} =$$

$$\frac{(-2)^2 - 2(-2) + 4}{(-2 - 2)\left[(-2)^2 + 4\right]} = \frac{12}{-32} = -\frac{3}{8}.$$

$\boxed{27}$ $\displaystyle\lim_{x \to 2} \frac{(1/x) - (1/2)}{x - 2} = \lim_{x \to 2} \frac{\frac{1}{x} - \frac{1}{2}}{x - 2}$ {equivalent form}

$$= \lim_{x \to 2} \frac{\frac{2}{2x} - \frac{x}{2x}}{x - 2} \qquad \{\text{lcd is } 2x\}$$

$$= \lim_{x \to 2} \frac{\frac{2 - x}{2x}}{x - 2} \qquad \{\text{combine terms}\}$$

$$= \lim_{x \to 2} \frac{-(x - 2)}{2x(x - 2)} \qquad \{\text{simplify, } -(x - 2) = 2 - x\}$$

$$= \lim_{x \to 2} \frac{-1}{2x} \qquad \{\text{cancel } x - 2\}$$

$$= -\frac{1}{4} \qquad \{\text{let } x = 2\}$$

$\boxed{29}$ $\displaystyle\lim_{x \to 1} \left(\frac{x^2}{x - 1} - \frac{1}{x - 1} \right) = \lim_{x \to 1} \frac{x^2 - 1}{x - 1}$ {combine terms}

$$= \lim_{x \to 1} \frac{(x - 1)(x + 1)}{x - 1} \qquad \{\text{factor } x^2 - 1\}$$

$$= \lim_{x \to 1} (x + 1) \qquad \{\text{cancel } x - 1\}$$

$$= 2 \qquad \{\text{let } x = 1\}$$

[37] A common strategy to change the form of a fraction is to rationalize the numerator or denominator. In this case, there is nothing to factor, and rationalizing the numerator is a reasonable attempt at changing the form of the given expression.

$$\lim_{h \to 0} \frac{4 - \sqrt{16 + h}}{h} = \lim_{h \to 0} \frac{4 - \sqrt{16 + h}}{h} \cdot \frac{4 + \sqrt{16 + h}}{4 + \sqrt{16 + h}}$$

$$= \lim_{h \to 0} \frac{16 - 4\sqrt{16 + h} + 4\sqrt{16 + h} - (\sqrt{16 + h})^2}{h(4 + \sqrt{16 + h})}$$

$$= \lim_{h \to 0} \frac{16 - (16 + h)}{h(4 + \sqrt{16 + h})} = \lim_{h \to 0} \frac{-h}{h(4 + \sqrt{16 + h})}$$

$$= \lim_{h \to 0} \frac{-1}{4 + \sqrt{16 + h}} = \frac{-1}{4 + \sqrt{16 + 0}} = -\frac{1}{8}$$

[39] If we substitute 1 for x into the expression, both numerator and denominator are equal to 0. This tells us that 1 is a root of both $x^2 + x - 2$ and $x^5 - 1$, i.e., $x - 1$ is a factor of each polynomial. Using synthetic or long division,

$\dfrac{x^5 - 1}{x - 1} = x^4 + x^3 + x^2 + x + 1$. Hence, $\displaystyle\lim_{x \to 1} \dfrac{x^2 + x - 2}{x^5 - 1} =$

$$\lim_{x \to 1} \frac{(x + 2)(x - 1)}{(x - 1)(x^4 + x^3 + x^2 + x + 1)} = \lim_{x \to 1} \frac{x + 2}{x^4 + x^3 + x^2 + x + 1} = \frac{3}{5}.$$

[45] From algebra, $\sqrt[n]{(\text{Expression})^n} = |\text{Expression}|$ if n is an even positive integer, and $\sqrt[n]{(\text{Expression})^n} = \text{Expression}$ if n is an odd positive integer.

Since x approaches 3 from the right, $x > 3$ and $x - 3 > 0$. Thus, $|x - 3| = x - 3$.

$$\lim_{x \to 3^+} \frac{\sqrt{(x - 3)^2}}{x - 3} = \lim_{x \to 3^+} \frac{|x - 3|}{x - 3} = \lim_{x \to 3^+} \frac{x - 3}{x - 3} = \lim_{x \to 3^+} (1) = 1.$$

[47] $x \to 5^+ \Rightarrow x > 5 \Rightarrow 2x > 10$. Thus, $2x - 10 > 0$ and $\sqrt{2x - 10}$ is always defined.

So, $\displaystyle\lim_{x \to 5^+} \dfrac{1 + \sqrt{2x - 10}}{x + 3} = \dfrac{1 + 0}{5 + 3} = \dfrac{1}{8}$. Note that *if* the original limit was

$$\lim_{x \to 5^-} \frac{1 + \sqrt{2x - 10}}{x + 3} \text{ or } \lim_{x \to 5} \frac{1 + \sqrt{2x - 10}}{x + 3}, \text{ the limit would not exist.}$$

[49] (a) $x \to 5^- \Rightarrow x < 5 \Rightarrow 5 > x \Rightarrow 5 - x > 0$ and $\sqrt{5 - x}$ is defined.

Thus, $\displaystyle\lim_{x \to 5^-} \sqrt{5 - x} = \sqrt{5 - 5} = 0$.

(b) $\displaystyle\lim_{x \to 5^+} \sqrt{5 - x}$ DNE since $5 - x < 0$ for $x > 5$.

(c) $\displaystyle\lim_{x \to 5} \sqrt{5 - x}$ DNE since the limit in (b) does not exist.

[53] When working with functions like f, it is important to discover a pattern of results. To find a general pattern, it is helpful to examine some specific values. Thus, we start by finding the value of f for a few intervals.

If $x \in [-2, -1)$, then $n = -2$. $f(x) = (-1)^{-2} = \dfrac{1}{(-1)^2} = 1$.

If $x \in [-1, 0)$, then $n = -1$. $f(x) = (-1)^{-1} = \dfrac{1}{(-1)^1} = -1$.

If $x \in [0, 1)$, then $n = 0$. $f(x) = (-1)^0 = 1$.

If $x \in [1, 2)$, then $n = 1$. $f(x) = (-1)^1 = -1$.

If $x \in [2, 3)$, then $n = 2$. $f(x) = (-1)^2 = 1$.

Hence, f assumes only two values, -1 and 1, and we obtain *Figure 53*. If $x \to n^-$, then x is in the interval preceding $[n, n + 1)$, namely, $[n - 1, n)$. The function value is (-1) raised to the left-endpoint of the interval, i.e., $\lim_{x \to n^-} f(x) = (-1)^{n-1}$. Similarly, if $x \to n^+$, then x is in the interval $[n, n + 1)$, and $\lim_{x \to n^+} f(x) = (-1)^n$.

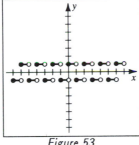

Figure 53

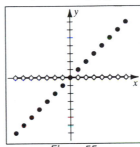

Figure 55

[55] $f(x)$ always equals 0 except when x is an integer. Thus, the graph in *Figure 55* is the line $y = 0$ (the x-axis) and the dots on the line $y = x$, which occur at every integer value. As $x \to n^-$ or $x \to n^+$, $\underline{x \ne n}$, and so $f(x) = 0$ in both of these cases. Hence, $\lim_{x \to n^-} f(x) = 0$ and $\lim_{x \to n^+} f(x) = 0$.

[57] See Figure 1.19 on page 18 of the text. In words, as x approaches any integer value from the right, then y is approaching the dot that is equal to that integer value. If x approaches any integer value from the left, then y is approaching the open circle that is equal to one less than that integer value. Mathematically, recall that if $n < x < n + 1$, then $[\![x]\!] = n$ and $\lim_{x \to n^+} [\![x]\!] = n$. Also, if $n - 1 < x < n$, then $[\![x]\!] = n - 1$ and $\lim_{x \to n^-} [\![x]\!] = n - 1$. Thus,

(a) $\lim_{x \to n^-} [\![x]\!] = n - 1$ (b) $\lim_{x \to n^+} [\![x]\!] = n$.

59 (a) $\qquad x \to n^- \Rightarrow n - 1 < x < n$ $\qquad \{ x \text{ is less than } n \}$

$\qquad\qquad \Rightarrow -(n - 1) > -x > -n$ $\qquad \{ \text{obtaining } -x \}$

$\qquad\qquad \Rightarrow -n < -x < -(n - 1)$ $\qquad \{ \text{equivalent inequality} \}$

$\qquad\qquad \Rightarrow [\![-x]\!] = -n$ $\qquad\qquad \{ [\![\ ]\!] \text{ is the integer to the left} \}$

Thus, $\lim\limits_{x \to n^-} (-[\![-x]\!]) = -\lim\limits_{x \to n^-} [\![-x]\!] = -(-n) = n$.

(b) Similar to the steps in part (a), $x \to n^+ \Rightarrow n < x < n + 1 \Rightarrow$

$\qquad -(n + 1) < -x < -n \Rightarrow [\![-x]\!] = -(n + 1) \Rightarrow$

$$\lim_{x \to n^+} (-[\![-x]\!]) = -\lim_{x \to n^+} [\![-x]\!] = -(-(n + 1)) = n + 1.$$

61 Considering the hint, we need to compare the values of x^2 and $|x|$ as x gets close to 0.

We know that if $-1 \le x \le 1$, then $0 \le x^2 \le |x|$ and $1 \le x^2 + 1 \le |x| + 1$.

But $\lim\limits_{x \to 0} 1 = 1$ and $\lim\limits_{x \to 0} (|x| + 1) = 1$ as given. By the sandwich theorem (2.15)

$\qquad$ with $f(x) = 1$, $g(x) = |x| + 1$, $h(x) = x^2 + 1$, and $L = 1$, $\lim\limits_{x \to 0} (x^2 + 1) = 1$.

63 For all $x \ne 0$, the absolute value of the sine is less than or equal to 1 and we have

$$0 \le \left| \sin (1/x) \right| \le 1.$$

We multiply this inequality by $|x|$ since we eventually want to obtain $x \sin (1/x)$.

$$|x| \cdot 0 \le |x| \left| \sin (1/x) \right| \le |x| \cdot 1$$

Simplifying, we obtain

$$0 \le \left| x \sin (1/x) \right| \le |x|.$$

If $|b| \le a$, then $-a \le b \le a$. Thus,

$$-|x| \le x \sin (1/x) \le |x|.$$

Since $\lim\limits_{x \to 0} (-|x|) = 0$ and $\lim\limits_{x \to 0} |x| = 0$, by the sandwich theorem with

$$f(x) = -|x|, \ g(x) = |x|, \ h(x) = x \sin (1/x), \text{ and } L = 0, \ \lim_{x \to 0} x \sin (1/x) = 0.$$

65 Since $0 \le f(x) \le c$ and $x^2 \ge 0$, $0 \le x^2 f(x) \le x^2 c$.

Now, $\lim\limits_{x \to 0} 0 = 0$ and $\lim\limits_{x \to 0} cx^2 = 0$. By the sandwich theorem with

$$f(x) = 0, \ g(x) = x^2 c, \ h(x) = x^2 f(x), \text{ and } L = 0, \ \lim_{x \to 0} x^2 f(x) = 0.$$

69 (a) $\lim\limits_{T \to -273^+} V = \lim\limits_{T \to -273^+} \left[V_0 (1 + \tfrac{1}{273} T) \right] = V_0 (1 + -1) = 0$

(b) If $T \to -273^-$, then $T < -273\,^\circ\text{C}$, and the volume V is negative, an absurdity.

71 (a) To investigate $\lim\limits_{p \to f^+} q$, we need to obtain an expression for q from the given

$\qquad$ relation. $\dfrac{1}{p} + \dfrac{1}{q} = \dfrac{1}{f} \Rightarrow \dfrac{1}{q} = \dfrac{1}{f} - \dfrac{1}{p} \Rightarrow q = \dfrac{pf}{p - f}$. $\lim\limits_{p \to f^+} q =$

$\qquad \lim\limits_{p \to f^+} \dfrac{pf}{p - f}$ DNE since the ratio becomes unbounded in the positive sense.

(b) As $p \to f^+$, q is becoming larger and the image is moving farther to the right and approaching an infinite distance from the lens.

Exercises 2.4

Note: Let LS denote $\lim\limits_{x \to a-} f(x)$, RS denote $\lim\limits_{x \to a+} f(x)$, and L denote $\lim\limits_{x \to a} f(x)$.

$\boxed{1}$ (a) As $x \to 4^-$, $x < 4$ and $(x - 4) \to 0^-$. In the ratio $\dfrac{5}{x - 4}$, the numerator is always positive and the denominator is always negative. Hence, the ratio is negative and becomes larger in the negative sense as $(x - 4) \to 0^-$. Thus, LS $= -\infty$.

(b) As $x \to 4^+$, $x > 4$ and $(x - 4) \to 0^+$. In the ratio $\dfrac{5}{x - 4}$, the numerator and denominator are always positive. Hence, the ratio is positive and becomes larger in the positive sense as $(x - 4) \to 0^+$. Thus, RS $= \infty$.

(c) Since the left-hand and right-hand limits are not equal, L DNE.

$\boxed{5}$ (a) As $x \to -8^-$, $3x \to -24$ and $(x + 8)^2 \to 0^+$. In the ratio $\dfrac{3x}{(x + 8)^2}$, the numerator is always negative and the denominator is always positive (since anything squared is nonnegative). Hence, the ratio is negative and becomes larger in the negative sense as $(x + 8)^2 \to 0^+$ and $3x \to -24$. Thus, LS $= -\infty$.

(b) As $x \to -8^+$, $3x \to -24$ and $(x + 8)^2 \to 0^+$. Thus, RS $= -\infty$.

(c) Since the left-hand and right-hand limits both equal $-\infty$, L $= -\infty$.

$\boxed{7}$ (a) As $x \to -1^-$, $2x^2 \to 2$. Factoring, $(x^2 - x - 2) = (x - 2)(x + 1)$. As $x \to -1^-$, $(x - 2) \to -3$ and $(x + 1) \to 0^-$ $(x < -1)$. Since $(x - 2)$ and $(x + 1)$ are both negative, their product, $(x - 2)(x + 1)$, is positive. The ratio $\dfrac{2x^2}{x^2 - x - 2}$ is positive and becomes larger as $(x^2 - x - 2) \to 0^+$ and $2x^2 \to 2$. Thus, LS $= \infty$.

(b) As $x \to -1^+$, $2x^2 \to 2$. As $x \to -1^+$, $(x - 2) \to -3$ and $(x + 1) \to 0^+$ $(x > -1)$. Since $(x - 2)$ is negative and $(x + 1)$ is positive, $(x^2 - x - 2) \to 0^-$, and is negative. The ratio $\dfrac{2x^2}{x^2 - x - 2}$ is negative and becomes larger in the negative sense as $(x^2 - x - 2) \to 0^-$ and $2x^2 \to 2$. Thus, RS $= -\infty$.

(c) Since the left-hand and right-hand limits are not equal, L DNE.

Note: The first step in 11–24 is the result of dividing the expression by the term containing the highest power of x in the denominator.

$\boxed{11}$ The highest power of x in the denominator is 2.

Divide each term in the numerator and denominator by x^2.

$$\lim_{x \to \infty} \frac{5x^2 - 3x + 1}{2x^2 + 4x - 7} = \lim_{x \to \infty} \frac{5x^2/x^2 - 3x/x^2 + 1/x^2}{2x^2/x^2 + 4x/x^2 - 7/x^2} =$$

$$\lim_{x \to \infty} \frac{5 - 3/x + 1/x^2}{2 + 4/x - 7/x^2} = \frac{5 - 0 + 0}{2 + 0 - 0} \, \{ \text{using (2.18)} \} = \frac{5}{2}$$

Note: Consider the general case of the limit of the quotient of two polynomials.

$$\lim_{x \to \infty} \frac{ax^m + \cdots + b}{cx^n + \cdots + d}$$

The leading terms (ax^m and cx^n) determine the behavior of the limit.

We investigate the three possible cases:

(i) $m < n$: The denominator is growing faster than the numerator and the function approaches the x-axis, $y = 0$.

(ii) $m = n$: The denominator and numerator are growing at similar rates and the function approaches the *ratio of leading coefficients*, $\frac{a}{c}$.

(iii) $m > n$: The numerator is growing faster than the denominator and the limit will be $+\infty$ or $-\infty$, depending on the signs of a and c. Using long division, we can determine the general behavior of the function as $x \to \infty$ and $x \to -\infty$.

This is a very important concept, and we will use this extensively in chapters 4, 5, 10, and 11 (among other places) to provide insight into solutions. You should, of course, solve the problems as your instructor requires, but this generalization will prove to be helpful for your intuitive sense of solving problems.

$\boxed{15}$ Intuitively, since the degree of the numerator, 2, is less than the degree of the denominator, 3, we think that the limit will be 0. Formally, the highest power of x in the denominator is 3. Divide each term in the numerator and denominator by x^3.

$$\lim_{x \to -\infty} \frac{2x^2 - 3}{4x^3 + 5x} = \lim_{x \to -\infty} \frac{2x^2/x^3 - 3/x^3}{4x^3/x^3 + 5x/x^3} =$$

$$\lim_{x \to -\infty} \frac{2/x - 3/x^3}{4 + 5/x^2} = \frac{0 - 0}{4 + 0} \, \{ \text{using (2.18)} \} = \frac{0}{4} = 0$$

$\boxed{17}$ Intuitively, the degree of the numerator, 3, is greater than the degree of the denominator, 2 — so the ratio increases without bound. Since the ratio of leading coefficients is $-\frac{1}{2}$, we think this limit will be $-\infty$. Formally, the highest power of x in the denominator is 2. Divide each term in the numerator and denominator by x^2.

$$\lim_{x \to \infty} \frac{-x^3 + 2x}{2x^2 - 3} = \lim_{x \to \infty} \frac{-x^3/x^2 + 2x/x^2}{2x^2/x^2 - 3/x^2} =$$

$$\lim_{x \to \infty} \frac{-x + 2/x}{2 - 3/x^2} = \frac{-\infty + 0}{2 - 0} = \frac{-\infty}{2} = -\infty$$

It is important to remember that $-\infty$ is *not* a real number. In this exercise the numerator *approaches* $-\infty$ while the denominator approaches 2. Thus, the ratio *approaches* $-\infty$.

$\boxed{21}$ Intuitively, since the degree of the numerator, 2, is equal to the degree of the denominator, 2, the limit will approach the cube root of the ratio of leading coefficients, i.e., $\sqrt[3]{1/1}$, or 1. Formally, the highest power of x in the denominator inside the cube root is 2. Divide each term inside the cube root by x^2. Actually, we have divided each term by $\sqrt[3]{x^2} = x^{2/3}$.

$$\lim_{x \to \infty} \sqrt[3]{\frac{8 + x^2}{x(x+1)}} = \lim_{x \to \infty} \sqrt[3]{\frac{8 + x^2}{x^2 + x}} =$$

$$\lim_{x \to \infty} \sqrt[3]{\frac{8/x^2 + x^2/x^2}{x^2/x^2 + x/x^2}} = \lim_{x \to \infty} \sqrt[3]{\frac{8/x^2 + 1}{1 + 1/x}} = \sqrt[3]{\frac{0 + 1}{1 + 0}} = \sqrt[3]{1} = 1$$

$\boxed{23}$ $\lim_{x \to \infty} \sin x$ DNE since $\sin x$ does not approach a real number L,

but rather oscillates between -1 and 1 as x increases without bound.

Note: Let VA and HA denote vertical and horizontal asymptote, respectively. The vertical asymptotes are found by finding the zeros of the denominator in the *reduced form* of $f(x)$.

The horizontal asymptote is found by finding $\lim_{x \to \infty} f(x)$ and $\lim_{x \to -\infty} f(x)$.

$\boxed{27}$ Since $f(x) = \dfrac{1}{x^2 - 4} = \dfrac{1}{(x+2)(x-2)}$, the denominator of f is zero at $x = \pm 2$ while the numerator is never zero. Thus, VA occur at $x = -2$ and $x = 2$.

$$\lim_{x \to \infty} \frac{1}{x^2 - 4} = \lim_{x \to \infty} \frac{1/x^2}{x^2/x^2 - 4/x^2} = \lim_{x \to \infty} \frac{1/x^2}{1 - 4/x^2} = \frac{0}{1 - 0} = 0. \qquad \text{Similarly,}$$

$\lim_{x \to -\infty} \dfrac{1}{x^2 - 4} = 0.$ There is one HA: $y = 0$. You will see later in the text that there are cases when two horizontal asymptotes exist for a function.

$\boxed{29}$ Since $x^2 + 1 \neq 0$ for any x, the denominator of f is never zero and there are no VA.

$$\lim_{x \to \infty} \frac{2x^2}{x^2 + 1} = \lim_{x \to \infty} \frac{2x^2/x^2}{x^2/x^2 + 1/x^2} = \lim_{x \to \infty} \frac{2}{1 + 1/x^2} = \frac{2}{1 + 0} = 2.$$

Similarly, $\lim_{x \to -\infty} \dfrac{2x^2}{x^2 + 1} = 2$. There is one HA: $y = 2$.

$\boxed{33}$ Since $f(x) = \dfrac{x^2 + 3x + 2}{x^2 + 2x - 3} = \dfrac{(x + 1)(x + 2)}{(x + 3)(x - 1)}$, the denominator of f is zero at

$x = -3, 1$, while the numerator is not zero at $x = -3, 1$. Thus, VA occur at

$x = -3$ and $x = 1$. $\displaystyle\lim_{x \to \infty} \frac{x^2 + 3x + 2}{x^2 + 2x - 3} = \lim_{x \to \infty} \frac{x^2/x^2 + 3x/x^2 + 2/x^2}{x^2/x^2 + 2x/x^2 - 3/x^2} =$

$\displaystyle\lim_{x \to \infty} \frac{1 + 3/x + 2/x^2}{1 + 2/x - 3/x^2} = \frac{1 + 0 + 0}{1 + 0 - 0} = 1.$ Similarly, $\displaystyle\lim_{x \to -\infty} \frac{x^2 + 3x + 2}{x^2 + 2x - 3} = 1.$

There is one HA: $y = 1$.

$\boxed{35}$ Since $f(x) = \dfrac{x + 4}{x^2 - 16} = \dfrac{x + 4}{(x + 4)(x - 4)} = \dfrac{1}{x - 4}$ if $x \neq -4$, the denominator of f is

zero at $x = 4$ and the numerator is nonzero. There is a VA at $x = 4$. No VA occurs

at $x = -4$ because $\displaystyle\lim_{x \to -4} f(x) = \lim_{x \to -4} \frac{1}{x - 4} = -\frac{1}{8}.$ Rather, there is a hole in the

graph of the function at $(-4, -\frac{1}{8})$. $\displaystyle\lim_{x \to \infty} \frac{x + 4}{x^2 - 4} = \lim_{x \to \infty} \frac{x/x^2 + 4/x^2}{x^2/x^2 - 4/x^2} =$

$\displaystyle\lim_{x \to \infty} \frac{1/x + 4/x^2}{1 - 4/x^2} = \frac{0 + 0}{1 - 0} = 0.$ Similarly, $\displaystyle\lim_{x \to -\infty} \frac{x + 4}{x^2 - 4} = 0.$

There is one HA: $y = 0$.

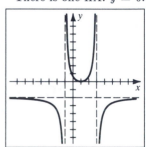

$\boxed{39}$ $\displaystyle\lim_{x \to -\infty} f(x) = \lim_{x \to \infty} f(x) = -2 \Rightarrow$ HA at $y = -2$.

$\displaystyle\lim_{x \to 3^-} f(x) = \infty$ and $\displaystyle\lim_{x \to 3^+} f(x) = -\infty \Rightarrow$

VA at $x = 3$.

$\displaystyle\lim_{x \to -1^-} f(x) = -\infty$ and $\displaystyle\lim_{x \to -1^+} f(x) = \infty \Rightarrow$

VA at $x = -1$.

First sketch in the asymptotes $y = -2$, $x = 3$, and

Figure 39

$x = -1$ as dashed lines. Next, try to draw portions of the graph that satisfy each

limit. It may take a few attempts — the answer is not unique. One possible sketch is

given in *Figure 39*.

$\boxed{41}$ (a) Since 5 gallons of water flow into the tank each minute and there is initially 50

gallons of water in the tank, the volume is 50 gallons plus 5 times the number of

minutes, i.e., $V(t) = 50 + 5t$. Since each additional gallon of water contains 0.1

lb of salt, $A(t) = 5(0.1)t = 0.5t$.

(b) The concentration is given by the amount of salt divided by the volume of water.

Hence, $c(t) = \dfrac{A(t)}{V(t)} = \dfrac{0.5t}{5t + 50} = \dfrac{t}{10t + 100}$ lb/gal.

(c) To see what happens over a long period of time, let $t \to \infty$. $\displaystyle\lim_{t \to \infty} c(t) =$

$\displaystyle\lim_{t \to \infty} \frac{t}{10t + 100} = \lim_{t \to \infty} \frac{t/t}{10t/t + 100/t} = \lim_{t \to \infty} \frac{1}{10 + 100/t} = \frac{1}{10} = 0.1$ lb/gal.

Thus, $c(t)$ approaches 0.1, which is the concentration of the entering salt water. This will make sense if we remember that no water is leaving the tank and only salt water is entering at a rate of 5 gallons per minute. After a long period of time, the 50 gallons of pure water will become insignificant compared to the large amount of salt water and the concentration will be almost 0.1. However, it will always be slightly less than 0.1 for any finite amount of time.

Exercises 2.5

$\boxed{1}$ f is discontinuous at $x = 2$ because $\displaystyle\lim_{x \to 2} f(x)$ does not exist.

It is a jump discontinuity since $\displaystyle\lim_{x \to 2^-} f(x) = 3$ and $\displaystyle\lim_{x \to 2^+} f(x) = 1$.

$\boxed{3}$ f is discontinuous at $x = 2$ because $\displaystyle\lim_{x \to 2} f(x) \neq f(2)$. The discontinuity is removable

since we could define $f(2)$ to equal 1 and then f would be continuous.

$\boxed{7}$ f is discontinuous at $x = 2$ because $f(2)$ is undefined.

The discontinuity is infinite since $\displaystyle\lim_{x \to 2^-} f(x) = -\infty$ and $\displaystyle\lim_{x \to 2^+} f(x) = \infty$.

$\boxed{11}$ See *Figure 41* in §2.1. Since $\displaystyle\lim_{x \to 1^-} f(x) = \lim_{x \to 1^-} (x^2 - 1) = 0$ and

$\displaystyle\lim_{x \to 1^+} f(x) = \lim_{x \to 1^+} (4 - x) = 3$, there is a jump discontinuity at 1.

$\boxed{13}$ Since $\displaystyle\lim_{x \to -2} f(x) = \lim_{x \to -2} |x + 3| = 1 \neq 2 = f(-2)$, the discontinuity is removable.

$\boxed{17}$ f has a discontinuity at $x = 0$ since $f(0)$ is undefined. To determine the type of discontinuity, we must approximate $\displaystyle\lim_{x \to 0^+} f(x)$ and $\displaystyle\lim_{x \to 0^-} f(x)$.

From trigonometry, $\cos\left(\frac{\pi}{2} - \alpha\right) = \sin \alpha$ is a cofunction identity.

Hence, we can rewrite $f(x) = x^{-1/3} \sin\left[\cos\left(\frac{\pi}{2} - x^2\right)\right]$ as $f(x) = \displaystyle\lim_{x \to 0} \frac{\sin\left[\sin\left(x^2\right)\right]}{\sqrt[3]{x}}$.

$$x = \pm 0.1 \Rightarrow f(x) = \pm 0.02154$$
$$x = \pm 0.01 \Rightarrow f(x) = \pm 0.00046$$
$$x = \pm 0.001 \Rightarrow f(x) = \pm 0.00001$$

It *appears* that $\displaystyle\lim_{x \to 0^+} f(x) = \lim_{x \to 0^-} f(x) = 0$. Thus, the discontinuity is removable.

19 Since $\lim_{x \to 4} \sqrt{2x - 5} = \sqrt{3}$ and $\lim_{x \to 4} 3x = 12$,

$\lim_{x \to 4} f(x) = \lim_{x \to 4} (\sqrt{2x - 5} + 3x) = \sqrt{3} + 12$. Also, $f(4) = \sqrt{3} + 12$.

Since $\lim_{x \to 4} f(x) = f(4)$, f is continuous at $x = 4$. *Note:* We only need to verify (iii) of

(2.20) to show that a function is continuous at a number c.

23 Since f is not defined at -2, (2.20)(i) is not satisfied.

25 $\lim_{x \to 3} f(x) = \lim_{x \to 3} \dfrac{x^2 - 9}{x - 3} = \lim_{x \to 3} \dfrac{(x - 3)(x + 3)}{x - 3} = \lim_{x \to 3} (x + 3) = 6$.

However, $f(3) = 4$. Since $\lim_{x \to 3} f(x) \neq f(3)$,

f is not continuous at $x = 3$ because (2.20)(iii) is not satisfied.

29 Refer to Figure 2.2 and the accompanying table in the text.

It *appears* that $\lim_{x \to 0} f(x) = \lim_{x \to 0} \dfrac{\sin x}{x} = 1$, but since $f(0) = 0$,

f is not continuous at $x = 0$ because (2.20)(iii) is not satisfied.

31 $f(x) = \dfrac{3}{x^2 + x - 6} = \dfrac{3}{(x + 3)(x - 2)}$. The denominator is zero if $x = -3$ or $x = 2$.

Thus, by (2.21)(ii), f is discontinuous at $x = -3$ and $x = 2$.

35 To show that a function f is continuous on a closed interval $[a, b]$,

we must show that the following three conditions are true:

    (i) f is continuous for all x in the *open* interval (a, b).

    (ii) f is continuous from the right at a.

    (iii) f is continuous from the left at b.

  (i)   If $4 < c < 8$, $\lim_{x \to c} f(x) = \lim_{x \to c} \sqrt{x - 4} = \sqrt{c - 4} = f(c)$.

Thus, f is continuous on $(4, 8)$.

  (ii) $\lim_{x \to 4^+} f(x) = \lim_{x \to 4^+} \sqrt{x - 4} = \sqrt{4 - 4} = 0 = f(4)$.

  (iii) $\lim_{x \to 8^-} f(x) = \lim_{x \to 8^-} \sqrt{x - 4} = \sqrt{8 - 4} = 2 = f(8)$.

Hence, f is continuous on $[4, 8]$ by (2.22).

37 If $c > 0$, $\lim_{x \to c} f(x) = \lim_{x \to c} (1/x^2) = 1/c^2 = f(c)$.

Hence, f is continuous on $(0, \infty)$ by (2.20).

Note: For 39–54, each function f is continuous on its domain.

39 $f(x) = \dfrac{3x - 5}{2x^2 - x - 3} = \dfrac{3x - 5}{(2x - 3)(x + 1)}$. By (2.21)(ii), f is continuous at every

number except when $(2x - 3)(x + 1) = 0$. f is continuous at $\{x : x \neq -1, \frac{3}{2}\}$.

43 f will be continuous when $\sqrt{x^2 - 1}$ is a nonzero real number. This occurs when

$x^2 - 1 > 0 \Rightarrow x^2 > 1 \Rightarrow |x| > 1$. Thus, f is continuous on $(-\infty, -1) \cup (1, \infty)$.

[49] f will be continuous when $\sqrt{x^2 - 9}$ and $\sqrt{25 - x^2}$ are both defined *and* $x \neq 4$.

$\sqrt{x^2 - 9}$ is defined when $x^2 \geq 9$ or $|x| \geq 3$. $\sqrt{25 - x^2}$ is defined when

$25 \geq x^2$ or $|x| \leq 5$. Thus, f is continuous on $[-5, -3] \cup [3, 4) \cup (4, 5]$.

[51] f will be continuous when $\tan 2x$ is defined. Recalling that the tangent function is

undefined at $x = \frac{\pi}{2}$ and has period π, we see that $\tan 2x$ is undefined when

$2x = \frac{\pi}{2} + \pi n$, or $x = \frac{\pi}{4} + \frac{\pi}{2}n$. Thus, $\tan 2x$ is continuous at $\{x : x \neq \frac{\pi}{4} + \frac{\pi}{2}n\}$.

[53] f will be continuous when $\csc \frac{1}{2}x$ is defined. Recalling that the cosecant function is

undefined when the sine function is zero, we see that $\csc \frac{1}{2}x$ is undefined when

$\frac{1}{2}x = \pi n$, or $x = 2\pi n$. Thus, f is continuous at $\{x : x \neq 2\pi n\}$.

[55] To verify (2.26), we need to find the value of c in $[a, b]$ such that $f(c) = w$, where w

is in $[f(a), f(b)]$. $f(-1) \leq w \leq f(2) \Leftrightarrow 0 \leq w \leq 9$; $f(c) = w \Rightarrow c^3 + 1 = w \Rightarrow$

$c^3 = w - 1 \Rightarrow c = \sqrt[3]{w - 1}$. Thus, if $0 \leq w \leq 9$, then $c \in [-1, 2]$.

[57] $f(1) \leq w \leq f(3) \Leftrightarrow 0 \leq w \leq 6$; $f(c) = w \Rightarrow c^2 - c = w \Rightarrow c^2 - c - w = 0 \Rightarrow$

$c = \dfrac{1 \pm \sqrt{1 + 4w}}{2}$ (Use the quadratic formula with $a = 1$, $b = -1$, and $c = -w$.)

$= \frac{1}{2} \pm \frac{1}{2}\sqrt{4w + 1}$. If c is going to be in the interval $[1, 3]$, then we must choose the

"+" to *add* to $\frac{1}{2}$. Thus, $c = \dfrac{1 + \sqrt{1 + 4w}}{2} = \frac{1}{2} + \frac{1}{2}\sqrt{4w + 1}$ for $1 \leq c \leq 3$.

Note: If $w = 0$, then $c = 1$, and if $w = 6$, then $c = 3$.

[59] We must find some number x_1 such that $f(x_1) < 100$ and some number x_2 such that

$f(x_2) > 100$. These numbers are not unique and may require some trial and error.

$x_1 = 0$ and $x_2 = 10$ are values that will satisfy the IVT. Hence, $f(0) = -9 < 100$

and $f(10) = 561 > 100$. Since f is continuous on $[0, 10]$, there is at least one number

a in $[0, 10]$ such that $f(a) = 100$.

| 2.6 Review Exercises |

[3] Since $\lim\limits_{x \to -2} (4x^2 + x) = 14 > 0$, $\lim\limits_{x \to -2} (2x - \sqrt{4x^2 + x}) = -4 - \sqrt{14}$.

[5] If we substitute $x = \frac{3}{2}$ into the expression, we obtain $\frac{0}{0}$, which is undefined. If we

factor both numerator and denominator, we can cancel the common factor, $2x - 3$.

$$\lim_{x \to 3/2} \frac{2x^2 + x - 6}{4x^2 - 4x - 3} = \lim_{x \to 3/2} \frac{(2x - 3)(x + 2)}{(2x - 3)(2x + 1)} = \lim_{x \to 3/2} \frac{x + 2}{2x + 1} =$$

$$\frac{\frac{3}{2} + 2}{2(\frac{3}{2}) + 1} = \frac{7}{8}$$

[9] As $x \to 0^+$, $\sqrt{x} \to 0^+$, so $\lim\limits_{x \to 0^+} \dfrac{1}{\sqrt{x}}$ DNE since the ratio becomes unbounded $(+\infty)$.

[11] If we substitute $x = \frac{1}{2}$ into the expression, we obtain $\frac{0}{0}$, which is undefined. If we factor both numerator and denominator, we can cancel the common factor, $2x - 1$.

$$\lim_{x \to 1/2} \frac{8x^3 - 1}{2x - 1} = \lim_{x \to 1/2} \frac{(2x - 1)(4x^2 + 2x + 1)}{2x - 1} = \lim_{x \to 1/2} (4x^2 + 2x + 1) =$$

$$4(\tfrac{1}{2})^2 + 2(\tfrac{1}{2}) + 1 = 3$$

[13] As $x \to 3^+$, $x > 3$, $0 > 3 - x$, or, equivalently, $3 - x < 0$.

$$\text{Thus, } \lim_{x \to 3^+} \frac{3 - x}{|3 - x|} = \lim_{x \to 3^+} \frac{3 - x}{-(3 - x)} = \lim_{x \to 3^+} (-1) = -1.$$

[15] Use the binomial theorem (located on the lower left corner of the back end papers of the text) to expand $(a + h)^4$.

$$(a + h)^4 = a^4 + \binom{4}{1} a^3 h^1 + \binom{4}{2} a^2 h^2 + \binom{4}{3} a^1 h^3 + h^4$$

$$= a^4 + \frac{4!}{1!\,3!} a^3 h + \frac{4!}{2!\,2!} a^2 h^2 + \frac{4!}{3!\,1!} ah^3 + h^4$$

$$= a^4 + 4a^3 h + 6a^2 h^2 + 4ah^3 + h^4$$

$$\lim_{h \to 0} \frac{(a + h)^4 - a^4}{h} = \lim_{h \to 0} \frac{(a^4 + 4a^3 h + 6a^2 h^2 + 4ah^3 + h^4) - a^4}{h} \quad \{\text{expand}\}$$

$$= \lim_{h \to 0} \frac{h(4a^3 + 6a^2 h + 4ah^2 + h^3)}{h} \quad \{\text{combine terms, factor } h\}$$

$$= \lim_{h \to 0} (4a^3 + 6a^2 h + 4ah^2 + h^3) \quad \{\text{cancel } h\}$$

$$= 4a^3 + 0 + 0 + 0 = 4a^3 \quad \{\text{let } h = 0\}$$

[17] If we substitute $x = -3$ into the expression, we obtain $\frac{0}{0}$, which is undefined.

Factor $x^3 + 27 = x^3 + (3)^3$ by using the sum of two cubes formula.

$$\lim_{x \to -3} \sqrt[3]{\frac{x + 3}{x^3 + 27}} = \lim_{x \to -3} \sqrt[3]{\frac{x + 3}{(x + 3)(x^2 - 3x + 9)}} = \lim_{x \to -3} \frac{1}{\sqrt[3]{x^2 - 3x + 9}} =$$

$$\frac{1}{\sqrt[3]{(-3)^2 - 3(-3) + 9}} = \frac{1}{\sqrt[3]{27}} = \frac{1}{3}$$

[19] Intuitively, leading terms of the numerator and denominator are $6x^2$ and $4x^2$, respectively. Since they have the same degree, 2, we think the limit will be the ratio of leading coefficients, $\frac{6}{4}$, or $\frac{3}{2}$. Formally, if we multiply out the denominator, the highest power of x is 2. Dividing the numerator and denominator by x^2 is equivalent to dividing each factor by x.

$$\lim_{x \to -\infty} \frac{(2x - 5)(3x + 1)}{(x + 7)(4x - 9)} = \lim_{x \to -\infty} \frac{(2x/x - 5/x)(3x/x + 1/x)}{(x/x + 7/x)(4x/x - 9/x)} =$$

$$\lim_{x \to -\infty} \frac{(2 - 5/x)(3 + 1/x)}{(1 + 7/x)(4 - 9/x)} = \frac{2 \cdot 3}{1 \cdot 4} = \frac{3}{2}$$

[23] As $x \to (\frac{2}{3})^+$, $x^2 \to (\frac{4}{9})^+$, and $4 - 9x^2 \to 0^-$. Thus, $\lim_{x \to 2/3^+} \frac{x^2}{4 - 9x^2} = -\infty$.

$\boxed{25}$ $\sqrt{x} - \frac{1}{\sqrt{x}} = \frac{x-1}{\sqrt{x}}$. As $x \to 0^+$, $x - 1 \to -1$ and $\sqrt{x} \to 0^+$. Thus, as $x \to 0^+$,

the ratio becomes unbounded in the negative sense and so $\lim\limits_{x \to 0^+} \left(\sqrt{x} - \frac{1}{\sqrt{x}} \right) = -\infty$.

$\boxed{29}$ From *Figure 29*, if $x < -3$, the graph of $f(x) = \frac{1}{2 - 3x}$ is increasing and positive.

There is a horizontal asymptote of $y = 0$.

(a) $\lim\limits_{x \to -3^-} f(x) = \lim\limits_{x \to -3^-} \frac{1}{2 - 3x} = \frac{1}{2 - 3(-3)} = \frac{1}{11}$

(b) $\lim\limits_{x \to -3^+} f(x) = \lim\limits_{x \to -3^+} \sqrt[3]{x + 2} = \sqrt[3]{-3 + 2} = \sqrt[3]{-1} = -1$

(c) Since the left-hand and right-hand limits are not equal, $\lim\limits_{x \to -3} f(x)$ DNE.

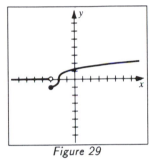

Figure 29

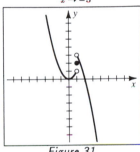

Figure 31

$\boxed{31}$ (a) If $x \to 1^-$, $x < 1$, and $\lim\limits_{x \to 1^-} f(x) = \lim\limits_{x \to 1^-} x^2 = 1$.

(b) If $x \to 1^+$, $x > 1$, and $\lim\limits_{x \to 1^+} f(x) = \lim\limits_{x \to 1^+} (4 - x^2) = 3$.

(c) $\lim\limits_{x \to 1} f(x)$ DNE since the left-hand and right-hand limits are not equal.

$\boxed{33}$ Here, $f(x) = 5x - 21$, $L = 9$, and $a = 6$. Then $|f(x) - L| = |(5x - 21) - 9|$

$= |5x - 30| = 5|x - 6|$. Thus, $|f(x) - L| < \epsilon \Leftrightarrow 5|x - 6| < \epsilon \Leftrightarrow |x - 6| < \frac{\epsilon}{5}$.

Hence, we may choose $\delta = \frac{\epsilon}{5}$. Now if $0 < |x - 6| < \delta$,

then $|f(x) - L| = 5|x - 6| < 5\delta = 5(\frac{\epsilon}{5}) = \epsilon$, as desired.

$\boxed{35}$ f will be discontinuous whenever f is undefined by (2.20)(i).

$$f(x) = \frac{|x^2 - 16|}{x^2 - 16} = \frac{|x^2 - 16|}{(x + 4)(x - 4)}; f \text{ is discontinuous at } -4 \text{ and } 4.$$

Note: For 39–42, since each f is a sum, difference, product, etc. of continuous functions,

each f is continuous on its domain.

$\boxed{41}$ f will be continuous whenever $\sqrt{9 - x^2}$ is defined *and* $x^4 - 16 \neq 0$.

$9 - x^2 \geq 0 \Rightarrow 9 \geq x^2 \Rightarrow |x| \leq 3$; $x^4 - 16 \neq 0 \Rightarrow x^2 \neq \pm 4 \Rightarrow x \neq \pm 2$.

f is continuous on $[-3, -2) \cup (-2, 2) \cup (2, 3]$.

43 By (2.20), f is continuous at $x = 8$ since

$$\lim_{x \to 8} f(x) = \lim_{x \to 8} \sqrt{5x + 9} = \sqrt{\lim_{x \to 8} (5x + 9)} = \sqrt{49} = 7 = f(8).$$

Chapter 3: The Derivative

1 (a) $m_a = \lim\limits_{h \to 0} \dfrac{f(a + h) - f(a)}{h}$ { Definition (3.1) }

$= \lim\limits_{h \to 0} \dfrac{\left[5(a + h)^2 - 4(a + h)\right] - (5a^2 - 4a)}{h}$

                        { use the definition of f, $f(x) = 5x^2 - 4x$ }

$= \lim\limits_{h \to 0} \dfrac{\left[5(a^2 + 2ah + h^2) - 4a - 4h\right] - (5a^2 - 4a)}{h}$ { expand }

$= \lim\limits_{h \to 0} \dfrac{5a^2 + 10ah + 5h^2 - 4a - 4h - 5a^2 + 4a}{h}$ { simplify }

$= \lim\limits_{h \to 0} \dfrac{10ah + 5h^2 - 4h}{h}$ { combine terms }

$= \lim\limits_{h \to 0} \dfrac{h(10a + 5h - 4)}{h}$ { factor out h }

$= \lim\limits_{h \to 0} (10a + 5h - 4)$ { cancel h }

$= 10a - 4$ { let $h = 0$ }

(b) $m_2 = 10(2) - 4 = 16$ and $f(2) = 12$. Using the point-slope form of a line (1.8)(ii), $y - y_1 = m(x - x_1)$, with $m = m_2 = 16$ and $(x_1, y_1) = (2, 12)$ yields $y - 12 = 16(x - 2)$, or $y = 16x - 20$.

3 (a) $m_a = \lim\limits_{h \to 0} \dfrac{f(a + h) - f(a)}{h}$ { Definition (3.1) }

$= \lim\limits_{h \to 0} \dfrac{(a + h)^3 - a^3}{h}$ { use the definition of f, $f(x) = x^3$ }

$= \lim\limits_{h \to 0} \dfrac{(a^3 + 3a^2h + 3ah^2 + h^3) - a^3}{h}$ { expand $(a + h)^3$ }

$= \lim\limits_{h \to 0} \dfrac{3a^2h + 3ah^2 + h^3}{h}$ { simplify }

$= \lim\limits_{h \to 0} \dfrac{h(3a^2 + 3ah + h^2)}{h}$ { factor out h }

$= \lim\limits_{h \to 0} (3a^2 + 3ah + h^2)$ { cancel h }

$= 3a^2$ { let $h = 0$ }

(b) $m_2 = 3(2)^2 = 12$ and $f(2) = 8$. $y - y_1 = m(x - x_1)$, with $m = m_2 = 12$ and $(x_1, y_1) = (2, 8)$ yields $y - 8 = 12(x - 2)$, or $y = 12x - 16$.

$\boxed{7}$ (a) Let $f(x) = \sqrt{x}$. Then,

$$m_a = \lim_{h \to 0} \frac{f(a + h) - f(a)}{h} \qquad \{\text{Definition (3.1)}\}$$

$$= \lim_{h \to 0} \frac{\sqrt{a + h} - \sqrt{a}}{h} \qquad \{\text{definition of } f, f(x) = \sqrt{x}\}$$

$$= \lim_{h \to 0} \frac{\sqrt{a + h} - \sqrt{a}}{h} \cdot \frac{\sqrt{a + h} + \sqrt{a}}{\sqrt{a + h} + \sqrt{a}} \qquad \{\text{multiply by the conjugate}\}$$

$$= \lim_{h \to 0} \frac{(\sqrt{a + h})^2 - \sqrt{a}\sqrt{a + h} + \sqrt{a}\sqrt{a + h} - (\sqrt{a})^2}{h(\sqrt{a + h} + \sqrt{a})} \qquad \{\text{expand}\}$$

$$= \lim_{h \to 0} \frac{(a + h) - a}{h(\sqrt{a + h} + \sqrt{a})} \qquad \{\text{simplify}\}$$

$$= \lim_{h \to 0} \frac{h}{h(\sqrt{a + h} + \sqrt{a})} \qquad \{\text{simplify}\}$$

$$= \lim_{h \to 0} \frac{1}{\sqrt{a + h} + \sqrt{a}} \qquad \{\text{cancel } h\}$$

$$= \frac{1}{2\sqrt{a}}. \qquad \{\text{let } h = 0\}$$

(b) $m_4 = \frac{1}{2\sqrt{4}} = \frac{1}{4}$. $y - y_1 = m(x - x_1)$, with $m = m_4 = \frac{1}{4}$ and $(x_1, y_1) = (4, 2)$

yields $y - 2 = \frac{1}{4}(x - 4)$, or $y = \frac{1}{4}x + 1$.

(c) Sketch the graph of $y = \sqrt{x}$ and plot the point $(4, 2)$. Try to sketch the tangent line that intersects the graph *only* at the point $(4, 2)$. This is the line given by $y = \frac{1}{4}x + 1$, which has y-intercept $(0, 1)$. See *Figure 7*.

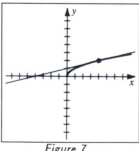

Figure 7

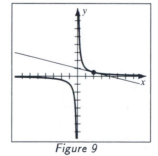

Figure 9

$\boxed{9}$ (a) Let $f(x) = 1/x$. Then,

$$m_a = \lim_{h \to 0} \frac{f(a + h) - f(a)}{h} \qquad \{\text{Definition (3.1)}\}$$

$$= \lim_{h \to 0} \frac{1/(a + h) - 1/a}{h} \qquad \{\text{definition of } f,\ f(x) = 1/x\}$$

$$= \lim_{h \to 0} \frac{1/(a + h) - 1/a}{h} \cdot \frac{(a + h)a}{(a + h)a} \qquad \{\text{multiply by lcd}\}$$

$$= \lim_{h \to 0} \frac{a - (a + h)}{h(a + h)a} \qquad \{\text{simplify}\}$$

$$= \lim_{h \to 0} \frac{-h}{h(a + h)a} \qquad \{\text{simplify}\}$$

$$= \lim_{h \to 0} \frac{-1}{(a + h)a} \qquad \{\text{cancel } h\}$$

$$= -\frac{1}{a^2} \qquad \{\text{let } h = 0\}$$

(b) $m_2 = -\dfrac{1}{(2)^2} = -\frac{1}{4}$. $y - y_1 = m(x - x_1)$, with $m = m_2 = -\frac{1}{4}$ and

$(x_1,\ y_1) = (2,\ \frac{1}{2})$ yields $y - \frac{1}{2} = -\frac{1}{4}(x - 2)$, or $y = -\frac{1}{4}x + 1$.

(c) Sketch the graph of $y = 1/x$ for $x > 0$ first. (The graph is symmetric with respect to the origin.) Plot the point $(2,\ \frac{1}{2})$. Try to sketch the tangent line that intersects the graph *only* at the point $(2,\ \frac{1}{2})$. This is the line given by $y = -\frac{1}{4}x + 1$, which has y-intercept $(0,\ 1)$. See *Figure 9*.

$\boxed{13}$ (a) The average velocity of P is equal to the average rate of change of the position function $s(t)$. Using (3.2), $v_{\text{av}} = \dfrac{d}{t} = \dfrac{s(a + h) - s(a)}{h}$. With $s(t) = 4t^2 + 3t$,

the average velocities (in cm/sec) for each interval are as follows.

$[1,\ 1.2]$: $\dfrac{s(1.2) - s(1)}{0.2} = \dfrac{9.36 - 7}{0.2} = 11.8$

$[1,\ 1.1]$: $\dfrac{s(1.1) - s(1)}{0.1} = \dfrac{8.14 - 7}{0.1} = 11.4$

$[1,\ 1.01]$: $\dfrac{s(1.01) - s(1)}{0.01} = \dfrac{7.1104 - 7}{0.01} = 11.04$

(b) The velocity of P at $t = 1$ is equal to the instantaneous rate of change of the position function s with respect to t at $t = 1$. We will find the velocity at time $t = a$ and then find v_1.

$$v_a = \lim_{h \to 0} \frac{s(a + h) - s(a)}{h} \qquad \{\text{Definition (3.3)}\}$$

$$= \lim_{h \to 0} \frac{\left[4(a + h)^2 + 3(a + h)\right] - (4a^2 + 3a)}{h} \qquad \{\text{definition of } s\}$$

$$= \lim_{h \to 0} \frac{4a^2 + 8ah + 4h^2 + 3a + 3h - 4a^2 - 3a}{h} \qquad \{\text{simplify}\}$$

$$= \lim_{h \to 0} \frac{8ah + 4h^2 + 3h}{h} \qquad \{\text{simplify}\}$$

$$= \lim_{h \to 0} \frac{h(8a + 4h + 3)}{h} \qquad \{\text{factor out } h\}$$

$$= \lim_{h \to 0} (8a + 4h + 3) \qquad \{\text{cancel } h\}$$

$$= 8a + 3 \qquad \{\text{let } h = 0\}$$

$v_1 = 8(1) + 3 = 11$ cm/sec.

$\boxed{15}$ (a) Let $s(t) = 160 - 16t^2$ be the position function of the sandbag. The velocity at time $t = a$ is

$$v_a = \lim_{h \to 0} \frac{s(a + h) - s(a)}{h} \qquad \{\text{Definition (3.3)}\}$$

$$= \lim_{h \to 0} \frac{\left[160 - 16(a + h)^2\right] - (160 - 16a^2)}{h} \qquad \{\text{definition of } s\}$$

$$= \lim_{h \to 0} \frac{160 - 16a^2 - 32ah - 16h^2 - 160 + 16a^2}{h} \qquad \{\text{simplify}\}$$

$$= \lim_{h \to 0} \frac{-32ah - 16h^2}{h} \qquad \{\text{simplify}\}$$

$$= \lim_{h \to 0} \frac{h(-32a - 16h)}{h} \qquad \{\text{factor out } h\}$$

$$= \lim_{h \to 0} (-32a - 16h) \qquad \{\text{cancel } h\}$$

$$= -32a \qquad \{\text{let } h = 0\}$$

$v_1 = -32(1) = -32$ ft/sec. (The negative sign indicates a downward direction.)

(b) The sandbag strikes the ground when $s(t) = 0 \Rightarrow 160 - 16t^2 = 0 \Rightarrow$ $16(10 - t^2) = 0 \Rightarrow t^2 = 10 \Rightarrow t = \sqrt{10}$ for $t \geq 0$. Thus, $v_a = -32\sqrt{10}$ ft/sec. *Note:* $s(t)$ is also 0 if $t = -\sqrt{10}$. However, the formula is valid only for $t \geq 0$.

17 First, we must determine the slope of the tangent line to the graph of $y = 1 + 1/x$.

With this slope and the points P and Q, we can then find the equations of the

tangent lines and their x-intercepts. The slope is $m_a = \lim\limits_{h \to 0} \dfrac{f(a + h) - f(a)}{h} =$

$\lim\limits_{h \to 0} \dfrac{\left[1 + 1/(a + h)\right] - (1 + 1/a)}{h} = \lim\limits_{h \to 0} \dfrac{-h}{h(a + h)a} = \lim\limits_{h \to 0} \dfrac{-1}{(a + h)a} = -\dfrac{1}{a^2}.$

(a) At $P(1, 2)$, $m_1 = -\dfrac{1}{(1)^2} = -1$, and the equation of the tangent line is

$(y - 2) = -1(x - 1)$ or $y = -x + 3$. If we let $y = 0$, we see that this line has

x-intercept 3. The creature at $x = 3$ will be hit.

(b) At $Q(\frac{3}{2}, \frac{5}{3})$, the equation of the tangent line is $(y - \frac{5}{3}) = -\frac{4}{9}(x - \frac{3}{2})$ or

$y = -\frac{4}{9}x + \frac{7}{3}$. If we let $y = 0$, then $\frac{4}{9}x = \frac{7}{3}$, or $x = \frac{21}{4}$,

and this line has x-intercept $\frac{21}{4}$. No creature is hit.

19 (a) Using (3.4)(i) with $f(x) = x^2 + 2$, $a = 3$, and $h = 3.5 - 3 = 0.5$,

we have $y_{av} = \dfrac{f(3.5) - f(3)}{0.5} = \dfrac{14.25 - 11}{0.5} = 6.5$.

(b) Using (3.4)(ii), $y_a = \lim\limits_{h \to 0} \dfrac{\left[(a + h)^2 + 2\right] - (a^2 + 2)}{h} = \lim\limits_{h \to 0} \dfrac{2ah + h^2}{h} =$

$\lim\limits_{h \to 0} (2a + h) = 2a$. Thus, $y_3 = 2(3) = 6$.

Remember, an average rate of change is a ratio of differences,

and an instantaneous rate of change is the *limit* of that same ratio.

23 The graph of $f(x) = \sin \pi x$ over $[0, 2]$ will have zeros

at $x = 0, 1, 2$, and be similar in shape to $y = \sin x$.

(a) At $x = 1.4$, the graph of f is sloping from left to

right (downward) at an angle of approximately

$45°$. Thus, $m_{1.4} \approx -1$.

(b) If h is small, then $\dfrac{f(a + h) - f(a)}{h} \approx m_a$.

Let $h = 0.0001$, $a = 1.4$, and $f(x) = \sin \pi x$.

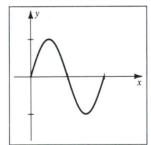

Figure 23

Then, $\dfrac{f(a + h) - f(a)}{h} = \dfrac{f(1.4 + 0.0001) - f(1.4)}{0.0001} = \dfrac{f(1.4001) - f(1.4)}{0.0001} \approx$

$\dfrac{-0.951154 - (-0.951057)}{0.0001} \approx -0.9703$ and similarly, $\dfrac{f(1.3999) - f(1.4)}{-0.0001} \approx$

-0.9713.

Exercises 3.2

$\boxed{3}$ (a) $\begin{aligned} f'(x) &= \lim_{h \to 0} \dfrac{f(x + h) - f(x)}{h} \end{aligned}$ $\{\text{Definition (3.5)}\}$

$= \lim_{h \to 0} \dfrac{\left[(x + h)^3 + (x + h)\right] - (x^3 + x)}{h}$ $\{\text{definition of } f\}$

$= \lim_{h \to 0} \dfrac{(x^3 + 3x^2h + 3xh^2 + h^3) + x + h - x^3 - x}{h}$ $\{\text{expand}\}$

$= \lim_{h \to 0} \dfrac{3x^2h + 3xh^2 + h^3 + h}{h}$ $\{\text{simplify}\}$

$= \lim_{h \to 0} \dfrac{h(3x^2 + 3xh + h^2 + 1)}{h}$ $\{\text{factor out } h\}$

$= \lim_{h \to 0} (3x^2 + 3xh + h^2 + 1)$ $\{\text{cancel } h\}$

$= 3x^2 + 1$ $\{\text{let } h = 0\}$

    (b) Since $f'(x) = 3x^2 + 1$ is always defined, its domain is all real numbers, or $\mathbb{R}$.

    (c) The slope of the tangent line at $P(1, 2)$ is $f'(1) = 4$.

           The equation of the tangent line is $y - 2 = 4(x - 1)$, or $y = 4x - 2$.

    (d) Since the tangent line is horizontal when $f'(x) = 0$ and $f'(x) \neq 0$ for any real value of x, there are no points on the graph where the tangent line is horizontal.

$\boxed{5}$ (a) Using (3.12) with $m = 9$ and $b = -2$, we have $f'(x) = 9$.

    (b) Since $f'(x) = 9$ is always defined, its domain is all real numbers: $\mathbb{R}$.

    (c) $y - y_1 = m(x - x_1)$ with $m = 9$ and $(x_1, y_1) = (3, 25) \Rightarrow y - 25 = 9(x - 3)$ $\Rightarrow y = 9x - 2$.   Note that since f is linear, the tangent line always coincides with f.

    (d) Since $f'(x) = 9 \neq 0$ for any x,

           there are no points on the graph at where the tangent line is horizontal.

$\boxed{7}$ (d) Since $f'(x) = 0$ for every x, the tangent line is horizontal at *all* points on the graph (there was an incorrect answer in the text).

$\boxed{9}$ (a) $f(x) = 1/x^3 = x^{-3} \Rightarrow f'(x) = -3x^{-3-1} = -3x^{-4} = -3/x^4$ using (3.14).

    (b) $f'(x) = -3/x^4$ is defined except when $x = 0$.

           Thus, the domain of f' is $(-\infty, 0) \cup (0, \infty)$.

    (c) The slope of the tangent line at $(2, \frac{1}{8})$ is $f'(2) = -\frac{3}{16}$.

           The tangent line is given by $y - \frac{1}{8} = -\frac{3}{16}(x - 2)$, or $y = -\frac{3}{16}x + \frac{1}{2}$.

    (d) Since $f'(x) = -3/x^4 \neq 0$ for any x, f has no horizontal tangent lines.

$\boxed{11}$ (a) $f(x) = 4x^{1/4} \Rightarrow f'(x) = 4(\frac{1}{4}x^{(1/4)-1}) = x^{-3/4} = 1/x^{3/4}$ using (3.15).

(b) $f'(x) = 1/x^{3/4}$ is defined when $x > 0$ { we cannot take a fourth root of a negative value }. Thus, the domain of f' is $(0, \infty)$.

(c) The slope of the tangent line at $(81, 12)$ is $f'(81) = \frac{1}{27}$.

The tangent line is given by $y - 12 = \frac{1}{27}(x - 81)$, or $y = \frac{1}{27}x + 9$.

(d) Since $f'(x) = 1/x^{3/4} \neq 0$ for any x, f has no horizontal tangent lines.

$\boxed{15}$ $f(x) = 9\sqrt[3]{x^2} = 9x^{2/3} \Rightarrow f'(x) = 9(\frac{2}{3}x^{(2/3)-1}) = 6x^{-1/3}$,

$f''(x) = 6(-\frac{1}{3}x^{(-1/3)-1}) = -2x^{-4/3}, f'''(x) = -2(-\frac{4}{3}x^{(-4/3)-1}) = \frac{8}{3}x^{-7/3}$

$\boxed{17}$ The notation $D_t^2 z$ means to differentiate z twice with respect to t.

$z = 25t^{9/5} \Rightarrow D_t z = 25(\frac{9}{5}t^{(9/5)-1}) = 45t^{4/5}$ and $D_t^2 z = 45(\frac{4}{5}t^{(4/5)-1}) = 36t^{-1/5}$

$\boxed{21}$ $f(x) = 1/x = x^{-1} \Rightarrow f'(x) = -1x^{-1-1} = -x^{-2} = -1/x^2$.

(a) No, because f is not differentiable at $x = 0$

(b) Yes, because f' exists for every number in $[1, 3)$

$\boxed{23}$ (a) From the graph we see that f is not differentiable at $x = 4$ because the slope of the tangent line approaches $-\infty$ as $x \to 4^-$.

(b) f' exists for every number in $[-5, 0]$ because the tangent line exists and has finite slope for each value of $x \in [-5, 0]$.

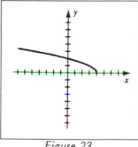

Figure 23

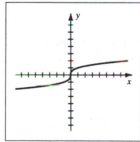

Figure 25

$\boxed{25}$ (a) Using (3.9), $\lim\limits_{x \to 0}|f'(x)| = \lim\limits_{x \to 0}\left|\frac{1}{3}x^{(1/3)-1}\right| = \lim\limits_{x \to 0}\left|\frac{1}{3}x^{-2/3}\right| = \lim\limits_{x \to 0}\left|\frac{1}{3x^{2/3}}\right| = \infty$

$\Rightarrow$ there is a vertical tangent line at $(0, 0)$.

(b) $\lim\limits_{x \to 0^+}\frac{1}{3x^{2/3}} = \infty$ and $\lim\limits_{x \to 0^-}\frac{1}{3x^{2/3}} = \infty$. By (3.10), f does not have a cusp at $(0, 0)$. (Note that $f'(x)$ is always positive.) See *Figure 25*.

$\boxed{27}$ (a) Using (3.9), $\lim\limits_{x \to 0} |f'(x)| = \lim\limits_{x \to 0} \left| \frac{2}{5} x^{-3/5} \right| = \lim\limits_{x \to 0} \left| \frac{2}{5x^{3/5}} \right| = \infty \Rightarrow$

there is a vertical tangent line at $(0, 0)$.

(b) If $x < 0$, then $f'(x) < 0$, and if $x > 0$, then $f'(x) > 0$. As $x \to 0^-$, $f'(x) \to -\infty$.

As $x \to 0^+$, $f'(x) \to \infty$. By (3.10), there is a cusp at $(0, 0)$. See *Figure 27*.

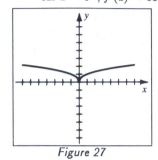

Figure 27

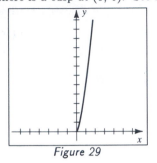

Figure 29

$\boxed{29}$ (a) Using (3.9), $\lim\limits_{x \to 0^+} |f'(x)| = \lim\limits_{x \to 0^+} \left| \frac{15}{2} x^{1/2} \right| = \lim\limits_{x \to 0^+} \left| \frac{15}{2} \sqrt{x} \right| = 0 \Rightarrow$

no vertical tangent line at $(0, 0)$.

(b) Since f' does not exist for negative values of x,

$$\lim\limits_{x \to 0^-} \left(\tfrac{15}{2} \sqrt{x} \right) \neq \pm \infty \text{ and no cusp is formed at } (0, 0). \text{ See *Figure 29*.}$$

$\boxed{31}$ To approximate $f'(x)$ using the figure, we must approximate the slope of the tangent line at each value of x. From the figure, it appears that $f'(-1) = 1$ since the tangent line is about $45°$ above the horizontal; $f'(1) = 0$ since the tangent line is horizontal; $f'(2)$ is undefined since there is a cusp at $x = 2$; and $f'(3) = -1$ since the tangent line is about $45°$ below the horizontal.

$\boxed{35}$ The right-hand derivative of $f(x) = [\![x - 2]\!]$ at $a = 2$ is equal to

$$\lim\limits_{h \to 0^+} \frac{f(2 + h) - f(2)}{h} = \lim\limits_{h \to 0^+} \frac{[\![2 + h - 2]\!] - [\![2 - 2]\!]}{h} = \lim\limits_{h \to 0^+} \frac{[\![h]\!] - 0}{h} =$$

$$\lim\limits_{h \to 0^+} \frac{0 - 0}{h} \{ \text{since } 0 < h < 1 \Rightarrow [\![h]\!] = 0 \} = \lim\limits_{h \to 0^+} \frac{0}{h} = \lim\limits_{h \to 0^+} 0 = 0.$$

The left-hand derivative of f at $a = 2$ is equal to

$$\lim\limits_{h \to 0^-} \frac{f(2 + h) - f(2)}{h} = \lim\limits_{h \to 0^-} \frac{[\![2 + h - 2]\!] - [\![2 - 2]\!]}{h} = \lim\limits_{h \to 0^-} \frac{[\![h]\!]}{h} =$$

$$\lim\limits_{h \to 0^-} \frac{-1}{h} \{ \text{since } -1 < h < 0 \Rightarrow [\![h]\!] = -1 \}, \text{ which DNE.}$$

Since the left-hand derivative fails to exist, the derivative fails to exist at $a = 2$.

39 See *Figure 39.* If $x < -1$, $f(x) = -x^2 \Rightarrow f'(x) = -2x$ and if $x > -1$, $f(x) = 2x + 3 \Rightarrow f'(x) = 2$. The right-hand derivative, 2, is equal to the left-hand derivative, $-2x$, at $x = -1$. However, the function is not continuous at $x = -1$ and hence, by the contrapositive* of Theorem (3.11), cannot be differentiable at that point. The domain of f' is $\{x : x \neq -1\}$.

* The contrapositive of "If P, then Q" is "If not Q, then not P" and is a logically equivalent statement. For (3.11), the contrapositive is "If a function f is not continuous at a, then f is not differentiable at a."

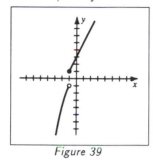

Figure 39

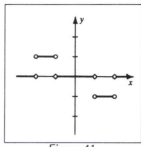

Figure 41

41 We can determine the values of f' by determining the slopes of the tangent lines to the graph of f. On $(-\infty, -2) \cup (-1, 1) \cup (2, \infty)$, the slope is 0. On $(-2, -1)$, the slope is 1 and on $(1, 2)$, the slope is -1. f is differentiable at all of these points, but f is not differentiable at $x = \pm 1, \pm 2$, since f has a corner at these points.

43 $v(t) = s'(t) = 2t^{-1/3}$; $v(t) = 4 \Rightarrow 2t^{-1/3} = 4 \Rightarrow t^{-1/3} = 2 \Rightarrow$
$$(t^{-1/3})^{-3} = (2)^{-3} \Rightarrow t = 2^{-3} = \tfrac{1}{8}.$$

45 $C = \tfrac{5}{9}(F - 32) \Rightarrow \tfrac{9}{5}C = F - 32 \Rightarrow F = \tfrac{9}{5}C + 32$. The rate of change of F with respect to C is equal to $F'(C) = F_C = \tfrac{9}{5}$. This means that a degree of increase on the Celsius scale corresponds to an increase of $\tfrac{9}{5}$ of a degree on the Fahrenheit scale.

51 (a) The formula gives an approximation of the slope of the tangent line at $(a, f(a))$ by using the slope of the secant line through $P(a - h, f(a - h))$ and $Q(a + h, f(a + h))$. To depict this situation, we have sketched $f(x) = \tfrac{1}{4}x^3$ over $[-1, 2.5]$.

If we wanted to approximate $f'(1)$, we would let $a = 1$ and then let $h = 1$ (just one possibility). The points P and Q are then $(0, 0)$ and $(2, 2)$. The slope of the line between P and Q is 1. The actual value of $f'(1)$ is $\tfrac{3}{4}$. We could have made our approximation better by choosing a smaller value for h.

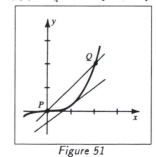

Figure 51

(b) $\lim\limits_{h \to 0} \dfrac{f(a + h) - f(a - h)}{2h}$ { given formula }

$= \lim\limits_{h \to 0} \dfrac{f(a + h) - f(a) - f(a - h) + f(a)}{2h}$ { subtract and add $f(a)$ }

$= \dfrac{1}{2} \lim\limits_{h \to 0} \dfrac{f(a + h) - f(a)}{h} - \dfrac{1}{2} \lim\limits_{h \to 0} \dfrac{f(a - h) - f(a)}{h}$ { split up }

$= \dfrac{1}{2} f'(a) + \dfrac{1}{2} \lim\limits_{k \to 0} \dfrac{f(a + k) - f(a)}{k}$ (where $k = -h$)   { definition of $f'(a)$ }

$= \dfrac{1}{2} f'(a) + \dfrac{1}{2} f'(a) = f'(a)$ { definition of $f'(a)$ }

(c) $h = 0.1$: $f'(1) \approx \dfrac{f(1.1) - f(0.9)}{0.2} = \dfrac{1/(1.1)^2 - 1/(0.9)^2}{0.2} \approx -2.0406$

  $h = 0.01$: $f'(1) \approx \dfrac{f(1.01) - f(0.99)}{0.02} = \dfrac{1/(1.01)^2 - 1/(0.99)^2}{0.02} \approx -2.0004$

  $h = 0.001$: $f'(1) \approx \dfrac{f(1.001) - f(0.999)}{0.002} = \dfrac{1/(1.001)^2 - 1/(0.999)^2}{0.002} \approx -2.0000$

(d) $f(x) = 1/x^2 \Rightarrow f'(x) = -2/x^3; \; f'(1) = -2$

$\boxed{53}$ From Exercise 51, $v(a) = s'(a) \approx \dfrac{s(a + h) - s(a - h)}{2h}$.

(a) $a = 3, \; h = 1$ { because the measurements were taken at 1 second intervals }:

$$v(3) = s'(3) \approx \dfrac{s(3 + 1) - s(3 - 1)}{2(1)} = \dfrac{149.0 - 42.6}{2} = 53.2 \text{ ft/sec}$$

(b) $a = 6, \; h = 1$:

$$v(6) = s'(6) \approx \dfrac{s(6 + 1) - s(6 - 1)}{2(1)} = \dfrac{396.7 - 220.1}{2} = 88.3 \text{ ft/sec}$$

$\boxed{\text{Exercises 3.3}}$

$\boxed{1}$ Using (3.18)(iii), $g(t) = 6t^{5/3} \Rightarrow g'(t) = 6 \cdot \dfrac{5}{3} \cdot t^{2/3} = 10t^{2/3}$.

$\boxed{7}$ Using (3.19), $g'(x) = (x^3 - 7) D_x (2x^2 + 3) + (2x^2 + 3) D_x (x^3 - 7) =$

  $(x^3 - 7)(4x) + (2x^2 + 3)(3x^2) = 4x^4 - 28x + 6x^4 + 9x^2 = 10x^4 + 9x^2 - 28x$.

$\boxed{9}$ We could use the product rule as in Exercise 7, but we choose to multiply the terms
  together first, and then differentiate. $f(x) = x^{1/2}(x^2 + x - 4) =$

$$x^{5/2} + x^{3/2} - 4x^{1/2} \Rightarrow f'(x) = \dfrac{5}{2}x^{3/2} + \dfrac{3}{2}x^{1/2} - 2x^{-1/2}.$$

$\boxed{15}$ Using (3.20), $f'(x) = \dfrac{(3x + 2) D_x (4x - 5) - (4x - 5) D_x (3x + 2)}{(3x + 2)^2} =$

$$\dfrac{(3x + 2)(4) - (4x - 5)(3)}{(3x + 2)^2} = \dfrac{12x + 8 - 12x + 15}{(3x + 2)^2} = \dfrac{23}{(3x + 2)^2}.$$

Note: If you think of a quotient as $\dfrac{\text{Hi}}{\text{Lo}}$, then a mnemonic device that may help you

remember the derivative formula for a quotient is $\dfrac{\text{Lo } d(\text{Hi}) - \text{Hi } d(\text{Lo})}{(\text{Lo})^2}$.

17 Using (3.20),

$$h(z) = \frac{8 - z + 3z^2}{2 - 9z} \Rightarrow h'(z) = \frac{(2 - 9z)(-1 + 6z) - (8 - z + 3z^2)(-9)}{(2 - 9z)^2} =$$

$$\frac{-2 + 9z + 12z - 54z^2 + 72 - 9z + 27z^2}{(2 - 9z)^2} = \frac{70 + 12z - 27z^2}{(2 - 9z)^2}.$$

21 Before solving Exercise 21, we would like to remind you of a factoring property that will often be used in future solutions. In the past, if you factored $x^4 + x^2$, you factored out the greatest common factor, x^2, and wrote the factored form as $x^2(x^2 + 1)$. If you were to factor $x^{-4} + x^{-2}$ in a similar fashion, you would factor out x^{-4} since -4 is the smallest exponent appearing on x (just as 2 was in $x^4 + x^2$). Hence, $x^{-4} + x^{-2} = x^{-4}(x^{-4-(-4)} + x^{-2-(-4)}) = x^{-4}(1 + x^2)$. We use this idea in the next solution. $g(t) = \dfrac{\sqrt[3]{t^2}}{3t - 5} = \dfrac{t^{2/3}}{3t - 5} \Rightarrow$

$$g'(t) = \frac{(3t - 5)(\frac{2}{3}t^{-1/3}) - (t^{2/3})(3)}{(3t - 5)^2} \qquad \{\text{using } (3.20)\}$$

$$= \frac{(3t - 5)(2t^{-1/3}) - 3 \cdot 3(t^{2/3})}{3(3t - 5)^2} \qquad \{\text{factor out } \tfrac{1}{3} \text{ (or multiply by } \tfrac{3}{3})\}$$

$$= \frac{t^{-1/3}\left[2(3t - 5) - 9t\right]}{3(3t - 5)^2} \qquad \{\text{factor out } t^{-1/3}\}$$

$$= -\frac{3t + 10}{3\sqrt[3]{t}(3t - 5)^2} \qquad \{\text{simplify}\}$$

23 By the reciprocal rule (3.21),

$$f'(x) = -\frac{D_x(1 + x + x^2 + x^3)}{(1 + x + x^2 + x^3)^2} = -\frac{1 + 2x + 3x^2}{(1 + x + x^2 + x^3)^2}.$$

29 $K'(s) = D_s(3^{-4}s^{-4}) = 3^{-4}D_s(s^{-4})$ { since 3^{-4} is a constant, (3.18)(iv) } $=$

$$3^{-4}(-4s^{-5}) = -\frac{4}{3^4}s^{-5} = -\frac{4}{81}s^{-5}.$$

33 $g(r) = (5r - 4)^{-2} = \dfrac{1}{(5r - 4)^2} = \dfrac{1}{25r^2 - 40r + 16}.$

Using the reciprocal rule (3.21) gives

$$g'(r) = -\frac{50r - 40}{(25r^2 - 40r + 16)^2} = -\frac{10(5r - 4)}{\left[(5r - 4)^2\right]^2} = -\frac{10(5r - 4)}{(5r - 4)^4} = -\frac{10}{(5r - 4)^3}.$$

35 We first change the form of $f(t)$ by multiplying the numerator and denominator by

the lcd of the entire fraction, $5t^2$. $f(t) = \dfrac{3/(5t) - 1}{(2/t^2) + 7} \cdot \dfrac{5t^2}{5t^2} = \dfrac{3t - 5t^2}{5(2 + 7t^2)}$.

Using (3.20), $f'(t) = \frac{1}{5} \cdot D_t\left(\dfrac{3t - 5t^2}{2 + 7t^2}\right) = \frac{1}{5} \cdot \dfrac{(2 + 7t^2)(3 - 10t) - (3t - 5t^2)(14t)}{(2 + 7t^2)^2} =$

$$\dfrac{6 + 21t^2 - 20t - 70t^3 - 42t^2 + 70t^3}{5(2 + 7t^2)^2} = \dfrac{-21t^2 - 20t + 6}{5(2 + 7t^2)^2}.$$

37 $M(x)$ can be simplified first by dividing the denominator, x^2, into each term of the

numerator. $M(x) = \dfrac{2x^3}{x^2} - \dfrac{7x^2}{x^2} + \dfrac{4x}{x^2} + \dfrac{3}{x^2} = 2x - 7 + 4x^{-1} + 3x^{-2}$.

Thus, $M'(x) = 2 - 4x^{-2} - 6x^{-3} = 2 - \dfrac{4}{x^2} - \dfrac{6}{x^3}$.

43 First, we differentiate y using the quotient rule (3.20).

$y = \dfrac{2x^2 + 3x - 6}{x - 2} \Rightarrow D_x\, y = \dfrac{(x - 2)(4x + 3) - (2x^2 + 3x - 6)(1)}{(x - 2)^2} =$

$\dfrac{4x^2 - 8x + 3x - 6 - 2x^2 - 3x + 6}{(x - 2)^2} = \dfrac{2x^2 - 8x}{(x - 2)^2} = \dfrac{2x(x - 4)}{(x - 2)^2}.$

$D_x\, y = 0$ when its numerator is 0. Thus, $D_x\, y = 0$ when $x = 0$ or $x = 4$.

45 First, we differentiate y twice. $y = 6x^4 + 24x^3 - 540x^2 + 7 \Rightarrow$

$D_x\, y = 24x^3 + 72x^2 - 1080x \Rightarrow D_x^2\, y = 72x^2 + 144x - 1080 =$

$72(x^2 + 2x - 15) = 72(x + 5)(x - 3). \ D_x^2\, y = 0 \Rightarrow x = -5, 3.$

49 (a) Using the quotient rule, $\dfrac{dy}{dx} = \dfrac{x^{2/3}(2x - 3) - (x^2 - 3x)(\frac{2}{3}x^{-1/3})}{(x^{2/3})^2} =$

$\dfrac{3x^{2/3}(2x - 3) - (x^2 - 3x)(2x^{-1/3})}{3(x^{2/3})^2} = \dfrac{x^{-1/3}\Big[3x(2x - 3) - 2(x^2 - 3x)\Big]}{3x^{4/3}} =$

$\dfrac{6x^2 - 9x - 2x^2 + 6x}{x^{1/3}(3x^{4/3})} = \dfrac{4x^2 - 3x}{3x^{5/3}} = \dfrac{x(4x - 3)}{x(3x^{2/3})} = \dfrac{4x - 3}{3\sqrt[3]{x^2}}.$

(b) First, rewrite y as $y = (x^2 - 3x)(x^{-2/3})$ and then apply the product rule.

$\dfrac{dy}{dx} = D_x\Big[(x^2 - 3x)(x^{-2/3})\Big] = (x^2 - 3x)(-\frac{2}{3}x^{-5/3}) + (x^{-2/3})(2x - 3) =$

$\dfrac{-2(x^2 - 3x)}{3x^{5/3}} + \dfrac{2x - 3}{x^{2/3}} \cdot \dfrac{3x}{3x} = \dfrac{-2x^2 + 6x}{3x^{5/3}} + \dfrac{6x^2 - 9x}{3x^{5/3}} =$

$\dfrac{4x^2 - 3x}{3x^{5/3}} = \dfrac{x(4x - 3)}{x(3x^{2/3})} = \dfrac{4x - 3}{3\sqrt[3]{x^2}}.$

(c) First, rewrite y as $y = \dfrac{x^2}{x^{2/3}} - \dfrac{3x}{x^{2/3}} = x^{4/3} - 3x^{1/3}$.

Then, $\dfrac{dy}{dx} = \frac{4}{3}x^{1/3} - x^{-2/3} = \dfrac{4x^{1/3}}{3} \cdot \dfrac{x^{2/3}}{x^{2/3}} - \dfrac{1}{x^{2/3}} \cdot \dfrac{3}{3} = \dfrac{4x - 3}{3\sqrt[3]{x^2}}.$

51 Using the quotient rule,

$$\frac{dy}{dx} = \frac{(x + 1)(3) - (3x + 4)(1)}{(x + 1)^2} = -\frac{1}{(x + 1)^2} = -\frac{1}{x^2 + 2x + 1}.$$ Applying the

reciprocal rule to $\frac{dy}{dx}$ yields $\frac{d^2y}{dx^2} = -\left[-\frac{2x + 2}{(x + 1)^4}\right] = \frac{2(x + 1)}{(x + 1)^4} = \frac{2}{(x + 1)^3}.$

53 The slope of the tangent line at $P(-2, 1)$ will be given by $f'(-2)$.

By the reciprocal rule, $f'(x) = 5\left[-\frac{2x}{(1 + x^2)^2}\right] = \frac{-10x}{(1 + x^2)^2}.$ $f'(-2) = \frac{20}{25} = \frac{4}{5}.$

The tangent line is $y - 1 = \frac{4}{5}(x + 2)$, or $y = \frac{4}{5}x + \frac{13}{5}.$

55 (a) Let $y = f(x)$. The tangent line is horizontal when $f'(x) = 0$.

$$f'(x) = 3x^2 + 4x - 4 = (3x - 2)(x + 2); f'(x) = 0 \Rightarrow x = \frac{2}{3}, -2.$$

(b) $2y + 8x = 5 \Leftrightarrow 2y = -8x + 5 \Leftrightarrow y = -4x + \frac{5}{2}.$

Hence, the given line has slope -4. $f'(x) = -4 \Rightarrow 3x^2 + 4x - 4 = -4 \Rightarrow$

$$3x^2 + 4x = 0 \Rightarrow x(3x + 4) = 0 \Rightarrow x = 0, -\frac{4}{3}.$$

59 We use (3.9) to find vertical tangent lines. Note that

$x = 0$ is a left endpoint of the domain and so we let

$x \rightarrow 0^+$ only. $y' = \frac{1}{2}x^{-1/2} = \frac{1}{2\sqrt{x}}; \lim\limits_{x \to 0^+}\left|\frac{1}{2\sqrt{x}}\right| = \infty$

$\Rightarrow$ vertical tangent line at $x = 0$. To graph

$y = \sqrt{x} - 4$, translate the graph of $y = \sqrt{x}$ down 4

units on the y-axis.

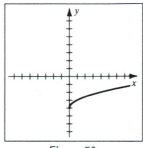

Figure 59

61 (a) The velocity of the balloon, $v(t)$, is the derivative of the position function, $s(t)$.

$$v(t) = s'(t) = 2 + 2t; v(1) = 4, v(4) = 10, \text{ and } v(8) = 18 \text{ (ft/sec)}$$

(b) We must first find the *time* when the balloon is 50 feet above the ground.

$$s(t) = 50 \Rightarrow t^2 + 2t - 44 = 0 \Rightarrow t = -1 \pm 3\sqrt{5} = -1 + 3\sqrt{5} \text{ for } t \geq 0.$$

The velocity at this time is $v(-1 + 3\sqrt{5}) = 6\sqrt{5}$ ft/sec ≈ 13.4 ft/sec.

63 The slope of the tangent line is given by y' at $x = a$.

The reciprocal rule gives $y' = a^3\left[-\frac{2x}{(a^2 + x^2)^2}\right] = \frac{-2a^3x}{(a^2 + x^2)^2}.$

At $x = a$, $y' = \frac{-2a^3(a)}{(2a^2)^2} = \frac{-2a^4}{4a^4} = -\frac{1}{2}.$

A common mistake is to differentiate $(a^2 + x^2)$ with respect to x, and write $2a + 2x$

rather than $2x$. Remember, a^2 is a constant, and its derivative with respect to x is 0.

65 We must find a line passing through $P(5, 9)$ and a point Q lying on the graph of $y = x^2$ such that $\overline{PQ}$ is the tangent line to the graph at point Q. Since Q is on the graph of $y = x^2$, its coordinates will have the form $Q(a, a^2)$. The slope of the tangent line at Q is $2a$. Now the points P and Q determine the line $(a^2 - 9) = 2a(a - 5) \Rightarrow a^2 - 10a + 9 = 0 \Rightarrow a = 1,\ 9$. Thus, there are two lines, and their slopes are 2 and 18. Their equations are: $(y - 9) = 2(x - 5)$ and $(y - 9) = 18(x - 5)$, or equivalently, $y = 2x - 1$ and $y = 18x - 81$.

67 (a) Using (3.18)(v) with $x = 2$, $(f + g)'(2) = f'(2) + g'(2) = -1 + 2 = 1$.

(b) Using (3.18)(vi) with $x = 2$, $(f - g)'(2) = f'(2) - g'(2) = -1 - 2 = -3$.

(c) Using (3.18)(iv) with $c = 4$ and $x = 2$, $(4f)'(2) = 4 f'(2) = 4(-1) = -4$.

(d) Using (3.19) with $x = 2$,

$$(fg)'(2) = f(2)\, g'(2) + g(2)\, f'(2) = (3)(2) + (-5)(-1) = 11.$$

(e) Using (3.20) with $x = 2$,

$$\left(\frac{f}{g}\right)'(2) = \frac{g(2)\, f'(2) - f(2)\, g'(2)}{\left[g(2)\right]^2} = \frac{(-5)(-1) - (3)(2)}{(-5)^2} = -\frac{1}{25}.$$

(f) Using (3.21) with $x = 2$, $(1/f)'(2) = -\dfrac{f'(2)}{\left[f(2)\right]^2} = -\dfrac{-1}{3^2} = \dfrac{1}{9}$.

69 (a) $(2f - g)'(2) = 2 f'(2) - g'(2) = 2(-1) - 2 = -4$

(b) $(5f + 3g)'(2) = 5 f'(2) + 3\, g'(2) = (5)(-1) + (3)(2) = 1$

(c) $(gg)'(2) = g(2)\, g'(2) + g(2)\, g'(2) = 2\, g(2)\, g'(2) = 2(-5)(2) = -20$

(d) Using the reciprocal rule,

$$\left(\frac{1}{f + g}\right)'(2) = -\frac{(f + g)'(2)}{\left[(f + g)(2)\right]^2} = -\frac{f'(2) + g'(2)}{\left[f(2) + g(2)\right]^2} = -\frac{-1 + 2}{(-2)^2} = -\frac{1}{4}.$$

71 For brevity, we suppress the argument of x for each function.

$D_x(fgh) = D_x[(fg)h] = (fg)h' + h\, D_x(fg) = fgh' + h(fg' + gf') = fgh' + fg'h + f'gh$.

With $f = g = h$, we have $D_x(f^3) = D_x(fff) =$

$$fff' + ff'f + f'ff = f^2 f' + f^2 f' + f^2 f' = 3 f^2 f'.$$

75 $dy/dx = (x)\ (2x^3 - 5x - 1)\ \ \ \ (12x)$

$ + (x)\ \ \ \ \ (6x^2 - 5)\ \ \ \ \ (6x^2 + 7)$

$ + (1)\ (2x^3 - 5x - 1)\ \ (6x^2 + 7)$

77 (a) $r(t) = 3t^{1/3}$; $r'(t) = 3(\frac{1}{3} t^{-2/3}) = t^{-2/3}$; $r'(8) = 8^{-2/3} = \frac{1}{4}$ cm/min

(b) $V = \frac{4}{3}\pi r^3$; $V(t) = \frac{4}{3}\pi (3\sqrt[3]{t})^3 = 36\pi t$; $V'(t) = 36\pi$; $V'(8) = 36\pi$ cm^3/min

(c) $S = 4\pi r^2$; $S(t) = 4\pi(3\sqrt[3]{t})^2 = 36\pi t^{2/3}$; $S'(t) = 36\pi \cdot \frac{2}{3} t^{-1/3} = \dfrac{24\pi}{\sqrt[3]{t}}$;

$$S'(8) = 12\pi \text{ cm}^2/\text{min}$$

$\boxed{79}$ $A = \pi r^2$ and $r(t) = 40t \Rightarrow A(t) = \pi(40t)^2 = 1600\pi t^2$; $A'(t) = 3200\pi t$;

(a) $A'(1) = 3200\pi$ (b) $A'(2) = 6400\pi$ (c) $A'(3) = 9600\pi$ (cm^2/sec)

Exercises 3.4

$\boxed{3}$ We recognize $G(v)$ as a product of two functions, v and $\csc v$. We will always "take out" constants before applying any of the differentiation rules. By the product rule,

$$G(v) = 5v \csc v \Rightarrow G'(v) = 5\big[v(-\csc v \cot v) + \csc v \cdot 1\big] = 5\csc v(1 - v \cot v).$$

$\boxed{7}$ By the quotient rule,

$$f(\theta) = \frac{\sin\theta}{\theta} \Rightarrow f'(\theta) = \frac{(\theta)(\cos\theta) - (\sin\theta)(1)}{(\theta)^2} = \frac{\theta\cos\theta - \sin\theta}{\theta^2}.$$

$\boxed{11}$ Using the product rule for each term, $f(x) = 2x \cot x + x^2 \tan x \Rightarrow$

$$f'(x) = 2x(-\csc^2 x) + 2\cot x + x^2 \sec^2 x + 2x \tan x =$$

$$-2x \csc^2 x + 2\cot x + x^2 \sec^2 x + 2x \tan x.$$

$\boxed{13}$ By the quotient rule,

$$h(z) = \frac{1 - \cos z}{1 + \cos z} \Rightarrow h'(z) = \frac{(1 + \cos z)(\sin z) - (1 - \cos z)(-\sin z)}{(1 + \cos z)^2} =$$

$$\frac{\sin z + \cos z \sin z + \sin z - \cos z \sin z}{(1 + \cos z)^2} = \frac{2\sin z}{(1 + \cos z)^2}.$$

$\boxed{15}$ First, rewrite g as $g(x) = \dfrac{1}{\sin x \tan x} = \dfrac{1}{\sin x} \cdot \dfrac{1}{\tan x} = \csc x \cot x$. Then,

$$\begin{aligned}
g'(x) &= \csc x(-\csc^2 x) + \cot x(-\csc x \cot x) & \{\text{product rule}\}\\
&= -\csc x(\csc^2 x + \cot^2 x) & \{\text{factor out } -\csc x\}\\
&= -\csc x\big[(1 + \cot^2 x) + \cot^2 x\big] & \{\csc^2 x = 1 + \cot^2 x\}\\
&= -\csc x(1 + 2\cot^2 x) & \{\text{simplify}\}
\end{aligned}$$

$\boxed{17}$ Using the product rule, $g(x) = (x + \csc x)\cot x \Rightarrow$

$$g'(x) = (x + \csc x)(-\csc^2 x) + \cot x(1 - \csc x \cot x) =$$

$$-x \csc^2 x - \csc^3 x + \cot x - \csc x \cot^2 x$$

$\boxed{21}$ $f(x) = \dfrac{\tan x}{1 + x^2} \Rightarrow$

$$f'(x) = \frac{(1 + x^2)(\sec^2 x) - (\tan x)(2x)}{(1 + x^2)^2} = \frac{\sec^2 x + x^2 \sec^2 x - 2x \tan x}{(1 + x^2)^2}$$

$\boxed{27}$ First, rewrite H as $H(\phi) = (\cot\phi + \csc\phi)(\tan\phi - \sin\phi)$

$$= \cot\phi \tan\phi - \cot\phi \sin\phi + \csc\phi \tan\phi - \csc\phi \sin\phi$$

$$= \frac{1}{\tan\phi} \cdot \tan\phi - \frac{\cos\phi}{\sin\phi} \cdot \sin\phi + \frac{1}{\sin\phi} \cdot \frac{\sin\phi}{\cos\phi} - \frac{1}{\sin\phi} \cdot \sin\phi$$

$$= 1 - \cos\phi + \sec\phi - 1 = -\cos\phi + \sec\phi. \text{ Thus, } H'(\phi) = \sin\phi + \sec\phi \tan\phi.$$

29 First, find the slope of the tangent line by calculating $f'(\frac{\pi}{4})$. The normal line will then have a slope of $-1/f'(\frac{\pi}{4})$ { the negative reciprocal } since it is perpendicular to the tangent line. $f(x) = \sec x \Rightarrow f'(x) = \sec x \tan x \Rightarrow f'(\frac{\pi}{4}) = \sqrt{2}(1) = \sqrt{2}$. Using the point-slope form with the point $(\frac{\pi}{4}, f(\frac{\pi}{4})) = (\frac{\pi}{4}, \sqrt{2})$, we obtain the following two equations:

tangent: $(y - \sqrt{2}) = \sqrt{2}(x - \frac{\pi}{4})$; $\qquad$ normal: $(y - \sqrt{2}) = -\frac{1}{\sqrt{2}}(x - \frac{\pi}{4})$.

33 On $(0, \frac{\pi}{2})$, $f(x) = \csc x + \sec x \Rightarrow f'(x) = -\csc x \cot x + \sec x \tan x = 0 \Rightarrow$

$$-\frac{1}{\sin x} \cdot \frac{\cos x}{\sin x} + \frac{1}{\cos x} \cdot \frac{\sin x}{\cos x} = 0 \Rightarrow \frac{\sin x}{\cos^2 x} = \frac{\cos x}{\sin^2 x} \Rightarrow$$

$\sin^3 x = \cos^3 x \{ \sin x \neq 0, \cos x \neq 0 \} \Rightarrow \sin x = \cos x \Rightarrow \frac{\sin x}{\cos x} = 1 \Rightarrow \tan x = 1 \Rightarrow$

$$x = \frac{\pi}{4}. \quad f(\frac{\pi}{4}) = \csc \frac{\pi}{4} + \sec \frac{\pi}{4} = \sqrt{2} + \sqrt{2} = 2\sqrt{2}. \text{ The point is } (\frac{\pi}{4}, 2\sqrt{2}).$$

35 (a) $f(x) = x + 2 \cos x \Rightarrow f'(x) = 1 - 2 \sin x = 0 \Rightarrow \sin x = \frac{1}{2} \Rightarrow$

$x = \frac{\pi}{6} + 2\pi n, \frac{5\pi}{6} + 2\pi n$. There are an infinite number of points at which the

$\qquad\qquad\qquad\qquad\qquad\qquad\qquad\qquad\qquad\qquad\qquad$ tangent line is horizontal.

(b) $P(0, f(0)) = P(0, 2)$ and $f'(0) = 1$.

$\qquad$ The equation of the tangent line of f at P is $(y - 2) = 1(x - 0)$, or $y = x + 2$.

37 (a) $y = 3 + 2 \sin x \Rightarrow y' = 2 \cos x$. The slope of the given line, $y = \sqrt{2}\,x - 5$, is $\sqrt{2}$.

$\qquad$ Setting the slopes equal, we have $y' = \sqrt{2} \Rightarrow 2 \cos x = \sqrt{2} \Rightarrow \cos x = \frac{\sqrt{2}}{2} \Rightarrow$

$$x = \frac{\pi}{4} + 2\pi n, \frac{7\pi}{4} + 2\pi n.$$

(b) At $x = \frac{\pi}{6}$, $y = 4$ and $y' = \sqrt{3}$.

$$\text{The equation of the tangent line is } (y - 4) = \sqrt{3}(x - \frac{\pi}{6}).$$

41 Point P will have a velocity of zero whenever $s'(t) = 0$.

$$v(t) = s'(t) = 1 - 2 \sin t = 0 \Rightarrow \sin t = \frac{1}{2} \Rightarrow t = \frac{\pi}{6} + 2\pi n, \frac{5\pi}{6} + 2\pi n.$$

43 $y = x^{3/2} + 2x \Rightarrow y' = \frac{3}{2} x^{1/2} + 2$. We want to solve the equation $y' = 8$.

Thus, $\frac{3}{2} x^{1/2} + 2 = 8 \Rightarrow x^{1/2} = 4 \Rightarrow x = 16$.

$$y = (16)^{3/2} + 2(16) = 64 + 32 = 96. \quad (16, 96)$$

47 (i) $\quad y = \tan x \Rightarrow D_x y = \sec^2 x = \sec x \cdot \sec x$.

(ii) $\quad D_x^2 y = \sec x(\sec x \tan x) + \sec x(\sec x \tan x) = 2 \sec^2 x \tan x$.

(iii) $\quad D_x^3 y = 2\left[\sec^2 x(\sec^2 x) + \tan x(2 \sec^2 x \tan x) \right]$

$$\{ D_x \sec^2 x = 2 \sec^2 x \tan x \text{ from (i) and (ii)} \}$$

$\qquad = 2 \sec^2 x(\sec^2 x + 2 \tan^2 x)$

$\qquad = 2 \sec^2 x\left[(1 + \tan^2 x) + 2 \tan^2 x \right]$

$\qquad = 2 \sec^2 x(3 \tan^2 x + 1)$.

49 $D_x \cot x = D_x \left(\dfrac{\cos x}{\sin x} \right)$ { definition of $\cot x$ }

$$= \frac{(\sin x)(-\sin x) - (\cos x)(\cos x)}{\sin^2 x} \quad \{ \text{quotient rule} \}$$

$$= \frac{-1(\sin^2 x + \cos^2 x)}{\sin^2 x} \quad \{ \text{simplify} \}$$

$$= -\frac{1}{\sin^2 x} \quad \{ \sin^2 x + \cos^2 x = 1 \}$$

$$= -\csc^2 x \quad \{ \text{reciprocal identity} \}$$

51 $D_x \sin 2x = D_x (2 \sin x \cos x)$ { using the hint }

$$= 2 \Big[\sin x(-\sin x) + \cos x \cos x \Big] \quad \{ \text{product rule} \}$$

$$= 2(\cos^2 x - \sin^2 x) \quad \{ \text{simplify} \}$$

$$= 2 \cos 2x \quad \{ \text{double angle formula for cosine} \}$$

Exercises 3.5

1 (a) Let $y = f(x) = 2x^2 - 4x + 5$.

$$\Delta y = f(x + \Delta x) - f(x)$$

$$= \Big[2(x + \Delta x)^2 - 4(x + \Delta x) + 5 \Big] - (2x^2 - 4x + 5)$$

$$= 2x^2 + 4x\Delta x + 2(\Delta x)^2 - 4x - 4\Delta x + 5 - 2x^2 + 4x - 5$$

$$= 4x\Delta x + 2(\Delta x)^2 - 4\Delta x$$

$$= (4x - 4)\Delta x + 2(\Delta x)^2$$

$dy = f'(x)\, dx = (4x - 4)\, dx$.

        Remember, Δy is the *exact* change in y and dy is an *approximation* of Δy.

(b) $x = 2$ and $\Delta x = -0.2 \Rightarrow$

$$\Delta y = (4x - 4)\Delta x + 2(\Delta x)^2 = 4(-0.2) + 2(-0.2)^2 = -0.72 \text{ and}$$

$$dy = (4x - 4)\, dx = 4(-0.2) = -0.8.$$

3 (a) Let $y = f(x) = 1/x^2$. $\Delta y = f(x + \Delta x) - f(x) = \left[\dfrac{1}{(x + \Delta x)^2} - \dfrac{1}{x^2} \right] =$

$$\frac{x^2 - (x + \Delta x)^2}{x^2(x + \Delta x)^2} = \frac{x^2 - x^2 - 2x\Delta x - (\Delta x)^2}{x^2(x + \Delta x)^2} = \frac{-(2x + \Delta x)\Delta x}{x^2(x + \Delta x)^2}.$$

$dy = f'(x)\, dx = (-2/x^3)\, dx$.

(b) $x = 3$ and $\Delta x = 0.3 \Rightarrow$

$$\Delta y = \frac{-(2x + \Delta x)\Delta x}{x^2(x + \Delta x)^2} = \frac{-(6.3)(0.3)}{9(3.3)^2} = \frac{-1.89}{98.01} = -\frac{7}{363} \approx -0.01928 \text{ and}$$

$$dy = (-2/x^3)\, dx = -\frac{2}{27}(0.3) = -\frac{1}{45} = -0.0\overline{2}.$$

$\boxed{7}$ (a) Let $y = f(x) = 3x^2 + 5x - 2$. $\Delta y = f(x + \Delta x) - f(x)$

$$= \left[3(x + \Delta x)^2 + 5(x + \Delta x) - 2 \right] - (3x^2 + 5x - 2)$$

$$= 3x^2 + 6x\Delta x + 3(\Delta x)^2 + 5x + 5\Delta x - 2 - 3x^2 - 5x + 2$$

$$= (6x + 5)\Delta x + 3(\Delta x)^2$$

(b) $dy = f'(x)\,dx = (6x + 5)\,dx$.

(c) Since $dx = \Delta x$, $dy - \Delta y = -3(\Delta x)^2$.

$\boxed{9}$ (a) Let $y = f(x) = 1/x$.

$$\Delta y = f(x + \Delta x) - f(x) = \frac{1}{x + \Delta x} - \frac{1}{x} = \frac{x - (x + \Delta x)}{x(x + \Delta x)} = \frac{-\Delta x}{x(x + \Delta x)}.$$

(b) $dy = f'(x)\,dx = -\frac{1}{x^2}\,dx$.

(c) Since $dx = \Delta x$, $dy - \Delta y = -\frac{1}{x^2}\,dx - \left[\frac{-\Delta x}{x(x + \Delta x)} \right] = -\frac{\Delta x}{x^2} + \frac{\Delta x}{x(x + \Delta x)} =$

$$\frac{-\Delta x}{x^2} \cdot \frac{x + \Delta x}{x + \Delta x} + \frac{\Delta x}{x(x + \Delta x)} \cdot \frac{x}{x} = \frac{-x\Delta x - (\Delta x)^2 + x\Delta x}{x^2(x + \Delta x)} = \frac{-(\Delta x)^2}{x^2(x + \Delta x)}.$$

$\boxed{11}$ $dy = f'(x)\,dx = (20x^4 - 24x^3 + 6x)\Delta x$.

$x = 1$ and $\Delta x = b - a = 0.03$ yield $dy = (20 - 24 + 6)(0.03) = 2(0.03) = 0.06$.

$$f(x + \Delta x) \approx f(x) + dy \Rightarrow f(1.03) \approx f(1) + dy = -4 + 0.06 = -3.94.$$

$\boxed{15}$ $dy = f'(\theta)\,\Delta\theta = (2\cos\theta - \sin\theta)\,\Delta\theta$.

$\theta = 30°$ and $\Delta\theta = b - a = -3°$ yield $dy = (2\cos 30° - \sin 30°)(\Delta\theta) =$

$(\sqrt{3} - \frac{1}{2})(-3 \cdot \frac{\pi}{180})$ {radian measure must be used} ≈ -0.0645. $f(x + \Delta x) \approx$

$$f(x) + dy \Rightarrow f(27°) \approx f(30°) + dy \approx (1 + \sqrt{3}/2) + (-0.0645) \approx 1.8015.$$

$\boxed{19}$ (a) With $h = 0.001$, $f'(2.5) \approx \dfrac{f(2.501) - f(2.499)}{2(0.001)} \approx -0.27315$.

$$y - f(2.5) = f'(2.5)(x - 2.5) \Rightarrow y \approx -0.98451 - 0.27315(x - 2.5)$$

(b) When $x = 2.6$, $y \approx -1.011825$.

(c) $f(2.6) \approx f(2.5) + dy =$

$$f(2.5) + f'(2.5)\Delta x \approx -0.98451 - 0.27315(0.1) = -1.011825$$

(d) They are equal because the tangent line approximation is equivalent to using

$$(3.31).$$

$\boxed{23}$ Let $y = f(x)$. The average error is $\dfrac{\Delta y}{y} \approx \dfrac{dy}{y} = \dfrac{f'(x)\,dx}{f(x)} = \dfrac{\left[4(\frac{1}{2}x^{-1/2}) + 3 \right]dx}{4\sqrt{x} + 3x} =$

$\dfrac{(2/\sqrt{x} + 3)\,dx}{4\sqrt{x} + 3x} = \dfrac{4(\pm 0.2)}{20} = \pm 0.04$ {when $x = 4$ and $\Delta x = \pm 0.2$}.

The percentage error is (average error) $\times$ 100%, i.e., $(\pm 0.04)(100\%)$ or $\pm 4\%$.

$\boxed{25}$ $A = 3x^2 - x \Rightarrow dA = A'(x)\,dx = (6x - 1)\,dx$.

$$x = 2 \text{ and } dx = 0.1 \Rightarrow dA = 11(0.1) = 1.1.$$

$\boxed{27}$ In this case, we are given the error in the "input" variable (the independent variable, x) and we want to find the error that will result in the "output" variable (the dependent variable, y). Average error =

$$\frac{\Delta y}{y} \approx \frac{dy}{y} = \frac{12x^2\, dx}{4x^3} = 3\left(\frac{dx}{x}\right) = 3\left(\frac{\Delta x}{x}\right) = 3(\pm 15\%) = \pm 45\%.$$

$\boxed{29}$ In this case, we know what the desired output error is $\left(\frac{dA}{A} = \pm 0.04\right)$, and we need a formula for the input error, $\frac{ds}{s}$. Average error = $\frac{\Delta A}{A} \approx \frac{dA}{A} = \frac{A'(s)\, ds}{A(s)} =$

$$\frac{15\left(\frac{2}{3}s^{-1/3}\right) ds}{15s^{2/3}} = \frac{10s^{-1/3}\, ds}{15s^{2/3}} = \frac{2}{3}\left(\frac{ds}{s}\right) \Rightarrow \frac{ds}{s} = \frac{3}{2}\frac{dA}{A} = \frac{3}{2}(\pm 0.04) = \pm 0.06.$$

$\boxed{35}$ For "word problems" involving differentials (and later, related rates), we introduce a "Find when knowing" statement that we have found to be helpful to students. As with other types of word problems, *getting started is the problem.* After carefully reading a problem, try to complete the statement

Find $\underline{what_1}$ when $\underline{what_2}$ knowing $\underline{what_3}$.

Typically, "$what_1$" will be dy, $\frac{dy}{y}$, $\frac{dy}{y} \times 100$, etc. "$What_2$" will be a particular condition pertaining to the problem, for example: $x = 2$, $\theta = 30°$, $h = s$, etc. "$What_3$" will usually be a condition related to "$what_1$", for example: $dx = \pm 0.01$, $\frac{dx}{x} = \pm 0.2$, etc. Sometimes $what_2$ and $what_3$ may not be given completely in the problem. You may have to determine them first.

Once you have determined *what* you need to find, and at *what* particular instant, and knowing *what* related information, you then need to set up an equation relating the variables used in the 3 *what's.*

For this problem, we let x denote the length of the base and A the area of one side. We need to *find* the maximum error in the area of the side (dA) *when* the length of the base is 48 feet *knowing* that the error in the measurement of the length of the base (dx) is ± 1 inch. Our "Find when knowing" statement becomes

Find $\underline{dA}$ when $\underline{x = 48'}$ knowing $\underline{dx = \pm 1''}$.

This statement indicates that we need to relate A with x. The area of the square is x^2 and the area of the equilateral triangle is $\frac{\sqrt{3}}{4}x^2$ (see inside cover). Thus, $A = f(x) = x^2 + \frac{\sqrt{3}}{4}x^2$. By (3.28), $dA = f'(x)\, dx = \left(2x + \frac{\sqrt{3}}{2}x\right) dx$. Note that we differentiated *before* substituting any *particular* values. A common mistake is to substitute particular values before differentiating.

(cont.)

Now we let $x = 48$ and $dx = \pm\frac{1}{12}$ ft to obtain $dA = (96 + 24\sqrt{3})(\pm\frac{1}{12}) = 8 + 2\sqrt{3} \approx \pm 11.4641$ ft. This is an estimate of the maximum error in the calculation of the area of one side.

Next, we want the average error in the calculation. This is $\frac{dA}{A}$, and the value for A is obtained by using our relationship $A = x^2 + \frac{\sqrt{3}}{4}x^2$, and letting $x = 48$, as if there was *no error* in the measurement of x. Thus, $A = 48^2 + \frac{\sqrt{3}}{4} \cdot 48^2 = 2304 + 576\sqrt{3} \approx 3301.66$ ft^2. $\frac{dA}{A} = \frac{f'(48)(\pm\frac{1}{12})}{f(48)} \approx \frac{\pm 11.4641}{3301.66} \approx \pm 0.00347$.

Finally, the percentage error is just 100 times the average error, or $\pm 0.347\%$.

$\boxed{37}$ For a cone, the volume V is given by $V = \frac{1}{3}\pi r^2 h$, where r denotes the radius of the cone and h its altitude. Try to formulate your own "Find when knowing" statement before continuing.

Our statement is:

Find __dr__ when __$r = 10$ cm__ knowing __$dV = 2$ cm^3__ and $h = r$.

$V = \frac{1}{3}\pi r^2 h$ and $h = r \Rightarrow V = \frac{1}{3}\pi r^3$; $\Delta V \approx dV = \pi r^2\, dr$;

$$2 = \pi(10)^2\, dr \Rightarrow dr = \tfrac{1}{50\pi} \text{ cm} = 0.00637 \text{ cm}.$$

$\boxed{39}$ Find __ds__ when __$s = 20$ cm__ knowing $\frac{dF}{F} = 10\%$.

$F(s) = \frac{Gm_1 m_2}{s^2} = (Gm_1 m_2)s^{-2} \Rightarrow F'(s) = -2(Gm_1 m_2)s^{-3}$. The average change in

F is $\frac{\Delta F}{F} \approx \frac{dF}{F} = \frac{F'(s)\,\Delta s}{F} = \frac{(-2Gm_1 m_2/s^3)\,\Delta s}{Gm_1 m_2/s^2} = \frac{-2\,\Delta s}{s}$. Since

$\frac{dF}{F} = 10\% \,\{0.1\}$, we have $0.1 = \frac{-2\,\Delta s}{s} \Rightarrow \Delta s = -\frac{1}{20}s = -\frac{1}{20}(20) = -1$ cm.

Hence, if we *decrease* the distance between two particles, we *increase* the force of attraction between them.

$\boxed{41}$ Let r denote the radius of the arteriole, P the pressure difference, and c the proportionality constant. Since P is inversely proportional to r^4, we have $P = \frac{c}{r^4}$. (If we had stated *directly* proportional, then the equation would be $P = cr^4$.) Our statement for this problem is:

Find __$\frac{\Delta P}{P}$__ when __$P = \frac{c}{r^4}$__ knowing __$\frac{dr}{r} = -10\%$__ .

Note here that we don't have any particular condition other than the general formula, $P = cr^{-4}$. $\frac{\Delta P}{P} \approx \frac{dP}{P} = \frac{-4cr^{-5}\,dr}{cr^{-4}} = -4\left(\frac{dr}{r}\right)$.

$\frac{dr}{r} = -0.1 \,\{10\% \text{ decrease}\} \Rightarrow \frac{\Delta P}{P} = +0.4$, an increase of 40%.

43 *Hint*: Find $\underline{dF}$ when $\underline{\theta = 45°}$ knowing $\underline{d\theta = 1 \cdot \frac{\pi}{180} \text{ rad}}$.

45 *Hint*: Find $\underline{dh}$ when $\underline{\theta = 60°}$ knowing $\underline{d\theta = \pm 15' = \pm \frac{\pi}{720} \text{ rad}}$.

47 Find $\underline{d\phi}$ when $\underline{\phi = 52°}$ knowing $\underline{dh = \pm 1'}$. From the text's figure,

$$\tan\phi = \frac{h}{\frac{1}{2} \cdot 230} \Rightarrow h = 115\tan\phi \text{ and } dh = 115\sec^2\phi\, d\phi \Rightarrow d\phi = \tfrac{1}{115}\cos^2\phi\, dh.$$

For $dh = \pm 1$ and $\phi = 52°$, we need $d\phi = \frac{1}{115}(\cos 52°)^2(\pm 1) \approx$

$$\pm 0.0033 \,\{\text{radians}\} = \tfrac{180}{\pi}(\pm 0.0033)\,\{\text{degrees}\} \approx \pm 0.19°.$$

51 $dA = A'(s)\,\Delta s = 2s\,\Delta s$ and

$$\Delta A = A(s + \Delta s) - A(s)$$
$$= (s + \Delta s)^2 - s^2$$
$$= 2s\,\Delta s + (\Delta s)^2.$$

dA is the shaded area and

$\Delta A - dA = (\Delta s)^2$ is as labeled.

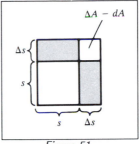

Figure 51

Exercises 3.6

1   Using (3.33) with $y = f(u) = u^2$ and $u = g(x) = x^3 - 4$,

$$\frac{dy}{dx} = \frac{dy}{du}\frac{du}{dx} = f'(u)\,g'(x) = (2u)(3x^2) = 6x^2 u = 6x^2(x^3 - 4).$$

3   $\frac{dy}{du} = D_u(u^{-1}) = -u^{-2} = -\frac{1}{u^2}$. To find $\frac{du}{dx} = D_x(\sqrt{3x - 2})$, we must use the

chain rule. $D_x\left[(3x - 2)^{1/2}\right] = \frac{1}{2}(3x - 2)^{-1/2} D_x(3x - 2) = \dfrac{3}{2(3x - 2)^{1/2}}.$

Thus, $\dfrac{dy}{dx} = \dfrac{dy}{du}\dfrac{du}{dx} =$

$$-\frac{1}{u^2} \cdot \frac{3}{2(3x - 2)^{1/2}} = -\frac{1}{\left[(3x - 2)^{1/2}\right]^2} \cdot \frac{3}{2(3x - 2)^{1/2}} = -\frac{3}{2(3x - 2)^{3/2}}.$$

7   Let $f(u) = u^3$ and $u = g(x) = x^2 - 3x + 8$.

Using the chain rule, $f'(x) = f'(u)\,g'(x) = 3u^2(2x - 3) = 3(x^2 - 3x + 8)^2(2x - 3)$.

11 Using the quotient rule to find $f'(x)$ yields $f'(x) = \dfrac{(x^2 - 1)^4 D_x(x) - x D_x(x^2 - 1)^4}{\left[(x^2 - 1)^4\right]^2}.$

We use the chain rule to calculate $D_x(x^2 - 1)^4$.

$D_x(x^2 - 1)^4 = 4(x^2 - 1)^3 D_x(x^2 - 1) = 4(x^2 - 1)^3(2x) = 8x(x^2 - 1)^3$. Thus,

$$f'(x) = \frac{(x^2 - 1)^4(1) - 8x^2(x^2 - 1)^3}{(x^2 - 1)^8} = \frac{(x^2 - 1)^3\left[(x^2 - 1) - 8x^2\right]}{(x^2 - 1)^8} = -\frac{7x^2 + 1}{(x^2 - 1)^5}.$$

17 Using the product rule to find $N'(x)$ yields

$$N'(x) = (6x - 7)^3 D_x\left[(8x^2 + 9)^2\right] + D_x\left[(6x - 7)^3\right](8x^2 + 9)^2.$$

Using the chain rule to find the two derivatives, we have

$$D_x\left[(8x^2 + 9)^2\right] = (2)(8x^2 + 9) D_x (8x^2 + 9) = 2(8x^2 + 9)(16x)$$

and $D_x\left[(6x - 7)^3\right] = 3(6x - 7)^2 D_x (6x - 7) = 3(6x - 7)^2(6).$

Hence, $N'(x) = (6x - 7)^3(2)(8x^2 + 9)(16x) + 3(6x - 7)^2(6)(8x^2 + 9)^2$

$$= (6x - 7)^2(8x^2 + 9)\left[32x(6x - 7) + 18(8x^2 + 9)\right]$$

$$= (6x - 7)^2(8x^2 + 9)(336x^2 - 224x + 162)$$

$$= 2(6x - 7)^2(8x^2 + 9)(168x^2 - 112x + 81).$$

21 $k(r) = \sqrt[3]{8r^3 + 27} = (8r^3 + 27)^{1/3} \Rightarrow k'(r) = \frac{1}{3}(8r^3 + 27)^{-2/3} D_r (8r^3 + 27) =$

$$\frac{1}{3}(8r^3 + 27)^{-2/3}(24r^2) = 8r^2(8r^3 + 27)^{-2/3}.$$

23 Using the reciprocal rule (3.21), $F(v) = \dfrac{5}{\sqrt[5]{v^5 - 32}} = \dfrac{5}{(v^5 - 32)^{1/5}} \Rightarrow$

$$F'(v) = -5\frac{D_v\left[(v^5 - 32)^{1/5}\right]}{\left[(v^5 - 32)^{1/5}\right]^2} = -5\frac{\frac{1}{5}(v^5 - 32)^{-4/5}(5v^4)}{(v^5 - 32)^{2/5}} = -\frac{5v^4}{(v^5 - 32)^{6/5}}.$$

25 We first change the form of $g(w)$. $g(w) = \dfrac{w^2 - 4w + 3}{w^{3/2}} = \dfrac{w^2}{w^{3/2}} - \dfrac{4w}{w^{3/2}} + \dfrac{3}{w^{3/2}} =$

$w^{1/2} - 4w^{-1/2} + 3w^{-3/2}$. Thus, $g'(w) = \frac{1}{2}w^{-1/2} + 2w^{-3/2} - \frac{9}{2}w^{-5/2} =$

$$\frac{1}{2}w^{-5/2}\left[w^2 + 2 \cdot 2w^1 - 9\right] = \frac{w^2 + 4w - 9}{2w^{5/2}}.$$

27 $H(x) = \dfrac{2x + 3}{\sqrt{4x^2 + 9}} = \dfrac{2x + 3}{(4x^2 + 9)^{1/2}}.$ The quotient rule gives us

$$H'(x) = \frac{(4x^2 + 9)^{1/2} D_x (2x + 3) - (2x + 3) D_x\left[(4x^2 + 9)^{1/2}\right]}{\left[(4x^2 + 9)^{1/2}\right]^2}$$

$$= \frac{(4x^2 + 9)^{1/2}(2) - (2x + 3)(\frac{1}{2})(4x^2 + 9)^{-1/2}(8x)}{4x^2 + 9}$$

$$= \frac{2(4x^2 + 9)^{-1/2}\left[(4x^2 + 9) - 2x(2x + 3)\right]}{4x^2 + 9} = \frac{6(3 - 2x)}{(4x^2 + 9)^{3/2}}.$$

31 Since $H(\theta) = \cos^5 3\theta = (\cos 3\theta)^5$, $H'(\theta) = 5(\cos 3\theta)^4 D_\theta (\cos 3\theta) =$

$$5(\cos 3\theta)^4(-\sin 3\theta) D_\theta (3\theta) = -5(\cos 3\theta)^4(\sin 3\theta)3 = -15\cos^4 3\theta \sin 3\theta.$$

33 $g'(z) = D_z\left[\sec (2z + 1)^2\right] = \sec (2z + 1)^2 \tan (2z + 1)^2 D_z\left[(2z + 1)^2\right]$

$$= \sec (2z + 1)^2 \tan (2z + 1)^2 (2)(2z + 1) D_z (2z)$$

$$= \sec (2z + 1)^2 \tan (2z + 1)^2 (2)(2z + 1)(2)$$

$$= 4(2z + 1) \sec (2z + 1)^2 \tan (2z + 1)^2$$

$\boxed{37}$ $f(x) = \cos(3x^2) + \cos^2 3x = \cos(3x^2) + (\cos 3x)^2 \Rightarrow$

$$f'(x) = -\sin(3x^2)\, D_x\,(3x^2) + 2(\cos 3x)^1\, D_x\,(\cos 3x)$$

$$= -\sin(3x^2) \cdot (6x) + 2(\cos 3x)(-\sin 3x)\, D_x\,(3x)$$

$$= -6x\sin(3x^2) - 2(\cos 3x)(\sin 3x)(3)$$

$$= -6x\sin(3x^2) - 6\cos 3x \sin 3x$$

$\boxed{39}$ $F(\phi) = \csc^2 2\phi = (\csc 2\phi)^2 \Rightarrow F'(\phi) = 2(\csc 2\phi)\, D_\phi\,(\csc 2\phi) =$

$$2(\csc 2\phi)(-\csc 2\phi \cot 2\phi)\, D_\phi\,(2\phi) = -2(\csc^2 2\phi \cot 2\phi)(2) = -4\csc^2 2\phi \cot 2\phi$$

$\boxed{43}$ $h(\theta) = \tan^2\theta\, \sec^3\theta = (\tan\theta)^2(\sec\theta)^3 \Rightarrow$

$$h'(\theta) = (\tan\theta)^2\, D_\theta\Big[(\sec\theta)^3\Big] + (\sec\theta)^3\, D_\theta\Big[(\tan\theta)^2\Big]$$

$$= (\tan\theta)^2\, 3(\sec\theta)^2\, D_\theta\,(\sec\theta) + (\sec\theta)^3\, 2(\tan\theta)\, D_\theta\,(\tan\theta)$$

$$= 3\tan^2\theta\, \sec^2\theta\,(\sec\theta\tan\theta) + 2(\sec^3\theta)(\tan\theta)(\sec^2\theta)$$

$$= 3\tan^3\theta\, \sec^3\theta + 2\tan\theta\, \sec^5\theta$$

$\boxed{45}$ $N(x) = (\sin 5x - \cos 5x)^5 \Rightarrow$

$$N'(x) = 5(\sin 5x - \cos 5x)^4\, D_x\,(\sin 5x - \cos 5x)$$

$$= 5(\sin 5x - \cos 5x)^4\Big[\cos 5x \cdot D_x\,(5x) - (-\sin 5x)\, D_x\,(5x)\Big]$$

$$= 5(\sin 5x - \cos 5x)^4\,(5\cos 5x + 5\sin 5x)$$

$$= 25(\sin 5x - \cos 5x)^4\,(\cos 5x + \sin 5x)$$

$\boxed{49}$ $h(w) = \dfrac{\cos 4w}{1 - \sin 4w} \Rightarrow$

$$h'(w) = \frac{(1 - \sin 4w)\, D_w\,(\cos 4w) - (\cos 4w)\, D_w\,(1 - \sin 4w)}{(1 - \sin 4w)^2}$$

$$= \frac{(1 - \sin 4w)(-4\sin 4w) - (\cos 4w)(-4\cos 4w)}{(1 - \sin 4w)^2}$$

$$= \frac{-4\sin 4w + 4\sin^2 4w + 4\cos^2 4w}{(1 - \sin 4w)^2}$$

$$= \frac{-4\sin 4w + 4(\sin^2 4w + \cos^2 4w)}{(1 - \sin 4w)^2}$$

$$= \frac{-4\sin 4w + 4(1)}{(1 - \sin 4w)^2} = \frac{4(1 - \sin 4w)}{(1 - \sin 4w)^2} = \frac{4}{1 - \sin 4w}$$

$\boxed{53}$ $f(x) = \sin\sqrt{x} + \sqrt{\sin x} = \sin x^{1/2} + (\sin x)^{1/2} \Rightarrow$

$$f'(x) = (\cos x^{1/2})\, D_x\,(x^{1/2}) + \tfrac{1}{2}(\sin x)^{-1/2}\, D_x\,(\sin x)$$

$$= (\cos\sqrt{x})(\tfrac{1}{2}x^{-1/2}) + \tfrac{1}{2}(\sin x)^{-1/2}\cos x = \frac{\cos\sqrt{x}}{2\sqrt{x}} + \frac{\cos x}{2\sqrt{\sin x}}$$

$\boxed{57}$ $g(x) = \sqrt{x^2 + 1}\, \tan \sqrt{x^2 + 1} \Rightarrow g'(x)$

$$= \sqrt{x^2 + 1}\, D_x \left(\tan \sqrt{x^2 + 1} \right) + \tan \sqrt{x^2 + 1}\, D_x \left(\sqrt{x^2 + 1} \right)$$

$$= \sqrt{x^2 + 1} \left(\sec^2 \sqrt{x^2 + 1} \right) D_x \left(\sqrt{x^2 + 1} \right) + \tan \sqrt{x^2 + 1} \left[\tfrac{1}{2}(x^2 + 1)^{-1/2} D_x \left(x^2 \right) \right]$$

$$= \sqrt{x^2 + 1} \left(\sec^2 \sqrt{x^2 + 1} \right) \left[\tfrac{1}{2}(x^2 + 1)^{-1/2}(2x) \right] + \tan \sqrt{x^2 + 1} \left[\tfrac{1}{2}(x^2 + 1)^{-1/2}(2x) \right]$$

$$= x \sec^2 \sqrt{x^2 + 1} + \frac{x \tan \sqrt{x^2 + 1}}{\sqrt{x^2 + 1}}$$

$\boxed{61}$ $h(x) = \sqrt{4 + \csc^2 3x} = \left(4 + \csc^2 3x \right)^{1/2} \Rightarrow$

$$h'(x) = \tfrac{1}{2}\left(4 + \csc^2 3x \right)^{-1/2} D_x \left[4 + \left(\csc 3x \right)^2 \right]$$

$$= \tfrac{1}{2}\left(4 + \csc^2 3x \right)^{-1/2} \left(2 \csc 3x \right) D_x \left(\csc 3x \right)$$

$$= \left(4 + \csc^2 3x \right)^{-1/2} \left(\csc 3x \right) \left(-\csc 3x \cot 3x \right) D_x \left(3x \right)$$

$$= -\left(4 + \csc^2 3x \right)^{-1/2} \left(\csc^2 3x \cot 3x \right)(3) = -\frac{3 \csc^2 3x \cot 3x}{\sqrt{4 + \csc^2 3x}}$$

$\boxed{63}$ (a) $y = f(x) = (4x^2 - 8x + 3)^4 \Rightarrow f'(x) = 4(4x^2 - 8x + 3)^3(8x - 8) =$

$4\left[(2x - 1)(2x - 3) \right]^3 (8)(x - 1) = 32(x - 1)(2x - 1)^3(2x - 3)^3.$ $f'(2) = 864.$

The equation of the tangent line is $y - 81 = 864(x - 2)$.

The equation of the normal line is $y - 81 = -\frac{1}{864}(x - 2)$.

(b) $f'(x) = 32(x - 1)(2x - 1)^3(2x - 3)^3 = 0 \Rightarrow x = 1, \frac{1}{2}, \frac{3}{2}$

$\boxed{67}$ (a) $y = f(x) = 3x + \sin 3x \Rightarrow f'(x) = 3 + 3\cos 3x.$ $f'(0) = 3 + 3 = 6.$

The equation of the tangent line is $y - 0 = 6(x - 0)$, or $y = 6x$.

The equation of the normal line is $y - 0 = -\frac{1}{6}(x - 0)$, or $y = -\frac{1}{6}x$.

(b) $f'(x) = 3 + 3\cos 3x = 3(1 + \cos 3x) = 0 \Rightarrow$

$$\cos 3x = -1 \Rightarrow 3x = \pi + 2\pi n \Rightarrow x = \tfrac{\pi}{3} + \tfrac{2\pi}{3} n.$$

$\boxed{69}$ $g(z) = \sqrt{3z + 1} = (3z + 1)^{1/2} \Rightarrow g'(z) = \tfrac{1}{2}(3z + 1)^{-1/2}(3) = \dfrac{3}{2(3z + 1)^{1/2}}.$

Using the reciprocal rule (3.21),

$$g''(z) = -\frac{3}{2}\left[\frac{D_z\left[(3z + 1)^{1/2} \right]}{\left[(3z + 1)^{1/2} \right]^2} \right] = -\frac{3}{2}\frac{\tfrac{1}{2}(3z + 1)^{-1/2}(3)}{(3z + 1)^1} = -\frac{9}{4(3z + 1)^{3/2}}.$$

$\boxed{73}$ $f(x) = \sin^3 x = (\sin x)^3 \Rightarrow f'(x) = 3(\sin x)^2 D_x (\sin x) = 3 \sin^2 x \cos x.$

$$f''(x) = 3\left[(\sin x)^2 D_x (\cos x) + (\cos x) D_x (\sin x)^2 \right]$$

$$= 3\left[\sin^2 x(-\sin x) + (\cos x) 2 \sin x\, D_x (\sin x) \right]$$

$$= 3\left[\sin^2 x \cdot (-\sin x) + (\cos x) 2 \sin x \cos x \right]$$

$$= 6 \sin x \cos^2 x - 3 \sin^3 x$$

75 Using the hint, we let $y = f(x) = \sqrt[3]{x}$. Thus, $dy = f'(x)\,dx = \frac{1}{3}x^{-2/3}\,dx =$

$\dfrac{1}{3\left(\sqrt[3]{x}\right)^2}\,dx$. When $x = 64$ and $dx = 1$, $dy = f'(64)(1) = \dfrac{1}{3\left(\sqrt[3]{64}\right)^2}(1) = \dfrac{1}{48}$.

Using (3.31), $f(65) \approx f(64) + dy = 4 + \frac{1}{48} = 4\frac{1}{48}$.

77 We wish to find a formula for dK/dt. K is not a function of t, but it is a function of v, which in turn is a function of t. Thus, to differentiate K with respect to t, we must "go through" v. Hence, we can use the chain rule to find dK/dt.

$$K = \tfrac{1}{2}mv^2 \Rightarrow \frac{dK}{dt} = \frac{dK}{dv}\frac{dv}{dt} = mv\frac{dv}{dt}.$$

79 Find $\underline{dW/dt}$ when $\underline{x = 1000}$ knowing $\underline{dx/dt = 6}$.

$$\frac{dW}{dt} = \frac{dW}{dx}\frac{dx}{dt} = 150\,(2)\left(\frac{6400}{6400 + x}\right)\left[-\frac{6400}{(6400 + x)^2}\right]\frac{dx}{dt} = -\frac{300\,(6400)^2}{(6400 + x)^3}\frac{dx}{dt}.$$

At $x = 1000$ with $\frac{dx}{dt} = 6$, $\dfrac{dW}{dt} = -\dfrac{300\,(6400)^2}{(7400)^3}\cdot 6 \approx -0.1819$ lbs/sec.

81 $k(2) = f(g(2)) = f(2) = -4$.

Using the chain rule, $k'(2) = f'(g(2))\,g'(2) = f'(2)\,g'(2) = 3\cdot 5 = 15$.

85 Since $h(x) = f(g(x))$, $h'(1.12) = f'(g(1.12))\cdot g'(1.12) = f'(2.232)\cdot g'(1.12)$.

By 3.2.51, $f'(2.232) \approx \dfrac{f(2.232 + h) - f(2.232 - h)}{2h}$, where h is the difference

between consecutive values of x, i.e., $h = 2.2320 - 2.2210 = 0.011$.

$g'(1.12) \approx \dfrac{g(1.12 + h) - g(1.12 - h)}{2h}$, where $h = 1.1200 - 1.1100 = 0.01$.

Thus, $f'(2.232)\cdot g'(1.12) \approx \left[\dfrac{f(2.243) - f(2.221)}{2(0.011)}\right]\left[\dfrac{g(1.13) - g(1.11)}{2(0.01)}\right] =$

$$\left(\frac{5.0310 - 4.9328}{0.022}\right)\left(\frac{2.243 - 2.221}{0.02}\right) = 4.91.$$

87 (a) Since f is even, $f(-x) = f(x)$. Differentiating both sides yields:

$D_x\big[f(-x)\big] = D_x\big[f(x)\big] \Rightarrow f'(-x)\cdot D_x\,(-x) = f'(x) \Rightarrow$

$f'(-x)(-1) = f'(x) \Rightarrow f'(-x) = -f'(x) \Rightarrow f'$ is odd.

Example : If $f(x) = x^4$ (an even function), then $f'(x) = 4x^3$ (an odd function).

(b) Since f is odd, $f(-x) = -f(x)$. Differentiating both sides yields:

$D_x\big[f(-x)\big] = D_x\big[-f(x)\big] \Rightarrow f'(-x)\cdot D_x\,(-x) = -f'(x) \Rightarrow$

$f'(-x)(-1) = -f'(x) \Rightarrow f'(-x) = f'(x) \Rightarrow f'$ is even.

Example: If $f(x) = x^3$ (an odd function), then $f'(x) = 3x^2$ (an even function).

89 (a) $W = (6 \times 10^{-5}) L^{2.74} \Rightarrow$

$$\frac{dW}{dt} = \frac{dW}{dL}\frac{dL}{dt} = 2.74(6 \times 10^{-5})L^{1.74}\frac{dL}{dt} = (1.644 \times 10^{-4})L^{1.74}\frac{dL}{dt}.$$

(b) We are given the weight W and $\frac{dW}{dt}$, and we are trying to find $\frac{dL}{dt}$. Since the

formula in part (a) relates $\frac{dW}{dt}$, L, and $\frac{dL}{dt}$, we must find the length L for the

given W. $W = (6 \times 10^{-5}) L^{2.74} = 0.5 \Rightarrow L^{2.74} = \dfrac{0.5}{6 \times 10^{-5}} = \frac{1}{12} \times 10^5 \Rightarrow$

$L = (\frac{1}{12} \times 10^5)^{1/2.74} \Rightarrow L \approx 26.975.$ $\frac{dW}{dt}$ is the change in weight with respect

to time. This is given as 0.4 kg/month. Since $\frac{dW}{dt} = \frac{dW}{dL}\frac{dL}{dt}$,

$$\text{we have } \frac{dL}{dt} = \frac{dW/dt}{dW/dL} \approx \frac{0.4}{(0.0001644)(26.975)^{1.74}} \approx 7.876 \text{ cm/month.}$$

91 (a) In this exercise, the measurement of r is assumed to be exact. Thus,

$S(h) = 6\pi\sqrt{36 + h^2}$. If there were no error in the measurement of h, then the

surface area would be $S(8) = 6\pi\sqrt{36 + 8^2} = 60\pi$. The maximum error in the

last calculation is $\Delta S \approx dS = S'(h)\, dh = 6\pi(\frac{1}{2})(36 + h^2)^{-1/2}(2h)\, dh =$

$$\frac{6\pi h}{\sqrt{36 + h^2}}\, dh. \text{ Thus, } \Delta S \approx \frac{6\pi(8)}{\sqrt{36 + 64}}(\pm 0.1) = \pm 0.48\pi \approx \pm 1.508 \text{ cm}^2.$$

(b) % error $\approx \dfrac{\pm 0.48\pi}{60\pi} \times 100\% = \pm 0.8\%$

Exercises 3.7

Note: For all exercises, we assume is that the denominators are nonzero.

3
$$D_x\left(2x^3 + x^2y + y^3\right) = D_x\left(1\right)$$
$$D_x\left(2x^3\right) + D_x\left(x^2y\right) + D_x\left(y^3\right) = 0$$
$$6x^2 + \left[x^2\, D_x\, y + y\, D_x\left(x^2\right)\right] + D_x\left(y^3\right) = 0$$
$$6x^2 + \left(x^2\, y' + y(2x)\right) + 3y^2\, D_x\, y = 0$$
$$6x^2 + \left(x^2 y' + 2xy\right) + 3y^2 y' = 0$$
$$x^2 y' + 3y^2 y' = -6x^2 - 2xy$$
$$(x^2 + 3y^2)y' = -(6x^2 + 2xy) \Rightarrow y' = -\frac{6x^2 + 2xy}{x^2 + 3y^2}$$

7
$$D_x\left(\sqrt{x} + \sqrt{y}\right) = D_x\left(100\right)$$
$$D_x\left(x^{1/2}\right) + D_x\left(y^{1/2}\right) = 0$$
$$\tfrac{1}{2}x^{-1/2} + \tfrac{1}{2}y^{-1/2}\, D_x\, y = 0$$
$$\tfrac{1}{2}x^{-1/2} + \tfrac{1}{2}y^{-1/2} y' = 0$$
$$\tfrac{1}{2}y^{-1/2} y' = -\tfrac{1}{2}x^{-1/2}$$
$$\frac{y'}{\sqrt{y}} = -\frac{1}{\sqrt{x}} \Rightarrow y' = -\sqrt{\frac{y}{x}}$$

⑨
$$D_x\left(x^2 + \sqrt{xy}\right) = D_x\left(7\right)$$

$$D_x\left(x^2\right) + D_x\left[(xy)^{1/2}\right] = 0$$

$$2x + \tfrac{1}{2}(xy)^{-1/2}\,D_x\left(xy\right) = 0$$

$$2x + \tfrac{1}{2}(xy)^{-1/2}\left[x\,D_x\,y + y\,D_x\left(x\right)\right] = 0$$

$$2x + \tfrac{1}{2}(xy)^{-1/2}(xy' + y) = 0$$

$$\frac{xy' + y}{2\sqrt{xy}} = -2x$$

$$xy' + y = -4x\sqrt{xy} \Rightarrow y' = \frac{-4x\sqrt{xy} - y}{x}$$

⑪
$$D_x\left[(\sin 3y)^2\right] = D_x\left(x + y - 1\right)$$

$$2(\sin 3y)\,D_x\left(\sin 3y\right) = D_x\left(x\right) + D_x\,y - D_x\left(1\right)$$

$$2(\sin 3y)(\cos 3y)\,D_x\left(3y\right) = 1 + y' - 0$$

$$2(\sin 3y)(\cos 3y)(3y') = 1 + y'$$

$$(6\sin 3y \cos 3y)y' - y' = 1$$

$$y'(6\sin 3y \cos 3y - 1) = 1 \Rightarrow y' = \frac{1}{6\sin 3y \cos 3y - 1} = \frac{1}{3\sin 6y - 1}$$

Note: $6\sin 3y \cos 3y = 3(2\sin 3y \cos 3y) = 3\sin\left(2\cdot 3y\right) = 3\,\sin 6y$

⑮
$$D_x\left(y^2\right) = D_x\left(x \cos y\right)$$

$$2y\,D_x\,y = x\,D_x\left(\cos y\right) + (\cos y)\,D_x\left(x\right)$$

$$2yy' = x(-\sin y)\,D_x\,y + (\cos y)(1)$$

$$2yy' = x(-\sin y)\,y' + \cos y$$

$$(x\sin y + 2y)y' = \cos y \Rightarrow y' = \frac{\cos y}{x\sin y + 2y}$$

⑰
$$D_x\left(x^2 + \sqrt{\sin y} - y^2\right) = D_x\left(1\right)$$

$$D_x\left(x^2\right) + D_x\left[(\sin y)^{1/2}\right] - D_x\left(y^2\right) = 0$$

$$2x + \tfrac{1}{2}(\sin y)^{-1/2}\,D_x\left(\sin y\right) - 2y\,D_x\,y = 0$$

$$2x + \tfrac{1}{2}(\sin y)^{-1/2}(\cos y)\,D_x\,y - 2yy' = 0$$

$$2x + \tfrac{1}{2}(\sin y)^{-1/2}(\cos y)\,y' - 2yy' = 0 \Rightarrow$$

$$2x = 2yy' - \frac{(\cos y)\,y'}{2\sqrt{\sin y}} \Rightarrow 2x = \left(2y - \frac{\cos y}{2\sqrt{\sin y}}\right)y' \Rightarrow$$

$$y' = \frac{2x}{2y - \dfrac{\cos y}{2\sqrt{\sin y}}} \cdot \frac{2\sqrt{\sin y}}{2\sqrt{\sin y}} \Rightarrow y' = \frac{4x\sqrt{\sin y}}{4y\sqrt{\sin y} - \cos y}$$

19 *Note:* Remember to treat a and b as constants when differentiating.

$$D_x\left[(x^2 + y^2 + a^2)^2\right] - D_x(4a^2x^2) = D_x(b^4)$$

$$2(x^2 + y^2 + a^2)\,D_x(x^2 + y^2 + a^2) - 4a^2\,D_x(x^2) = 0$$

$$2(x^2 + y^2 + a^2)(2x + 2y\,D_x\,y + 0) - 4a^2(2x) = 0$$

$$2(x^2 + y^2 + a^2)(2x + 2yy') - 8a^2x = 0$$

$$4y(x^2 + y^2 + a^2)y' = 8a^2x - 4x(x^2 + y^2 + a^2)$$

$$y' = \frac{2a^2x - x(x^2 + y^2 + a^2)}{y(x^2 + y^2 + a^2)}$$

At $P(2, \sqrt{2})$ with $a = 2$, $b = \sqrt{6}$, $y' = \dfrac{2(2)^2(2) - 2(2^2 + (\sqrt{2})^2 + 2^2)}{\sqrt{2}(2^2 + (\sqrt{2})^2 + 2^2)} = -\dfrac{\sqrt{2}}{5}.$

25

$$D_x(2x^3 - x^2y + y^3 - 1) = D_x(0)$$

$$D_x(2x^3) - D_x(x^2y) + D_x(y^3) - D_x(1) = 0$$

$$6x^2 - \left[x^2\,D_x\,y + y\,D_x(x^2)\right] + 3y^2\,D_x\,y - 0 = 0$$

$$6x^2 - (x^2y' + 2xy) + 3y^2y' = 0$$

$$3y^2y' - x^2y' = 2xy - 6x^2$$

$$y'(3y^2 - x^2) = 2xy - 6x^2 \Rightarrow y' = \frac{2xy - 6x^2}{3y^2 - x^2}$$

At $P(2, -3)$, $y' = \dfrac{2(2)(-3) - 6(2)^2}{3(-3)^2 - 2^2} = -\dfrac{36}{23},$

which is equal to the slope of the tangent line.

27

$$D_x(x^2y + \sin y) = D_x(2\pi)$$

$$D_x(x^2y) + D_x(\sin y) = 0$$

$$\left[x^2\,D_x\,y + y\,D_x(x^2)\right] + (\cos y)\,D_x\,y = 0$$

$$x^2y' + y(2x) + (\cos y)y' = 0$$

$$(x^2 + \cos y)y' = -2xy$$

$$y' = -\frac{2xy}{x^2 + \cos y}$$

At $P(1, 2\pi)$, $y' = -\dfrac{2(1)(2\pi)}{1^2 + \cos 2\pi} = -2\pi,$

which is equal to the slope of the tangent line.

$\boxed{29}$ $D_x\left(3x^2\right) + D_x\left(4y^2\right) = D_x\left(4\right) \Rightarrow 6x + 8y\,D_x\,y = 0 \Rightarrow 8yy' = -6x \Rightarrow y' = -\dfrac{3x}{4y}$.

$$y'' = -\frac{(4y)\,D_x\,(3x) - (3x)\,D_x\,(4y)}{(4y)^2} = -\frac{(4y)(3) - (3x)(4y')}{16y^2}$$

$$= -\frac{12y - 12x\left(-\frac{3x}{4y}\right)}{16y^2}\left(\text{since } y' = -\frac{3x}{4y}\right) = -\frac{12y - 12x\left(-\frac{3x}{4y}\right)}{16y^2}\cdot\frac{4y}{4y}$$

$$= -\frac{48y^2 + 36x^2}{64y^3} = -\frac{12(3x^2 + 4y^2)}{64y^3} = -\frac{3(4)}{16y^3}\left(\text{since } 3x^2 + 4y^2 = 4\right) = -\frac{3}{4y^3}$$

$\boxed{33}$ $D_x\left(\sin y\right) + D_x\,y = D_x\left(x\right) \Rightarrow (\cos y)\,D_x\,y + y' = 1 \Rightarrow (\cos y)y' + y' = 1 \Rightarrow$

$y' = \dfrac{1}{1 + \cos y}$. Using the reciprocal rule (3.21),

$$y'' = -\frac{D_x\left(1 + \cos y\right)}{(1 + \cos y)^2} = -\frac{(-\sin y)y'}{(1 + \cos y)^2} = \frac{\sin y}{(1 + \cos y)^3}\left(\text{since } y' = \frac{1}{1 + \cos y}\right).$$

$\boxed{35}$ If we solve the given equation for y, we find that $x^4 + y^4 - 1 = 0 \Leftrightarrow y^4 = 1 - x^4$

$\Leftrightarrow y = \pm\sqrt[4]{1 - x^4}$. At first glance it might appear that the answer is two.

However, we could define a function $f_c(x)$ as follows:

Let $f_c(x) = \begin{cases} \left(1 - x^4\right)^{1/4} & \text{if } -1 \le x \le c \\ -\left(1 - x^4\right)^{1/4} & \text{if } c < x \le 1 \end{cases}$ for any c in $[-1,\, 1]$. For each c in

$[-1,\, 1]$, we obtain a different function determined by the two formulas for y. Thus, there are an infinite number of solutions.

$\boxed{37}$ None; $x^2 + y^2 + 1 = 0 \Leftrightarrow x^2 + y^2 = -1$, which is not possible because $x^2 + y^2 \ge 0$. Since no values of x and y satisfy the equation, it cannot determine any functions implicitly.

$\boxed{41}$ $3x^2 - x^2y^3 + 4y = 12 \Rightarrow$

$$D_x\left(3x^2\right) - \left[x^2\,D_x\left(y^3\right) + y^3\,D_x\left(x^2\right)\right] + D_x\left(4y\right) = D_x\left(12\right) \Rightarrow$$

$$6x - \left(x^2\cdot 3y^2y' + 2xy^3\right) + 4y' = 0 \Rightarrow 4y' - 3x^2y^2y' = 2xy^3 - 6x \Rightarrow$$

$y' = \dfrac{2xy^3 - 6x}{4 - 3x^2y^2}$. At $(2,\, 0)$, $y' = \dfrac{2(2)(0)^3 - 6(2)}{4 - 3(2)^2(0)^2} = -3$.

Then, $\Delta y \approx dy = f'(2)\Delta x = (-3)(-0.03) = 0.09$.

Exercises 3.8

$\boxed{1}$ $A = x^2 \Rightarrow \dfrac{d}{dt}(A) = \dfrac{d}{dt}(x^2) \Rightarrow \dfrac{dA}{dt} = 2x\dfrac{dx}{dt} = 2(10)(3) = 60.$

3   $V = -5p^{3/2} \Rightarrow \frac{d}{dt}(V) = \frac{d}{dt}(-5p^{3/2}) \Rightarrow \frac{dV}{dt} = -\frac{15}{2}p^{1/2}\frac{dp}{dt} \Rightarrow \frac{dp}{dt} = -\frac{2}{15\sqrt{p}}\frac{dV}{dt}.$

We are given $\frac{dV}{dt}$, but not p. Solving $V = -5p^{3/2}$ for p, we have $p = \left[-\frac{1}{5}V\right]^{2/3}$.

When $V = -40$, $p = \left[-\frac{1}{5}(-40)\right]^{2/3} = \left[8^{1/3}\right]^2 = 2^2 = 4.$

$$\text{Thus, } \frac{dp}{dt} = -\frac{2}{15\sqrt{4}}(-4) = \frac{4}{15}.$$

5

$$x^2 + 3y^2 + 2y = 10$$

$$\frac{d}{dt}(x^2) + \frac{d}{dt}(3y^2) + \frac{d}{dt}(2y) = \frac{d}{dt}(10)$$

$$2x\frac{dx}{dt} + 6y\frac{dy}{dt} + 2\frac{dy}{dt} = 0$$

$$x\frac{dx}{dt} + (3y+1)\frac{dy}{dt} = 0$$

If $x = 3$, $y = -1$, and $\frac{dx}{dt} = 2$, then $(3)(2) + (-2)\frac{dy}{dt} = 0 \Rightarrow \frac{dy}{dt} = 3.$

11 In this problem, the volume of the balloon is changing. The formula for the volume of a sphere is $V = \frac{4}{3}\pi r^3$. We will use our "Find when knowing" statement for the related rates applications. Find dr/dt when diameter = 18 in. knowing $dV/dt = 5$ ft^3/min . Note that we need to change our "when" condition from diameter = 18 in. to radius = 9 in. = $\frac{9}{12}$ ft = $\frac{3}{4}$ ft so that we have matching units.

$V = \frac{4}{3}\pi r^3 \Rightarrow \frac{d}{dt}(V) = \frac{d}{dt}(\frac{4}{3}\pi r^3) \Rightarrow \frac{dV}{dt} = 4\pi r^2 \frac{dr}{dt} \Rightarrow$

$$\frac{dr}{dt} = \frac{1}{4\pi r^2}\frac{dV}{dt} = \frac{1}{4\pi(\frac{3}{4})^2}(5) = \frac{20}{9\pi} \approx 0.707 \text{ ft/min.}$$

15 Let x denote the distance between the tip of the shadow and the base of the pole, y the distance between the boy and the base, and z the length of the shadow. $\frac{dx}{dt}$ is the rate at which the tip of the shadow is moving and $\frac{dz}{dt}$ is the rate at which the length of the shadow is increasing.

Find dx/dt when $y = 18$ ft knowing $dy/dt = 4$ ft/sec .

To find $\frac{dx}{dt}$, we must find a formula that relates x and y. Using similar triangles in the figure from the text gives us $\frac{x}{16} = \frac{z}{5} \Rightarrow \frac{x}{16} = \frac{x-y}{5}$ $\{x = y + z \Rightarrow$ $z = x - y\} \Rightarrow \frac{x}{16} = \frac{x}{5} - \frac{y}{5}$ {multiply by 80} $\Rightarrow 5x = 16x - 16y \Rightarrow x = \frac{16}{11}y \Rightarrow$ $\frac{d}{dt}(x) = \frac{d}{dt}(\frac{16}{11}y) \Rightarrow \frac{dx}{dt} = \frac{16}{11}\frac{dy}{dt} = \frac{16}{11}(4) = \frac{64}{11} \approx 5.82$ ft/sec. To answer the second question, we need to find $\frac{dz}{dt}$. $z = x - y \Rightarrow \frac{dz}{dt} = \frac{dx}{dt} - \frac{dy}{dt} = \frac{64}{11} - 4 = \frac{20}{11} \approx 1.82$ ft/sec. Note that the 18 ft measurement was not used to answer either question.

$\boxed{17}$ Let T denote the thickness of the ice and note that the radius is 120 in. We must find a formula for the volume V of the ice and then find dV/dt when $T = 2$ in. knowing $dT/dt = -\frac{1}{4}$ in./hr . The volume of the ice (outer hemisphere $-$ inner hemisphere) is $V = \frac{2}{3}\pi r_{outer}^3 - \frac{2}{3}\pi r_{inner}^3 = \frac{2}{3}\pi(120 + T)^3 - \frac{2}{3}\pi(120)^3 \Rightarrow \frac{dV}{dt} =$ $2\pi(120 + T)^2 \frac{dT}{dt} \left\{ \frac{d}{dt}\left[\frac{2}{3}\pi(120)^3\right] = 0 \text{ since it is a constant} \right\} = 2\pi(120 + 2)^2(-\frac{1}{4})$ $= -7442\pi \approx -23{,}380$ in.^{3}/hr.

$\boxed{23}$ Let h denote the depth of the water and V the volume of the water in the trough. Find dh/dt when $h = 8$ in. knowing $dV/dt = 5$ ft^3/min . The area of the submerged triangular portion is $A = \frac{1}{2}\cdot\text{base}\cdot\text{height} = \frac{1}{2}\left(\frac{2h}{\sqrt{3}}\right)h \left\{ h = \frac{\sqrt{3}}{2}s \Rightarrow \right.$ $s = \frac{2h}{\sqrt{3}}$ from the front end papers of the text $\left.\right\} = \frac{h^2}{\sqrt{3}}$. The volume of water in the trough is just the area of the submerged triangular portion times the length of the trough. $V = 8A = \frac{8h^2}{\sqrt{3}} \Rightarrow \frac{dV}{dt} = \frac{16h}{\sqrt{3}}\frac{dh}{dt} \Rightarrow \frac{dh}{dt} = \frac{1}{16h}\sqrt{3}\frac{dV}{dt} = \frac{1}{16(\frac{2}{3})}\sqrt{3}(5)$ { 8 in. $= \frac{2}{3}$ ft } $= \frac{15}{32}\sqrt{3} \approx 0.81$ ft/min.

$\boxed{25}$ Let x denote the length of a side. Find dx/dt when $A = 200$ cm^2 knowing $dA/dt = -4$ cm^2/min . From the front end papers of the text, the area of the equilateral triangle is $A = \frac{\sqrt{3}}{4}x^2$. Thus, $\frac{dA}{dt} = \frac{\sqrt{3}}{2}x\frac{dx}{dt} \Rightarrow \frac{dx}{dt} = \frac{2}{\sqrt{3}\,x}\frac{dA}{dt}$. We need to find the length of x when A is 200. $A = 200 \Rightarrow 200 = \frac{\sqrt{3}}{4}x^2 \Rightarrow x = \left(\frac{800}{\sqrt{3}}\right)^{1/2}$.

Thus, $x = \left(\frac{800}{\sqrt{3}}\right)^{1/2}$ and $\frac{dA}{dt} = -4 \Rightarrow \frac{dx}{dt} = \frac{2}{\sqrt{3}\left(\frac{800}{\sqrt{3}}\right)^{1/2}}(-4) = -\frac{8\sqrt[4]{3}}{\sqrt{3}\,(20\sqrt{2})} =$ $-\frac{\sqrt{2}}{5\sqrt[4]{3}} \approx -0.2149$ cm/min.

$\boxed{29}$ Find dR/dt when $R_1 = 30$ ohms and $R_2 = 90$ ohms knowing

$dR_1/dt = 0.01$ ohm/sec and $dR_2/dt = 0.02$ ohm/sec . $\frac{1}{R} = \frac{1}{R_1} + \frac{1}{R_2} \Rightarrow$

$\frac{d}{dt}\left(\frac{1}{R}\right) = \frac{d}{dt}\left(\frac{1}{R_1}\right) + \frac{d}{dt}\left(\frac{1}{R_2}\right) \Rightarrow -\frac{1}{R^2}\frac{dR}{dt} = -\frac{1}{R_1^2}\frac{dR_1}{dt} - \frac{1}{R_2^2}\frac{dR_2}{dt}.$

$\frac{dR}{dt} = -R^2\left[-\frac{1}{R_1^2}\frac{dR_1}{dt} - \frac{1}{R_2^2}\frac{dR_2}{dt}\right]. \ R_1 = 30$ and $R_2 = 90 \Rightarrow \frac{1}{R} = \frac{1}{30} + \frac{1}{90} \Rightarrow$

$\frac{1}{R} = \frac{4}{90} \Rightarrow R = \frac{45}{2}. \ \frac{dR_1}{dt} = 0.01$ and $\frac{dR_2}{dt} = 0.02 \Rightarrow$

$$\frac{dR}{dt} = -\left(\frac{45}{2}\right)^2\left[-\frac{1}{(30)^2}\left(\frac{1}{100}\right) - \frac{1}{(90)^2}\left(\frac{2}{100}\right)\right] = \frac{11}{1600} = 0.006875 \text{ ohm/sec.}$$

$\boxed{31}$ Find dh/dt when $h = 4$ ft knowing $dV/dt = 100$ gal/min .

Using the given conversion factor, $\dfrac{dV}{dt} = \dfrac{100 \text{ gal}}{\text{min}} \times \dfrac{0.1337 \text{ ft}^3}{1 \text{ gal}} = 13.37 \text{ ft}^3/\text{min}$.

Since $a = 16$, $V = \frac{1}{3}\pi h^2 \big[3(16) - h\big] \Rightarrow \dfrac{dV}{dt} = \pi h(32 - h)\dfrac{dh}{dt} \Rightarrow$

$$\dfrac{dh}{dt} = \dfrac{1}{\pi h(32 - h)}\dfrac{dV}{dt} \approx \dfrac{1}{\pi(4)(28)}(13.37) = \dfrac{13.37}{112\pi} \approx 0.038 \text{ ft/min}.$$

$\boxed{35}$ Orient the plane as in *Figure 35*. The plane is located at the origin, traveling upward at a 45° angle. The control tower is located at P. If the plane travels a distance z, then its coordinates will be $\left(\frac{z}{\sqrt{2}}, \frac{z}{\sqrt{2}}\right)$ on the coordinate plane since the sides of a triangle with interior angles of 45°, 45°, and 90° are in the proportion 1–1–$\sqrt{2}$. In this case, the plane is traveling 360 mi/hr, or equivalently, 6 mi/min. L is the distance between the plane and the control tower. We must find dL/dt when $z = 6$ mi (one minute in flight) knowing $dz/dt = 360$ mi/hr . Using the distance formula $\{ 10{,}560 \text{ ft} = 2 \text{ mi} \}$,

$$L = \sqrt{\left(\tfrac{z}{\sqrt{2}} - 0\right)^2 + \left(\tfrac{z}{\sqrt{2}} - (-2)\right)^2} = \sqrt{\tfrac{1}{2}z^2 + \left(\tfrac{1}{2}z^2 + \tfrac{4}{\sqrt{2}}z + 4\right)} = \sqrt{z^2 + 2\sqrt{2}\,z + 4}$$

$\Rightarrow \dfrac{dL}{dt} = \frac{1}{2}(z^2 + 2\sqrt{2}\,z + 4)^{-1/2}(2z + 2\sqrt{2})\dfrac{dz}{dt}$. $z = 6$ and $\dfrac{dz}{dt} = 360 \Rightarrow$

$$\dfrac{dL}{dt} = \dfrac{z + \sqrt{2}}{\sqrt{z^2 + 2\sqrt{2}\,z + 4}}\dfrac{dz}{dt} = \dfrac{6 + \sqrt{2}}{\sqrt{40 + 12\sqrt{2}}}(360) = \dfrac{180(6 + \sqrt{2})}{\sqrt{10 + 3\sqrt{2}}} \approx 353.6 \text{ mi/hr}.$$

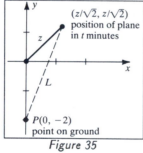

Figure 35

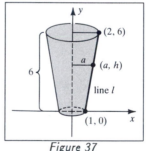

Figure 37

$\boxed{37}$ Consider the cup as shown in *Figure 37*. Let $b = 1$ and a vary with the depth of the water. The equation of line l is $y = 6(x - 1)$ $\{ l$ passes through $(1, 0)$ and $(2, 6) \}$ and thus, $h = 6(a - 1) \Rightarrow a = \frac{1}{6}(h + 6)$. We need to write the volume V in terms of h and then find dh/dt when $h = 4$ in. knowing $dV/dt = -3$ in.^{3}/hr .

$$V = \tfrac{1}{3}\pi h(a^2 + b^2 + ab) = \tfrac{1}{3}\pi h\Big[\tfrac{1}{36}(h^2 + 12h + 36) + 1 + \tfrac{1}{6}(h + 6)\Big] =$$

$$\tfrac{\pi}{108}h(h^2 + 18h + 108) = \tfrac{\pi}{108}h^3 + \tfrac{\pi}{6}h^2 + \pi h \Rightarrow \dfrac{dV}{dt} = \left(\tfrac{\pi}{36}h^2 + \tfrac{\pi}{3}h + \pi\right)\dfrac{dh}{dt}.$$

$$\dfrac{dV}{dt} = -3, \ h = 4 \Rightarrow -3 = \left(\tfrac{4}{9}\pi + \tfrac{4}{3}\pi + \pi\right)\dfrac{dh}{dt} \Rightarrow \dfrac{dh}{dt} = -\tfrac{27}{25\pi} \approx -0.3438 \text{ in./hr}.$$

$\boxed{41}$ Using $A = \frac{1}{2}bc \sin \alpha$ for any triangle, where α is the angle between sides b and c,

we have $A = \frac{1}{2}(6)(6) \sin \theta = 18 \sin \theta$. Find $\underline{dA/dt}$ when $\underline{\theta = 30°}$ knowing

$\underline{d\theta/dt = 2°/\text{min}}$. $A = 18 \sin \theta \Rightarrow \frac{dA}{dt} = 18 \cos \theta \frac{d\theta}{dt} =$

$18(\frac{\sqrt{3}}{2})(2 \cdot \frac{\pi}{180})$ { convert $2°$ to radians; $\cos 30° = \frac{\sqrt{3}}{2}$ } $= \frac{\pi}{10}\sqrt{3} \approx 0.54$ in.^{2}/min.

$\boxed{43}$ Let s denote the distance between the top of the control tower and the airplane and x

the distance the airplane is down the runway. Find $\underline{ds/dt}$ when $\underline{x = 300 \text{ ft}}$

knowing $\underline{dx/dt = 100 \text{ mi/hr}}$. $s^2 = x^2 + 300^2 + 20^2 \Rightarrow 2s\frac{ds}{dt} = 2x\frac{dx}{dt} \Rightarrow$

$\frac{ds}{dt} = \frac{x}{s}\frac{dx}{dt}$. When $x = 300$ ft, $s^2 = 300^2 + 300^2 + 20^2 \Rightarrow s = \sqrt{180,400}$ ft.

$\frac{dx}{dt} = \frac{100 \text{ mi}}{\text{hr}} \times \frac{5280 \text{ ft}}{1 \text{ mi}} \times \frac{1 \text{ hr}}{60 \text{ min}} = \frac{8800 \text{ ft}}{\text{min}}$. Thus, $\frac{ds}{dt} = \frac{300}{\sqrt{180,400}}(8800) =$

$\frac{2,640,000}{\sqrt{180,400}} \approx 6215.6$ ft/min ≈ 70.63 mi/hr. $\left\{ \frac{1 \text{ ft}}{\text{min}} \times \frac{1 \text{ mi}}{5280 \text{ ft}} \times \frac{60 \text{ min}}{1 \text{ hr}} \right\}$

$\boxed{45}$ Let x denote the horizontal distance between the plane and the observer, and θ the

angle of elevation. When the plane is 60,000 ft from the observer $\{ \theta = 30° \}$, the

constant height is 30,000 ft. Find $\underline{dx/dt}$ when $\underline{\theta = 30°}$ knowing

$\underline{d\theta/dt = 0.5°/\text{sec}}$. $\tan \theta = \frac{30,000}{x} \Rightarrow x = 30,000 \cot \theta \Rightarrow$

$\frac{dx}{dt} = -30,000 \csc^2\theta \frac{d\theta}{dt} = -30,000(2)^2(0.5 \cdot \frac{\pi}{180})$ { convert $0.5°$ to radians;

$\csc 30° = 2$ } $= -\frac{1000\pi}{3}$ ft/sec ≈ -714.0 mi/hr (toward the observer).

$$\left\{ \frac{1 \text{ ft}}{\text{sec}} \times \frac{1 \text{ mi}}{5280 \text{ ft}} \times \frac{3600 \text{ sec}}{1 \text{ hr}} \right\}$$

$\boxed{47}$ Let ϕ represent the angle the wheel turns through when the pedals are rotated

through an angle θ. Using the arc length formula $\{ s = r\theta \}$, $s_1 = 5\theta$ and $s_2 = 2\phi$.

Assuming no slippage in the chain, the length of chain moving around each sprocket

must be equal, i.e., $s_1 = s_2 \Rightarrow 5\theta = 2\phi \Rightarrow \phi = \frac{5}{2}\theta$. The length x that the wheel

travels is 14ϕ since the wheel has a radius of 14 and rotates through an angle ϕ.

$x = 14\phi = 14(\frac{5}{2}\theta) = 35\theta \Rightarrow \frac{dx}{dt} = 35\frac{d\theta}{dt}$ in./sec $=$

$$\frac{175}{88}\frac{d\theta}{dt} \text{ mi/hr} \left\{ \frac{35 \text{ in.}}{\text{sec}} \times \frac{1 \text{ ft}}{12 \text{ in.}} \times \frac{1 \text{ mi}}{5280 \text{ ft}} \times \frac{3600 \text{ sec}}{1 \text{ hr}} \right\}.$$

49 (a) From the figure in the text, $\sin\theta = \frac{r}{l}$, or $r = l\sin\theta$.

$r = l\sin\theta$ and $v^2 = rg\tan\theta \Rightarrow v^2 = gl\sin\theta\tan\theta$.

$\frac{d}{dt}(v^2) = gl\frac{d}{dt}(\sin\theta\tan\theta)$ { g and l are constants } $\Rightarrow$

$$2v\frac{dv}{dt} = gl(\sin\theta\sec^2\theta + \tan\theta\cos\theta)\frac{d\theta}{dt}$$

$$= gl(\sin\theta\sec^2\theta + \sin\theta)\frac{d\theta}{dt}$$

$$= g(l\sin\theta)(\sec^2\theta + 1)\frac{d\theta}{dt} = gr(\sec^2\theta + 1)\frac{d\theta}{dt}$$

(b) $r = l\sin\theta \Rightarrow \frac{dr}{dt} = l\cos\theta\frac{d\theta}{dt} \Rightarrow \frac{d\theta}{dt} = \frac{1}{l\cos\theta}\frac{dr}{dt}$.

Substituting this expression for $\frac{d\theta}{dt}$ into the result from part (a) yields

$$2v\frac{dv}{dt} = gr(\sec^2\theta + 1)\frac{1}{l\cos\theta}\frac{dr}{dt}$$

$$= g(l\sin\theta)(\sec^2\theta + 1)\frac{1}{l\cos\theta}\frac{dr}{dt}$$

$$= g\tan\theta\,(\sec^2\theta + 1)\frac{dr}{dt}.$$

3.9 Review Exercises

1   Using Definition (3.5) with $f(x) = \dfrac{4}{3x^2 + 2}$,

$f'(x) = \lim\limits_{h\to0}\dfrac{f(x+h) - f(x)}{h}$ { Definition (3.5) }

$= \lim\limits_{h\to0}\dfrac{1}{h}\left[\dfrac{4}{3(x+h)^2 + 2} - \dfrac{4}{3x^2 + 2}\right]$ { definition of f }

$= \lim\limits_{h\to0}\dfrac{4}{h}\left[\dfrac{3x^2 + 2 - 3(x+h)^2 - 2}{\left[3(x+h)^2 + 2\right](3x^2 + 2)}\right]$ { combine fractions }

$= \lim\limits_{h\to0}\dfrac{4}{h}\left[\dfrac{-6xh - 3h^2}{\left[3(x+h)^2 + 2\right](3x^2 + 2)}\right]$ { simplify }

$= \lim\limits_{h\to0}4\left[\dfrac{-6x - 3h}{\left[3(x+h)^2 + 2\right](3x^2 + 2)}\right]$ { factor and cancel h }

$= -\dfrac{24x}{(3x^2 + 2)^2}$ { let $h = 0$ }

7   $F(z) = \sqrt[3]{7z^2 - 4z + 3} = (7z^2 - 4z + 3)^{1/3} \Rightarrow$

$$F'(z) = \tfrac{1}{3}(7z^2 - 4z + 3)^{-2/3}(14z - 4) = \dfrac{2(7z - 2)}{3(7z^2 - 4z + 3)^{2/3}}.$$

9   Using the reciprocal rule (3.21),

$$G(x) = \dfrac{6}{(3x^2 - 1)^4} \Rightarrow G'(x) = (-6)\dfrac{4(3x^2 - 1)^3(6x)}{\left[(3x^2 - 1)^4\right]^2} = -\dfrac{144x}{(3x^2 - 1)^5}.$$

13   $g(x) = \sqrt[5]{(3x + 2)^4} = (3x + 2)^{4/5} \Rightarrow g'(x) = \tfrac{4}{5}(3x + 2)^{-1/5}(3) = \dfrac{12}{5(3x + 2)^{1/5}}.$

$\boxed{15}$ $r'(s) = 4\left(\dfrac{8s^2 - 4}{1 - 9s^3}\right)^3 D_s\left(\dfrac{8s^2 - 4}{1 - 9s^3}\right)$

$= 4\left(\dfrac{8s^2 - 4}{1 - 9s^3}\right)^3 \dfrac{(1 - 9s^3)\, D_s\,(8s^2 - 4) - (8s^2 - 4)\, D_s\,(1 - 9s^3)}{(1 - 9s^3)^2}$

$= 4\left(\dfrac{8s^2 - 4}{1 - 9s^3}\right)^3 \dfrac{(1 - 9s^3)(16s) - (8s^2 - 4)(-27s^2)}{(1 - 9s^3)^2}$

$= 4\left(\dfrac{8s^2 - 4}{1 - 9s^3}\right)^3 \dfrac{16s - 144s^4 + 216s^4 - 108s^2}{(1 - 9s^3)^2}$

$= \dfrac{4(8s^2 - 4)^3(72s^4 - 108s^2 + 16s)}{(1 - 9s^3)^5} = \dfrac{1024s(2s^2 - 1)^3(18s^3 - 27s + 4)}{(1 - 9s^3)^5}$

$\boxed{17}$ $F(x) = (x^6 + 1)^5(3x + 2)^3 \Rightarrow$

$F'(x) = (x^6 + 1)^5 D_x\,(3x + 2)^3 + (3x + 2)^3 D_x\,(x^6 + 1)^5$

$= (x^6 + 1)^5(3)(3x + 2)^2(3) + (3x + 2)^3(5)(x^6 + 1)^4(6x^5)$

$= 9(x^6 + 1)^5(3x + 2)^2 + 30x^5(x^6 + 1)^4(3x + 2)^3$

$= (x^6 + 1)^4(3x + 2)^2\Big[9(x^6 + 1) + 30x^5(3x + 2)\Big]$

$= (x^6 + 1)^4(3x + 2)^2\Big[9x^6 + 9 + 90x^6 + 60x^5\Big]$

$= 3(x^6 + 1)^4(3x + 2)^2(33x^6 + 20x^5 + 3)$

$\boxed{23}$ $f(w) = \sqrt{\dfrac{2w + 5}{7w - 9}} = \left(\dfrac{2w + 5}{7w - 9}\right)^{1/2} \Rightarrow$

$f'(w) = \dfrac{1}{2}\left(\dfrac{2w + 5}{7w - 9}\right)^{-1/2} D_w\left(\dfrac{2w + 5}{7w - 9}\right)$

$= \dfrac{1}{2}\left(\dfrac{2w + 5}{7w - 9}\right)^{-1/2} \dfrac{(7w - 9)(2) - (2w + 5)(7)}{(7w - 9)^2}$

$= \dfrac{1}{2}\left(\dfrac{7w - 9}{2w + 5}\right)^{1/2} \dfrac{-53}{(7w - 9)^2}$

$= -\dfrac{53}{2(2w + 5)^{1/2}(7w - 9)^{3/2}} = -\dfrac{53}{2\sqrt{(2w + 5)(7w - 9)^3}}$

$\boxed{27}$ $f(x) = \sin^2(4x^3) = \Big[\sin(4x^3)\Big]^2 \Rightarrow$

$f'(x) = 2\sin(4x^3)\, D_x\Big[\sin(4x^3)\Big]$

$= 2\sin(4x^3)\,\cos(4x^3)\, D_x\,(4x^3)$

$= \Big[2(\sin 4x^3)(\cos 4x^3)\Big](12x^2) = 12x^2\sin(2\cdot 4x^3) = 12x^2\sin 8x^3$

$\boxed{29}$ $h(x) = (\sec x + \tan x)^5 \Rightarrow$

$$h'(x) = 5(\sec x + \tan x)^4 \, D_x \, (\sec x + \tan x)$$

$$= 5(\sec x + \tan x)^4 (\sec x \tan x + \sec^2 x)$$

$$= 5(\sec x + \tan x)^4 (\sec x)(\tan x + \sec x) = 5 \sec x (\sec x + \tan x)^5$$

$\boxed{33}$ $K(\theta) = \dfrac{\sin 2\theta}{1 + \cos 2\theta} \Rightarrow$

$$K'(\theta) = \frac{(1 + \cos 2\theta) \, D_\theta \, (\sin 2\theta) - (\sin 2\theta) \, D_\theta \, (1 + \cos 2\theta)}{(1 + \cos 2\theta)^2}$$

$$= \frac{(1 + \cos 2\theta)(2 \cos 2\theta) - (\sin 2\theta)(-2 \sin 2\theta)}{(1 + \cos 2\theta)^2}$$

$$= \frac{2 \cos 2\theta + 2 \cos^2 2\theta + 2 \sin^2 2\theta}{(1 + \cos 2\theta)^2}$$

$$= \frac{2 \cos 2\theta + 2(\cos^2 2\theta + \sin^2 2\theta)}{(1 + \cos 2\theta)^2} = \frac{2(1 + \cos 2\theta)}{(1 + \cos 2\theta)^2} = \frac{2}{1 + \cos 2\theta}$$

$\boxed{35}$ $g(x) = (\cos \sqrt[3]{x} - \sin \sqrt[3]{x})^3 = \left[\cos(x^{1/3}) - \sin(x^{1/3})\right]^3 \Rightarrow$

$$g'(x) = 3\left[\cos(x^{1/3}) - \sin(x^{1/3})\right]^2 D_x\left[\cos(x^{1/3}) - \sin(x^{1/3})\right]$$

$$= 3\left[\cos(x^{1/3}) - \sin(x^{1/3})\right]^2 \left[-\sin(x^{1/3}) \, D_x \, (x^{1/3}) - \cos(x^{1/3}) \, D_x \, (x^{1/3})\right]$$

$$= 3(\cos \sqrt[3]{x} - \sin \sqrt[3]{x})^2 \left[-\sin \sqrt[3]{x} \, (\tfrac{1}{3})(x^{-2/3}) - \cos \sqrt[3]{x} \, (\tfrac{1}{3})(x^{-2/3})\right]$$

$$= -\frac{(\cos \sqrt[3]{x} - \sin \sqrt[3]{x})^2 (\cos \sqrt[3]{x} + \sin \sqrt[3]{x})}{\sqrt[3]{x^2}}$$

$\boxed{37}$ $G(u) = \dfrac{\csc u + 1}{\cot u + 1} \Rightarrow$

$$G'(u) = \frac{(\cot u + 1) \, D_u \, (\csc u + 1) - (\csc u + 1) \, D_u \, (\cot u + 1)}{(\cot u + 1)^2}$$

$$= \frac{(\cot u + 1)(-\csc u \cot u) - (\csc u + 1)(-\csc^2 u)}{(\cot u + 1)^2}$$

$$= \frac{-\csc u \cot^2 u - \csc u \cot u + \csc^3 u + \csc^2 u}{(\cot u + 1)^2}$$

$$= \frac{\csc u(-\cot^2 u - \cot u + \csc^2 u + \csc u)}{(\cot u + 1)^2}$$

$$= \frac{\csc u(1 - \cot u + \csc u)}{(\cot u + 1)^2} \quad \{\csc^2 u - \cot^2 u = 1\}$$

$\boxed{45}$ $\dfrac{\sqrt{x}+1}{\sqrt{y}+1} = y \Rightarrow y(\sqrt{y}+1) = \sqrt{x}+1 \Rightarrow y^{3/2} + y = x^{1/2} + 1 \Rightarrow$

$\frac{3}{2}y^{1/2}\, D_x\, y + D_x\, y = \frac{1}{2}x^{-1/2} + D_x\,(1) \Rightarrow (\frac{3}{2}y^{1/2} + 1)y' = \frac{1}{2}x^{-1/2} \Rightarrow$

$$(3\sqrt{y}+2)y' = \frac{1}{\sqrt{x}} \Rightarrow y' = \frac{1}{\sqrt{x}\,(3\sqrt{y}+2)}$$

$\boxed{47}$ $xy^2 = \sin(x+2y) \Rightarrow D_x\,(xy^2) = D_x\Big[\sin(x+2y)\Big] \Rightarrow$

$\qquad x\,D_x\,(y^2) + y^2\,D_x\,(x) = \cos(x+2y)\,D_x\,(x+2y)$

$\qquad x(2yy') + y^2 = \cos(x+2y)\big[1 + 2y'\big]$

$\qquad 2xyy' - 2\cos(x+2y)y' = \cos(x+2y) - y^2 \Rightarrow y' = \dfrac{\cos(x+2y) - y^2}{2xy - 2\cos(x+2y)}$

$\boxed{49}$ $y = 2x - 4x^{-1/2} \Rightarrow y' = 2 + 2x^{-3/2} = 2 + 2/\sqrt{x^3}$. At $P(4, 6)$, $y' = 2 + \dfrac{2}{(\sqrt{4})^3} =$

$\frac{9}{4}$. The slope of the tangent line is $\frac{9}{4}$ and the slope of the normal line is $-\frac{1}{9/4} = -\frac{4}{9}$.

The equation of the tangent line is $y - 6 = \frac{9}{4}(x - 4)$, or $y = \frac{9}{4}x - 3$.

$\qquad$ The equation of the normal line is $y - 6 = -\frac{4}{9}(x - 4)$, or $y = -\frac{4}{9}x + \frac{70}{9}$.

$\boxed{51}$ $2x + 4y = 5 \Leftrightarrow y = -\frac{1}{2}x + \frac{5}{4}$. The slope of the given line is $-\frac{1}{2}$. The slope of any

line perpendicular to the given line is 2. $\quad y = 3x - \cos 2x \Rightarrow y' = 3 + 2\sin 2x$.

$y' = 2 \Rightarrow 3 + 2\sin 2x = 2 \Rightarrow \sin 2x = -\frac{1}{2} \Rightarrow 2x = \frac{7\pi}{6} + 2\pi n \quad$ or $\quad \frac{11\pi}{6} + 2\pi n \Rightarrow$

$x = \frac{7\pi}{12} + \pi n$ or $\frac{11\pi}{12} + \pi n$.

$\boxed{53}$ $y = 5x^3 + 4\sqrt{x} = 5x^3 + 4x^{1/2} \Rightarrow y' = 15x^2 + 4(\frac{1}{2}x^{-1/2}) = 15x^2 + \dfrac{2}{\sqrt{x}} \Rightarrow$

$\qquad y'' = 30x + 2(-\frac{1}{2}x^{-3/2}) = 30x - \dfrac{1}{\sqrt{x^3}} \Rightarrow y''' = 30 - (-\frac{3}{2}x^{-5/2}) = 30 + \dfrac{3}{2\sqrt{x^5}}$

$\boxed{55}$ $x^2 + 4xy - y^2 = 8 \Rightarrow D_x(x^2) + 4x D_x y + y D_x(4x) - D_x(y^2) = D_x(8) \Rightarrow$

$2x + 4xy' + 4y - 2yy' = 0 \Rightarrow 4xy' - 2yy' = -(2x + 4y) \Rightarrow$

$y' = -\dfrac{2x + 4y}{4x - 2y} = \dfrac{x + 2y}{y - 2x}.$ $y'' = \dfrac{(y - 2x) D_x(x + 2y) - (x + 2y) D_x(y - 2x)}{(y - 2x)^2} =$

$\dfrac{(y - 2x)(1 + 2y') - (x + 2y)(y' - 2)}{(y - 2x)^2}.$ Substituting for y', the <u>numerator</u> becomes:

$(y - 2x)\left(1 + 2 \cdot \dfrac{x + 2y}{y - 2x}\right) - (x + 2y)\left(\dfrac{x + 2y}{y - 2x} - 2\right) =$

$(y - 2x) + 2(x + 2y) - \left[\dfrac{(x + 2y)^2}{y - 2x}\right] + 2(x + 2y) = (2x + 9y) - \dfrac{(x + 2y)^2}{y - 2x}.$

It now follows that $y'' = \dfrac{2x + 9y - \dfrac{(x + 2y)^2}{y - 2x}}{(y - 2x)^2} \cdot \dfrac{y - 2x}{y - 2x} =$

$\dfrac{(2x + 9y)(y - 2x) - (x + 2y)^2}{(y - 2x)^3} = \dfrac{2xy - 4x^2 + 9y^2 - 18xy - x^2 - 4xy - 4y^2}{(y - 2x)^3} =$

$\dfrac{5y^2 - 20xy - 5x^2}{(y - 2x)^3} = \dfrac{5(y^2 - 4xy - x^2)}{(y - 2x)^3} =$

$\dfrac{5(-8)}{(y - 2x)^3} \{x^2 + 4xy - y^2 = 8 \Rightarrow y^2 - 4xy - x^2 = -8\} = -\dfrac{40}{(y - 2x)^3}.$

$\boxed{57}$ (a) $\Delta y = f(x + \Delta x) - f(x) = \left[3(x + \Delta x)^2 - 7\right] - (3x^2 - 7) = 6x\Delta x + 3(\Delta x)^2$

(b) $dy = f'(x)\, dx = 6x\, dx$

(c) $dy - \Delta y = 6x\, dx - \left[6x\Delta x + 3(\Delta x)^2\right] = -3(\Delta x)^2$

$\boxed{59}$ $A = \frac{1}{4}\sqrt{3}\, x^2$, where x is the length of a side. The maximum error ΔA can be

estimated by using dA. Hence, find <u>dA</u> when <u>$x = 4$ in.</u> knowing

<u>$dx = \pm 0.03$ in.</u> . $dA = \frac{1}{2}\sqrt{3}\, x\, dx = \frac{1}{2}\sqrt{3}(4)(\pm 0.03) = \pm 0.06\sqrt{3} \approx \pm 0.104$ in.^{2}.

The average error is $\dfrac{\Delta A}{A} \approx \dfrac{dA}{A} = \dfrac{\pm 0.06\sqrt{3}}{4\sqrt{3}} = (\pm 0.015)$ and

the percentage error is $\pm 1.5\%$.

$\boxed{63}$ (a) $(2f - 3g)'(2) = 2f'(2) - 3g'(2) = 2(4) - 3(2) = 2.$

(b) $(2f - 3g)''(2) = 2f''(2) - 3g''(2) = 2(-2) - 3(1) = -7.$

(c) $(fg)'(2) = f(2)\, g'(2) + g(2) f'(2) = (-1)(2) + (-3)(4) = -14.$

(d) $(fg)'' = (fg' + gf')' = (fg'' + g'f') + (gf'' + f'g') = fg'' + 2f'g' + gf'';$

at $x = 2$, the value is $(-1)(1) + 2(4)(2) + (-3)(-2) = 21.$

(e) $\left(\dfrac{f}{g}\right)'(2) = \dfrac{g(2) f'(2) - f(2)\, g'(2)}{[g(2)]^2} = \dfrac{(-3)(4) - (-1)(2)}{(-3)^2} = -\dfrac{10}{9}.$

(f) $\left(\dfrac{f}{g}\right)'' = \left(\dfrac{gf' - fg'}{g^2}\right)' = \dfrac{g^2(gf' - fg')' - (gf' - fg')(g^2)'}{g^4} =$

$$\dfrac{g^2\big[(gf'' + g'f') - (fg'' + f'g')\big] - 2gg'(gf' - fg')}{g^4};$$

at $x = 2$, the value is $\dfrac{9\big[(6 + 8) - (-1 + 8)\big] + 12(-12 + 2)}{81} = -\dfrac{57}{81} = -\dfrac{19}{27}.$

$\boxed{65}$ (a) $f(x) = 3(x + 1)^{1/3} - 4 \Rightarrow f'(x) = \dfrac{1}{(x + 1)^{2/3}}$, undefined at $x = -1$. Since

$(x + 1)^{2/3} \to 0^+$ as $x \to -1$, $\displaystyle\lim_{x \to -1^+} f'(x) = \infty$ and $\displaystyle\lim_{x \to -1^-} f'(x) = \infty \Rightarrow f$ has a

vertical tangent line at $(-1, -4)$. There is no cusp since $f'(x)$ is always positive.

(b) $f(x) = 2(x - 8)^{2/3} - 1 \Rightarrow f'(x) = \dfrac{4}{3(x - 8)^{1/3}}$, undefined at $x = 8$.

As $x \to 8^+$, $x > 8$ and $(x - 8)^{1/3} \to 0^+$. As $x \to 8^-$, $x < 8$ and $(x - 8)^{1/3} \to 0^-$.

Thus, $\displaystyle\lim_{x \to 8^+} f'(x) = \infty$ and $\displaystyle\lim_{x \to 8^-} f'(x) = -\infty \Rightarrow f$ has a cusp at $(8, -1)$.

Note: If f has a cusp, then it must also have a vertical tangent line.

$\boxed{67}$ Find dR/R when $R = kT^4$ knowing $dT/T = 0.5\%$.

$dR = 4kT^3\, dT \Rightarrow \dfrac{dR}{R} = \dfrac{4kT^3}{kT^4}\, dT = 4\dfrac{dT}{T}.$

Since $\dfrac{dT}{T} \approx \dfrac{\Delta T}{T} = 0.5\%$, we have $\dfrac{\Delta R}{R} \approx \dfrac{dR}{R} = 4(0.5\%) = 2\%.$

$\boxed{71}$ Let h denote the depth of the water and w the width of the surface of the water.

Find dV/dt when $h = 1$ ft knowing $dh/dt = \frac{1}{4}$ in./min. Using similar triangles

and *Figure 71*, we see that the right triangles on the end of the trapezoid have a base

equal to 1 and a height equal to 2. Thus, $\frac{x}{h} = \frac{1}{2} \Rightarrow h = 2x$ and hence,

$w = x + 3 + x = 3 + 2x = 3 + 2x = 3 + h$. The cross-sectional area of the

water at one end of the trough is

$A = \frac{1}{2}(\text{lower base} + \text{upper base})h$

$= \frac{1}{2}\big[3 + (3 + h)\big]h = 3h + \frac{1}{2}h^2.$

Hence, the volume V of water is

$V = 10A = 30h + 5h^2.$

$\dfrac{dV}{dt} = (30 + 10h)\dfrac{dh}{dt} = \big[30 + 10(1)\big]\cdot\big[\frac{1}{4}(\frac{1}{12})\big]$

$\{\frac{1}{4}$ in./min $= \frac{1}{4}(\frac{1}{12})$ ft/min $\} = \frac{5}{6}$ ft^3/min.

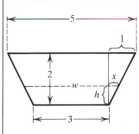

Figure 71

$\boxed{75}$ (a) Consider the ferris wheel to be in the xy-plane with the origin centered at the rotating axis of the ferris wheel. The (x, y) coordinates of any passenger can be given by $(50\cos\theta,\ 50\sin\theta)$, where θ is the angle in standard position with terminal side passing through the point (x, y). Since ground level is at $y = -60$, $h = y + 60 = 50\sin\theta + 60$. The wheel rotates 2π radians every 30 seconds, so $\theta = \frac{2\pi}{30}t - \frac{\pi}{2}$. Note that at $t = 0$, $\theta = -\frac{\pi}{2}$ and the seat is at the bottom. Thus, $h(t) = 50\sin\left(\frac{2\pi}{30}t - \frac{\pi}{2}\right) + 60 = 60 - 50\cos\frac{\pi}{15}t$

$$\left\{\sin\left(x - \tfrac{\pi}{2}\right) = \sin\left[-\left(\tfrac{\pi}{2} - x\right)\right] = -\sin\left(\tfrac{\pi}{2} - x\right) = -\cos x\right\}.$$

(b) $h = 55 \Rightarrow 60 - 50\cos\frac{\pi}{15}t = 55 \Rightarrow \cos\frac{\pi}{15}t = 0.1 \Rightarrow t_0 = \frac{15}{\pi}\cos^{-1}0.1 \approx 7.02$.

$h'(t) = -50\left(\frac{\pi}{15}\right)\left(-\sin\frac{\pi}{15}t\right) = \frac{10\pi}{3}\sin\frac{\pi}{15}t$ and $h'(t_0) = \frac{10\pi}{3}\sin\left(\frac{\pi}{15}\cdot\frac{15}{\pi}\cos^{-1}\frac{1}{10}\right) =$

$\frac{10\pi}{3}\sin\left(\cos^{-1}\frac{1}{10}\right) = \frac{10\pi}{3}\left(\frac{\sqrt{99}}{10}\right) = \sqrt{11}\,\pi \approx 10.4$ ft/sec. To see that $\sin\left(\cos^{-1}\frac{1}{10}\right) = $

$\frac{\sqrt{99}}{10}$, draw a right triangle with angle α having adjacent side 1 and hypotenuse 10. By the Pythagorean theorem, the opposite side is $\sqrt{10^2 - 1^2} = \sqrt{99}$. Hence, $\sin\alpha = \frac{\sqrt{99}}{10}$. Alternatively, using a calculator, $h'(7.02) \approx 10.4$ ft/sec.

Chapter 4: Applications of the Derivative

Note: In this chapter and subsequent chapters, we will use the following notation:

  MAX for absolute maximum (maxima), *MIN* for absolute minimum (minima),

  LMAX for local maximum (maxima), *LMIN* for local minimum (minima),

  and *CN* for critical number(s).

1 The absolute maximum and absolute minimum are simply the largest and smallest function values (or y values). In this text, "estimating the extrema" refers to the *absolute* extrema. See the discussion after Figure 4.7. *MAX* of 4 at 2; *MIN* of 0 at 4; *LMAX* at $x = 2$, $6 \le x \le 8$; *LMIN* at $x = 4$, $6 < x < 8$, $x = 10$. *Note:* There is *not* a local minimum at $x = 6$ since there is no *open* interval containing 6 such that $f(x) \ge f(6)$ for every x in the interval. The same holds for $x = 8$. However, there are local maxima at $x = 6$ and $x = 8$.

3 (a) On $[-3, 3)$, the *MIN* of -6 occurs at the left endpoint, $x = -3$. If $x = 3$ was included in the interval, the *MAX* would occur there. Since 3 is not included, no *specific* value of x less than 3 will give a *MAX*.   *MIN:* $f(-3) = -6$; *MAX:* none

(b) There is a *LMIN* of $-\frac{2}{3}$ at $x = 1$. However, as $x \to -3^+$, there are values less than $-\frac{2}{3}$. If $x = -3$ *were* included, then the *MIN* would occur there. Since -3 is not included, no *specific* value of x greater than -3 will give a *MIN*. On $(-3, \sqrt{3})$, the *MAX* of $\frac{2}{3}$ occurs at $x = -1$.   *MIN:* none; *MAX:* $f(-1) = \frac{2}{3}$

(c) If we had the interval $[-\sqrt{3}, 1]$, the *MIN* of f would occur at $x = 1$. Since 1 is not included, no *MIN* exists. The *MAX* of $\frac{2}{3}$ occurs at $x = -1$.

MIN: none; *MAX:* $f(-1) = \frac{2}{3}$

(d) Since $[0, 3]$ is a closed interval and f is continuous, both a *MAX* and *MIN* must exist by (4.3). The *MAX* of 6 occurs at the right endpoint, $x = 3$.

MIN: $f(1) = -\frac{2}{3}$; *MAX:* $f(3) = 6$

⑤ We first find any critical numbers of f.

$f(x) = \frac{1}{2}x^2 - 2x \Rightarrow f'(x) = x - 2$ and $f'(x) = 0 \Rightarrow$
$x = 2$. $f(2) = -2$. The graph of f is a parabola
opening upward with its vertex at the point $(2, -2)$.
The y value of the vertex is the *MIN* of f when it is in
the interval. For parts (a), (b), and (c), we examine
the associated *closed* interval, and make our
conclusion based on whether or not the endpoints of the
interval are included.

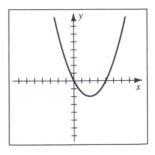

Figure 5

(a) On $[0, 5)$, the *MAX* of $\frac{5}{2}$ occurs at $x = 5$. Since 5 is not included, no *MAX* exists.

MIN: $f(2) = -2$; *MAX*: none

(b) On $(0, 2)$, the *MAX* occurs at $x = 0$ and the *MIN* occurs at $x = 2$. Since 0 and 2 are not included, no *MAX* nor *MIN* exists.   MIN: none; *MAX*: none

(c) On $(0, 4]$, the *MAX* occurs at $x = 4$. Since 4 is not included, no *MAX* exists.

MIN: $f(2) = -2$; *MAX*: none

(d) Since $[2, 5]$ is a closed interval and f is continuous, both a *MAX* and *MIN* must exist by (4.3).   MIN: $f(2) = -2$; *MAX*: $f(5) = \frac{5}{2}$

⑦ Since $[-3, 1]$ is a closed interval, we will follow the Guidelines for finding the extrema of a continuous function f on $[a, b]$ (4.9).

(1) From (4.8), we need to find all x values such that $f'(x) = 0$ or $f'(x)$ does not exist. $f(x) = 5 - 6x^2 - 2x^3 \Rightarrow f'(x) = -12x - 6x^2 = -6x(x + 2) = 0 \Rightarrow$ $x = 0, -2$.

(2) The corresponding y values for each critical number in part (a) are $f(0) = 5$ and $f(-2) = -3$.

(3) The endpoint values are $f(-3) = 5$ and $f(1) = -3$.

(4) The possible candidates for the *MAX* and *MIN* are only the numbers -3 and 5. Thus, 5 is the *MAX* and -3 is the *MIN*. A common mistake is to confuse the x values $(-3, -2, 0, 1)$ with the y values.

⑨ Following the guidelines in (4.9), we find the critical numbers first. $f(x) = 1 - x^{2/3}$

$\Rightarrow f'(x) = -\frac{2}{3}x^{-1/3} = -\frac{2}{3\sqrt[3]{x}}$. $f'(x) \neq 0$ since the numerator is never zero.
However, $f'(x)$ does not exist at $x = 0$ since the denominator is zero when $x = 0$.
Thus, the only critical number is 0. Finding the function values for 0 and the
endpoints, we have $f(0) = 1$, $f(-1) = 0$, and $f(8) = -3$.   *MAX*: 1, *MIN*: -3

$\boxed{13}$ $s(t) = 2t^3 + t^2 - 20t + 4 \Rightarrow$

$s'(t) = 6t^2 + 2t - 20 = 2(3t^2 + t - 10) = 2(3t - 5)(t + 2) = 0 \Leftrightarrow t = \frac{5}{3}, -2.$

Since $s'(t)$ is always defined, the only *CN* of s are -2 and $\frac{5}{3}$.

$\boxed{17}$ $f(z) = \sqrt{z^2 - 16} = (z^2 - 16)^{1/2} \Rightarrow f'(z) = \frac{1}{2}(z^2 - 16)^{-1/2}(2z) = \dfrac{z}{\sqrt{z^2 - 16}} = 0$ if

$z = 0$, but 0 is not in the domain of f, which is $|z| \ge 4$. $f'(z)$ is undefined when

$z = \pm 4$ — these are endpoint extrema. $f'(z)$ is undefined for $|z| < 4$,

but these values are not in the domain of f. Thus, the only *CN* are $z = \pm 4$.

$\boxed{19}$ $h(x) = (2x - 5)\sqrt{x^2 - 4} \Rightarrow h'(x) = (2x - 5)\frac{1}{2}(x^2 - 4)^{-1/2}(2x) + 2(x^2 - 4)^{1/2} =$

$(x^2 - 4)^{-1/2}\Big[(2x - 5)x + 2(x^2 - 4)\Big] = \dfrac{4x^2 - 5x - 8}{\sqrt{x^2 - 4}}.$

$h'(x) = 0 \Rightarrow x = \dfrac{-(-5) \pm \sqrt{(-5)^2 - 4(4)(-8)}}{2(4)} = \dfrac{5 \pm \sqrt{153}}{8} \approx 2.17, -0.92.$

Only $\dfrac{5 + \sqrt{153}}{8}$ is in the domain of h, which is $|x| \ge 2$.

$h'(x)$ is undefined at $x = \pm 2$. Thus, the *CN* are $\dfrac{5 + \sqrt{153}}{8}$ and ± 2.

$\boxed{21}$ $g(t) = t^2 \sqrt[3]{2t - 5} \Rightarrow g'(t) = t^2(\frac{1}{3})(2t - 5)^{-2/3}(2) + (2t - 5)^{1/3}(2t) =$

$\frac{2}{3}(2t - 5)^{-2/3}\Big[t^2 + 3t(2t - 5)\Big] = \dfrac{2(7t^2 - 15t)}{3(2t - 5)^{2/3}} = \dfrac{2t(7t - 15)}{3(2t - 5)^{2/3}} = 0$ at $t = 0, \frac{15}{7},$

and fails to exist if $t = \frac{5}{2}$. Thus, the *CN* are $0, \frac{15}{7},$ and $\frac{5}{2}$.

$\boxed{23}$ $G(x) = \dfrac{2x - 3}{x^2 - 9} \Rightarrow G'(x) = \dfrac{(x^2 - 9)(2) - (2x - 3)(2x)}{(x^2 - 9)^2} = \dfrac{-2x^2 + 6x - 18}{(x^2 - 9)^2}.$

To determine if $G'(x) = 0$, we use the quadratic formula. In this case, we check the

value of the discriminant, $b^2 - 4ac$, and find out that it is $(6)^2 - 4(-2)(-18) =$

$-108 < 0$. Hence, the roots are non-real, complex numbers and $G'(x) \ne 0$. $G'(x)$

fails to exist at $x = \pm 3$. Since ± 3 are not in the domain of G, there are no *CN*.

$\boxed{25}$ $f(t) = \sin^2 t - \cos t \Rightarrow f'(t) = 2(\sin t)^1 D_t (\sin t) - (-\sin t) =$

$2\sin t \cos t + \sin t = \sin t(2\cos t + 1) = 0 \Leftrightarrow \sin t = 0$ or $\cos t = -\frac{1}{2} \Leftrightarrow$

$t = \pi n, t = \frac{2\pi}{3} + 2\pi n, t = \frac{4\pi}{3} + 2\pi n.$ All of these values are *CN*.

$\boxed{29}$ $f(x) = \dfrac{1 + \sin x}{1 - \sin x} \Rightarrow f'(x) = \dfrac{(1 - \sin x)\cos x - (1 + \sin x)(-\cos x)}{(1 - \sin x)^2} = \dfrac{2\cos x}{(1 - \sin x)^2} =$

$0 \Leftrightarrow \cos x = 0 \Leftrightarrow x = \frac{\pi}{2} + 2\pi n$ or $\frac{3\pi}{2} + 2\pi n.$ Note that $x = \frac{\pi}{2} + 2\pi n$ are not in

the domain of f and both f and f' are undefined at these values. Remember, if x is a

CN, then x *must* be in the domain of f. Thus, the *CN* are $x = \frac{3\pi}{2} + 2\pi n.$

$\boxed{33}$ $H(\phi) = \cot\phi + \csc\phi \Rightarrow H'(\phi) = -\csc^2\phi - \csc\phi\cot\phi = 0 \Leftrightarrow$

$\csc\phi(\csc\phi + \cot\phi) = 0 \Leftrightarrow \csc\phi = 0$ or $\csc\phi = -\cot\phi$. Now, $\csc\phi \neq 0$ for all ϕ

and $\csc\phi = -\cot\phi \Leftrightarrow \dfrac{1}{\sin\phi} = -\dfrac{\cos\phi}{\sin\phi} \Leftrightarrow \sin\phi = -\cos\phi\sin\phi \Leftrightarrow$

$\sin\phi(1 + \cos\phi) = 0 \Leftrightarrow \cos\phi = -1$ or $\sin\phi = 0$. $\cos\phi = -1 \Rightarrow \phi = \pi + 2\pi n$,

but these values are not in the domain of H, which is $\mathbb{R} - \{\pi n\}$. $\sin\phi = 0 \Rightarrow$

$\phi = \pi n$, and these values are not in the domain of H. Thus, there are no CN.

$\boxed{35}$ $f(x) = \sec(x^2 + 1) \Rightarrow f'(x) = 2x\sec(x^2 + 1)\tan(x^2 + 1) = 0$ when $x = 0$ or

$\tan(x^2 + 1) = 0$ since $\sec(x^2 + 1) \neq 0$ for any x. $\tan(x^2 + 1) = 0 \Rightarrow$

$x^2 + 1 = \pi k$ { where k must be a *positive* integer since $x^2 + 1 \geq 1$ } $\Rightarrow$

$x = \pm\sqrt{\pi k - 1}$. Thus, the CN are 0 and $\pm\sqrt{\pi k - 1}$ for $k = 1, 2, 3, \ldots$.

$\boxed{37}$ (a) Since $f(x) = x^{1/3} \Rightarrow f'(x) = \frac{1}{3}x^{-2/3} = \dfrac{1}{3\sqrt[3]{x^2}}$, $f'(0)$ does not exist. If $a \neq 0$,

then $f'(a) \neq 0$. Hence 0 is the only critical number of f. The number $f(0) = 0$

is not a local extremum, since $f(x) < 0$ if $x < 0$ and $f(x) > 0$ if $x > 0$.

(b) $f(x) = x^{2/3} \Rightarrow f'(x) = \frac{2}{3}x^{-1/3} = \dfrac{2}{3\sqrt[3]{x}}$. The only critical number is 0 for the

same reasons given in part (a). The number $f(0) = 0$ is a local minimum, since

$f(x) > 0$ if $x \neq 0$.

$\boxed{39}$ (a) $f(x) = x^3 + 1 \Rightarrow f'(x) = 3x^2$. $f'(x) = 0 \Rightarrow$

     $x = 0$.   There is a critical number, 0, but

     $f(0) = 1$ is not a local extremum, since

     $f(x) < f(0)$ if $x < 0$ and $f(x) > f(0)$ if $x > 0$.

(c) By (2.21)(i), the function is continuous at every

     number a. If $0 < x_1 < x_2 < 1$, then $f(x_1) < f(x_2)$

     and hence there is neither a maximum nor a

     minimum on $(0, 1)$.

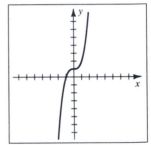

Figure 39

(d) This does not contradict Theorem (4.3) because the interval $(0, 1)$ is *open*, and

the theorem assumes that the interval is *closed*.

$\boxed{43}$ If $x = n$ is an integer, then $f'(n)$ does not exist because f is not continuous at $x = n$.

Remember, if f is not continuous, then it cannot be differentiable. Otherwise,

$f'(x) = 0$ for every $x \neq n$. Thus, for every value of x, either $f'(x) = 0$ or $f'(x)$ is

undefined and every x is a CN of f.

$\boxed{45}$ If $f(x) = ax^2 + bx + c$ and $a \neq 0$, then $f'(x) = 2ax + b$.

                 $f'(x) = 0 \Rightarrow x = -b/(2a)$. This is the only critical number of f.

49–52 *Note*: The exercises in the text will often ask you to estimate coordinates from a graph. This is especially true of the calculator exercises. When reading graphs it is generally difficult to estimate coordinates beyond $\pm\frac{1}{2}$ of the smallest grid marking. For example, if a graph is marked in units of 0.2, then we will usually have difficulty estimating coordinates beyond the accuracy of ± 0.1. The answers given carry more accuracy than you will usually be able to achieve without some additional aid such as a graphics calculator or computer.

Exercises 4.2

1 Draw a line between the two endpoints of the graph of f. This is the secant line. Now examine various tangent lines to the graph. When the tangent line to the graph is parallel to the secant line, the corresponding x value satisfies the mean value theorem. It appears that these values are approximately 3 and 7.

Note: In Exercises 3–8, since the given functions are continuous and differentiable everywhere, we only need to verify that $f(a) = f(b)$ and solve $f'(c) = 0$.

5 $f(x) = x^4 + 4x^2 + 1 \Rightarrow f'(x) = 4x^3 + 8x = 4x(x^2 + 2)$. $f(-3) = f(3) = 118$.

$\qquad f'(c) = 4c(c^2 + 2) = 0 \Rightarrow c = 0$ since $c^2 + 2 \neq 0$ for any c.

7 $f(x) = \sin 2x \Rightarrow f'(x) = 2\cos 2x$. $f(0) = f(\pi) = 0$.

$f'(c) = 2\cos 2c = 0 \Rightarrow \cos 2c = 0 \Rightarrow 2c = \frac{\pi}{2} + \pi n \Rightarrow c = \frac{\pi}{4} + \frac{\pi}{2}n$.

$\qquad\qquad\qquad$ In the open interval $(0, \pi)$, $c = \frac{\pi}{4}$ and $\frac{3\pi}{4}$.

Note: In Exercises 9–24, unless otherwise specified, the functions are continuous on the indicated closed interval and differentiable on the associated open interval, thereby satisfying the hypotheses of the mean value theorem (MVT).

11 $f(1)$ is undefined $\Rightarrow f$ is not continuous on $[0, 2]$ by (2.20).

13 $f(x) = x^{2/3} \Rightarrow f'(x) = \frac{2}{3}x^{-1/3} = \frac{2}{3\sqrt[3]{x}}$.

$\qquad\qquad f'(0)$ does not exist $\Rightarrow f$ is not differentiable at every $x \in (-8, 8)$.

17 $f(x) = x^3 - 2x^2 + x + 3$ is continuous on $[-1, 1]$ and $f'(x) = 3x^2 - 4x + 1$ exists for each x in $(-1, 1)$. Thus, there exists a c in $(-1, 1)$ such that $f(1) - f(-1) = f'(c)\big[1 - (-1)\big] \Rightarrow 3 - (-1) = (3c^2 - 4c + 1)(2) \Rightarrow 3c^2 - 4c - 1 = 0 \Rightarrow c = \frac{1}{3}(2 - \sqrt{7}) \approx -0.22 \in (-1, 1)$. *Note*: The value $c = \frac{1}{3}(2 + \sqrt{7}) \approx 1.55 \notin (-1, 1)$ and therefore does not satisfy the MVT.

19 $f(x) = 4 + \sqrt{x - 1}$ is continuous on $[1, 5]$ and $f'(x) = \frac{1}{2}(x - 1)^{-1/2} = \frac{1}{2\sqrt{x - 1}}$ exists for each x in $(1, 5)$. Thus, there exists a c in $(1, 5)$ such that

$f(5) - f(1) = f'(c)(5 - 1) \Rightarrow 6 - 4 = \frac{1}{2\sqrt{c - 1}}(4) \Rightarrow 1 = \frac{1}{\sqrt{c - 1}} \Rightarrow$

$\qquad\qquad\qquad\qquad\qquad\qquad \sqrt{c - 1} = 1 \Rightarrow c = 2.$

23 $f(x) = \sin x$ is continuous on $[0, \frac{\pi}{2}]$ and $f'(x) = \cos x$ exists for each x in $(0, \frac{\pi}{2})$. Thus, there exists a c in $(0, \frac{\pi}{2})$ such that $f(\frac{\pi}{2}) - f(0) = f'(c)(\frac{\pi}{2} - 0) \Rightarrow 1 - 0 = (\cos c)(\frac{\pi}{2})$

$$\Rightarrow \cos c = \tfrac{2}{\pi} \Rightarrow c \approx 0.88 \in (0, \tfrac{\pi}{2}).$$

27 $f(4) - f(-1) = f'(c)\big[4 - (-1)\big] \Rightarrow 1 - (-4) = (-4/c^2)(5) \Rightarrow 1 = (-4/c^2) \Rightarrow$ $c^2 = -4$. The last equation has no real solutions. This is not a contradiction,

because $f(0)$ is undefined and so f is not continuous on the interval $[-1, 4]$.

31 If f has degree 3, then $f'(x)$ is a polynomial of degree 2. Consequently the equation $f(b) - f(a) = f'(x)(b - a)$ has at most two solutions since a, b, $f(a)$, and $f(b)$ are all constants, and could be solved using the quadratic formula. If f has degree n, then $f'(x)$ is a polynomial of degree $n - 1$ and there are at most $n - 1$ solutions. Recall that a polynomial of degree n has at most n zeros.

33 Let x be any number in $(a, b]$. Applying the MVT to the interval $[a, x]$ yields

$$
\begin{aligned}
f(x) - f(a) &= f'(c)(x - a) && \{\, c \text{ is in } (a, x)\,\} \\
&= 0(x - a) && \{\, f'(c) = 0\,\} \\
&= 0 && \{\, \text{simplify}\,\}
\end{aligned}
$$

Thus, $f(x) - f(a) = 0 \Rightarrow f(x) = f(a)$, and hence f is a constant function, i.e.,

$$f(x) = k \text{ for some real number } k.$$

35 Let $s(t)$ be the distance traveled, $s'(t)$ the instantaneous velocity, t_A the time of departure, and t_B the time of arrival. By the mean value theorem,

$$s'(c) = \frac{s(t_B) - s(t_A)}{t_B - t_A} = \frac{50}{1} = 50 \text{ mi/hr for some time } c \text{ in } (0, 1).$$

Rephrasing this result, the instantaneous velocity must equal

the average velocity at least once over a specific time interval.

37 Let t_1 be the time when the person weighed 487 pounds and t_2 the time when the person weighed 130 pounds. By the MVT, for some time t between t_1 and t_2,

$$\frac{dW}{dt} = \frac{W(t_2) - W(t_1)}{t_2 - t_1} = -\frac{357}{8} = -44.625 \text{ lb/mo} < -44 \text{ lb/mo}.$$

The weight *loss* was more than 44 lb/mo at some time $t \in (t_1, t_2)$.

39 Let $f(x) = \sin x$ on $[u, v]$. By the MVT, where $c \in (u, v)$,

$$
\begin{aligned}
\big|f(v) - f(u)\big| &= \big|f'(c)(v - u)\big| && \{\, \text{absolute value of both sides of MVT}\,\} \\
|\sin v - \sin u| &= \big|f'(c)\big|\,|v - u| && \{\, \text{definition of } f;\ |ab| = |a|\,|b|\,\} \\
|\sin u - \sin v| &= |\cos c|\,|u - v| && \{\, |a - b| = |b - a|;\ f'(x) = \cos x\,\} \\
&\leq (1)\,|u - v| && \{\, |\cos c| \leq 1\,\}
\end{aligned}
$$

The proof would be similar if $u > v$ and $c \in (v, u)$.

Note: We will use the notation ↑ for increasing and ↓ for decreasing in this chapter and subsequent chapters.

3   $f(x) = 2x^3 + x^2 - 20x + 1 \Rightarrow f'(x) = 6x^2 + 2x - 20 = 2(3x - 5)(x + 2) = 0 \Leftrightarrow$ $x = \frac{5}{3}, -2$. Similar to the charts in the text, we construct a brief first derivative chart. The columns consist of the intervals, determined by the critical numbers; the sign of the first derivative, determined by substituting any value from the associated interval into the derivative; and the conclusion, ↑ if the sign is + and ↓ if the sign is − . If you are unsure of how to determine the sign of the derivative, review the solutions in §1.1 that used the sign chart.

Interval	Sign	Conclusion
$(-\infty, -2)$	+	↑ on $(-\infty, -2]$
$(-2, \frac{5}{3})$	−	↓ on $[-2, \frac{5}{3}]$
$(\frac{5}{3}, \infty)$	+	↑ on $[\frac{5}{3}, \infty)$

Summarizing the table information, on $(-\infty, -2) \cup (\frac{5}{3}, \infty)$, $f'(x) > 0$ and f is ↑ on $(-\infty, -2] \cup [\frac{5}{3}, \infty)$. On $(-2, \frac{5}{3})$, $f'(x) < 0$ and f is ↓ on $[-2, \frac{5}{3}]$. The chart indicates that the function increases, decreases, and then increases. This makes sense since the function is a cubic polynomial with a positive leading coefficient. Thus, by (4.14)(i), $f(-2) = 29$ is a *LMAX* and by (4.14)(ii), $f(\frac{5}{3}) = -\frac{548}{27}$ is a *LMIN*. When graphing f, plot the *LMAX* at $(-2, 29)$ and the *LMIN* at $(\frac{5}{3}, -\frac{548}{27}) \approx (1.7, -20)$ first. Another easy point to plot is the y-intercept. Letting $x = 0$ yields $y = 1$. Sketch a smooth curve that clearly shows horizontal tangent lines at the extrema points.

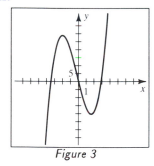
Figure 3

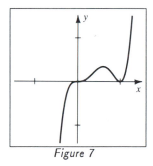
Figure 7

7   $f(x) = 10x^3(x - 1)^2 \Rightarrow f'(x) = 30x^2(x - 1)^2 + 2(x - 1)(10x^3) =$ $10x^2(x - 1)\big[3(x - 1) + 2x\big] = 10x^2(x - 1)(5x - 3) = 0 \Leftrightarrow x = 0, \frac{3}{5}, 1.$ Since $10x^2 \geq 0$, the sign of f' is determined by the sign of the quadratic $(x - 1)(5x - 3)$.   (cont.)

Interval	Sign	Conclusion
$(-\infty, \frac{3}{5})$	$+$	$\uparrow$ on $(-\infty, \frac{3}{5}]$
$(\frac{3}{5}, 1)$	$-$	$\downarrow$ on $[\frac{3}{5}, 1]$
$(1, \infty)$	$+$	$\uparrow$ on $[1, \infty)$

Thus, by (4.14)(i), $f(\frac{3}{5}) = \frac{216}{625} \approx 0.35$ is a *LMAX* and by (4.14)(ii), $f(1) = 0$ is a *LMIN*. When graphing f, plot the *LMAX* at $(\frac{3}{5}, \frac{216}{625})$, the *LMIN* at $(1, 0)$, and the zero at $(0, 0)$ first. Notice that f' is zero at $(0, 0)$, but f is increasing throughout $(-\infty, \frac{5}{3}]$. This indicates that there is a horizontal tangent line at $(0, 0)$.

$\boxed{9}$ $f(x) = x^{4/3} + 4x^{1/3} \Rightarrow f'(x) = \frac{4}{3}x^{1/3} + \frac{4}{3}x^{-2/3} = \frac{4}{3}x^{-2/3}(x + 1) = \dfrac{4(x + 1)}{3x^{2/3}} = 0$

$\Leftrightarrow x = -1$. See the solution to Exercise 21 in §3.3 for a discussion on factoring expressions with negative exponents. f' fails to exist at $x = 0$.

Since $3x^{2/3} \geq 0$, the sign of f' is determined by the factor $(x + 1)$.

Interval	Sign	Conclusion
$(-\infty, -1)$	$-$	$\downarrow$ on $(-\infty, -1]$
$(-1, \infty)$	$+$	$\uparrow$ on $[-1, \infty)$

$f(0) = 0$ is neither a *LMIN* nor a *LMAX* since f' does not change sign, by (4.14)(iii). Thus, $f(-1) = -3$ is a *LMIN*. There is a vertical tangent line at $x = 0$,

since f' fails to exist there.

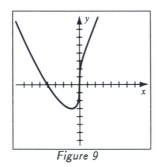

Figure 9

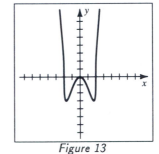

Figure 13

$\boxed{13}$ $f(x) = x^2 \sqrt[3]{x^2 - 4} \Rightarrow f'(x) = x^2(\frac{1}{3})(x^2 - 4)^{-2/3}(2x) + 2x(x^2 - 4)^{1/3} =$

$\frac{2}{3}x(x^2 - 4)^{-2/3}\left[x^2 + 3(x^2 - 4)\right] = \dfrac{8x(x^2 - 3)}{3(x^2 - 4)^{2/3}} = 0 \Leftrightarrow x = 0, \pm\sqrt{3}$.

f' fails to exist at $x = \pm 2$.

Interval	Sign	Conclusion
$(-\infty, -2)$	$-$	$\downarrow$ on $(-\infty, -2]$
$(-2, -\sqrt{3})$	$-$	$\downarrow$ on $[-2, -\sqrt{3}]$
$(-\sqrt{3}, 0)$	$+$	$\uparrow$ on $[-\sqrt{3}, 0]$
$(0, \sqrt{3})$	$-$	$\downarrow$ on $[0, \sqrt{3}]$
$(\sqrt{3}, 2)$	$+$	$\uparrow$ on $[\sqrt{3}, 2]$
$(2, \infty)$	$+$	$\uparrow$ on $[2, \infty)$

Thus, f is $\uparrow$ on $[-\sqrt{3}, 0] \cup [\sqrt{3}, \infty)$ and f is $\downarrow$ on $(-\infty, -\sqrt{3}] \cup [0, \sqrt{3}]$. $f(0) = 0$ is a *LMAX* and $f(\pm\sqrt{3}) = -3$ are *LMIN*. There are no extrema at $x = \pm 2$ since f' does not change sign, but there are vertical tangent lines at these values. Also,

$$\lim_{x \to \infty} f(x) = \lim_{x \to -\infty} f(x) = \infty.$$

$\boxed{15}$ $f(x) = x\sqrt{x^2 - 9} \Rightarrow f'(x) = x \cdot \frac{1}{2}(x^2 - 9)^{-1/2}(2x) + (x^2 - 9)^{1/2} =$

$(x^2 - 9)^{-1/2}\left[x^2 + (x^2 - 9)\right] = \dfrac{2x^2 - 9}{\sqrt{x^2 - 9}} = 0 \Leftrightarrow x = \pm\sqrt{\frac{9}{2}} \approx \pm 2.12$. These values

are not in the domain of f, which is $|x| \geq 3$. f' fails to exist at ± 3. f' is positive throughout its domain, and hence f is $\uparrow$ on $(-\infty, -3] \cup [3, \infty)$. There are no extrema. There are vertical tangent lines at $x = \pm 3$. $\lim\limits_{x \to \infty} f(x) = \infty$ and $\lim\limits_{x \to -\infty} f(x) = -\infty$.

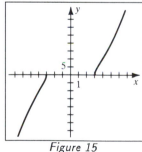

Figure 15

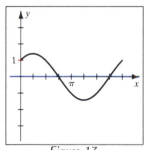

Figure 17

$\boxed{17}$ $f(x) = \cos x + \sin x \Rightarrow f'(x) = \cos x - \sin x$. $f'(x) = 0 \Rightarrow \cos x = \sin x \Rightarrow$ $x = \frac{\pi}{4}, \frac{5\pi}{4}$. Since $\cos x > \sin x$ on $[0, \frac{\pi}{4}) \cup (\frac{5\pi}{4}, 2\pi]$,

it follows that $f'(x) > 0$ on these intervals.

Interval	Sign	Conclusion
$(0, \frac{\pi}{4})$	$+$	$\uparrow$ on $[0, \frac{\pi}{4}]$
$(\frac{\pi}{4}, \frac{5\pi}{4})$	$-$	$\downarrow$ on $[\frac{\pi}{4}, \frac{5\pi}{4}]$
$(\frac{5\pi}{4}, 2\pi)$	$+$	$\uparrow$ on $[\frac{5\pi}{4}, 2\pi]$

Thus, $f(\frac{\pi}{4}) = \sqrt{2}$ is a *LMAX* and $f(\frac{5\pi}{4}) = -\sqrt{2}$ is a *LMIN*.

$\boxed{21}$ $f(x) = 2\cos x + \sin 2x \Rightarrow f'(x) = -2\sin x + 2\cos 2x$

$= -2\sin x + 2(1 - 2\sin^2 x) = 2 - 2\sin x - 4\sin^2 x$

$= 2(1 - 2\sin x)(\sin x + 1) = 0 \quad \Leftrightarrow \quad \sin x = \frac{1}{2}$ or

$\sin x = -1$ if $x = \frac{\pi}{6}, \frac{5\pi}{6}, \frac{3\pi}{2}$. Since $(\sin x + 1) \geq 0$ for

all x, the sign of $f'(x)$ is determined by $(1 - 2\sin x)$.

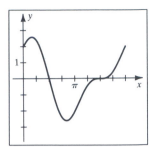

Figure 21

Interval	Sign	Conclusion
$(0, \frac{\pi}{6})$	$+$	$\uparrow$ on $[0, \frac{\pi}{6}]$
$(\frac{\pi}{6}, \frac{5\pi}{6})$	$-$	$\downarrow$ on $[\frac{\pi}{6}, \frac{5\pi}{6}]$
$(\frac{5\pi}{6}, 2\pi)$	$+$	$\uparrow$ on $[\frac{5\pi}{6}, 2\pi]$

Thus, $f(\frac{\pi}{6}) = \frac{3}{2}\sqrt{3}$ is a *LMAX* and $f(\frac{5\pi}{6}) = -\frac{3}{2}\sqrt{3}$ is a *LMIN*.

At $f(\frac{3\pi}{2}) = 0$, $f' = 0$ but there is no local extrema since f is $\uparrow$ throughout $[\frac{5\pi}{6}, 2\pi]$.

$\boxed{23}$ $f(x) = \sqrt[3]{x^3 - 9x} \Rightarrow f'(x) = \frac{1}{3}(x^3 - 9x)^{-2/3}(3x^2 - 9) = \dfrac{x^2 - 3}{\left[x(x-3)(x+3)\right]^{2/3}}$.

f' is undefined at $x = 0, \pm 3$. Otherwise the denominator is always positive and so

the sign of f' is determined by $(x^2 - 3)$.

Interval	Sign	Conclusion
$(-\infty, -\sqrt{3})$	$+$	$\uparrow$ on $(-\infty, -\sqrt{3}]$
$(-\sqrt{3}, \sqrt{3})$	$-$	$\downarrow$ on $[-\sqrt{3}, \sqrt{3}]$
$(\sqrt{3}, \infty)$	$+$	$\uparrow$ on $[\sqrt{3}, \infty)$

Thus, $f(-\sqrt{3}) = (6\sqrt{3})^{1/3} \approx 2.18$ is a *LMAX* and $f(\sqrt{3}) = -(6\sqrt{3})^{1/3}$ is a *LMIN*.

f' does not change sign when $x = 0, \pm 3$,

and so no local extrema exist at these values of f.

$\boxed{27}$ $f(x) = \dfrac{\sqrt{x - 3}}{x^2} \Rightarrow f'(x) = \dfrac{x^2(\frac{1}{2})(x-3)^{-1/2} - (x-3)^{1/2}(2x)}{(x^2)^2} =$

$$\dfrac{x(x-3)^{-1/2}\left[x - 4(x-3)\right]}{2x^4} = \dfrac{-3x + 12}{2x^3\sqrt{x-3}} = 0 \Leftrightarrow x = 4.$$

f' fails to exist at $x = 0$ and 3, but 0 is not in the domain of f, which is $x \geq 3$.

Interval	Sign	Conclusion
$(3, 4)$	$+$	$\uparrow$ on $[3, 4]$
$(4, \infty)$	$-$	$\downarrow$ on $[4, \infty)$

Thus, $f(4) = \frac{1}{16}$ is a *LMAX*. $x = 3$ is an endpoint of the domain of f and cannot be a

local extrema since there is no *open* interval in the domain of f containing 3.

31 $f(x) = 2\tan x - \tan^2 x \Rightarrow f'(x) = 2\sec^2 x - 2\tan x \sec^2 x =$
    $2\sec^2 x(1 - \tan x) = 0$ on $\left[-\frac{\pi}{3}, \frac{\pi}{3}\right] \Rightarrow x = \frac{\pi}{4}$ since $\sec^2 x \neq 0$.

The sign of f' is determined by $(1 - \tan x)$ since $\sec^2 x > 0$.

Interval	Sign	Conclusion
$\left(-\frac{\pi}{3}, \frac{\pi}{4}\right)$	$+$	$\uparrow$ on $\left[-\frac{\pi}{3}, \frac{\pi}{4}\right]$
$\left(\frac{\pi}{4}, \frac{\pi}{3}\right)$	$-$	$\downarrow$ on $\left[\frac{\pi}{4}, \frac{\pi}{3}\right]$

Since f' changes from positive to negative at $x = \frac{\pi}{4}$, $f(\frac{\pi}{4}) = 1$ is a *LMAX*.

33 $y = \frac{1}{2}x + \cos x \Rightarrow y' = \frac{1}{2} - \sin x = 0$ on $[-2\pi, 2\pi] \Rightarrow x = -\frac{11\pi}{6}, -\frac{7\pi}{6}, \frac{\pi}{6}, \frac{5\pi}{6}$.

Interval	Sign	Conclusion
$\left(-2\pi, -\frac{11\pi}{6}\right)$	$+$	$\uparrow$ on $\left[-2\pi, -\frac{11\pi}{6}\right]$
$\left(-\frac{11\pi}{6}, -\frac{7\pi}{6}\right)$	$-$	$\downarrow$ on $\left[-\frac{11\pi}{6}, -\frac{7\pi}{6}\right]$
$\left(-\frac{7\pi}{6}, \frac{\pi}{6}\right)$	$+$	$\uparrow$ on $\left[-\frac{7\pi}{6}, \frac{\pi}{6}\right]$
$\left(\frac{\pi}{6}, \frac{5\pi}{6}\right)$	$-$	$\downarrow$ on $\left[\frac{\pi}{6}, \frac{5\pi}{6}\right]$
$\left(\frac{5\pi}{6}, 2\pi\right)$	$+$	$\uparrow$ on $\left[\frac{5\pi}{6}, 2\pi\right]$

Thus, there are *LMIN* at $x = -\frac{7\pi}{6}, \frac{5\pi}{6}$ and *LMAX* at $x = -\frac{11\pi}{6}, \frac{\pi}{6}$.

35 First, plot the points $(\pm 2, -4)$ and $(0, 3)$. Since $f'(-2) = f'(2) = 0$, the graph has
    horizontal tangent lines at $(\pm 2, -4)$. Since $f'(0)$ is undefined, the graph has a
    vertical tangent line at $(0, 3)$. $f'(x) > 0$ if $-2 < x < 0$ or $x > 2 \Rightarrow f$ is increasing on
    $[-2, 0]$ and $[2, \infty)$. $f'(x) < 0$ if $x < -2$ or $0 < x < 2 \Rightarrow$ decreasing on $(-\infty, -2]$
    and $[0, 2]$. $(\pm 2, -4)$ are *LMIN* and $(0, 3)$ is a *LMAX*. Sketch a smooth curve
    satisfying these conditions.

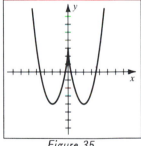

Figure 35

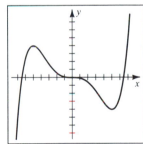

Figure 39

39 First, plot the points $(-5, 4)$, $(0, 0)$, and $(5, -4)$. Since $f'(-5) = f'(0) = f'(5) =$
    0, f has horizontal tangent lines at these points. $f'(x) > 0$ if $|x| > 5 \Rightarrow f$ is $\uparrow$ on
    $(-\infty, -5] \cup [5, \infty)$. $f'(x) < 0$ if $0 < |x| < 5 \Rightarrow f$ is decreasing on $[-5, 5]$. $(-5, 4)$
    is a *LMAX* and $(5, -4)$ is a *LMIN*. Since f is decreasing on $[-5, 5]$, $(0, 0)$ is not a
    local extrema. However, the graph of f "levels off" at $(0, 0)$. Sketch a smooth curve
    satisfying these conditions.

[43] At $x \approx -0.51$, $f'(x) = 0$ and changes sign from positive to negative. At $x \approx 0.49$, $f'(x) = 0$ and changes sign from negative to positive. There is a $LMAX$ at $x \approx -0.51$ and a $LMIN$ at $x \approx 0.49$. The $LMAX$ and $LMIN$ are for f, Figure 43 is a sketch of f'.

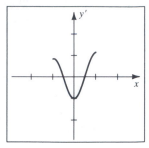

Figure 43

| Exercises 4.4 |

Note: Throughout this section and all subsequent sections, CU denotes concave up, CD, concave down, and PI, point(s) of inflection.

[3] $f(x) = 3x^4 - 4x^3 + 6 \Rightarrow f'(x) = 12x^3 - 12x^2 = 12x^2(x - 1) = 0 \Leftrightarrow x = 0, 1$.
$f''(x) = 36x^2 - 24x = 12x(3x - 2)$. We now apply the second derivative test (4.18) to all critical numbers c such that $f'(c) = 0$. $f''(1) = 12 > 0 \Rightarrow f(1) = 5$ is a $LMIN$ by (4.18)(ii). Note that the value 12 is not important in terms of magnitude, only in terms of its sign. In fact, sometimes you will not even compute this value, rather, just determine if it is positive or negative. $f''(0) = 0$ gives no information. By the first derivative test, $x = 0$ is not an extremum because f' does not change sign at $x = 0$. The graph of f levels off at $(0, 6)$.

Next, we wish to determine the intervals where f is CU and CD. We start by finding the critical numbers of the *first derivative*, i.e., the zeros of f'' and the numbers at which f'' does not exist. In this case, f'' always exists and $f''(x) = 0$ at $x = 0, \frac{2}{3}$. We will use a chart similar to those in the text. The intervals are determined by the values we just found and the domain of the function. The term "Sign" in the heading of the second column refers to the sign of $f''(x)$. The third column indicates that the graph is CU on the interval if the sign of f'' is $+$ and CD on the interval if the sign of f'' is $-$.

Interval	Sign	Concavity
$(-\infty, 0)$	$+$	CU
$(0, \frac{2}{3})$	$-$	CD
$(\frac{2}{3}, \infty)$	$+$	CU

Lastly, we need to determine the x-coordinates of the points of inflection. To do this, we merely need to examine the table and determine if a change in concavity has occurred. In this case, there are two changes in concavity. Thus, there are PI at $x = 0, \frac{2}{3}$. See Figure 3.

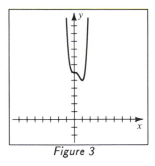

Figure 3

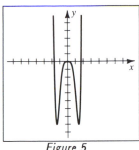

Figure 5

$\boxed{5}$ $f(x) = 2x^6 - 6x^4 \Rightarrow f'(x) = 12x^5 - 24x^3 = 12x^3(x^2 - 2) = 0 \Leftrightarrow x = 0, \pm\sqrt{2}.$

$f''(x) = 60x^4 - 72x^2 = 60x^2(x^2 - \frac{6}{5}).$ $f''(\pm\sqrt{2}) = 96 > 0 \Rightarrow f(\pm\sqrt{2}) = -8$ are

LMIN. $f''(0) = 0$ gives no information. By the first derivative test, $f(0) = 0$ is a

LMAX since f' changes from positive to negative at $x = 0$. Since $60x^2 \geq 0$, the sign

of f'' is determined by $(x^2 - \frac{6}{5})$.

Interval	Sign	Concavity
$(-\infty, -\sqrt{6/5})$	$+$	*CU*
$(-\sqrt{6/5}, 0)$	$-$	*CD*
$(0, \sqrt{6/5})$	$-$	*CD*
$(\sqrt{6/5}, \infty)$	$+$	*CU*

Thus, there are *PI* at $x = \pm\sqrt{6/5}$. Since f' changes from positive to negative at

$x = 0$ and $f'(0) = 0$, f' is *decreasing* at $x = 0$.

Thus, it is *CD* by (4.15)(ii) on $\left(-\sqrt{\frac{6}{5}}, \sqrt{\frac{6}{5}}\right)$.

$\boxed{11}$ $f(x) = \sqrt[3]{x^2}(3x + 10) \Rightarrow f'(x) = x^{2/3}(3) + (3x + 10)\frac{2}{3}x^{-1/3} = $

$\frac{1}{3}x^{-1/3}\left[9x + 2(3x + 10)\right] = \frac{15x + 20}{3x^{1/3}} = \frac{5(3x + 4)}{3x^{1/3}} = 0 \Leftrightarrow x = -\frac{4}{3}.$

f' fails to exist at $x = 0$. $f''(x) = \frac{5}{3} \cdot \dfrac{x^{1/3}(3) - (3x + 4)\frac{1}{3}x^{-2/3}}{(x^{1/3})^2} = $

$\frac{5}{3} \cdot \dfrac{x^{-2/3}\left[9x - (3x + 4)\right]}{3x^{2/3}} = \dfrac{10(3x - 2)}{9x^{4/3}}.$ $f''(-\frac{4}{3}) < 0$ and $f(-\frac{4}{3}) = 4\sqrt[3]{6} \approx 7.27$ is

a *LMAX* by (4.18)(i). Since $f''(0)$ is undefined, use the first derivative test to show

that $f(0) = 0$ is a *LMIN*. This is true since f' changes sign from negative to positive

at $x = 0$. Since $9x^{4/3} \geq 0$, the sign of f'' is determined by $(3x - 2)$.

Interval	Sign	Concavity
$(-\infty, 0)$	$-$	*CD*
$(0, \frac{2}{3})$	$-$	*CD*
$(\frac{2}{3}, \infty)$	$+$	*CU*

Thus, there is a *PI* at $x = \frac{2}{3}$.   (cont.)

Note: f is not *CD* at $x = 0$ since no tangent line exists at $x = 0$.

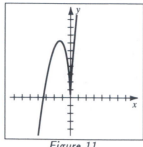

Figure 11   Figure 13

13 $f(x) = x^2(3x - 5)^{1/3} \Rightarrow f'(x) = x^2(\frac{1}{3})(3x - 5)^{-2/3}(3) + (3x - 5)^{1/3}(2x) =$

$x(3x - 5)^{-2/3}\left[x + 2(3x - 5)\right] = \dfrac{x(7x - 10)}{(3x - 5)^{2/3}} = 0 \Leftrightarrow x = 0, \frac{10}{7}$. f' fails to exist at

$x = \frac{5}{3}$. $f''(x) = \dfrac{(3x - 5)^{2/3}(14x - 10) - (7x^2 - 10x)(\frac{2}{3})(3x - 5)^{-1/3}(3)}{\left[(3x - 5)^{2/3}\right]^2} =$

$\dfrac{(3x - 5)^{-1/3}\left[(3x - 5)(14x - 10) - 2(7x^2 - 10x)\right]}{(3x - 5)^{4/3}} = \dfrac{2(14x^2 - 40x + 25)}{(3x - 5)^{5/3}}.$

$f''(0) = -2\sqrt[3]{5} < 0 \Rightarrow f(0) = 0$ is a *LMAX*. $f''(\frac{10}{7}) = 2\sqrt[3]{245} > 0 \Rightarrow f(\frac{10}{7}) =$

$-\frac{100}{343}\sqrt[3]{245} \approx -1.82$ is a *LMIN*.   By the first derivative test, $f(\frac{5}{3}) = 0$ is not an

extremum since f' does not change sign at $x = \frac{5}{3}$. $f''(x) = 0 \Rightarrow 14x^2 - 40x + 25 =$

$0 \Rightarrow x = \dfrac{20 \pm 5\sqrt{2}}{14}.$ Let $a = \dfrac{20 - 5\sqrt{2}}{14} \approx 0.92$ and $b = \dfrac{20 + 5\sqrt{2}}{14} \approx 1.93.$ The

numerator of f'' is negative between a and b and positive outside these values. The

denominator of f'' is negative when $x < \frac{5}{3}$ and positive when $x > \frac{5}{3}$.

Interval	Sign	Concavity
$(-\infty, a)$	$-$	*CD*
$(a, \frac{5}{3})$	$+$	*CU*
$(\frac{5}{3}, b)$	$-$	*CD*
(b, ∞)	$+$	*CU*

Thus, there are *PI* at $x = a, \frac{5}{3}$, and b.

15 $f(x) = 8x^{1/3} + x^{4/3} \Rightarrow f'(x) = \frac{8}{3}x^{-2/3} + \frac{4}{3}x^{1/3} = \frac{4}{3}x^{-2/3}(2 + x) = \dfrac{4(2 + x)}{3x^{2/3}} = 0$

$\Leftrightarrow x = -2$. f' fails to exist at $x = 0$.

$f''(x) = \dfrac{4}{3} \cdot \dfrac{x^{2/3}(1) - (2 + x)(\frac{2}{3}x^{-1/3})}{(x^{2/3})^2} = \dfrac{4}{3} \cdot \dfrac{x^{-1/3}\left[3x - 2(2 + x)\right]}{3x^{4/3}} = \dfrac{4(x - 4)}{9x^{5/3}} \Rightarrow$

$f''(-2) = \frac{2}{3}\sqrt[3]{2} > 0$ and $f(-2) = -6\sqrt[3]{2} \approx -7.55$ is a *LMIN*. $f''(0)$ is undefined.

By the first derivative test,

$f(0) = 0$ is not a local extremum since f' does not change sign at $x = 0$.

Interval	Sign	Concavity
$(-\infty, 0)$	$+$	CU
$(0, 4)$	$-$	CD
$(4, \infty)$	$+$	CU

Thus, there are *PI* at $x = 0, 4$.

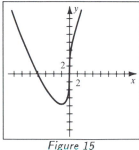

Figure 15   Figure 17

[17] $f(x) = x^2\sqrt{9 - x^2} \Rightarrow f'(x) = x^2(\tfrac{1}{2})(9 - x^2)^{-1/2}(-2x) + (9 - x^2)^{1/2}(2x) =$

$x(9 - x^2)^{-1/2}\left[-x^2 + 2(9 - x^2)\right] = \dfrac{3x(6 - x^2)}{\sqrt{9 - x^2}} = 0 \Leftrightarrow x = 0, \pm\sqrt{6}$.

f' fails to exist at $x = \pm 3$, which are endpoints of the domain.

$f''(x) = 3 \cdot \dfrac{(9 - x^2)^{1/2}(6 - 3x^2) - (6x - x^3)(\tfrac{1}{2})(9 - x^2)^{-1/2}(-2x)}{\left[(9 - x^2)^{1/2}\right]^2} =$

$3 \cdot \dfrac{(9 - x^2)^{-1/2}\left[(9 - x^2)(6 - 3x^2) + x(6x - x^3)\right]}{9 - x^2} = \dfrac{3(54 - 27x^2 + 2x^4)}{(9 - x^2)^{3/2}}.$

$f''(\pm\sqrt{6}) = -12\sqrt{3} < 0 \Rightarrow f(\pm\sqrt{6}) = 6\sqrt{3} \approx 10.4$ are *LMAX*. $f''(0) = 6 > 0 \Rightarrow$

$f(0) = 0$ is a *LMIN*. $f''(x) = 0 \Rightarrow 2x^4 - 27x^2 + 54 = 0 \Rightarrow x^2 = \dfrac{27 \pm \sqrt{297}}{4} \Rightarrow$

$x^2 = \dfrac{27 - 3\sqrt{33}}{4}$ since x^2 must be less than or equal to 9 to be in the domain of f.

Thus, $x = \pm\tfrac{1}{2}\sqrt{27 - 3\sqrt{33}}$ for $|x| < 3$. Let $a = -\tfrac{1}{2}\sqrt{27 - 3\sqrt{33}} \approx -1.56$ and

$b = -a$. The sign of f'' is determined by its numerator.

$54 - 27x^2 + 2x^4$ is positive on (a, b) and negative on $[-3, a) \cup (b, 3]$.

Interval	Sign	Concavity
$(-3, a)$	$-$	CD
(a, b)	$+$	CU
$(b, 3)$	$-$	CD

x-coordinates of *PI* are a and b. There are vertical tangents at $x = \pm 3$.

Note: The solutions for Exercises 19 & 23 first appeared in the solutions of Exercises
17 & 21 in §4.3, where they were solved using the first derivative test.

$\boxed{19}$ The *CN* are $x = \frac{\pi}{4}, \frac{5\pi}{4}$. $f''(x) = -\cos x - \sin x$.

$$f''(\tfrac{\pi}{4}) = -\sqrt{2} < 0 \Rightarrow f(\tfrac{\pi}{4}) = \sqrt{2} \text{ is a } LMAX.$$

$$f''(\tfrac{5\pi}{4}) = \sqrt{2} > 0 \Rightarrow f(\tfrac{5\pi}{4}) = -\sqrt{2} \text{ is a } LMIN.$$

$\boxed{23}$ The *CN* are $x = \frac{\pi}{6}, \frac{5\pi}{6}, \frac{3\pi}{2}$. $f''(x) = -2\cos x - 4\sin 2x$.

$$f''(\tfrac{\pi}{6}) = -3\sqrt{3} < 0 \Rightarrow f(\tfrac{\pi}{6}) = \tfrac{3\sqrt{3}}{2} \text{ is a } LMAX.$$

$$f''(\tfrac{5\pi}{6}) = 3\sqrt{3} > 0 \Rightarrow f(\tfrac{5\pi}{6}) = -\tfrac{3\sqrt{3}}{2} \text{ is a } LMIN.$$

$\frac{3\pi}{2}$ is also a *CN*, but $f''(\frac{3\pi}{2}) = 0$ gives no information.

$\boxed{27}$ *Note:* See Exercise 31 in §4.3. The only *CN* in $(-\frac{\pi}{3}, \frac{\pi}{3})$ is $x = \frac{\pi}{4}$.

$f'(x) = 2\sec^2 x(1 - \tan x) \Rightarrow$

$$f''(x) = 4\sec x(\sec x \tan x)(1 - \tan x) + 2\sec^2 x(-\sec^2 x)$$
$$= 4\sec^2 x \tan x(1 - \tan x) - 2\sec^4 x$$

$f''(\frac{\pi}{4}) = -8 < 0 \Rightarrow f(\frac{\pi}{4}) = 1$ is a *LMAX*.

$\boxed{29}$ The *CN* on $(-2\pi, 2\pi)$ are $x = -\frac{11\pi}{6}, -\frac{7\pi}{6}, \frac{\pi}{6}, \frac{5\pi}{6}$. $f'(x) = \frac{1}{2} - \sin x \Rightarrow$

$f''(x) = -\cos x$. Since $f''(-\frac{11\pi}{6}) = f''(\frac{\pi}{6}) = -\frac{\sqrt{3}}{2} < 0$,

$f(-\frac{11\pi}{6}) = \frac{\sqrt{3}}{2} - \frac{11\pi}{12} \approx -2.01$ and $f(\frac{\pi}{6}) = \frac{\sqrt{3}}{2} + \frac{\pi}{12} \approx 1.13$ are *LMAX*.

Since $f''(-\frac{7\pi}{6}) = f''(\frac{5\pi}{6}) = \frac{\sqrt{3}}{2} > 0$,

$$f(-\tfrac{7\pi}{6}) = -\tfrac{\sqrt{3}}{2} - \tfrac{7\pi}{12} \approx -2.70 \text{ and } f(\tfrac{5\pi}{6}) = \tfrac{5\pi}{12} - \tfrac{\sqrt{3}}{2} \approx 0.44 \text{ are } LMIN.$$

$\boxed{31}$ First, plot the points $(0, 1)$ and $(2, 3)$. Since $f'(0) = f'(2) = 0$, there are horizontal
tangent lines at these points. $f'(x) < 0$ if $|x - 1| > 1 \Rightarrow f$ is $\downarrow$ on $(-\infty, 0] \cup [2, \infty)$.
$f'(x) > 0$ if $|x - 1| < 1 \Rightarrow f$ is $\uparrow$ on $[0, 2]$. $f''(x) > 0$ if $x < 1 \Rightarrow f$ is *CU* on $(-\infty, 1)$.
$f''(x) > 0$ if $x > 1 \Rightarrow f$ is *CD* on $(1, \infty)$. There is a *PI* at $x = 1$. It may be helpful
to set up first and second derivative charts as below.

Interval	Sign	Conclusion		Interval	Sign	Concavity
$(-\infty, 0)$	$-$	$\downarrow$ on $(-\infty, 0]$		$(-\infty, 1)$	$+$	*CU*
$(0, 2)$	$+$	$\uparrow$ on $[0, 2]$		$(1, \infty)$	$-$	*CD*
$(2, \infty)$	$-$	$\downarrow$ on $[2, \infty)$				

See *Figure 31*.

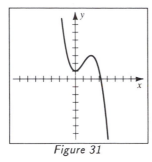

Figure 31

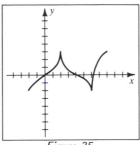

Figure 35

35 First, plot the points $(-2, -2)$, $(6, -2)$, $(0, 0)$, $(4, 0)$, $(2, 3)$, and $(8, 3)$. Since f' is undefined at 2 and 6, f has a vertical tangent line at $x = 2$ and $x = 6$. Since $f'(0) = 1$, f has a tangent line with a slope of 1 as it crosses the origin. $f'(x) > 0$ throughout $(-\infty, 2)$ and $(6, \infty) \Rightarrow f$ is $\uparrow$ on $(-\infty, 2] \cup [6, \infty)$. $f'(x) < 0$ if $|x - 4| < 2 \Rightarrow f$ is $\downarrow$ on $[2, 6]$. $f''(x) < 0$ throughout $(-\infty, 0)$, $(4, 6)$, and $(6, \infty) \Rightarrow f$ is CD on $(-\infty, 0) \cup (4, 6) \cup (6, \infty)$. $f''(x) > 0$ throughout $(0, 2)$ and $(2, 4) \Rightarrow f$ is CU on $(0, 2) \cup \quad (2, 4)$. There are PI at $x = 0, 4$. It may take a couple attempts to sketch a continuous curve satisfying all these conditions.

Interval	Sign	Conclusion
$(-\infty, 2)$	$+$	$\uparrow$ on $(-\infty, 2]$
$(2, 6)$	$-$	$\downarrow$ on $[2, 6]$
$(6, \infty)$	$+$	$\uparrow$ on $[6, \infty)$

Interval	Sign	Concavity
$(-\infty, 0)$	$-$	CD
$(0, 2)$	$+$	CU
$(2, 4)$	$+$	CU
$(4, 6)$	$-$	CD
$(6, \infty)$	$-$	CD

39 If $f(x) = ax^2 + bx + c$, where a, b, and c are constants, then $f''(x) = 2a$, which does not change sign. Thus, there is no point of inflection.

(a) $f''(x) > 0 \Leftrightarrow a > 0$. Hence, f is CU if $a > 0$.

(b) $f''(x) < 0 \Leftrightarrow a < 0$. Hence, f is CD if $a < 0$.

43 Estimating from the graph, we see that $f''(x) = 0$ at $x = \frac{3}{10}, \frac{11}{5}$. However, f'' does not change sign at these values. Since $f''(x) \geq 0$ on $[0, 3]$, the graph of f does not change concavity and f is CU on $(0, 3)$ with no PI.

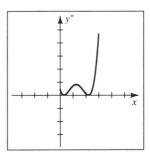

Figure 43

Exercises 4.5

Note: It is helpful to know where the function intersects the horizontal or oblique asymptote to determine how the function is approaching the asymptote. None of the functions in the exercises have more than one point of intersection and it will be denoted by $I(x, y)$. We include this in guideline 7 for Exercises 1, 3, 9, and 11, and then mention it when appropriate. We will use Guidelines (4.19) for Exercises 1, 3, 9, and 11—after these we will use a shorter style to summarize the graph. Also, we will include fewer steps on *how* to get the first and second derivatives, and concentrate on their implications for the graphs.

1 $f(x) = \dfrac{2x - 5}{x + 3}$.

1) Since f is undefined when $x = -3$, the domain of f is $\mathbb{R} - \{-3\}$.

2) f has an infinite discontinuity at $x = -3$ and is continuous at all other numbers.

3) The x-intercepts will be the solutions to the equation $f(x) = 0$, i.e., when the numerator is zero. For this function, we will have an x-intercept at $x = \frac{5}{2}$. The y-intercept is $f(0)$. Note that this will always be the ratio of the constant terms in the numerator and the denominator for a rational function. In this problem, $f(0) = -\frac{5}{3}$ is the y-intercept.

4) $f(-x) = \dfrac{-2x - 5}{-x + 3} \neq f(x) \Rightarrow f$ is not even.

$-f(x) = -\dfrac{2x - 5}{x + 3} \neq f(-x) \Rightarrow f$ is not odd. Since f is neither even nor odd,

the graph is not symmetric with respect to the y-axis or to the origin.

5) $f'(x) = \dfrac{(x + 3)(2) - (2x - 5)(1)}{(x + 3)^2} = \dfrac{11}{(x + 3)^2}$. f' is never equal to zero and is undefined only when $x = -3$. However, this is not in the domain of f. Hence, there are no critical numbers and f has no extrema.

6) $f''(x) = -11 \cdot \dfrac{2(x + 3)}{\left[(x + 3)^2\right]^2} = -\dfrac{22}{(x + 3)^3}$. f'' is never equal to zero and is undefined only when $x = -3$, again, not in the domain of f. Hence, there are no points of inflection. If $x < -3$, then $f'' > 0$ and f is *CU*. If $x > -3$, then $f'' < 0$ and f is *CD*.

7) Reread the *Note* in §2.4, Exercise 11. This should help you determine the horizontal asymptote without much work. In this case, the degree of the numerator, 1, is equal to the degree of the denominator. Hence, the horizontal asymptote will equal the ratio of leading coefficients, $\frac{2}{1}$.

Next, we determine where, if at all, the function intersects the horizontal asymptote. We set the function equal to the value of the horizontal asymptote and then solve for x.

$$\frac{2x - 5}{x + 3} = 2 \Rightarrow 2x - 5 = 2x + 6 \Rightarrow -5 = 6.$$

The last equation is a contradiction since it is always false. This indicates that the function doesn't cross the horizontal asymptote. This is helpful information because we now have another "boundary" to help us sketch the function.

The vertical asymptotes occur where the function is undefined. In this case, there is one vertical asymptote at $x = -3$ since $\lim\limits_{x \to -3^-} f(x) = \infty$ and $\lim\limits_{x \to -3^+} f(x) = -\infty$.

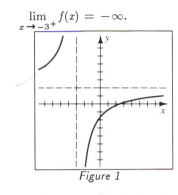

Figure 1

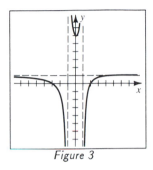

Figure 3

3 $f(x) = \dfrac{x^2 + x - 6}{x^2 - 1} = \dfrac{(x + 3)(x - 2)}{(x + 1)(x - 1)}.$

1) The domain of f is $\mathbb{R} - \{\pm 1\}$.

2) f is continuous at all numbers except ± 1.

3) x-intercepts: $f(x) = 0 \Rightarrow (x + 3)(x - 2) = 0 \Rightarrow x = -3, 2.$

   y-intercept: $f(0) = \frac{-6}{-1} = 6.$

4) Since f is neither even nor odd,

   the graph is not symmetric with respect to the y-axis or to the origin.

5) $f'(x) = \dfrac{(x^2 - 1)(2x + 1) - (x^2 + x - 6)(2x)}{(x^2 - 1)^2} = -\dfrac{x^2 - 10x + 1}{(x^2 - 1)^2} = 0 \Leftrightarrow$

   $x^2 - 10x + 1 = 0 \Leftrightarrow x = 5 \pm 2\sqrt{6}.$ Since $(x^2 - 1)^2 \geq 0$, the sign of f' is determined by $-(x^2 - 10x + 1)$. This is positive between $x = 5 - 2\sqrt{6}$ and $x = 5 + 2\sqrt{6}$, and negative outside these values. Thus, f' changes from negative to positive at $x = 5 - 2\sqrt{6} \Rightarrow$ *LMAX* of $f(5 + 2\sqrt{6}) \approx 1.05$;   and f' changes from positive to negative at $x = 5 + 2\sqrt{6} \Rightarrow$ *LMIN* of $f(5 - 2\sqrt{6}) \approx 5.95.$

6) $f''(x) = -\dfrac{(x^2 - 1)^2(2x - 10) - (x^2 - 10x + 1)(2)(x^2 - 1)(2x)}{\left[(x^2 - 1)^2\right]^2} =$

   $\dfrac{2(x^2 - 1)\left[(x^2 - 1)(x - 5) - (x^2 - 10x + 1)(2x)\right]}{(x^2 - 1)^4} = \dfrac{2(x^3 - 15x^2 + 3x - 5)}{(x^2 - 1)^3}.$

(cont.)

You are not expected to be able to solve $f''(x) = 0$ without the aid of some type of computing equipment. We will examine points of inflection *intuitively* in guideline 7, even though it is not required.

7) The degree of the numerator, 2, is equal to the degree of the denominator. Hence, the horizontal asymptote is the ratio of leading coefficients, i.e., $y = \frac{1}{1} = 1$. Setting the function equal to 1 and solving for x yields:

$$\frac{x^2 + x - 6}{x^2 - 1} = 1 \Rightarrow x^2 + x - 6 = x^2 - 1 \Rightarrow x = 5.$$

This indicates that when $x = 5$, $y = 1$, and the graph will cross the horizontal asymptote at $I(5, 1)$. Note in *Figure 3* that this is barely noticeable. In fact, if you didn't know that this point was on the graph, you might think that the figure was incorrect. However, the graph goes through $(5, 1)$ and rises to the maximum at $x = 5 + 2\sqrt{6}$ (≈ 1.05) and then starts to decrease. The graph is *CD* here, but cannot continue to be *CD*, because it would have to cross the horizontal asymptote—but it can't, because we know it only crosses at $x = 5$. The function changes concavity between $x = 14$ and $x = 15$, continues to decrease, and approaches the horizontal asymptote as x increases without bound.

If you examine the graph of $y = x^3 - 15x^2 + 3x - 5$, you will see that there is only the one root mentioned above. Hence, this is the only solution for $f''(x) = 0$ and is the only *PI* for the graph.

There are vertical asymptotes at $x = \pm 1$.

⑨ $f(x) = \dfrac{x + 4}{\sqrt{x}}$.

1) Since the denominator, $\sqrt{x}$, is defined and nonzero when $x > 0$, the domain of f is $x > 0$.

2) f is continuous on its domain, $(0, \infty)$.

3) x-intercepts: $f(x) = 0 \Rightarrow x + 4 = 0 \Rightarrow x = -4$, however,

    this value is not in the domain of f.

y-intercept: $f(0) = \frac{4}{0}$ is not defined. There is no y-intercept.

4) Since f is neither even nor odd,

    the graph is not symmetric with respect to the y-axis or to the origin.

5) $f'(x) = \dfrac{x^{1/2}(1) - (x + 4)(\frac{1}{2})(x^{-1/2})}{(\sqrt{x})^2} = \dfrac{x^{-1/2}\left[2x - (x + 4)\right]}{2x} = \dfrac{x - 4}{2x^{3/2}} = 0 \Leftrightarrow$

$x = 4$. f' is undefined at $x = 0$, but $x = 0$ is not in the domain of f.

$f'(x) < 0$ on $(0, 4)$ and $f'(x) > 0$ on $(4, \infty)$. f is $\downarrow$ on $(0, 4]$ and f is $\uparrow$ on $[4, \infty)$.

6) $f''(x) = \dfrac{2x^{3/2}(1) - (x-4)(3x^{1/2})}{(2x^{3/2})^2} = \dfrac{x^{1/2}\left[2x - 3(x-4)\right]}{4x^3} = -\dfrac{x - 12}{4x^{5/2}}.$

The sign of f'' is determined by $12 - x$. On $(0, 12)$, $f'' > 0$ and f is *CU*. On $(12, \infty)$, $f'' < 0$ and f is *CD*. $f''(4) > 0 \Rightarrow$ *LMIN* of $f(4) = 4$. {We could have used the first derivative test in step 5 to show this.}

7) Since the degree of the numerator, 1, is greater than the degree of the denominator, $\frac{1}{2}$, we will use division to determine the behavior of f as $x \to \infty$. In this case, we can rewrite f as $f(x) = \sqrt{x} + (4/\sqrt{x})$ and see that as $x \to \infty$, $f(x) \to \sqrt{x}$. This means that the graph will approach the curve $y = \sqrt{x}$ as $x \to \infty$.

To determine if the function crosses the curve $y = \sqrt{x}$, we set the function equal to $\sqrt{x}$ and solve for x.

$$\frac{x + 4}{\sqrt{x}} = \sqrt{x} \Rightarrow x + 4 = x \Rightarrow 4 = 0.$$

This is a contradiction, and hence the function will not cross $y = \sqrt{x}$.

There is a vertical asymptote at $x = 0$.

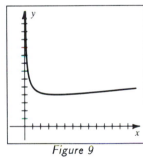

Figure 9

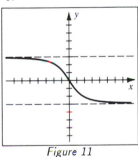

Figure 11

11 $f(x) = \dfrac{-3x}{\sqrt{x^2 + 4}}.$

1) Since $\sqrt{x^2 + 4}$ never equals zero and is always defined, the domain of f is $\mathbb{R}$.

2) f is continuous on $\mathbb{R}$.

3) x-intercepts: $f(x) = 0 \Rightarrow -3x = 0 \Rightarrow x = 0$. y-intercept: $f(0) = 0$.

4) $f(-x) = \dfrac{-3(-x)}{\sqrt{(-x)^2 + 4}} = \dfrac{3x}{\sqrt{x^2 + 4}} = -\left(\dfrac{-3x}{\sqrt{x^2 + 4}}\right) = -f(x) \Rightarrow$

f is an odd function and is symmetric with respect to the origin.

5) $f'(x) = \dfrac{(x^2 + 4)^{1/2}(-3) - (-3x)(\frac{1}{2})(x^2 + 4)^{-1/2}(2x)}{(\sqrt{x^2 + 4})^2} =$

$$\dfrac{(x^2 + 4)^{-1/2}\left[-3(x^2 + 4) + 3x^2\right]}{x^2 + 4} = -\dfrac{12}{(x^2 + 4)^{3/2}} \neq 0.$$

$f'(x) < 0$ for all $x \Rightarrow f$ is $\downarrow$ on $(-\infty, \infty)$. There are no extrema.

6) $f''(x) = -12(-\frac{3}{2})(x^2 + 4)^{-5/2}(2x) = \dfrac{36x}{(x^2 + 4)^{5/2}}.$

When $x < 0$, $f''(x) < 0$ and f is CD on $(-\infty, 0)$.

When $x > 0$, $f''(x) > 0$ and f is CU on $(0, \infty)$. There is a PI at $x = 0$.

7) $\lim\limits_{x \to \infty} f(x) = -3 \Rightarrow$ horizontal asymptote at $y = -3$. $\lim\limits_{x \to -\infty} f(x) = 3 \Rightarrow$

horizontal asymptote at $y = 3$. {Note that it is possible for f to decrease on $(-\infty, \infty)$ and still have its graph remain between $y = -3$ and $y = 3$.} Setting $f(x)$ equal to ± 3 yields $0 = 36$ and indicates that the function never crosses either horizontal asymptote.

There are no vertical asymptotes.

[13] $f(x) = \dfrac{x^2 - x - 6}{x + 1} = \dfrac{(x - 3)(x + 2)}{x + 1}.$ The domain of f is $\mathbb{R} - \{-1\}$.

x-intercepts: $f(x) = 0 \Rightarrow x = 3, -2.$ y-intercept: $f(0) = -6.$

$f'(x) = \dfrac{(x + 1)(2x - 1) - (x^2 - x - 6)(1)}{(x + 1)^2} = \dfrac{x^2 + 2x + 5}{(x + 1)^2} \neq 0$ since

$x^2 + 2x + 5 > 0.$ $f'(x) > 0$ whenever it is defined and

f is $\uparrow$ on $(-\infty, -1) \cup (-1, \infty)$. There are no extrema.

$f''(x) = \dfrac{(x + 1)^2(2x + 2) - (x^2 + 2x + 5)(2)(x + 1)}{\left[(x + 1)^2\right]^2} =$

$$\dfrac{2(x + 1)\left[x^2 + 2x + 1 - x^2 - 2x - 5\right]}{(x + 1)^4} = -\dfrac{8}{(x + 1)^3}.$$

If $x < -1$, $f''(x) > 0$ and f is CU on $(-\infty, -1)$.

If $x > -1$, $f''(x) < 0$ and f is CD on $(-1, \infty)$. There are no PI.

Dividing $x^2 - x - 6$ by $x + 1$, we obtain an alternate form of $f(x)$, i.e.,

$f(x) = x - 2 - \dfrac{4}{x + 1}.$ We see that as $|x| \to \infty$, $f(x) \to x - 2$,

an oblique asymptote. There is a vertical asymptote at $x = -1$. See *Figure 13*.

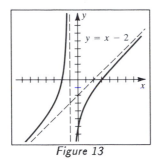

Figure 13

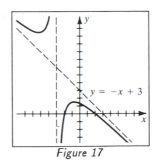

Figure 17

$\boxed{17}$ $f(x) = \dfrac{4 - x^2}{x + 3} = \dfrac{(2 + x)(2 - x)}{x + 3}$. The domain of f is $\mathbb{R} - \{-3\}$.

x-intercepts: $f(x) = 0 \Rightarrow x = \pm 2$. y-intercept: $f(0) = \frac{4}{3}$.

$$f'(x) = \frac{(x + 3)(-2x) - (4 - x^2)(1)}{(x + 3)^2} = -\frac{x^2 + 6x + 4}{(x + 3)^2} = 0 \Leftrightarrow x^2 + 6x + 4 = 0$$

$\Leftrightarrow x = -3 \pm \sqrt{5}$. The sign of f' is determined by $-(x^2 + 6x + 4)$. $f' > 0$ between

$x = -3 - \sqrt{5}$ and $x = -3 + \sqrt{5}$ ($x \neq -3$) and negative otherwise. f is $\uparrow$ on

$\quad [-3 - \sqrt{5}, -3) \cup (-3, -3 + \sqrt{5}]$ and $\downarrow$ on $(-\infty, -3 - \sqrt{5}] \cup [-3 + \sqrt{5}, \infty)$.

$$f''(x) = -\frac{(x + 3)^2(2x + 6) - (x^2 + 6x + 4)(2)(x + 3)}{\left[(x + 3)^2\right]^2} =$$

$$-\frac{2(x + 3)\left[x^2 + 6x + 9 - x^2 - 6x - 4\right]}{(x + 3)^4} = -\frac{10}{(x + 3)^3}.$$

$f''(-3 + \sqrt{5}) = -\dfrac{10}{(\sqrt{5})^3} < 0 \Rightarrow LMAX$ of $f(-3 + \sqrt{5}) \approx 1.53$.

$f''(-3 - \sqrt{5}) = -\dfrac{10}{(-\sqrt{5})^3} > 0 \Rightarrow LMIN$ of $f(-3 - \sqrt{5}) \approx 10.47$.

If $x < -3$, $f'' > 0$ and f is CU on $(-\infty, -3)$.

$\qquad$ If $x > -3$, $f'' < 0$ and f is CD on $(-3, \infty)$. There are no PI.

Dividing $4 - x^2$ by $x + 3$, we obtain an alternate form of $f(x)$, i.e.,

$f(x) = -x + 3 - \dfrac{5}{x + 3}$. We see that as $|x| \to \infty$, $f(x) \to -x + 3$,

$\qquad$ an oblique asymptote. There is a vertical asymptote at $x = -3$.

$\boxed{19}$ $f(x) = \dfrac{3x}{(x + 8)^2}$. The domain of f is $\mathbb{R} - \{-8\}$. Since the degree of the numerator is less than the degree of the denominator, $y = 0$ is the horizontal asymptote. $f(0) = 0$ is the y-intercept and is also the only x-intercept. The function crosses the horizontal asymptote at $I(0, 0)$.

$$f'(x) = \frac{(x + 8)^2(3) - 3x(2)(x + 8)}{\left[(x + 8)^2\right]^2} = \frac{3(x + 8)\left[x + 8 - 2x\right]}{(x + 8)^4} = -\frac{3(x - 8)}{(x + 8)^3} = 0$$

$$\Leftrightarrow x = 8. \ f \text{ is } \downarrow \text{ on } (-\infty, -8) \cup (8, \infty). \ f \text{ is } \uparrow \text{ on } (-8, 8).$$

$$f''(x) = -3 \cdot \frac{(x + 8)^3(1) - (x - 8)(3)(x + 8)^2}{\left[(x + 8)^3\right]^2} = -3 \cdot \frac{(x + 8)^2\left[x + 8 - 3(x - 8)\right]}{(x + 8)^6}$$

$$= \frac{6(x - 16)}{(x + 8)^4}. \ f \text{ is } CU \text{ on } (16, \infty) \text{ and } CD \text{ on } (-\infty, 16) \ \{x \ne -8\}.$$

$$f''(8) < 0 \Rightarrow LMAX \text{ of } f(8) = \tfrac{3}{32}. \ \text{There is a } PI \text{ at } (16, \tfrac{1}{12}).$$

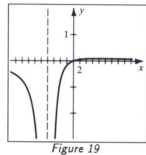

Figure 19   Figure 21

$\boxed{21}$ $f(x) = \dfrac{3x}{x^2 + 1}$ intersects the horizontal asymptote, $y = 0$, at $I(0, 0)$.

$$f'(x) = \frac{-3(x^2 - 1)}{(x^2 + 1)^2} = 0 \Rightarrow x = \pm 1. \ f \text{ is } \downarrow \text{ when } |x| \ge 1 \text{ and } f \text{ is } \uparrow \text{ when } |x| \le 1.$$

$$f''(x) = \frac{6x(x^2 - 3)}{(x^2 + 1)^3} = 0 \Rightarrow x = 0, \pm\sqrt{3}. \ f''(1) < 0 \Rightarrow LMAX \text{ of } f(1) = \tfrac{3}{2}.$$

$$f''(-1) > 0 \Rightarrow LMIN \text{ of } f(-1) = -\tfrac{3}{2}.$$

$$f'' \text{ changes sign at } x = 0, \pm\sqrt{3} \Rightarrow PI: (0, 0), (\pm\sqrt{3}, \pm\tfrac{3}{4}\sqrt{3}).$$

$\boxed{23}$ $f(x) = x^2 - \dfrac{27}{x^2} = \dfrac{x^4 - 27}{x^2}. \ f'(x) = 2x + \dfrac{54}{x^3} = \dfrac{2(x^4 + 27)}{x^3} \ne 0 \Rightarrow$

there are no extrema. f is $\downarrow$ on $(-\infty, 0)$ and $\uparrow$ on $(0, \infty)$.

$$f''(x) = 2 - \frac{162}{x^4} = \frac{2(x - 3)(x + 3)(x^2 + 9)}{x^4} = 0 \Leftrightarrow x = \pm 3.$$

f is CU on $(-\infty, -3) \cup (3, \infty)$ and CD on $(-3, 0) \cup (0, 3)$. Since f'' changes sign at $x = \pm 3$, there are PI at $(\pm 3, 6)$. Note that from the original form of the function, $x^2 - (27/x^2)$, as $|x| \to \infty$, f approaches the graph of $y = x^2$. This is the shape of the function as $|x|$ increases without bound. See *Figure 23*.

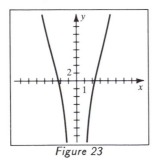

Figure 23

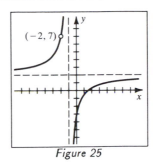

Figure 25

$\boxed{25}$ $f(x) = \dfrac{2x^2 + x - 6}{x^2 + 3x + 2} = \dfrac{(x + 2)(2x - 3)}{(x + 2)(x + 1)} = \dfrac{2x - 3}{x + 1}$ for $x \neq -2$.

To determine the value of y when $x = -2$, substitute -2 into $\dfrac{2x - 3}{x + 1}$ to get 7.

There is a hole (open circle) in the graph at $(-2, 7)$. Hence, we need only graph

$\quad y = \dfrac{2x - 3}{x + 1}$, and add the hole. This is similar to the solution of Exercise 1.

$\boxed{29}$ $f(x) = \left| x^2 - 6x + 5 \right| = \left| (x - 1)(x - 5) \right|$. $\quad y = x^2 - 6x + 5$ is a parabola opening upward with zeros at $x = 1$ and $x = 5$. It is negative on the interval $(1, 5)$. The graph of $f(x) = \left| x^2 - 6x + 5 \right|$ has the portion of the graph of $y = x^2 - 6x + 5$ between $x = 1$ and $x = 5$ reflected through the x-axis.

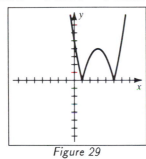

Figure 29

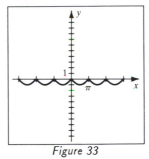

Figure 33

$\boxed{33}$ $f(x) = -\left| \sin x \right|$. First, graph $y = \left| \sin x \right|$ by reflecting the graph of $y = \sin x$ through the x-axis whenever $\sin x < 0$. Now reflect this entire graph through the x-axis to obtain the graph of $y = -\left| \sin x \right|$.

$\boxed{35}$ We will analyze each piece of information separately, and then draw some general conclusions about the graph of f.

1) $f(x) = \dfrac{x - 3}{x^2 - 1} = \dfrac{x - 3}{(x + 1)(x - 1)}$. From this information, we know that the x-intercept is 3, the y-intercept is 3, the horizontal asymptote is $y = 0$, and the vertical asymptotes are $x = \pm 1$. Also, the function $only$ crosses the horizontal asymptote (the x-axis) at $x = 3$.

2) $f'(x) = \dfrac{-x^2 + 6x - 1}{(x^2 - 1)^2}$; $f'(x) = 0$ if $x = 3 \pm 2\sqrt{2}$ (approximately 5.83, 0.17).

This piece of information indicates that the only critical numbers are $x = 3 \pm 2\sqrt{2}$, and these are the *only* candidates for extrema.

$$f' > 0 \text{ on } (0.17,\ 5.83) \ \{\, x \neq 1 \,\} \text{ and } f' < 0 \text{ otherwise } \{\, x \neq -1 \,\}.$$

3) $f''(x) = \dfrac{2x^3 - 18x^2 + 6x - 6}{(x^2 - 1)^3}$; $f''(x) = 0$ if $x = 2\sqrt[3]{2} + 2\sqrt[3]{4} + 3 \approx 8.69$.

This indicates that the *only* possible *PI* is at $x \approx 8.69$.

4) $f(5.83) \approx 0.09$; $f(0.17) \approx 2.91$; and $f(8.69) \approx 0.08$. These are the y values of the x values listed in (3) and (4). Start sketching the graph by drawing the asymptotes and plotting the points already discussed. Sketch the graph by regions that are determined by the vertical asymptotes.

On $(-\infty,\ -1)$, either f is ↑, above the x-axis, and approaches ∞ as $x \to -1^-$, or f is ↓, below the x-axis, and approaches $-\infty$ as $x \to -1^-$. Test any value less than -1 in the function to determine if f is positive or negative. This test gives us a negative numerator and a positive denominator, i.e., a negative value. Thus, $f < 0$ for every $x < -1$, and hence, f is ↓ on $(-\infty,\ -1)$.

On $(-1,\ 1)$, we have the points $(0,\ 3)$ and $(0.17,\ 2.91)$. Since the function has no other x-intercepts, it must approach the vertical asymptotes through positive values, and $(0.17,\ 2.91)$ must be a *LMIN*.

On $(1,\ \infty)$, we have an x-intercept at 3, the point $(5.83,\ 0.09)$, and the point $(8.69,\ 0.08)$. The function crosses the horizontal asymptote at $x = 3$, reaches a *LMAX* at $x \approx 5.83$, starts to decrease, changes concavity from *CD* to *CU* at $x \approx 8.69$, and approaches the horizontal asymptote ($y = 0$) as $x \to \infty$.

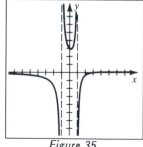

Figure 35

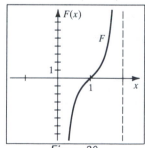

Figure 39

39 (a) Let $k > 0$ be a constant of proportionality. The attraction to the particle at 0

is $F_1 = \dfrac{k(1)(-1)}{x^2} = -\dfrac{k}{x^2}$. In this case, the negative sign indicates a force to the

left. The attraction to the particle at 2 is $F_2 = -\dfrac{k(1)(-1)}{(2-x)^2} = \dfrac{k}{(x-2)^2}$.

The additional negative sign is used to indicate that F_2 is in the opposite

direction as F_1. Since $F = F_1 + F_2$, the result follows.

(b) See *Figure 39*. $k = 1 \Rightarrow F(x) = -\dfrac{1}{x^2} + \dfrac{1}{(x-2)^2} = \dfrac{4(x-1)}{x^2(x-2)^2}$.

F has an x-intercept at $x = 1$ and vertical asymptotes at $x = 0$ and $x = 2$.

$$F'(x) = \frac{-4(3x^2 - 6x + 4)}{x^3(x-2)^3} \neq 0. \quad F'(x) > 0 \text{ on } (0, 2) \Rightarrow F \text{ is } \uparrow \text{ on } (0, 2).$$

Exercises 4.6

Note: Many of the application problems fit into this form: Maximize or minimize a

quantity that is determined by two variables. These two variables are then

related by another equation called the constraint. The general solution method is

to solve the constraint for one of the variables and substitute that expression into

the equation for the quantity. This is then just a function of one variable and the

problem can be solved using methods already covered.

Example: Find the maximum value of S if $S = xy$ and $x + y = 20$.

Solution: S is in terms of two variables, so we solve $x + y = 20$ for y and substitute this

expression into $S = xy$. $x + y = 20 \Rightarrow y = 20 - x \Rightarrow S = x(20 - x)$. Now

S is just a function of x, and to maximize S we need the first derivative. $S =$

$20x - x^2 \Rightarrow S' = 20 - 2x$. $S' = 0 \Rightarrow x = 10 \Rightarrow y = 20 - 10 = 10$, and S

$= xy = 10(10) = 100$. Thus, the maximum value of S is 100 when x and y

are both 10.

This example is fairly simple, and it is our goal to help reduce the

exercises to problems of this type when possible.

1 We wish to minimize the surface area of the box, and the constraint is that the

volume can only be 4 ft^3. Let x denote the length of a side of the square base and y

the height of the box. The volume of the box is the area of the base of the box times

its height, i.e., $V = x^2 y$. The constraining equation is $V = 4$, or $x^2 y = 4$. The

equation for the surface area is the sum of the four sides and the base, i.e.,

$S = 4xy + x^2$. Solving $V = x^2 y$ for y and substituting into the surface area equation

yields

$$S = 4xy + x^2 = 4x\left(\frac{4}{x^2}\right) + x^2 = \frac{16}{x} + x^2. \qquad \text{(cont.)}$$

Getting to this point is the tough part of the problem. To find the minimum of S, we differentiate and solve $S' = 0$. $S' = -\frac{16}{x^2} + 2x$. $S' = 0 \Rightarrow 2x = \frac{16}{x^2} \Rightarrow 2x^3 = 16 \Rightarrow x^3 = 8 \Rightarrow x = 2$. The height is $y = 4/2^2 = 1$. Thus, the dimensions are 2 ft × 2 ft × 1 ft. Note that the problem did not ask for the minimum surface area, only the dimensions. Since $S'' = 32/x^3 + 2 > 0$ for $x > 0$, this will be a minimum value.

[5] We wish to maximize the area enclosed by the fence, and the constraint is the amount of fence. The length of the fence used is $4y + 3x = 1000 \Rightarrow y = 250 - \frac{3}{4}x$. $A = xy = x(250 - \frac{3}{4}x) = 250x - \frac{3}{4}x^2$, where $x \in [0, \frac{1000}{3}]$. The values 0 and $\frac{1000}{3}$ are the smallest and largest values for x—obtained from the constraint $4y + 3x = 1000$. $A'(x) = 250 - \frac{3}{2}x = 0 \Rightarrow x = 166\frac{2}{3}$. Evaluating the endpoints and the critical number, we find that $A(0) = A(\frac{1000}{3}) = 0$, and $A(166\frac{2}{3}) = 20{,}833\frac{1}{3}$.

Thus, $x = 166\frac{2}{3}$ ft and $y = 125$ ft give the maximum area.

[9] Let x and y denote the distances shown in *Figure 9*, and L the length of the ladder. Using similar triangles, $\frac{y}{x+1} = \frac{8}{x}$, or $y = \frac{8(x+1)}{x}$. We wish to minimize L subject to $y = 8(x+1)/x$. Using the Pythagorean theorem, we have

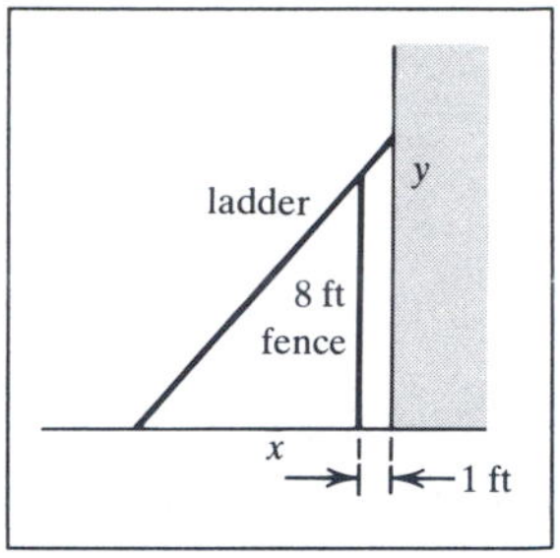

$L^2 = (x+1)^2 + y^2$

$= (x^2 + 2x + 1) + \frac{64(x^2 + 2x + 1)}{x^2}$

$= (x^2 + 2x + 1) + \left(64 + \frac{128}{x} + \frac{64}{x^2}\right)$

$= x^2 + 2x + 65 + (128/x) + (64/x^2) = f(x)$. We can minimize L by minimizing $f(x) = L^2$ since a positive expression will be minimized if its square is minimized. $f'(x) = 2x + 2 - (128/x^2) - (128/x^3) = 2 \cdot \frac{x^4 + x^3 - 64x - 64}{x^3} =$

$2 \cdot \frac{x^3(x+1) - 64(x+1)}{x^3} = \frac{2(x^3 - 64)(x+1)}{x^3} = 0$ if $x = -1, 4$ { disregard

$x = -1$}. $x = 4 \Rightarrow y = 10$ and $L = 5\sqrt{5} \approx 11.18$ ft.

This is a minimum, since $f'' = (2 + 256/x^3 + 384/x^4) > 0$ for $x > 0$.

Figure 9

[11] Let w denote the width, l the length, and h the height. Note that $w = \frac{3}{4}l$. We wish to minimize the cost subject to the constraint that the volume is 900 ft^3. Hence, $V = lwh = \frac{3}{4}l^2h = 900 \Rightarrow h = 1200/l^2$. To determine the cost, we add the cost of all 6 sides (the floor, the roof, 2 sides with dimensions length × height, and 2 sides

with dimensions width × height). The cost of each side is its cost per square foot times its area. Thus, the cost is given by $C = 4(wl) + 6(2lh) + 6(2wh) + 3(wl)$. Substituting for w and h yields

$$C = 4(\tfrac{3}{4}l \cdot l) + 6\left(2l \cdot \frac{1200}{l^2}\right) + 6\left(2 \cdot \tfrac{3}{4}l \cdot \frac{1200}{l^2}\right) + 3(\tfrac{3}{4}l \cdot l)$$

$$= 3l^2 + (14{,}400/l) + (10{,}800/l) + \tfrac{9}{4}l^2 = \tfrac{21}{4}l^2 + (25{,}200/l).$$

$C' = \tfrac{21}{2}l - (25{,}200/l^2) = 0 \Rightarrow 21l^3 - 50{,}400 = 0 \Rightarrow$

$l = \sqrt[3]{2400} = 2(\sqrt[3]{300}) \approx 13.38$ ft, $w = \tfrac{3}{2}(\sqrt[3]{300}) \approx 10.04$ ft, $h = \sqrt[3]{300} \approx 6.69$ ft.

Since $C'' = \tfrac{21}{2} + (50{,}400/l^3) > 0$ for $l > 0$, this gives a minimum.

⬛13 Let x denote the width of the field, y the length, and k, a constant, the length of the barn, where $0 < k < y$. The area of the field is given by $A = xy$ and the amount of fence used is given by $2x + y + (y - k) = 500 \Rightarrow y = \dfrac{500 + k}{2} - x$.

Thus, $A = \left(\dfrac{500 + k}{2}\right)x - x^2$. $A' = \dfrac{500 + k}{2} - 2x = 0 \Rightarrow x = \dfrac{500 + k}{4} = y$.

These values give a maximum since $A'' = -2 < 0$.

$$x = y \Rightarrow \text{the rectangle is a square.}$$

⬛15 Let x denote the number of rooms reserved, where $x \in [30, 60]$. The price for each room is given by $p(x) = \left[80 - (x - 30)\right] = 110 - x$. The revenue received is $R(x) = xp(x) = x(110 - x) = 110x - x^2$. Now, $R'(x) = 110 - 2x = 0 \Rightarrow x = 55$. Evaluating the endpoints and the critical number we find that $R(30) = 2400$, $R(60) = 3000$, and $R(55) = 3025$. Thus, maximum revenue occurs when 55 rooms are rented. Note that the maximum will not occur on $[0, 30]$ since $R(x)$ is maximum when $x = 30$ on this interval.

⬛17 The volume of the cylindrical part is $\pi r^2 h$, and the volume of each hemisphere is $\tfrac{1}{2}(\tfrac{4}{3})\pi r^3$. The total volume is $V = \pi r^2 h + \tfrac{4}{3}\pi r^3 = 10\pi \Rightarrow$

$$h = \frac{10\pi - \tfrac{4}{3}\pi r^3}{\pi r^2} = \frac{30 - 4r^3}{3r^2}.$$

The total surface area is $S = 2\pi r h + 4\pi r^2 = 2\pi r\left(\dfrac{30 - 4r^3}{3r^2}\right) + 4\pi r^2$.

Since the construction of the end piece is twice as expensive as the cylinder, the cost function is $C(r) = 2\pi r\left(\dfrac{30 - 4r^3}{3r^2}\right) + 2(4\pi r^2) = \left(\dfrac{60\pi}{3r} - \dfrac{8\pi r^2}{3}\right) + 8\pi r^2 = $

$(20\pi/r) + \tfrac{16}{3}\pi r^2$. $C'(r) = (-20\pi/r^2) + \tfrac{32}{3}\pi r$ and $C''(r) = (40\pi/r^3) + \tfrac{32}{3}\pi$.

$C'(r) = \dfrac{-20\pi + \tfrac{32}{3}\pi r^3}{r^2} = \dfrac{-60\pi + 32\pi r^3}{3r^2} = 0$ if $r^3 = \tfrac{60}{32} = \tfrac{15}{8} \Rightarrow r = \tfrac{1}{2}\sqrt[3]{15}$.

Since $C'' > 0$ for $r > 0$, this value for r will give a minimum.

$$h = 2\sqrt[3]{15} \text{ for this value of } r.$$

$\boxed{21}$ If V is the volume of the cone, then $V = \frac{1}{3}\pi r^2 h$. *Figure 21* shows a cross section of
the sphere and cone. To relate r and h, consider the small right triangle in *Figure 21*.
The vertical leg is $h - a$ if $h > a$ (shown) or $a - h$ if $h < a$.
$r^2 + (h - a)^2 = a^2 \Rightarrow r^2 = 2ah - h^2$ (in either case).
$V = \frac{1}{3}\pi(2ah^2 - h^3) \Rightarrow V' = \frac{\pi}{3}(4a - 3h)h = 0$ if $h = \frac{4}{3}a$ { $h = 0$ is an endpoint }
and the volume is $V = \frac{\pi}{3}\left[2a(\frac{4}{3}a)^2 - (\frac{4}{3}a)^3\right] = \frac{\pi}{3}\left[\frac{32}{9}a^3 - \frac{64}{27}a^3\right] = \frac{32}{81}\pi a^3$.

This is a maximum since the endpoints of $h = 0, 2a$ yield $V = 0$.

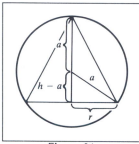

Figure 21

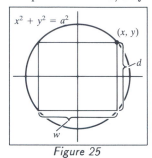

Figure 25

$\boxed{23}$ An arbitrary point on the graph of $y = x^2 + 1$ has coordinates $(x, x^2 + 1)$.
If $f(x)$ is the square of the distance of a point $(x, x^2 + 1)$ on the parabola from $(3, 1)$,
then $f(x) = (x - 3)^2 + \left[(x^2 + 1) - 1\right]^2 = (x - 3)^2 + x^4$. If we minimize f we will
minimize the distance. $f'(x) = 2(x - 3) + 4x^3 = 4x^3 + 2x - 6 = 0$ if $x = 1$.
{ Since the sum of the coefficients of f' is zero, 1 is a root. }

The point $(1, 2)$ gives a minimum, since $f''(x) = 12x^2 + 2 > 0$.

$\boxed{25}$ If S denotes the strength, w the width, and d the depth, then $S = kwd^2$, where $k > 0$
is a proportionality constant. With the circular cross section placed as shown in
Figure 25, we have $w = 2x$, $d = 2y$, and $y^2 = a^2 - x^2$. $S = k(2x)4y^2 =$
$k(2x)4(a^2 - x^2) = 8k(a^2 x - x^3)$. $S' = 8k(a^2 - 3x^2) = 0$ if $x = a/\sqrt{3}$.   Thus,
$w = 2a/\sqrt{3}$ and $d = 2y = 2\sqrt{a^2 - x^2} = 2\sqrt{a^2 - (a^2/3)} = 2\sqrt{2}a/\sqrt{3}$, which is $\sqrt{2}$
times the width. This is a maximum, since the endpoints of $x = 0, a$ yield $S = 0$.

$\boxed{29}$ (a) Let $3y$ denote the length of the piece of wire bent into the equilateral triangle and
$6x$ { width x and length $2x$ } the amount bent into the rectangle. The triangle's
area is $\frac{1}{4}\sqrt{3}\,y^2$ { inside cover of text }. Now, $3y + 6x = 36 \Rightarrow y = 12 - 2x$.
The total area is $A(x) = \frac{1}{4}\sqrt{3}(12 - 2x)^2 + 2x^2 = \sqrt{3}(6 - x)^2 + 2x^2$.

$A' = (4 + 2\sqrt{3})x - 12\sqrt{3} = 0 \Leftrightarrow x = \dfrac{12\sqrt{3}}{4 + 2\sqrt{3}} \approx 2.785$. Use $6x \approx 16.71$ cm

for the rectangle. This is a minimum, since $A'' = 4 + 2\sqrt{3} > 0$.

(b) The maximum must occur at an endpoint. $A(0) = 36\sqrt{3} < A(6) = 72$.

Use all the wire for the rectangle to obtain a maximum area.

35 Let x denote the number of trees planted in *excess* of 24 per acre. The total yield is the number of trees per acre times the yield per tree. To obtain expressions for the number of trees per acre and the yield per tree, it is helpful to write some specific values in a table format.

x	0	1	2	$\cdots$
number of trees	24	25	26	$\cdots$
yield per tree	600	588	576	$\cdots$

Thus, $(24 + x)$ is the number of trees per acre and $(600 - 12x)$ is the yield per tree. The total yield per acre is $f(x) = (24 + x)(600 - 12x) = -12x^2 + 312x + 14{,}400$. $f'(x) = -24x + 312 = 0 \Leftrightarrow x = 13$. $f''(x) = -24 < 0 \Rightarrow$ this is a maximum.

Plant $24 + 13 = 37$ trees per acre.

39 Let S_1 denote the amount of smoke from factory A and S_2 the amount of smoke from factory B. The pollution at a distance x from factory A and a distance $4 - x$ from factory B on a straight line path between A and B is $P = \dfrac{kS_1}{x^3} + \dfrac{kS_2}{(4 - x)^3}$, where $k > 0$ is a constant of proportionality. Since factory A emits twice as much smoke as factory B, $S_1 = 2S_2$. Thus, $P = \dfrac{2kS_2}{x^3} + \dfrac{kS_2}{(4 - x)^3} \Rightarrow P' = -\dfrac{6kS_2}{x^4} + \dfrac{3kS_2}{(4 - x)^4}$.

$P' = 0 \Rightarrow \dfrac{3kS_2}{(4 - x)^4} = \dfrac{6kS_2}{x^4} \Rightarrow \dfrac{3kS_2}{6kS_2} = \dfrac{(4 - x)^4}{x^4} \Rightarrow \left(\dfrac{4 - x}{x}\right)^4 = \dfrac{1}{2} \Rightarrow$

$\dfrac{4 - x}{x} = \sqrt[4]{\tfrac{1}{2}} \Rightarrow 4 - x = \sqrt[4]{\tfrac{1}{2}}\,x \Rightarrow 4 = x + \sqrt[4]{\tfrac{1}{2}}\,x \Rightarrow 4 = (1 + \sqrt[4]{\tfrac{1}{2}})x \Rightarrow$

$x = \dfrac{4}{1 + \sqrt[4]{1/2}} \approx 2.17$ mi from A. Since $P'' = \dfrac{24kS_2}{x^5} + \dfrac{12kS_2}{(4 - x)^5} > 0$ for

$x \in (0, 4)$, this gives a minimum.

41 (a) The volume of the pyramid is $V = \tfrac{1}{3}x^2 h$. Let y denote its slant height. Then, we have a right triangle with sides h and $\tfrac{1}{2}x$, and hypotenuse y. By the Pythagorean theorem, $y^2 = h^2 + (\tfrac{1}{2}x)^2$. The surface area of one of the triangular sides is $\tfrac{1}{2}xy = \tfrac{1}{2}x\sqrt{h^2 + \tfrac{1}{4}x^2}$. Since there are 4 such sides, $S = 2x\sqrt{h^2 + \tfrac{1}{4}x^2}$. Solving for h yields $S^2 = 4x^2(h^2 + \tfrac{1}{4}x^2) = 4x^2 h^2 + x^4 \Rightarrow$

$h^2 = \dfrac{S^2 - x^4}{4x^2} \Rightarrow h = \dfrac{\sqrt{S^2 - x^4}}{2x}$. Thus, $V = \tfrac{1}{3}x^2 \dfrac{\sqrt{S^2 - x^4}}{2x} = \tfrac{1}{6}x\sqrt{S^2 - x^4}$.

(b) $V'(x) = \frac{1}{6}\left[x(\frac{1}{2})(S^2 - x^4)^{-1/2}(-4x^3) + (S^2 - x^4)^{1/2}(1)\right]$

$\qquad = \frac{1}{6}(S^2 - x^4)^{-1/2}\left[-2x^4 + (S^2 - x^4)\right]$

$\qquad = \frac{1}{6}(4x^2h^2)^{-1/2}(4x^2h^2 - 2x^4)\ \{S^2 = 4x^2h^2 + x^4 \Rightarrow S^2 - x^4 = 4x^2h^2\}$

$\qquad = \dfrac{2x^2(2h^2 - x^2)}{6(2xh)} = \dfrac{x^2(2h^2 - x^2)}{6xh} = 0 \Rightarrow x = \sqrt{2}\,h.$

(V' is undefined at $x = 0$.) Since $x > 0$ and $h > 0$,

the sign of $V' = \dfrac{x^2(2h^2 - x^2)}{6xh}$ is determined by the sign of $(2h^2 - x^2)$.

$\qquad V' > 0$ for $0 < x < \sqrt{2}\,h$ and $V' < 0$ for $x > \sqrt{2}\,h \Rightarrow$ maximum.

43 (a) Since 1 mi = 5,280 ft and each car requires $(12 + d)$ ft,

it follows directly that the bridge can hold $[\![5280/(12 + d)]\!]$ cars.

The greatest integer function is necessary since a fraction of a car is not allowed.

(b) Since the bridge is 1 mile long, the car "density" is $\dfrac{5280}{12 + d}$ cars/mi. If each car

is moving at v mi/hr, then the flow rate is $F = [\![5280v/(12 + d)]\!]$ cars/hr.

(c) Since $d = 0.025v^2$, $F = \dfrac{5280v}{12 + 0.025v^2}$.

$F' = \dfrac{(12 + 0.025v^2)(5280) - (5280v)(0.05v)}{(12 + 0.025v^2)^2} = \dfrac{63{,}360 - 132v^2}{(12 + 0.025v^2)^2} = 0 \Rightarrow$

$63{,}360 - 132v^2 = 0 \Rightarrow v = 4\sqrt{30} \approx 21.9$ mi/hr. $F' > 0$ for $0 < v < 4\sqrt{30}$ and

$\qquad\qquad\qquad F' < 0$ for $v > 4\sqrt{30} \Rightarrow 21.9$ mi/hr is a maximum value.

45 Let x denote the distance from A to B. Then, the distance from B to C is

$\sqrt{(40 - x)^2 + 20^2}$ and the cost (times 10,000) is $C(x) = 5x + 10\sqrt{(40 - x)^2 + 400}$

$\Rightarrow C'(x) = 5 + 10 \cdot \frac{1}{2}\left[(40 - x)^2 + 400\right]^{-1/2}(2)(40 - x)(-1) =$

$5 - \dfrac{10(40 - x)}{\left[(40 - x)^2 + 400\right]^{1/2}} = 0 \Rightarrow 5\sqrt{(40 - x)^2 + 400} = 10(40 - x) \Rightarrow$

$25\left[(40 - x)^2 + 400\right] = 100(40 - x)^2 \Rightarrow 10{,}000 = 75(40 - x)^2 \Rightarrow 40 - x = \dfrac{20}{\sqrt{3}}.$

Now, $\tan\theta = \dfrac{20}{40 - x} = \dfrac{20}{20/\sqrt{3}} = \sqrt{3} \Rightarrow \theta = 60°$. Evaluating the endpoints and the

critical number, we find that $C(20) \approx 383$, $C(40) = 400$, and $C(20/\sqrt{3}) \approx 289$;

$\qquad\qquad\qquad\qquad\qquad\qquad\qquad$ the minimum occurs at $\theta = 60°$.

49 Let $x = \overline{DC}$ and $h = \overline{BC}$. The volume of a cylinder is $V = \pi r^2 h$.

We need to express V in terms of θ. Now,

$$\sin\theta = \tfrac{h}{L} \Rightarrow h = L\sin\theta \quad \text{and} \quad \cos\theta = \tfrac{x}{L} \Rightarrow x = L\cos\theta.$$

Since x is the circumference of the cylinder, $x = 2\pi r \Rightarrow r = \frac{x}{2\pi}$. Thus, $r = \frac{L\cos\theta}{2\pi}$.

We can now write V in terms of θ. $V(\theta) = \pi\left(\frac{L\cos\theta}{2\pi}\right)^2(L\sin\theta) = \frac{L^3}{4\pi}(\cos^2\theta\,\sin\theta)$.

$V'(\theta) = \frac{L^3}{4\pi}(-2\cos\theta\,\sin^2\theta + \cos^3\theta) = 0 \Leftrightarrow \cos\theta\,(\cos^2\theta - 2\sin^2\theta) = 0 \Leftrightarrow$

$\cos\theta = 0$ or $\dfrac{\sin^2\theta}{\cos^2\theta} = \frac{1}{2} \Rightarrow \tan\theta = \pm\frac{1}{2}\sqrt{2}$. Since θ must be acute, we have

$\theta_0 = \tan^{-1}(\frac{1}{2}\sqrt{2}) \approx 35.3°$. Since $V' > 0$ for $0° < \theta < \theta_0$ and

$\hfill V' < 0$ for $\theta_0 < \theta < 90°$, θ_0 gives a maximum value.

$\boxed{53}$ For each $0 < \theta < \frac{\pi}{2}$, there is exactly one rod of length L that touches the corner *and* both walls. We would like to find the *minimum* of all such rods. If the rod is longer than this minimum, then it will not be able to pass around the corner. Divide the rod in the figure in the text at the corner into two parts of length A and B.

$$\sin\theta = \frac{4}{A} \Rightarrow A = \frac{4}{\sin\theta} \quad \text{and} \quad \cos\theta = \frac{3}{B} \Rightarrow B = \frac{3}{\cos\theta}.$$

Then, $L = A + B = \dfrac{4}{\sin\theta} + \dfrac{3}{\cos\theta}$. Using the reciprocal rule twice,

$L' = -\dfrac{4\cos\theta}{\sin^2\theta} + \dfrac{3\sin\theta}{\cos^2\theta} = \dfrac{3\sin^3\theta - 4\cos^3\theta}{\cos^2\theta\,\sin^2\theta} = 0 \Rightarrow 3\sin^3\theta = 4\cos^3\theta \Rightarrow$

$\tan^3\theta = \frac{4}{3} \Rightarrow \tan\theta = \sqrt[3]{\frac{4}{3}} \Rightarrow \theta_0 \approx 47.74°$ and $L = A + B = \dfrac{4}{\sin\theta_0} + \dfrac{3}{\cos\theta_0} \approx$

$5.4044 + 4.4612 \approx 9.87$ ft. Since for $0° < \theta < \theta_0$, $L' < 0$ and for $\theta_0 < \theta < 90°$,

$\hfill L' > 0$, this will be a minimum value for L.

$\boxed{55}$ (a) The volume of the cone and the cylinder is given by $V = \frac{1}{3}\pi R^2 h_1 + \pi R^2 h_2$, where h_1 is the height of the cone and h_2 is the height of the cylinder. The total surface area is $S = 2\pi R h_2 + \pi R\sqrt{R^2 + h_1^2}$ {inside front cover of text}.

We need to express h_1 and h_2 in terms of θ. From the volume equation,

$$V - \frac{1}{3}\pi R^2 h_1 = \pi R^2 h_2 \Rightarrow h_2 = \frac{V}{\pi R^2} - \frac{\pi R^2 h_1}{3\pi R^2} = \frac{V}{\pi R^2} - \frac{h_1}{3} = \frac{V}{\pi R^2} - \frac{R\cot\theta}{3}$$

since $\tan\theta = \dfrac{R}{h_1}$ or $h_1 = R\cot\theta$. Thus,

$S = 2\pi R\left(\dfrac{V}{\pi R^2} - \dfrac{R\cot\theta}{3}\right) + \pi R\sqrt{R^2 + R^2\cot^2\theta}$

$\quad = \dfrac{2V}{R} - \frac{2}{3}\pi R^2\cot\theta + \pi R^2\sqrt{1 + \cot^2\theta}$

$\quad = \dfrac{2V}{R} + \pi R^2\left(\csc\theta - \frac{2}{3}\cot\theta\right)\{\sqrt{1 + \cot^2\theta} = \csc\theta$ since θ is acute$\}$.

(b) $S'(\theta) = 0 + \pi R^2(-\csc\theta\,\cot\theta + \frac{2}{3}\csc^2\theta) = 0 \Rightarrow \cot\theta - \frac{2}{3}\csc\theta = 0$

$\{\csc\theta \ne 0\} \Rightarrow \cos\theta = \frac{2}{3}$. Let $\theta_0 = \cos^{-1}\frac{2}{3} \approx 48.2°$. Since

$S' = \pi R^2\csc\theta(\frac{2}{3}\csc\theta - \cot\theta)$, the sign of S' is determined by $\frac{2}{3}\csc\theta - \cot\theta =$

$\dfrac{1}{\sin\theta}(\frac{2}{3} - \cos\theta)$ for $\theta \in (0, \frac{\pi}{2})$. $S' < 0$ for $0 < \theta < \theta_0$ and

$\hfill S' > 0$ for $\theta_0 < \theta < \frac{\pi}{2}$, this value of θ_0 gives a minimum.

Exercises 4.7

Note: In Exercises 1–8, $v(t) = s'(t)$ and $a(t) = v'(t) = s''(t)$.

The motion is to the *right* when $v(t) > 0$ and to the *left* when $v(t) < 0$.

$\boxed{1}$ $s(t) = 3t^2 - 12t + 1 \Rightarrow v(t) = 6(t - 2) \Rightarrow a(t) = 6.$ $v < 0$ on $[0, 2)$ and

$v > 0$ on $(2, 5]$; left in $[0, 2)$; right in $(2, 5]$. The point is initially at $s(0) = 1$.

It moves left to $s(2) = -11$ and then right to $s(5) = 16$.

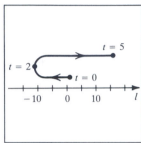

Figure 1

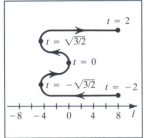

Figure 7

$\boxed{7}$ $s(t) = 2t^4 - 6t^2 \Rightarrow v(t) = 8t^3 - 12t.$ $v(t) = 4t(2t^2 - 3) = 0$ when $t = 0, \pm\sqrt{\frac{3}{2}}$.

$a(t) = 12(2t^2 - 1)$.

Time Interval	$(-2, -\sqrt{3/2})$	$(-\sqrt{3/2}, 0)$	$(0, \sqrt{3/2})$	$(\sqrt{3/2}, 2)$
k	-1.5	-1	1	1.5
Test value $v(k)$	-9	4	-4	9
Sign of $v(t)$	$-$	$+$	$-$	$+$
Direction of motion	left	right	left	right

The motion is left in $[-2, -\sqrt{\frac{3}{2}})$; right in $(-\sqrt{\frac{3}{2}}, 0)$; left in $(0, \sqrt{\frac{3}{2}})$; right in $(\sqrt{\frac{3}{2}}, 2]$.

The point is initially at $s(-2) = 8$. It moves left to $s(-\sqrt{\frac{3}{2}}) = -4.5$,

right to $s(0) = 0$, left to $s(\sqrt{\frac{3}{2}}) = -4.5$, and right to $s(2) = 8$.

$\boxed{9}$ (a) $s(t) = 5t^2 + 2 \Rightarrow v(t) = 10t$; $v(3) = 30$ ft/sec

(b) $v(t) = 10t = 28 \Rightarrow t = \frac{28}{10}$ or 2.8 sec

$\boxed{11}$ (a) $s(t) = 144t - 16t^2 \Rightarrow v(t) = 144 - 32t \Rightarrow a(t) = -32$

(b) The maximum height occurs when $v(t) = 16(9 - 2t) = 0 \Rightarrow t = \frac{9}{2}$.

This height is $s(\frac{9}{2}) = 324$ ft.

(c) The flight begins and ends when $s(t) = 0$. $s(t) = 144t - 16t^2 = 16t(9 - t)$.

$s(t) = 0 \Rightarrow t = 0, 9$ sec. $t = 0$ corresponds to the beginning of the flight and

$t = 9$, the end. The duration of the flight is 9 sec.

Note: In Exercises 13–16, consider the general position function $s(t) = a\sin\omega t$ {or cos}.

The amplitude is $|a|$, the period is $\frac{2\pi}{\omega}$, and the frequency is $\frac{\omega}{2\pi}$, where $\omega > 0$.

$\boxed{13}$ $s(t) = 5\cos\frac{\pi}{4}t \Rightarrow$ amplitude $= |a| = |5| = 5$; period $= \frac{2\pi}{\omega} = \frac{2\pi}{\pi/4} = 8$ sec.

The frequency is the reciprocal of the period, i.e., $\frac{1}{8}$ cycle/sec.

$\boxed{17}$ $V = 220\sin 360\pi t \Rightarrow V' = 220\,(360\pi\cos 360\pi t) = 79{,}200\pi\cos 360\pi t$;

$$V'(1) = 79{,}200\pi.$$

$I = 20\sin\left(360\pi t - \frac{\pi}{4}\right) \Rightarrow I' = 20\left[360\pi\cos\left(360\pi t - \frac{\pi}{4}\right)\right] = 7200\pi\cos\left(360\pi t - \frac{\pi}{4}\right)$;

$I'(1) = 7200\pi\cos\left(360\pi - \frac{\pi}{4}\right) = 7200\pi\cos\left(-\frac{\pi}{4}\right)$ {$\cos\left(360\pi + t\right) = \cos t$ for any t}

$$= 7200\pi\left(\tfrac{\sqrt{2}}{2}\right) = 3600\sqrt{2}\,\pi.$$

$\boxed{19}$ (a) The wave's amplitude is $\frac{1}{2}$(high point $-$ low point) $= \frac{1}{2}(12 - 3) = 4.5$. The period is the amount of time needed to make one complete cycle. From the graph, it appears that there is a minimum at about 7 A.M. and another at about 7 P.M. Thus, the period is 12. The "normal" position of a sine wave is to have a zero at the origin before immediately increasing to a maximum. This point appears to occur at about $(10, 7.5)$, and thus, the phase shift is 10. Now $\frac{2\pi}{b} = 12$ $\Rightarrow b = \frac{\pi}{6}$ and $-\frac{c}{b} = 10 \Rightarrow c = -\frac{5\pi}{3}$. The wave is shifted $\frac{3+12}{3} = 7.5$ units upward so $d = 7.5$ and $y = 4.5\sin\left[\frac{\pi}{6}(t - 10)\right] + 7.5$.

(b) $y' = \frac{3\pi}{4}\cos\left[\frac{\pi}{6}(t - 10)\right]$. At $t = 12$, $y' = \frac{3\pi}{4}\cos\frac{\pi}{3} = \frac{3\pi}{4}\cdot\frac{1}{2} = \frac{3\pi}{8} \approx 1.178$ ft/hr.

$\boxed{23}$ $s(t) = k\cos(\omega t + b) \Rightarrow s'(t) = -k\omega\sin(\omega t + b) \Rightarrow s''(t) = -k\omega^2\cos(\omega t + b)$.

Now, $s''(t) = -k\omega^2\cos(\omega t + b) = -\omega^2\left[k\cos(\omega t + b)\right] = -\omega^2 s(t)$.

$\boxed{25}$ Let (x, y) be a point on the circle $x^2 + y^2 = a^2$ at some time t, and for simplicity, assume that $(x, y) = (1, 0)$ at $t = 0$. Then, if the angle θ is drawn in standard position with its terminal side passing through (x, y), we have $\frac{x}{a} = \cos\theta$, or equivalently, $x = a\cos\theta$. This indicates that x is in simple harmonic motion by (4.22), where $k = a$, $\omega = 1$, and $b = 0$.

$\boxed{29}$ (a) $C(x) = 800 + 0.04x + 0.0002x^2$; $C(100) = 800 + 4 + 2 = 806$.

(b) The average cost function is $c(x) = C(x)/x = (800/x) + 0.04 + 0.0002x$.

The marginal cost function is $C'(x) = 0.04 + 0.0004x$.

$$c(100) = 8 + 0.04 + 0.02 = 8.06.\quad C'(100) = 0.04 + 0.04 = 0.08.$$

$\boxed{33}$ $C(x) = 100 + 50x + (100/x) \Rightarrow C'(x) = 50 - (100/x^2)$.

The marginal cost at $x = 5$ is $C'(5) = \$46$,

whereas the actual cost of the sixth motor is $C(6) - C(5) = 416\frac{2}{3} - 370 \approx \$46.67$.

35 (a) $p(x) = 50 - 0.1x \Rightarrow p'(x) = -0.1$

(b) $R(x) = x\,p(x) = x(50 - 0.1x) = 50x - 0.1x^2$

(c) $P(x) = R(x) - C(x) = (50x - 0.1x^2) - (10 + 2x) = -0.1x^2 + 48x - 10$

(d) $P(x) = -0.1x^2 + 48x - 10 \Rightarrow P'(x) = -0.2x + 48$

(e) $P'(x) = -0.2x + 48 = 0 \Leftrightarrow x = 240.$

$$P''(x) = -0.2 < 0 \Rightarrow P(240) = 5750 \text{ is a MAX.}$$

(f) $C'(x) = 2 \Rightarrow C'(10) = 2$, the marginal cost when the demand is 10 units.

37 $p(x) = 1800 - 2x,\ 1 \le x \le 100,\ C(x) = 1000 + x + 0.01x^2$

(a) $R(x) = x\,p(x) = x(1800 - 2x) = 1800x - 2x^2$

(b) $P(x) = R(x) - C(x) =$

$$(1800x - 2x^2) - (1000 + x + 0.01x^2) = -2.01x^2 + 1799x - 1000$$

(c) $P'(x) = -4.02x + 1799 = 0 \Leftrightarrow x \approx 447.51$, which is not in the domain.

Evaluating the endpoints of the domain, we find that

$$P(1) = 796.99 \text{ and } P(100) = 158,800; \text{ the maximum profit occurs at } x = 100.$$

(d) $P(100) = \$158,800$

Exercises 4.8

Note: Polynomials can be evaluated more easily (and accurately) using nested multiplication. Example: $3x^4 - 4x^3 + x - 6 = \{[(3x - 4)x]x + 1\}x - 6.$

Note: Numerical results may vary slightly. In the following, no rounding has been performed before the final result has been reached. This increases both accuracy and ease of calculation.

1 First, we must find a function f such that $f(\sqrt[3]{2}) = 0$. If $x = \sqrt[3]{2}$ is the solution of an equation, then $x^3 = 2$, and $x^3 - 2 = 0$. Thus, we will let $f(x) = x^3 - 2$.

$f(1) = -1 < 0$, $f(2) = 6 > 0 \Rightarrow$ there is a root in the interval $[1, 2]$.

$f(x_n) = x_n^3 - 2$ and $f'(x_n) = 3x_n^2$. Using (4.23), we have

$$x_{n+1} = x_n - \frac{f(x_n)}{f'(x_n)} \Rightarrow x_{n+1} = x_n - \frac{x_n^3 - 2}{3x_n^2}. \text{ Choosing } x_1 = 1 \text{ yields:}$$

$$x_2 = 1 - \frac{(1)^3 - 2}{3(1)^2} = \frac{4}{3} \approx 1.3333,\ x_3 = \frac{4}{3} - \frac{\left(\frac{4}{3}\right)^3 - 2}{3\left(\frac{4}{3}\right)^2} = \frac{91}{72} \approx 1.2639,$$

$x_4 = 1.2599$, and $x_5 = 1.2599$. Since x_4 and x_5 are the same (to four decimal places), we assign 1.2599 as the root value. If you pick a different x_1 in $[1, 2]$, you may find that the x_k differ. However, you should still obtain the correct root. The more accurately you can estimate x_1, the fewer x_k you will need to find.

Root: ≈ 1.2599

$\boxed{3}$ $x_{n+1} = x_n - \dfrac{x_n^4 + 2x_n^3 - 5x_n^2 + 1}{4x_n^3 + 6x_n^2 - 10x_n}$

$\quad\quad = x_n - \dfrac{\left[(x_n + 2)x_n - 5\right]x_n^2 + 1}{\left[(4x_n + 6)x_n - 10\right]x_n}$. Since $f'(1) = 0$, we cannot use $x_1 = 1$.

Instead, $x_1 = 2$ yields: $x_2 = 2 - \frac{13}{36} = \frac{59}{36} \approx 1.6389$, $x_3 = 1.4319$, $x_4 = 1.3472$,

$\quad\quad\quad\quad\quad x_5 = 1.3320$, $x_6 = 1.3315$, and $x_7 = 1.3315$. <u>Root: ≈ 1.3315</u>

$\boxed{7}$ Finding the largest zero of $f(x) = x^4 - 11x^2 - 44x - 24$ is equivalent to finding the largest zero of the equation $x^4 - 11x^2 - 44x - 24 = 0$. This equation can be written as $x^4 = 11x^2 + 44x + 24$, and each side can be thought of as a separate function. In this case, let $g(x) = x^4$ and $h(x) = 11x^2 + 44x + 24$. When the graphs of g and h intersect, then the function f is zero $\{$ since $f(x) = g(x) - h(x)\}$. Hence, any zeros will occur where $x^4 = 11x^2 + 44x + 24$. From *Figure 7*, the largest zero

is in $[4, 5]$. $x_{n+1} = x_n - \dfrac{x_n^4 - 11x_n^2 - 44x_n - 24}{4x_n^3 - 22x_n - 44} =$

$x_n - \dfrac{\left[(x_n^2 - 11)x_n - 44\right]x_n - 24}{(4x_n^2 - 22)x_n - 44}$ and $x_1 = 4$ yield: $x_2 = 4.9677$,

$\quad\quad\quad x_3 = 4.6860$, $x_4 = 4.6465$, $x_5 = 4.6458$, and $x_6 = 4.6458$. <u>Root: ≈ 4.6458</u>

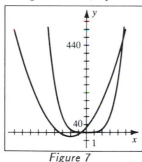

Figure 7

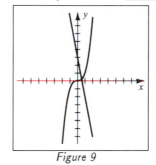

Figure 9

$\boxed{9}$ The roots of $x^3 + 5x - 3 = 0$ occur where the graphs of $y = x^3$ and $y = -5x + 3$ intersect. From *Figure 9*, the only root is in $[0, 1]$.

$x_{n+1} = x_n - \dfrac{x_n^3 + 5x_n - 3}{3x_n^2 + 5} = x_n - \dfrac{(x_n^2 + 5)x_n - 3}{3x_n^2 + 5}$ and $x_1 = 0$ yield: $x_2 = 0.60$,

$\quad\quad\quad\quad\quad\quad x_3 = 0.56$, and $x_4 = 0.56$. <u>Root: ≈ 0.56</u>

$\boxed{11}$ The root occurs where the graphs of $y = 2x$ and $y = 3\sin x$ intersect. From *Figure*

11, the positive root is in $[1, 2]$. Make sure your calculator is set in *radian* mode.

$$x_{n+1} = x_n - \frac{2x_n - 3\sin x_n}{2 - 3\cos x_n} \text{ and } x_1 = 1.5 \text{ yield: } x_2 = 1.50. \quad \underline{\text{Root:} \approx 1.50}$$

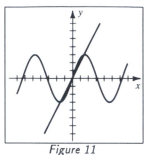

Figure 11

$\boxed{17}$ Any roots will occur where $x^3 = 3x - 1$.

From *Figure 17*, there are roots near -2, 0, and 2. $x_{n+1} = x_n - \dfrac{x_n^3 - 3x_n + 1}{3x_n^2 - 3}$.

Letting $x_1 = -2$ yields: $x_2 = -1.89$, $x_3 = -1.88$, and $x_4 = -1.88$.

Letting $x_1 = 0$ yields: $x_2 = 0.33$, $x_3 = 0.35$, and $x_4 = 0.35$.

Letting $x_1 = 2$ yields: $x_2 = 1.67$, $x_3 = 1.55$, $x_4 = 1.53$, and $x_5 = 1.53$.

$$\underline{\text{Roots:} \approx -1.88, \ 0.35, \text{ and } 1.53}$$

$\boxed{19}$ Any roots will occur where $\sin x = 2x - 5$. From *Figure 19*, there is a root near 3.

$$x_{n+1} = x_n - \frac{2x_n - 5 - \sin x_n}{2 - \cos x_n}.$$

Letting $x_1 = 3$ yields: $x_2 = 2.71$ and $x_3 = 2.71$. $\quad \underline{\text{Root:} \approx 2.71}$

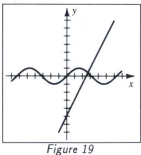

Figure 19

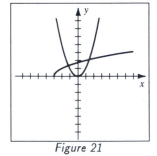

Figure 21

$\boxed{21}$ From *Figure 21*, the intersection points are in $[-2, -1]$ and $[1, 2]$.

Using (4.23) with $f(x) = x^2 - \sqrt{x + 3}$, $f'(x) = 2x - \frac{1}{2}(x + 3)^{-1/2}$,

and $x_1 = -1$ yields: $x_2 = -1.18$, $x_3 = -1.16$, and $x_4 = -1.16$.

Letting $x_1 = 1$ yields: $x_2 = 1.57$, $x_3 = 1.46$, $x_4 = 1.45$, and $x_5 = 1.45$.

23 From *Figure 23*, an intersection point is near $x = 3$.

Using (4.23) with $f(x) = \cos\frac{1}{2}x + x^2 - 9$,

$f'(x) = -\frac{1}{2}\sin\frac{1}{2}x + 2x$, and $x_1 = 3$ yields: $x_2 = 2.99$,

and $x_3 = 2.99$. By symmetry, the other point is near

$x = -2.99$.

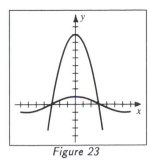

Figure 23

25 (a) $x_{n+1} = x_n - \dfrac{\sin x_n}{\cos x_n} = x_n - \tan x_n$ and $x_1 = 3$ yields:

$$x_2 = 3.1425465, \ x_3 = 3.1415927, \ x_4 = 3.1415926, \text{ and } x_5 = 3.1415926.$$

(b) They approach the nearest root, which is 2π.

29 (a) f: $x_{n+1} = x_n - \dfrac{(x_n - 1)^3(x_n^2 - 3x_n + 7)}{3(x_n - 1)^2(x_n^2 - 3x_n + 7) + (x_n - 1)^3(2x_n - 3)}$

$\qquad = x_n - \dfrac{(x_n - 1)(x_n^2 - 3x_n + 7)}{3(x_n^2 - 3x_n + 7) + (x_n - 1)(2x_n - 3)};$

$\qquad\qquad x_1 = 1.1, \ x_2 = 1.066485, \ x_3 = 1.044237, \ x_4 = 1.029451$

g: $x_{n+1} = x_n - \dfrac{(x_n - 1)(x_n^2 - 3x_n + 7)}{(x_n^2 - 3x_n + 7) + (x_n - 1)(2x_n - 3)};$

$\qquad\qquad x_1 = 1.1, \ x_2 = 0.9983437, \ x_3 = 0.9999995, \ x_4 = 1.000000$

(b) $x = 1$ is a multiple root of the equation $f(x) = 0$.

As a result, $f'(1) = 0$ and Newton's method will converge more slowly.

The higher the multiplicity, the slower the convergence.

4.9 Review Exercises

1 $f(x) = -x^2 + 6x - 8 \Rightarrow f'(x) = -2x + 6. \ f'(x) = 0 \Rightarrow x = 3.$

Evaluating the critical number and endpoints, we find that $f(3) = 1$, $f(1) = -3$,

and $f(6) = -8$. Max: $f(3) = 1$; min: $f(6) = -8$

3 The critical numbers of f occur when $f' = 0$ or when f' is undefined.

$f(x) = (x + 2)^3(3x - 1)^4 \Rightarrow$

$\quad f'(x) = (x + 2)^3(4)(3x - 1)^3(3) + 3(x + 2)^2(3x - 1)^4 \qquad \{\text{product rule}\}$

$\qquad = 3(x + 2)^2(3x - 1)^3\big[4(x + 2) + (3x - 1)\big] \qquad \{\text{factor out gcf}\}$

$\qquad = 3(x + 2)^2(3x - 1)^3(7x + 7) \qquad\qquad\qquad \{\text{simplify}\}$

f' is always defined. $f'(x) = 0 \Rightarrow x = -2, -1, \frac{1}{3}.$

$\boxed{7}$ $f(x) = (4 - x)x^{1/3} = 4x^{1/3} - x^{4/3} \Rightarrow$

$f'(x) = \frac{4}{3}x^{-2/3} - \frac{4}{3}x^{1/3} = \frac{4}{3}x^{-2/3}(1 - x) = \frac{4(1 - x)}{3x^{2/3}}.$ $f' = 0$ at $x = 1$.

f' is undefined at $x = 0$. The sign of f' is determined by $(1 - x)$ since $x^{2/3} \geq 0$.

f is $\uparrow$ on $(-\infty, 1]$ and $\downarrow$ on $[1, \infty)$. By the first derivative test,

we have a *LMAX* at $f(1) = 3$. There is a vertical tangent line at $(0, 0)$.

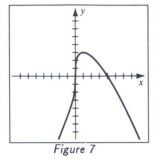   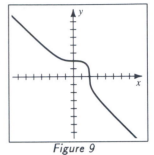

Figure 7   Figure 9

$\boxed{9}$ $f(x) = \sqrt[3]{8 - x^3} \Rightarrow f'(x) = \frac{1}{3}(8 - x^3)^{-2/3}(-3x^2) = -\frac{x^2}{(8 - x^3)^{2/3}}.$

$f''(x) = -\dfrac{(8 - x^3)^{2/3}(2x) - (x^2)(\frac{2}{3})(8 - x^3)^{-1/3}(-3x^2)}{\left[(8 - x^3)^{2/3}\right]^2}$

$= -\dfrac{2x(8 - x^3)^{-1/3}\left[(8 - x^3) + x^3\right]}{(8 - x^3)^{4/3}} = -\dfrac{16x}{(8 - x^3)^{5/3}}.$

The *CN* of f are 0 and 2. Since $f''(0) = 0$ and $f''(2)$ is undefined, the second derivative test is not applicable. Use the *first* derivative test to show that there are no extrema. $f'(x) < 0$ for all $x \neq 0, 2$. Hence, f is $\downarrow$ on $\mathbb{R}$ and there are no extrema. $f'' > 0$ when $x < 0$ or $x > 2$. $f'' < 0$ when $0 < x < 2$. Thus, f is *CU* on $(-\infty, 0)$ and $(2, \infty)$, and f is *CD* on $(0, 2)$. The x-coordinates of the *PI* are 0 and 2. Note that since $f(x) = \sqrt[3]{8 - x^3} = (-x)\sqrt[3]{1 - 8/x^3}$, as $|x|$ gets large, f has an oblique asymptote of $y = -x$.

13 $f(x) = 2\sin x - \cos 2x \Rightarrow f'(x) = 2\cos x + 2\sin 2x = 2\cos x + 4\sin x \cos x =$

$2\cos x(1 + 2\sin x)$. $f'(x) = 0 \Leftrightarrow \cos x = 0$ or $\sin x = -\frac{1}{2}$.

On $[0, 2\pi]$, the solutions are $\frac{\pi}{2}$, $\frac{3\pi}{2}$ and $\frac{7\pi}{6}$, $\frac{11\pi}{6}$. $f''(x) = -2\sin x + 4\cos 2x$.

$f''(\frac{\pi}{2}) = -6 < 0$ and $f''(\frac{3\pi}{2}) = -2 < 0 \Rightarrow$ LMAX at $f(\frac{\pi}{2}) = 3$ and $f(\frac{3\pi}{2}) = -1$.

$f''(\frac{7\pi}{6}) = f''(\frac{11\pi}{6}) = 3 > 0$. There are LMIN of $-\frac{3}{2}$ at both points.

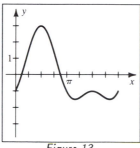

Figure 13

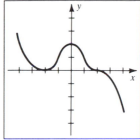

Figure 15

15 First, plot the points $(0, 2)$, $(-2, 0)$, and $(2, 0)$. $f'(-2) = f'(0) = f'(2) = 0 \Rightarrow$
there is a horizontal tangent line at each of these points. $f'(x) > 0$ if $-2 < x < 0 \Rightarrow$
f is $\uparrow$ on $[-2, 0]$. $f'(x) < 0$ if $x < -2$ or $x > 0 \Rightarrow f$ is decreasing on $(-\infty, -2] \cup$
$[0, \infty)$. Thus, $f(-2) = 0$ is a LMIN and $f(0) = 2$ is a LMAX. $f(2) = 0$ is neither.
$f''(x) > 0$ if $x < -1$ or $1 < x < 2 \Rightarrow f$ is CU on $(-\infty, -1) \cup (1, 2)$. $f''(x) < 0$ if
$-1 < x < 1$ or $x > 2 \Rightarrow f$ is CD on $(-1, 1) \cup (2, \infty)$.

Note: HA, OA, and VA denote horizontal, oblique, and vertical asymptotes, respectively.

$\boxed{17}$ $f(x) = \dfrac{3x^2}{9x^2 - 25} \Rightarrow f'(x) = \dfrac{(9x^2 - 25)(6x) - (3x^2)(18x)}{(9x^2 - 25)^2} = -\dfrac{150x}{(9x^2 - 25)^2}.$

$f''(x) = -\dfrac{(9x^2 - 25)^2(150) - (150x)(2)(9x^2 - 25)(18x)}{\left[(9x^2 - 25)^2\right]^2}$

$= -\dfrac{150(9x^2 - 25)\left[(9x^2 - 25) - 36x^2\right]}{(9x^2 - 25)^4} = \dfrac{150(27x^2 + 25)}{(9x^2 - 25)^3}.$

Since $f'(0) = 0$ and $f''(0) < 0$, there is a *LMAX* at $f(0) = 0$.

The denominator is zero when $x = \pm\frac{5}{3} \Rightarrow$ VA @ $x = \pm\frac{5}{3}$.

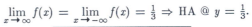
$\displaystyle\lim_{x \to \infty} f(x) = \lim_{x \to -\infty} f(x) = \tfrac{1}{3} \Rightarrow$ HA @ $y = \tfrac{1}{3}.$

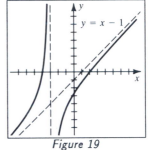

Figure 17   Figure 19

$\boxed{19}$ $f(x) = \dfrac{x^2 + 2x - 8}{x + 3} \Rightarrow f'(x) = \dfrac{x^2 + 6x + 14}{(x + 3)^2}$ and $f''(x) = -\dfrac{10}{(x + 3)^3}.$

$x^2 + 6x + 14 > 0 \Rightarrow f'(x) > 0$, and f is $\uparrow$ on $(-\infty, -3) \cup (-3, \infty)$.

There are no extrema. VA: $x = -3$.

Since $f(x) = x - 1 + \dfrac{-5}{x + 3}$, there is an OA of $y = x - 1$. No HA.

$\boxed{23}$ $f(4) - f(0) = f'(c)(4 - 0) \Rightarrow 85 - 1 = (3c^2 + 2c + 1)(4 - 0) \Rightarrow$

$3c^2 + 2c + 1 = 21 \Rightarrow 3c^2 + 2c - 20 = 0 \Rightarrow c = \tfrac{1}{3}(-1 + \sqrt{61}) \approx 2.3 \in (0, 4).$

Note that $\tfrac{1}{3}(-1 - \sqrt{61}) \approx -2.9 \notin (0, 4).$

27 See *Figure 27*. Let α be the angle between the two sheets as shown.

The volume will be maximized when the cross-sectional area is maximized.

$A = \frac{1}{2}bc\sin\alpha$ { from trigonometry } $= \frac{1}{2}(10)(10)\sin\alpha = 50\sin\alpha$.

This is maximum when $\sin\alpha = 1$, or $\alpha = \frac{\pi}{2}$. (No differentiation is necessary.)

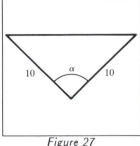

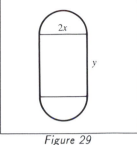

Figure 27   Figure 29

29 Let x denote the radius of the semicircles and y the length of the rectangle as shown in *Figure 29*. The perimeter is $2y + 2\pi x = \frac{1}{2} \Rightarrow y = \frac{1}{4} - \pi x$. The area of the rectangle is $A = (2x)y = \frac{1}{2}x - 2\pi x^2 \Rightarrow A' = \frac{1}{2} - 4\pi x$. $A' = 0$ if $x = \frac{1}{8\pi}$ mi.

Thus, $y = \frac{1}{4} - \pi(\frac{1}{8\pi}) = \frac{1}{8}$ mi. This is a maximum area for the rectangle, since $A'' = -4\pi < 0$ for all x.

33 $s(t) = \dfrac{t^2 + 3t + 1}{t^2 + 1} \Rightarrow$

$$v(t) = s'(t) = \frac{(t^2 + 1)(2t + 3) - (t^2 + 3t + 1)(2t)}{(t^2 + 1)^2} = \frac{3(1 - t^2)}{(t^2 + 1)^2}.$$

$$a(t) = v'(t) = 3 \cdot \frac{(t^2 + 1)^2(-2t) - (1 - t^2)(2)(t^2 + 1)(2t)}{\left[(t^2 + 1)^2\right]^2} =$$

$$3 \cdot \frac{2t(t^2 + 1)\left[-(t^2 + 1) - 2(1 - t^2)\right]}{(t^2 + 1)^4} = \frac{6t(t^2 - 3)}{(t^2 + 1)^3}.$$

The direction of the motion is determined by the sign of $(1 - t^2)$ in $v(t)$.

Since $v(t) > 0$ *only* for $|t| < 1$, the motion is to the left in $[-2, -1)$,

to the right in $(-1, 1)$, and to the left in $(1, 2]$.

39 Let $f(x) = \sin x - x\cos x$ and $f'(x) = \cos x - \left[x(-\sin x) - (\cos x)(1)\right] = x\sin x$.

Make sure your calculator is set in *radian* mode. Choose $x_1 = 4$, since $\pi < 4 < \frac{3\pi}{2}$.

$$x_{n+1} = x_n - \frac{\sin x_n - x_n\cos x_n}{x_n\sin x_n} \text{ and } x_1 = 4 \text{ yield:}$$

$$x_2 = 4.614, \; x_3 = 4.496, \; x_4 = 4.493, \text{ and } x_5 = 4.493. \; \underline{\text{Root} \approx 4.493}$$

Chapter 5: Integrals

$\boxed{3}$ $\quad \int (9t^2 - 4t + 3)\, dt = \int 9t^2\, dt - \int 4t\, dt + \int 3\, dt \qquad\qquad$ { apply (5.6)(ii) & (iii) }

$$= 9\int t^2\, dt - 4\int t\, dt + 3\int dt \qquad\qquad \text{\{ apply (5.6)(i) \}}$$

$$= 9\left[\frac{t^{2+1}}{2+1}\right] - 4\left[\frac{t^{1+1}}{1+1}\right] + 3t + C \qquad \text{\{ apply (5.4)(1) \& (2) \}}$$

$$= 9(\tfrac{1}{3}t^3) - 4(\tfrac{1}{2}t^2) + 3t + C \qquad\qquad \text{\{ simplify \}}$$

$$= 3t^3 - 2t^2 + 3t + C \qquad\qquad \text{\{ simplify \}}$$

$\boxed{5}$ $\quad \displaystyle\int\left(\frac{1}{z^3} - \frac{3}{z^2}\right) dz = \int (z^{-3} - 3z^{-2})\, dz \qquad\qquad$ { change to x^r form }

$$= \frac{z^{-3+1}}{-3+1} - 3\left[\frac{z^{-2+1}}{-2+1}\right] + C \qquad\qquad \text{\{ apply (5.6) and (5.4) \}}$$

$$= -\frac{z^{-2}}{-2} - 3\cdot\frac{z^{-1}}{-1} + C \qquad\qquad \text{\{ simplify \}}$$

$$= -\frac{1}{2z^2} + \frac{3}{z} + C \qquad\qquad \text{\{ simplify \}}$$

$\boxed{7}$ $\quad \displaystyle\int\left(3\sqrt{u} + \frac{1}{\sqrt{u}}\right) du = \int (3u^{1/2} + u^{-1/2})\, du \qquad\qquad$ { change to x^r form }

$$= 3\left[\frac{u^{(1/2)+1}}{\frac{1}{2}+1}\right] + \left[\frac{u^{(-1/2)+1}}{-\frac{1}{2}+1}\right] + C \qquad \text{\{ apply (5.6) \& (5.4) \}}$$

$$= 3\cdot\frac{u^{3/2}}{\frac{3}{2}} + \frac{u^{1/2}}{\frac{1}{2}} + C \qquad\qquad \text{\{ simplify \}}$$

$$= 2u^{3/2} + 2u^{1/2} + C \qquad\qquad \text{\{ simplify \}}$$

$\boxed{11}$ $\quad \int (3x - 1)^2\, dx = \int (9x^2 - 6x + 1)\, dx \qquad\qquad$ { expand }

$$= 9\cdot\frac{x^3}{3} - 6\cdot\frac{x^2}{2} + x + C \qquad \text{\{ apply (5.6) \& (5.4) \}}$$

$$= 3x^3 - 3x^2 + x + C \qquad\qquad \text{\{ simplify \}}$$

$\boxed{15}$ $\quad \displaystyle\int \frac{8x - 5}{\sqrt[3]{x}}\, dx = \int\left(\frac{8x}{\sqrt[3]{x}} - \frac{5}{\sqrt[3]{x}}\right) dx \qquad\qquad$ { break up integrand }

$$= \int (8x^{2/3} - 5x^{-1/3})\, dx \qquad\qquad \text{\{ change to } x^r \text{ form \}}$$

$$= 8\left[\frac{x^{(2/3)+1}}{\frac{2}{3}+1}\right] - 5\left[\frac{x^{(-1/3)+1}}{-\frac{1}{3}+1}\right] + C \qquad \text{\{ apply (5.6) \& (5.4) \}}$$

$$= 8\cdot\frac{x^{5/3}}{\frac{5}{3}} - 5\cdot\frac{x^{2/3}}{\frac{2}{3}} + C \qquad\qquad \text{\{ simplify \}}$$

$$= \tfrac{24}{5}x^{5/3} - \tfrac{15}{2}x^{2/3} + C \qquad\qquad \text{\{ simplify \}}$$

[17]
$$\int \frac{x^3 - 1}{x - 1}\, dx = \int \frac{(x - 1)(x^2 + x + 1)}{x - 1}\, dx \qquad \{\text{factor a difference of cubes}\}$$

$$= \int (x^2 + x + 1)\, dx \ (x \neq 1) \qquad \{\text{cancel } (x - 1)\}$$

$$= \tfrac{1}{3}x^3 + \tfrac{1}{2}x^2 + x + C \qquad \{\text{apply (5.6) \& (5.4)}\}$$

[19]
$$\int \frac{(t^2 + 3)^2}{t^6}\, dt = \int \frac{t^4 + 6t^2 + 9}{t^6}\, dt \qquad \{\text{expand}\}$$

$$= \int \left(\frac{t^4}{t^6} + 6 \cdot \frac{t^2}{t^6} + 9 \cdot \frac{1}{t^6} \right) dt \qquad \{\text{break up integrand}\}$$

$$= \int (t^{-2} + 6t^{-4} + 9t^{-6})\, dt \qquad \{\text{simplify and change form}\}$$

$$= \left[\frac{t^{-2+1}}{-2 + 1} \right] + 6 \left[\frac{t^{-4+1}}{-4 + 1} \right] + 9 \left[\frac{t^{-6+1}}{-6 + 1} \right] + C \quad \{\text{apply (5.6) \& (5.4)}\}$$

$$= \frac{t^{-1}}{-1} + 6 \cdot \frac{t^{-3}}{-3} + 9 \cdot \frac{t^{-5}}{-5} + C \qquad \{\text{simplify}\}$$

$$= -t^{-1} - 2t^{-3} - \tfrac{9}{5}t^{-5} + C \qquad \{\text{simplify}\}$$

[23]
$$\int \frac{7}{\csc x}\, dx = 7 \int \frac{1}{\csc x}\, dx = 7 \int \sin x\, dx = 7(-\cos x) + C = -7 \cos x + C$$

[27]
$$\int \frac{\sec t}{\cos t}\, dt = \int \sec t \cdot \frac{1}{\cos t}\, dt = \int \sec t \cdot \sec t\, dt = \int \sec^2 t\, dt = \tan t + C$$

[31]
$$\int \frac{\sec w \sin w}{\cos w}\, dw = \int \sec w \cdot \frac{\sin w}{\cos w}\, dw = \int \sec w \tan w\, dw = \sec w + C$$

[33]
$$\int \frac{(1 + \cot^2 z) \cot z}{\csc z}\, dz = \int \frac{\csc^2 z \cot z}{\csc z}\, dz = \int \csc z \cot z\, dz = -\csc z + C$$

[35] By Theorem (5.5)(i) with $f(x) = \sqrt{x^2 + 4}$, $\displaystyle \int D_x \sqrt{x^2 + 4}\, dx = \sqrt{x^2 + 4} + C.$

[37] By Theorem (5.5)(i) with $f(x) = \sin \sqrt[3]{x}$, $\displaystyle \int \frac{d}{dx}(\sin \sqrt[3]{x})\, dx = \sin \sqrt[3]{x} + C.$

[39] By Theorem (5.5)(ii) with $f(x) = x^3 \sqrt{x - 4}$, $\displaystyle D_x \int (x^3 \sqrt{x - 4})\, dx = x^3 \sqrt{x - 4}.$

[43] Exercises 43–48 are included to make sure that you understand the role of variables that are treated as constants if they are *not* the variable of integration. In this problem, a^2 is merely a constant since x is the variable of integration. Thus, applying (5.4)(1) yields $\int a^2\, dx = a^2 \int dx = a^2 x + C.$

[45] In this problem, a and b are constants, and t is the variable of integration.

$$\int (at + b)\, dt = a \cdot \frac{t^2}{2} + bt + C = \tfrac{1}{2}at^2 + bt + C$$

49 $f'(x) = 12x^2 - 6x + 1$ { given }

$\int f'(x)\,dx = \int (12x^2 - 6x + 1)\,dx$ { integrate both sides }

$f(x) = 12 \cdot \frac{x^3}{3} - 6 \cdot \frac{x^2}{2} + x + C$ { find antiderivative }

$= 4x^3 - 3x^2 + x + C$ { simplify }

We now have $f(x)$ and need to determine the constant C such that the condition $f(1) = 5$ is satisfied. $f(1) = 4(1)^3 - 3(1)^2 + 1 + C = 2 + C$.

$$f(1) = 5 \Rightarrow 2 + C = 5 \Rightarrow C = 3. \text{ Thus, } f(x) = 4x^3 - 3x^2 + x + 3.$$

53 $f''(x) = 4x - 1$ { given }

$\int f''(x)\,dx = \int (4x - 1)\,dx$ { integrate both sides }

$f'(x) = 4 \cdot \frac{x^2}{2} - x + C$ { find antiderivative }

$= 2x^2 - x + C$ { simplify }

$f'(2) = 2(2)^2 - 2 + C = 6 + C.$ $f'(2) = -2 \Rightarrow 6 + C = -2 \Rightarrow C = -8.$

Thus, $f'(x) = 2x^2 - x - 8$.

$\int f'(x)\,dx = \int (2x^2 - x - 8)\,dx$ { integrate both sides }

$f(x) = \frac{2}{3}x^3 - \frac{1}{2}x^2 - 8x + D$ { simplify }

$f(1) = \frac{2}{3}(1)^3 - \frac{1}{2}(1)^2 - 8(1) + D = -\frac{47}{6} + D.$

$$f(1) = 3 \Rightarrow -\frac{47}{6} + D = \frac{18}{6} \Rightarrow D = \frac{65}{6}. \text{ Thus, } f(x) = \frac{2}{3}x^3 - \frac{1}{2}x^2 - 8x + \frac{65}{6}.$$

55 $\dfrac{d^2 y}{dx^2} = 3\sin x - 4\cos x$ { given }

$\displaystyle\int \frac{d^2 y}{dx^2}\,dx = \int (3\sin x - 4\cos x)\,dx$ { integrate both sides }

$\dfrac{dy}{dx} = -3\cos x - 4\sin x + C$ { find antiderivative }

When $x = 0$, $y' = -3(1) - 4(0) + C = -3 + C$.

$$y' = 2 \Rightarrow -3 + C = 2 \Rightarrow C = 5. \text{ Thus, } \frac{dy}{dx} = -3\cos x - 4\sin x + 5.$$

$\displaystyle\int \frac{dy}{dx}\,dx = \int (-3\cos x - 4\sin x + 5)\,dx$ { integrate both sides }

$y = -3\sin x + 4\cos x + 5x + D$ { find antiderivative }

When $x = 0$, $y = -3(0) + 4(1) + 5(0) + D = 4 + D$.

$$y = 7 \Rightarrow 4 + D = 7 \Rightarrow D = 3. \text{ Thus, } y = -3\sin x + 4\cos x + 5x + 3.$$

57 Since $a(t) = v'(t) = 2 - 6t$, $v(t) = \int a(t)\,dt = 2t - 3t^2 + C$.

$v(0) = -5 \Rightarrow C = -5$. Thus, $v(t) = 2t - 3t^2 - 5$.

Since $v(t) = s'(t)$, $s(t) = \int v(t)\,dt = t^2 - t^3 - 5t + D$. $s(0) = 4 \Rightarrow D = 4$.

$$\text{Thus, } s(t) = t^2 - t^3 - 5t + 4.$$

$\boxed{59}$ (a) The acceleration due to gravity is -32 ft/sec^2. $a(t) = v'(t) = -32 \Rightarrow$

$v(t) = \int a(t)\, dt = -32t + C$. $v(0) = 1600 \Rightarrow C = 1600$.

Hence, $v(t) = s'(t) = -32t + 1600 \Rightarrow s(t) = \int v(t)\, dt = -16t^2 + 1600t + D$.

$s(0) = 0 \Rightarrow D = 0$. Thus, $s(t) = -16t^2 + 1600t$.

(b) The maximum height occurs when the projectile's upward velocity decreases to

zero. Thus, $v(t) = -32t + 1600 = 0 \Rightarrow t = 50$. After 50 seconds,

the actual maximum height is $s(50) = -16(50)^2 + 1600(50) = 40{,}000$ ft.

$\boxed{65}$ Let k denote the constant acceleration needed so that $a(t) = k$. If the car is to travel

500 feet in 10 seconds, then $s(10) = 500$ feet is a condition that must be satisfied.

Now $a(t) = k \Rightarrow v(t) = kt + C$. At this point in the solution, we need another

initial condition to help us determine C. Since the automobile is *starting at rest*, the

velocity at time $t = 0$ is zero, i.e., $v(0) = 0$. Thus, $v(0) = C \Rightarrow C = 0 \Rightarrow$

$v(t) = kt$.

Now, $s(t) = \int v(t)\, dt = \frac{1}{2}kt^2 + D$. We have two constants, k and D, that

need to be determined, but only one condition concerning $s(t)$. The position of the

car at time $t = 0$ is zero implies that $s(0) = 0$. Thus, $s(0) = D \Rightarrow D = 0 \Rightarrow s(t)$

$= \frac{1}{2}kt^2$. Finally, $s(10) = \frac{1}{2}k(10)^2 = 50k$. $s(10) = 500 \Rightarrow 50k = 500 \Rightarrow k = 10$

ft/sec^2.

Note that this is (10 feet per second) per second. At the end of 1 second, the

car is traveling with a velocity of 10 ft/sec; at the end of 2 seconds, 20 ft/sec; and so

on. At the end of 10 seconds, the car's velocity is 100 ft/sec, or, equivalently,

$\frac{100 \cdot 3600}{5280} = \frac{750}{11} \approx 68.2$ mi/hr.

$\boxed{67}$ The rate of consumption, dA/dt (or $A'(t)$), is given as $5 + 0.01t$. Hence, the total

amount of natural gas consumed at time t is $A(t) = \int A'(t)\, dt = 5t + 0.005t^2 + C$.

To determine C, we need to find a condition concerning A at some particular time.

Since none of the gas reserves have been used, this condition is that at time $t = 0$,

$A(0) = 0$. $A(0) = C \Rightarrow C = 0$ and $A(t) = 0.005t^2 + 5t$. The country's natural

gas reserves will be depleted when $A(t) = 100$.

$$A(t) = 100 \qquad \{\,\text{the given condition to satisfy}\,\}$$

$$0.005t^2 + 5t = 100 \qquad \{\,\text{definition of } A(t)\,\}$$

$$5t^2 + 5000t = 100{,}000 \qquad \{\,\text{multiply by } 1000\,\}$$

$$t^2 + 1000t - 20{,}000 = 0 \qquad \{\,\text{divide by 5 and set in standard form}\,\}$$

Using the quadratic formula to solve for t, we have

$$t = \frac{-(1000) \pm \sqrt{(1000)^2 - 4(1)(-20{,}000)}}{2(1)} = \frac{-1000 \pm \sqrt{2^2 \cdot 300^2 \cdot 3}}{2} =$$

$-500 \pm 300\sqrt{3}$. Since t must be positive, $t = -500 + 300\sqrt{3} \approx 19.62$ yr.

Exercises 5.2

Note: Let I denote the given integral.

1 Using (5.7) with $u = g(x) = 2x^2 + 3$ and $du = g'(x)\,dx = 4x\,dx$, we have

$$\int x(2x^2 + 3)^{10}\,dx = \tfrac{1}{4}\int (2x^2 + 3)^{10}(4x)\,dx \qquad \{\,\text{introduce } \tfrac{1}{4} \text{ and } 4\,\}$$

$$= \tfrac{1}{4}\int u^{10}\,du \qquad \{\,\text{substitute for } u \text{ and } du\,\}$$

$$= \tfrac{1}{4}(\tfrac{1}{11}u^{11}) + C \qquad \{\,\text{find antiderivative}\,\}$$

$$= \tfrac{1}{44}u^{11} + C \qquad \{\,\text{simplify}\,\}$$

$$= \tfrac{1}{44}(2x^2 + 3)^{11} + C \qquad \{\,\text{resubstitute for } u\,\}$$

5 $u = 1 + \sqrt{x}$, $du = \dfrac{1}{2\sqrt{x}}\,dx$ or, equivalently, $2\,du = \dfrac{1}{\sqrt{x}}\,dx \Rightarrow \displaystyle\int \dfrac{(1 + \sqrt{x})^3}{\sqrt{x}}\,dx =$

$$\int (1 + \sqrt{x})^3 \left(\tfrac{1}{\sqrt{x}}\,dx\right) = \int u^3\,(2\,du) = 2\int u^3\,du = \tfrac{1}{2}u^4 + C = \tfrac{1}{2}(1 + \sqrt{x})^4 + C.$$

7 $u = x^{3/2}$, $du = \tfrac{3}{2}x^{1/2}\,dx$ or $\tfrac{2}{3}\,du = \sqrt{x}\,dx \Rightarrow \displaystyle\int \sqrt{x}\,\cos\sqrt{x^3}\,dx = \int \cos\sqrt{x^3}\,(\sqrt{x}\,dx) =$

$$\int \cos u\,(\tfrac{2}{3}\,du) = \tfrac{2}{3}\int \cos u\,du = \tfrac{2}{3}\sin u + C = \tfrac{2}{3}\sin\sqrt{x^3} + C.$$

9 Since we know how to integrate $\sqrt{u}$, let $u = 3x - 2$.

$u = 3x - 2$, $du = 3\,dx$ or $\tfrac{1}{3}\,du = dx \Rightarrow \int \sqrt{3x - 2}\,dx = \int \sqrt{u}\,(\tfrac{1}{3})\,du = \tfrac{1}{3}\int u^{1/2}\,du =$

$$\tfrac{1}{3}\left[\frac{u^{3/2}}{\frac{3}{2}}\right] + C = \tfrac{2}{9}u^{3/2} + C = \tfrac{2}{9}(3x - 2)^{3/2} + C.$$

15 Let $u = v^3 - 1$ since we know how to integrate $\sqrt{u}$ and the variable part of du,

which is v^2, appears in the integrand. $u = v^3 - 1$, $du = 3v^2\,dv$ or $\tfrac{1}{3}\,du = v^2\,dv \Rightarrow$

$\int v^2\sqrt{v^3 - 1}\,dv = \int \sqrt{u}\,(\tfrac{1}{3})\,du = \tfrac{1}{3}\int u^{1/2}\,du = \tfrac{2}{9}u^{3/2} + C = \tfrac{2}{9}(v^3 - 1)^{3/2} + C.$

Note that if v^2 had been v or v^3, this substitution would not have worked so easily.

17 Let $u = 1 - 2x^2$ since we know how to integrate $\dfrac{1}{\sqrt[3]{u}} = u^{-1/3}$

and the variable part of du, which is x, appears in the integrand.

$$u = 1 - 2x^2,\; du = -4x\,dx \text{ or } -\tfrac{1}{4}\,du = x\,dx \Rightarrow \int \frac{x}{\sqrt[3]{1 - 2x^2}}\,dx = \int \frac{1}{\sqrt[3]{u}}\left(-\tfrac{1}{4}\right)du =$$

$$-\tfrac{1}{4}\int u^{-1/3}\,du = -\tfrac{3}{8}u^{2/3} + C = -\tfrac{3}{8}(1 - 2x^2)^{2/3} + C.$$

19 $\int (s^2 + 1)^2\,ds = \int (s^4 + 2s^2 + 1)\,ds = \tfrac{1}{5}s^5 + \tfrac{2}{3}s^3 + s + C.$

$\boxed{23}$ Since we know how to integrate $\dfrac{1}{u^3} = u^{-3}$, let $u = t^2 - 4t + 3$.

$u = t^2 - 4t + 3,\ du = (2t - 4)\,dt = 2(t - 2)\,dt,$ or $\frac{1}{2}\,du = (t - 2)\,dt \Rightarrow$

$$\int \frac{t - 2}{(t^2 - 4t + 3)^3}\,dt = \int \frac{1}{u^3}\left(\tfrac{1}{2}\right)du = \tfrac{1}{2}\int u^{-3}\,du = -\tfrac{1}{4}u^{-2} + C =$$

$-\dfrac{1}{4(t^2 - 4t + 3)^2} + C.$ Note that if the term $(t - 2)$ had not been in the

numerator of the integrand, this substitution would not have worked so easily.

$\boxed{27}$ Since we know how to integrate $\cos u$, let $u = 4x - 3$.

$u = 4x - 3,\ du = 4\,dx$ or $\frac{1}{4}\,du = dx \Rightarrow$

$\int \cos(4x - 3)\,dx = \int \cos u \left(\tfrac{1}{4}\right)du = \tfrac{1}{4}\int \cos u\,du = \tfrac{1}{4}\sin u + C = \tfrac{1}{4}\sin(4x - 3) + C.$

$\boxed{31}$ $u = \sin 3x,\ du = 3\cos 3x\,dx$ or $\frac{1}{3}\,du = \cos 3x\,dx \Rightarrow$

$\int \cos 3x \sqrt[3]{\sin 3x}\,dx = \int \sqrt[3]{u}\left(\tfrac{1}{3}\right)du = \tfrac{1}{3}\int u^{1/3}\,du = \tfrac{1}{4}u^{4/3} + C = \tfrac{1}{4}(\sin 3x)^{4/3} + C.$

$\boxed{33}$ $\int (\sin x + \cos x)^2\,dx = \int (\sin^2 x + 2\sin x \cos x + \cos^2 x)\,dx$

$$= \int \left[(\sin^2 x + \cos^2 x) + (2\sin x \cos x)\right]dx = \int (1 + \sin 2x)\,dx.$$

Now let $u = 2x$ and $\frac{1}{2}\,du = dx.$

$$\text{Then, I} = \int (1 + \sin u)\tfrac{1}{2}\,du = \tfrac{1}{2}u - \tfrac{1}{2}\cos u + C = x - \tfrac{1}{2}\cos 2x + C.$$

$\boxed{35}$ $u = 1 + \cos x,\ -du = \sin x\,dx \Rightarrow$

$$\int \sin x(1 + \cos x)^2\,dx = -\int u^2\,du = -\tfrac{1}{3}u^3 + D = -\tfrac{1}{3}(1 + \cos x)^3 + D.$$

This is one form of an antiderivative for I. To obtain the form in the answer section,

we expand $(1 + \cos x)^3$. Thus, $\text{I} = -\tfrac{1}{3}(1 + 3\cos x + 3\cos^2 x + \cos^3 x) + D =$

$-\cos x - \cos^2 x - \tfrac{1}{3}\cos^3 x + (D - \tfrac{1}{3}) = -\cos x - \cos^2 x - \tfrac{1}{3}\cos^3 x + C,$

where $C = D - \tfrac{1}{3}$ is another constant.

$\boxed{39}$ $u = 1 - \sin t,\ -du = \cos t\,dt \Rightarrow$

$$\int \frac{\cos t}{(1 - \sin t)^2}\,dt = \int \frac{1}{u^2}(-1)\,du = -\int u^{-2}\,du = u^{-1} + C = \frac{1}{1 - \sin t} + C.$$

$\boxed{43}$ $u = \sec 3x,\ \frac{1}{3}\,du = \sec 3x \tan 3x\,dx \Rightarrow \int \sec^2 3x \tan 3x\,dx =$

$$\int \sec 3x(\sec 3x \tan 3x)\,dx = \tfrac{1}{3}\int u\,du = \tfrac{1}{6}u^2 + C_1 = \tfrac{1}{6}\sec^2 3x + C_1.$$

Alternate solution: $u = \tan 3x,\ \frac{1}{3}\,du = \sec^2 3x\,dx \Rightarrow \int \sec^2 3x \tan 3x\,dx =$

$$\int \tan 3x(\sec^2 3x)\,dx = \tfrac{1}{3}\int u\,du = \tfrac{1}{6}u^2 + C_2 = \tfrac{1}{6}\tan^2 3x + C_2.$$

These answers are both correct. Note that they differ *only* by a constant since

$\tfrac{1}{6}\sec^2 3x + C_1 = \tfrac{1}{6}(1 + \tan^2 3x) + C_1 = \tfrac{1}{6}\tan^2 3x + \tfrac{1}{6} + C_1.$ Thus, $C_2 = \tfrac{1}{6} + C_1.$

$\boxed{47}$ $u = x^2,\ \frac{1}{2}\,du = x\,dx \Rightarrow$

$$\int x \cot(x^2) \csc(x^2)\,dx = \tfrac{1}{2}\int \cot u \csc u\,du = -\tfrac{1}{2}\csc u + C = -\tfrac{1}{2}\csc(x^2) + C.$$

49   $f'(x) = \sqrt[3]{3x + 2}$ { given differential equation }

$\int f'(x)\, dx = \int (3x + 2)^{1/3}\, dx$ { integrate both sides }

$f(x) = \frac{1}{3}\int u^{1/3}\, du$ { substitute $u = 3x + 2$, $\frac{1}{3}\, du = dx$ }

$= \frac{1}{4}u^{4/3} + C$ { find antiderivative }

$= \frac{1}{4}(3x + 2)^{4/3} + C$ { substitute for u }

$f(2) = \frac{1}{4}(8)^{4/3} + C = 4 + C.$ $f(2) = 9 \Rightarrow 4 + C = 9 \Rightarrow C = 5.$

Thus, $f(x) = \frac{1}{4}(3x + 2)^{4/3} + 5.$

51   $f''(x) = 16\cos 2x - 3\sin x$ { given differential equation }

$\int f''(x)\, dx = \int (16\cos 2x - 3\sin x)\, dx$ { integrate both sides }

$f'(x) = 8\sin 2x + 3\cos x + C$ { find antiderivative }

$f'(0) = 8(0) + 3(1) + C = 3 + C.$ $f'(0) = 4 \Rightarrow 3 + C = 4 \Rightarrow C = 1.$

$f'(x) = 8\sin 2x + 3\cos x + 1$ { replace C with 1 }

$\int f'(x)\, dx = \int (8\sin 2x + 3\cos x + 1)\, dx$ { integrate both sides }

$f(x) = -4\cos 2x + 3\sin x + x + D$ { find antiderivative }

$f(0) = -4(1) + 3(0) + (0) + D = -4 + D.$ $f(0) = -2 \Rightarrow -4 + D = -2 \Rightarrow$

$D = 2.$ Thus, $f(x) = 3\sin x - 4\cos 2x + x + 2.$

53   (a) $u = x + 4$, $du = dx \Rightarrow$

$I = \int u^2\, du = \frac{1}{3}u^3 + C_1 = \frac{1}{3}(x + 4)^3 + C_1 =$

$\frac{1}{3}\left[x^3 + 3(x^2)(4)^1 + 3(x)^1(4)^2 + 4^3\right] + C_1 = \frac{1}{3}x^3 + 4x^2 + 16x + \frac{64}{3} + C_1.$

(b) $I = \int (x^2 + 8x + 16)\, dx = \frac{1}{3}x^3 + 4x^2 + 16x + C_2;$ $C_2 = C_1 + \frac{64}{3}.$

59   $dA/dt = 4000 + 2000\sin\left(\frac{\pi}{90}t\right)$ { given demand rate }

$\int \frac{dA}{dt}\, dt = \int \left[4000 + 2000\sin\left(\frac{\pi}{90}t\right)\right] dt$ { integrate both sides }

$A(t) = \int (4000 + 2000\sin u)\frac{90}{\pi}\, du$ { substitute $u = \frac{\pi}{90}t$, $\frac{90}{\pi}\, du = dt$ }

$= \frac{90}{\pi}(4000u - 2000\cos u) + C$ { find antiderivative }

$= 4000t - \frac{180{,}000}{\pi}\cos\left(\frac{\pi}{90}t\right) + C$ { substitute for u }

The total water consumption during 90 days of summer can be estimated by finding the difference between the ninetieth day and the present time. $A(90) - A(0) =$

$(360{,}000 + \frac{180{,}000}{\pi} + C) - (-\frac{180{,}000}{\pi} + C) = 360{,}000 + \frac{360{,}000}{\pi} \approx 474{,}592 \text{ ft}^3.$

$\boxed{61}$ (a) $dV/dt = a\sin bt$. If the maximum of dV/dt is 0.6, then the amplitude $a = 0.6$.

Since the period is 5 seconds, $b = \frac{2\pi}{5}$. Thus, we have $dV/dt = 0.6\sin\left(\frac{2\pi}{5}t\right)$.

(b) $\int \frac{dV}{dt}\,dt = \int 0.6\sin\left(\frac{2\pi}{5}t\right)dt \Rightarrow V(t) = \int \frac{3}{5}\sin u\left(\frac{5}{2\pi}\right)du \ \{\, u = \frac{2\pi}{5}t,\ \frac{5}{2\pi}\,du = dt\,\} =$

$\frac{3}{2\pi}(-\cos u) + C \Rightarrow V(t) = -\frac{3}{2\pi}\cos\left(\frac{2\pi}{5}t\right) + C$.   Since the inhaling process

requires $\frac{5}{2}$ sec, we can assume that at $t = 0$ the volume starts to increase. A

person is inhaling while $dV/dt > 0$ from $t = 0$ to $t = \frac{5}{2}$. Thus, the volume

inhaled is given by $V(\frac{5}{2}) - V(0) = (\frac{3}{2\pi} + C) - (-\frac{3}{2\pi} + C) = \frac{3}{\pi} \approx 0.95$ L.

$\boxed{63}$ First way: $u = \sin x,\ du = \cos x\,dx \Rightarrow$

$\int \sin x \cos x\,dx = \int u\,du = \frac{1}{2}u^2 + C = \frac{1}{2}\sin^2 x + C.$

Second way: $u = \cos x,\ -du = \sin x\,dx \Rightarrow$

$\int \sin x \cos x\,dx = -\int u\,du = -\frac{1}{2}u^2 + D = -\frac{1}{2}\cos^2 x + D.$

Third way: $2\sin x \cos x = \sin 2x \Rightarrow \sin x \cos x = \frac{1}{2}\sin 2x \Rightarrow$

$\int \sin x \cos x\,dx = \frac{1}{2}\int \sin 2x\,dx = -\frac{1}{4}\cos 2x + E.$

The three answers are all antiderivatives of $\sin x \cos x$. Note that $\frac{1}{2}\sin^2 x = \frac{1}{2}\left(\frac{1 - \cos 2x}{2}\right) = \frac{1}{4} - \frac{1}{4}\cos 2x$. Comparing the first answer and the third answer, we

have $\frac{1}{4} + C = E$.   Also note that $-\frac{1}{2}\cos^2 x = -\frac{1}{2}\left(\frac{1 + \cos 2x}{2}\right) = -\frac{1}{4} - \frac{1}{4}\cos 2x$.

Comparing the second answer and the third answer, we have $-\frac{1}{4} + D = E$. Thus,

the three answers simply differ by a constant.

Exercises 5.3

$\boxed{1}$ $\displaystyle\sum_{j=1}^{4} (j^2 + 1) = (1^2 + 1) + (2^2 + 1) + (3^2 + 1) + (4^2 + 1) =$

$2 + 5 + 10 + 17 = 34$

$\boxed{3}$ $\displaystyle\sum_{k=0}^{5} k(k - 1)$

$= 0(0 - 1) + 1(1 - 1) + 2(2 - 1) + 3(3 - 1) + 4(4 - 1) + 5(5 - 1)$

$= 0 + 0 + 2 + 6 + 12 + 20 = 40$

$\boxed{7}$ $\displaystyle\sum_{i=1}^{50} 10 = 50(10) = 500$, by (5.10).

$\boxed{11}$ $\displaystyle\sum_{k=1}^{n} (k^3 + 2k^2 - k + 4)$

$$= \sum_{k=1}^{n} k^3 + 2 \sum_{k=1}^{n} k^2 - \sum_{k=1}^{n} k + \sum_{k=1}^{n} 4 \qquad \{\,\text{apply } (5.11)\,\}$$

$$= \left[\frac{n(n+1)}{2}\right]^2 + 2 \cdot \frac{n(n+1)(2n+1)}{6} - \frac{n(n+1)}{2} + 4n \quad \{\,\text{apply } (5.12) \ \& \ (5.10)\,\}$$

$$= \tfrac{1}{4}\left[n^2(n^2 + 2n + 1)\right] + 2 \cdot \tfrac{1}{6}\left[(n^2 + n)(2n+1)\right] - \tfrac{1}{2}(n^2 + n) + 4n \quad \{\,\text{simplify}\,\}$$

$$= (\tfrac{1}{4}n^4 + \tfrac{1}{2}n^3 + \tfrac{1}{4}n^2) + 2(\tfrac{1}{3}n^3 + \tfrac{1}{2}n^2 + \tfrac{1}{6}n) - \tfrac{1}{2}n^2 - \tfrac{1}{2}n + 4n \qquad \{\,\text{simplify}\,\}$$

$$= \tfrac{1}{4}n^4 + \tfrac{7}{6}n^3 + \tfrac{3}{4}n^2 + \tfrac{23}{6}n \qquad \{\,\text{combine terms}\,\}$$

$$= \tfrac{1}{12}n(3n^3 + 14n^2 + 9n + 46) \qquad \{\,\text{factor out } \tfrac{1}{12}n\,\}$$

$\boxed{13}$ $1 + 5 + 9 + 13 + 17$ • Each term is found by adding 4 to the previous term, or, equivalently, by adding $4(k-1)$ to the first term. The first term is 1 so each term is given by $1 + 4(k-1) = 4k - 3$. The summation can be written as $\displaystyle\sum_{k=1}^{5} (4k - 3)$. Another summation describing this pattern is $\displaystyle\sum_{j=0}^{4} (4j + 1)$, found by replacing k with $j + 1$.

Alternate Solution: Since the difference in terms is 4, the coefficient of the summation variable will be 4. Consider the term $4k + a$. When $k = 1$, $4k + a$ should be 1, $\therefore a = -3$.

$\boxed{15}$ $\tfrac{1}{2} + \tfrac{2}{5} + \tfrac{3}{8} + \tfrac{4}{11}$ • The numerator of the kth term is k. The denominator of the kth term is found by adding 3 to the previous denominator, or, equivalently, by adding $3(k-1)$ to the first denominator. The first denominator is 2 so the kth denominator is given by $2 + 3(k-1) = 3k - 1$. The summation can be written as $\displaystyle\sum_{k=1}^{4} \frac{k}{3k-1}$. Another summation describing this pattern is $\displaystyle\sum_{k=0}^{3} \frac{1+k}{2+3k}$.

Alternate Solution: Let the numerator be k and the denominator $3k + a$.

When $k = 1$, $3k + a$ should be 2, $\therefore a = -1$.

Note: If a function f is decreasing on an interval, use the right-hand endpoint for A_{IP} and the left-hand endpoint for A_{CP}. Reverse this choice if f is increasing.

$\boxed{19}$ (a) $f(x) = 3 - x \Rightarrow f'(x) = -1 < 0$ on $[-2, 2] \Rightarrow f$ is $\downarrow$.

$n = \dfrac{b - a}{\Delta x} = \dfrac{2 - (-2)}{1} = 4$. The minimum of f will be at the right-hand endpoint of the intervals $[-2, -1]$, $[-1, 0]$, $[0, 1]$, and $[1, 2]$.

$$A_{IP} = \sum_{k=1}^{4} f(u_k)\,\Delta x = \sum_{k=1}^{4} f(u_k) \ \{\text{since } \Delta x = 1\}$$

$$= f(-1) + f(0) + f(1) + f(2) = 4 + 3 + 2 + 1 = 10$$

(b) The maximum of f will be at the left-hand endpoint of the intervals in part (a).

$$A_{CP} = \sum_{k=1}^{4} f(v_k)\,\Delta x = \sum_{k=1}^{4} f(v_k) \ \{\text{since } \Delta x = 1\}$$

$$= f(-2) + f(-1) + f(0) + f(1) = 5 + 4 + 3 + 2 = 14$$

Also note that $A_{CP} = A_{IP} + f(-2)\,\Delta x - f(2)\,\Delta x = A_{IP} + f(-2) - f(2) = 10 + 5 - 1 = 14$. This last technique can be used on any increasing or decreasing function f. If there are several terms to add together, it can make finding A_{CP} much easier.

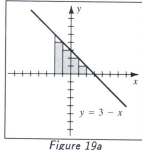

Figure 19a

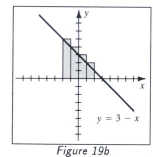

Figure 19b

$\boxed{21}$ (a) $f(x) = x^2 + 1 \Rightarrow f'(x) = 2x > 0$ on $[1, 3] \Rightarrow f$ is $\uparrow$. $n = \dfrac{b - a}{\Delta x} = \dfrac{3 - 1}{1/2} = 4$.

The minimum of f will be at the left-hand endpoint of the intervals $[1, \frac{3}{2}]$, $[\frac{3}{2}, 2]$, $[2, \frac{5}{2}]$, and $[\frac{5}{2}, 3]$. See *Figure 21a*.

$$A_{IP} = \sum_{k=1}^{4} f(u_k)\,\Delta x = \tfrac{1}{2}\sum_{k=1}^{4} f(u_k)$$

$$= \tfrac{1}{2}\left[f(1) + f(\tfrac{3}{2}) + f(2) + f(\tfrac{5}{2})\right] = \tfrac{1}{2}\left[2 + \tfrac{13}{4} + 5 + \tfrac{29}{4}\right] = \tfrac{35}{4}$$

(b) The maximum of f will be at the right-hand endpoint of the intervals in part (a).

$$A_{CP} = \sum_{k=1}^{4} f(v_k)\,\Delta x = A_{IP} + f(3)\,\Delta x - f(1)\,\Delta x = \tfrac{35}{4} + 10(\tfrac{1}{2}) - 2(\tfrac{1}{2}) = \tfrac{51}{4}$$

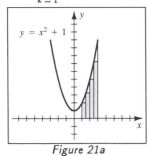

Figure 21a

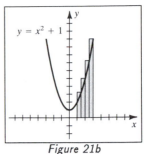

Figure 21b

$\boxed{23}$ (a) $f(x) = \sqrt{\sin x} \Rightarrow f'(x) = \dfrac{\cos x}{2\sqrt{\sin x}} > 0$ on $[0, 1.5] \Rightarrow f$ is $\uparrow$.

$n = \dfrac{b-a}{\Delta x} = \dfrac{1.5 - 0}{0.15} = 10$. The minimum of f will be at the left-hand

endpoint of the intervals $[0, 0.15]$, $[0.15, 0.30]$, ..., $[1.20, 1.35]$, and $[1.35, 1.5]$.

$$A_{IP} = \sum_{k=1}^{10} f(u_k)\,\Delta x$$

$$= \Big[f(0) + f(0.15) + f(0.3) + \cdots + f(1.2) + f(1.35)\Big]\Delta x$$

$$\{\text{be sure to use radians}\}$$

$$= \Big[0 + 0.38657 + 0.54362 + \cdots + 0.96542 + 0.98779\Big](0.15)$$

$$\approx (6.9364)(0.15) \approx 1.0405$$

(b) The maximum of f will be at the right-hand endpoint of the intervals in part (a).

$$A_{CP} = \sum_{k=1}^{10} f(v_k)\,\Delta x = A_{IP} + f(1.5)\,\Delta x - f(0)\,\Delta x$$

$$\approx 1.0405 + (0.9987)(0.15) - (0)(0.15) \approx 1.1903$$

$\boxed{25}$ (a) $b = 4 \Rightarrow \Delta x = \frac{4}{n}$ and $x_k = k\Delta x = \frac{4k}{n}$. Since f is $\uparrow$, f takes on its minimum value at the left-hand endpoint of $[x_{k-1}, x_k]$ and its maximum value at the right-hand endpoint. Thus, $u_k = x_{k-1} = \frac{4(k-1)}{n}$ and $v_k = x_k = \frac{4k}{n}$.

$$
\begin{aligned}
A_{\text{IP}} &= \sum_{k=1}^{n} f(u_k)\,\Delta x && \{\text{definition of } A_{\text{IP}}\} \\[2mm]
&= \sum_{k=1}^{n} (2u_k + 3)\,\Delta x && \{\text{definition of } f\} \\[2mm]
&= \sum_{k=1}^{n} \left[2 \cdot \frac{4(k-1)}{n} + 3 \right]\frac{4}{n} && \{\text{definition of } u_k \text{ and } \Delta x\} \\[2mm]
&= \frac{4}{n} \sum_{k=1}^{n} \left(\frac{8k}{n} - \frac{8}{n} + 3 \right) && \{\tfrac{4}{n} \text{ is a constant, } k \text{ is the variable}\} \\[2mm]
&= \frac{4}{n} \left(\sum_{k=1}^{n} \frac{8k}{n} - \sum_{k=1}^{n} \frac{8}{n} + \sum_{k=1}^{n} 3 \right) && \{\text{apply (5.11)(i) and (5.11)(iii)}\} \\[2mm]
&= \frac{32}{n^2} \sum_{k=1}^{n} k - \frac{32}{n^2} \sum_{k=1}^{n} 1 + \frac{12}{n} \sum_{k=1}^{n} 1 && \{\text{apply (5.11)(ii)}\} \\[2mm]
&= \frac{32}{n^2} \cdot \frac{(n)(n+1)}{2} - \frac{32}{n^2} \cdot n + \frac{12}{n} \cdot n && \{\text{apply (5.10) and (5.12)}\} \\[2mm]
&= \frac{16(n)(n+1)}{n^2} - \frac{32}{n} + 12. && \{\text{simplify}\}
\end{aligned}
$$

As in Chapter 2, we encourage you to intuitively evaluate the limit. In the first term, the highest degree on n in the numerator is 2, the same as in the denominator. Thus, the limit of the first term is the ratio of leading coefficients, $\frac{16}{1}$. $\lim\limits_{n \to \infty} A_{\text{IP}} = 16 - 0 + 12 = 28$.

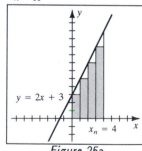

Figure 25a

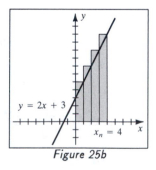

Figure 25b

(b) $\begin{aligned}[t]
A_{\text{CP}} &= \sum_{k=1}^{n} f(v_k)\,\Delta x && \{\text{definition of } A_{\text{CP}}\} \\[2mm]
&= \sum_{k=1}^{n} (2v_k + 3)\,\Delta x && \{\text{definition of } f\} \\[2mm]
&= \sum_{k=1}^{n} \left[2 \cdot \frac{4k}{n} + 3 \right]\frac{4}{n} && \{\text{definition of } v_k \text{ and } \Delta x\}
\end{aligned}$

$$= \frac{4}{n}\left(\sum_{k=1}^{n} \frac{8k}{n} + \sum_{k=1}^{n} 3 \right) \qquad \{ \frac{4}{n} \text{ is a constant, } k \text{ is the variable} \}$$

$$= \frac{32}{n^2} \sum_{k=1}^{n} k + \frac{12}{n} \sum_{k=1}^{n} 1 \qquad \{ \text{apply (5.11)} \}$$

$$= \frac{32}{n^2} \cdot \frac{n(n+1)}{2} + \frac{12}{n} \cdot n \qquad \{ \text{apply (5.10) and (5.12)} \}$$

$$= \frac{16n(n+1)}{n^2} + 12 \qquad \{ \text{simplify} \}$$

$$\lim_{n \to \infty} A_{CP} = 16 + 12 = 28.$$

$\boxed{27}$ (a) $b = 3 \Rightarrow \Delta x = \frac{3}{n}$ and $x_k = k\Delta x = \frac{3k}{n}$. Since f is $\downarrow$, f takes on its maximum value at the left-hand endpoint of $[x_{k-1}, x_k]$ and its minimum value at the right-hand endpoint. Thus, $u_k = x_k = \frac{3k}{n}$ and $v_k = x_{k-1} = \frac{3(k-1)}{n}$.

$$A_{IP} = \sum_{k=1}^{n} f(u_k) \Delta x = \sum_{k=1}^{n} \left(9 - \frac{9k^2}{n^2} \right)\left(\frac{3}{n} \right)$$

$$= \frac{27}{n}\left[\sum_{k=1}^{n} 1 - \sum_{k=1}^{n} \frac{k^2}{n^2} \right]$$

$$= \frac{27}{n} \sum_{k=1}^{n} 1 - \frac{27}{n^3} \sum_{k=1}^{n} k^2$$

$$= \frac{27n}{n} - \frac{27n(n+1)(2n+1)}{6n^3}. \quad \lim_{n \to \infty} A_{IP} = 27 - 9 = 18.$$

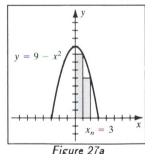

Figure 27a

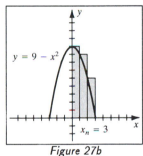

Figure 27b

(b) $A_{CP} = \sum_{k=1}^{n} f(v_k) \Delta x = \sum_{k=1}^{n} \left(9 - \frac{9(k-1)^2}{n^2} \right)\left(\frac{3}{n} \right)$

$$= \frac{27}{n}\left[\sum_{k=1}^{n} 1 - \sum_{k=1}^{n} \frac{k^2}{n^2} + \sum_{k=1}^{n} \frac{2k}{n^2} - \sum_{k=1}^{n} \frac{1}{n^2} \right]$$

$$= \frac{27}{n} \sum_{k=1}^{n} 1 - \frac{27}{n^3} \sum_{k=1}^{n} k^2 + \frac{54}{n^3} \sum_{k=1}^{n} k - \frac{27}{n^3} \sum_{k=1}^{n} 1$$

$$= \frac{27n}{n} - \frac{27n(n+1)(2n+1)}{6n^3} + \frac{54n(n+1)}{2n^3} - \frac{27n}{n^3}.$$

$$\lim_{n \to \infty} A_{CP} = 27 - 9 + 0 - 0 = 18.$$

$\boxed{31}$ Let A_k denote the area under the graph of $f(x) = x^3$ from 0 to k, $k > 0$.

From Example 7, $A_b = \frac{b^4}{4}$ and so $A_k = \frac{k^4}{4}$.

(a) $A = A_3 - A_1 = \frac{3^4}{4} - \frac{1^4}{4} = 20$

(b) $A = A_b - A_a = \frac{b^4}{4} - \frac{a^4}{4} = \frac{1}{4}(b^4 - a^4)$

Exercises 5.4

$\boxed{1}$ (a) $\Delta x_1 = (1.1 - 0) = 1.1$, $\Delta x_2 = (2.6 - 1.1) = 1.5$, $\Delta x_3 = (3.7 - 2.6) = 1.1$,

$$\Delta x_4 = (4.1 - 3.7) = 0.4, \; \Delta x_5 = (5 - 4.1) = 0.9.$$

(b) $\|P\| = $ maximum $\Delta x_k = \Delta x_2 = 1.5$.

$\boxed{5}$ Since there are four numbers in the partition, $n = 4 - 1 = 3$.

$$R_P = \sum_{k=1}^{3} f(w_k)\,\Delta x_k = \sum_{k=1}^{3} (2w_k + 3)\,\Delta x_k =$$

(a) $f(3) \cdot 2 + f(4) \cdot 1 + f(5) \cdot 1 = 18 + 11 + 13 = 42$

(b) $f(1) \cdot 2 + f(3) \cdot 1 + f(4) \cdot 1 = 10 + 9 + 11 = 30$

(c) $f(2) \cdot 2 + f(\frac{7}{2}) \cdot 1 + f(\frac{9}{2}) \cdot 1 = 14 + 10 + 12 = 36$

$\boxed{7}$ Since the partition is regular, $\Delta x_k = 1$, $\forall k$, and the intervals are $[0,\,1]$, $[1,\,2]$, $[2,\,3]$, $[3,\,4]$, $[4,\,5]$, and $[5,\,6]$.

$$R_P = \sum_{k=1}^{6} f(w_k)\,\Delta x_k$$

$$= f(\tfrac{1}{2}) + f(\tfrac{3}{2}) + f(\tfrac{5}{2}) + f(\tfrac{7}{2}) + f(\tfrac{9}{2}) + f(\tfrac{11}{2})$$

$$= \tfrac{63}{8} + \tfrac{55}{8} + \tfrac{39}{8} + \tfrac{15}{8} - \tfrac{17}{8} - \tfrac{57}{8} = \tfrac{49}{4}.$$

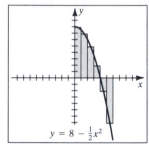

$y = 8 - \frac{1}{2}x^2$

Figure 7

$\boxed{13}$ $f(w_k) = 3w_k^2 - 2w_k + 5$, $a = -1$, and $b = 2$.

By (5.16), $\displaystyle\lim_{\|P\| \to 0} \sum_{k=1}^{n} (3w_k^2 - 2w_k + 5)\,\Delta x_k = \int_a^b f(x)\,dx = \int_{-1}^{2} (3x^2 - 2x + 5)\,dx.$

$\boxed{17}$ $\displaystyle\int_4^1 \sqrt{x}\,dx = -\int_1^4 \sqrt{x}\,dx = -\frac{14}{3}$, by (5.17).

$\boxed{21}$ $\displaystyle\int_4^4 \sqrt{x}\,dx + \int_4^1 \sqrt{x}\,dx = 0 + -\int_1^4 \sqrt{x}\,dx = -\frac{14}{3}$, by (5.18) and (5.17).

$\boxed{23}$ $5x + 4y = 20 \Leftrightarrow y = -\frac{5}{4}x + 5$. The integral that represents the area under the line from $x = 0$ to $x = 4$ is $\displaystyle\int_0^4 \left(-\frac{5}{4}x + 5\right) dx.$

$\boxed{25}$ $(x - 2)^2 + y^2 = 9 \; (y \geq 0) \Leftrightarrow y = \sqrt{9 - (x - 2)^2}$. The integral that represents the area under the semicircle from $x = -1$ to $x = 5$ is $\displaystyle\int_{-1}^{5} \sqrt{9 - (x - 2)^2}\,dx.$

$\boxed{29}$ First, graph the line $y = f(x) = 2x + 6$ between $x = -3$ and $x = 2$. Note that $f(-3) = 0$ and $f(2) = 10$. The area bounded by $x = -3$, $x = 2$, the x-axis, and $f(x) = 2x + 6$ is that of a triangle with base $2 - (-3) = 5$ and height $f(2) = 10$. Hence, the value is $\frac{1}{2}bh = \frac{1}{2}(5)(10) = 25$.

$\boxed{31}$ First, graph the "V-shape" $y = f(x) = |x - 1|$ between $x = 0$ and $x = 3$. Note that $f(0) = 1$, $f(1) = 0$, and $f(3) = 2$. The area bounded by $x = 0$, $x = 3$, the x-axis, and $f(x) = |x - 1|$ is that of two triangles. Value $= \frac{1}{2}(1)(1) + \frac{1}{2}(2)(2) = 2.5$.

$\boxed{33}$ The area bounded by $x = 0$, $x = 3$, the x-axis, and $f(x) = \sqrt{9 - x^2}$ is that of the first quadrant portion of a circle { quarter circle } centered at the origin with radius 3.

$$\text{Value} = \frac{1}{4}\pi r^2 = \frac{1}{4}\pi(3)^2 = \frac{9\pi}{4}.$$

$\boxed{35}$ $y = 3 + \sqrt{4 - x^2} \Rightarrow y - 3 = \sqrt{4 - x^2} \Rightarrow (y - 3)^2 = 4 - x^2 \Rightarrow$
$x^2 + (y - 3)^2 = 4$, where $y \geq 3$. The area bounded by $x = -2$, $x = 2$, the x-axis, and $f(x) = 3 + \sqrt{4 - x^2}$ is that of the top half of a circle centered at $(0, 3)$ with radius 2 and a rectangle below the circle of height 3 and base 4.

$$\text{Value} = \frac{1}{2}\pi(2)^2 + (3)(4) = 12 + 2\pi.$$

Exercises 5.5

$\boxed{1}$ By (5.21), $I = 5\big[4 - (-2)\big] = 30$.

$\boxed{7}$ $I = 3\int_1^4 x^2\, dx + \int_1^4 5\, dx = 3(21) + 5(4 - 1) = 78$

$\boxed{11}$ We want to use Corollary (5.27) with $f(x) = 3x^2 + 4$ and $g(x) = 2x^2 + 5$.

To use this corollary, we must have f and g integrable and $f \geq g$ on $[1, 2]$. Since f and g are polynomials, they are continuous. By (5.20), the functions are integrable. $f(x) \geq g(x) \Leftrightarrow 3x^2 + 4 \geq 2x^2 + 5 \Leftrightarrow x^2 \geq 1 \Leftrightarrow x \in (-\infty, -1] \cup [1, \infty)$. Since the interval $[1, 2]$ is in this set of values, we may apply (5.27), and hence, $\int_1^2 (3x^2 + 4)\, dx \geq \int_1^2 (2x^2 + 5)\, dx.$

$\boxed{13}$ The graph of $y = x^2 - 6x + 8 = (x - 2)(x - 4)$ is a parabola opening upwards with x-intercepts at 2 and 4. Thus, $x^2 - 6x + 8 \leq 0$ on $[2, 4]$ so $-(x^2 - 6x + 8) \geq 0$ on $[2, 4]$. By (5.26), $\int_2^4 -(x^2 - 6x + 8)\, dx = -\int_2^4 (x^2 - 6x + 8)\, dx \geq 0$, or, equivalently, $\int_2^4 (x^2 - 6x + 8)\, dx \leq 0$.

$\boxed{15}$ $-1 \leq \sin x \leq 1 \Rightarrow 0 \leq (1 + \sin x) \leq 2$. By (5.26), $\int_0^{2\pi} (1 + \sin x)\, dx \geq 0$.

$\boxed{17}$ $\int_5^1 f(x)\, dx + \int_{-3}^5 f(x)\, dx = \int_{-3}^5 f(x)\, dx + \int_5^1 f(x)\, dx = \int_{-3}^1 f(x)\, dx,$

$$\text{by (5.25) with } a = -3, \ b = 1, \text{ and } c = 5.$$

$\boxed{21}$ $\int_c^{c+h} f(x)\,dx - \int_c^h f(x)\,dx = \int_h^{c+h} f(x)\,dx + \int_h^c f(x)\,dx =$

$$\int_h^c f(x)\,dx + \int_c^{c+h} f(x)\,dx = \int_h^{c+h} f(x)\,dx, \text{ by (5.25) with } a = h \text{ and } b = c + h.$$

$\boxed{23}$ (a) $\int_0^3 3x^2\,dx = f(z)(3 - 0) \Rightarrow 27 = (3z^2)(3) \Rightarrow z^2 = 3 \Rightarrow z = \sqrt{3}$,

$$\text{since } -\sqrt{3} \notin (0, 3).$$

(b) By (5.29), $f_{av} = \frac{1}{3-0}\int_0^3 f(x)\,dx = \frac{1}{3-0}(27) = 9.$

$\boxed{27}$ (a) $\int_{-1}^8 3\sqrt{x+1}\,dx = f(z)\big[8 - (-1)\big] \Rightarrow 54 = 3\sqrt{z+1}\,(9) \Rightarrow \sqrt{z+1} = 2 \Rightarrow$

$$z + 1 = 4 \Rightarrow z = 3.$$

(b) $f_{av} = \frac{1}{8-(-1)}\int_{-1}^8 f(x)\,dx = \frac{1}{8-(-1)}(54) = 6.$

$\boxed{31}$ $\int_{-2}^3 (8x^3 + 3x - 1)\,dx = f(z)\big[3 - (-2)\big] \Rightarrow (8z^3 + 3z - 1)(5) = 132.5 \Leftrightarrow$

$40z^3 + 15z - 137.5 = 0.$ Let $f(x) = 40x^3 + 15x - 137.5$ and $f'(x) = 120x^2 + 15.$

The graphs of $y = 40x^3$ and $y = 137.5 - 15x$ intersect between 1 and 2, so we will

use $x_1 = 1.5$ as our initial guess, and $x_{n+1} = x_n - \dfrac{40x_n^3 + 15x_n - 137.5}{120x_n^2 + 15}$ for our

iteration formula for Newton's method. If $x_1 = 1.5$, $x_2 = 1.430$, $x_3 = 1.426$, and

$x_4 = 1.426.$

$\boxed{33}$ $\quad\int_a^b \big[cf(x) + dg(x)\big]\,dx = \int_a^b cf(x)\,dx + \int_a^b dg(x)\,dx \qquad \{\text{by } (5.23)(i)\}$

$$= c\int_a^b f(x)\,dx + d\int_a^b g(x)\,dx \qquad \{\text{by } (5.22)\}$$

$\boxed{\text{Exercises 5.6}}$

$\boxed{1}$ $\int_1^4 (x^2 - 4x - 3)\,dx = \big[\tfrac{1}{3}x^3 - 2x^2 - 3x\big]_1^4 =$

$$(\tfrac{64}{3} - 32 - 12) - (\tfrac{1}{3} - 2 - 3) = -18$$

$\boxed{7}$ $\int_1^2 \dfrac{5}{x^6}\,dx = 5\int_1^2 x^{-6}\,dx = 5\big[-\tfrac{1}{5}x^{-5}\big]_1^2 = -(\tfrac{1}{32} - 1) = \tfrac{31}{32}$

$\boxed{9}$ $\int_4^9 \dfrac{t-3}{\sqrt{t}}\,dt = \int_4^9 \Big(\dfrac{t}{\sqrt{t}} - \dfrac{3}{\sqrt{t}}\Big)\,dt = \int_4^9 (t^{1/2} - 3t^{-1/2})\,dt =$

$$\big[\tfrac{2}{3}t^{3/2} - 6t^{1/2}\big]_4^9 = \big[\tfrac{2}{3}(27) - 6(3)\big] - \big[\tfrac{2}{3}(8) - 6(2)\big] = 0 - (\tfrac{16}{3} - 12) = \tfrac{20}{3}$$

[11] Whenever the integration interval is of the form $[-a, a]$, it is advisable to examine the integrand to see if it (or part of it) is an odd or even function. See Example 7, part (c) for an example of how to work with an integrand that contains both odd and even portions. Let $f(s) = \sqrt[3]{s^2} + 2$. $f(-s) = \sqrt[3]{(-s)^2} + 2 = f(s) \Rightarrow f$ is an even function. By (5.34)(i),

$$\int_{-8}^{8} (\sqrt[3]{s^2} + 2)\, ds = 2\int_0^8 (s^{2/3} + 2)\, ds = 2\left[\tfrac{3}{5}s^{5/3} + 2s\right]_0^8 = 2\left[\tfrac{3}{5}(2^5) + 16\right] = \tfrac{352}{5}.$$

[15] Since $x \neq 1$, $\displaystyle\int_3^2 \frac{x^2 - 1}{x - 1}\, dx = \int_3^2 \frac{(x-1)(x+1)}{x-1}\, dx = \int_3^2 (x+1)\, dx =$

$$\left[\tfrac{1}{2}x^2 + x\right]_3^2 = 4 - \tfrac{15}{2} = -\tfrac{7}{2}.$$

[21] Before we can integrate, we must remove the absolute value symbol.

Since $|x - 4| = -(x - 4)$ if $x < 4$ and $|x - 4| = (x - 4)$ if $x \geq 4$, we have

$$\int_{-3}^{6} |x - 4|\, dx = \int_{-3}^{4} -(x - 4)\, dx + \int_4^6 (x - 4)\, dx =$$

$$\left[4x - \tfrac{1}{2}x^2\right]_{-3}^{4} + \left[\tfrac{1}{2}x^2 - 4x\right]_4^6 = \left[8 - (-\tfrac{33}{2})\right] + \left[-6 - (-8)\right] = \tfrac{53}{2}.$$

[25] Let $f(v) = (v^2 - 1)^3 v$. $f(-v) = \left[(-v)^2 - 1\right]^3 (-v) = -(v^2 - 1)^3 v = -f(v) \Rightarrow$

f is an odd function. By (5.34)(ii), $\displaystyle\int_{-1}^{1} (v^2 - 1)^3 v\, dv = 0.$

[27] $u = 3 - 2x \Rightarrow -\tfrac{1}{2}\, du = dx$. $x = 0, 1 \Rightarrow u = 3, 1$. Thus, $\displaystyle\int_0^1 \frac{1}{(3 - 2x)^2}\, dx =$

$-\tfrac{1}{2}\displaystyle\int_3^1 u^{-2}\, du = -\tfrac{1}{2}\left[-\tfrac{1}{u}\right]_3^1 = \tfrac{1}{2}(1 - \tfrac{1}{3}) = \tfrac{1}{3}$. A common mistake is to make the change of variables from x to u and then substitute the original expression in x back in for the expression in u. But remember, this is what you did for *indefinite* integrals. Once you've changed the integrand and the limits in terms of the substitution variable u for a *definite* integral, there is no need to use the expression in x again.

[29] $u = \sqrt{x} + 1 \Rightarrow 2\, du = \dfrac{dx}{\sqrt{x}}$. $x = 1, 4 \Rightarrow u = 2, 3$.

Thus, $\displaystyle\int_1^4 \frac{1}{\sqrt{x}\,(\sqrt{x} + 1)^3}\, dx = 2\int_2^3 u^{-3}\, du = 2\left[-\tfrac{1}{2}u^{-2}\right]_2^3 = -(\tfrac{1}{9} - \tfrac{1}{4}) = \tfrac{5}{36}.$

[33] $\displaystyle\int_{\pi/4}^{\pi/3} (4\sin 2\theta + 6\cos 3\theta)\, d\theta = \left[-2\cos 2\theta + 2\sin 3\theta\right]_{\pi/4}^{\pi/3} =$

$$(1 + 0) - (0 + \sqrt{2}) = 1 - \sqrt{2} \approx -0.41$$

[35] Let $f(x) = x + \sin 5x$. $f(-x) = -x + \sin(-5x) = -x - \sin 5x = -f(x) \Rightarrow$

f is an odd function. By (5.34)(ii), $\displaystyle\int_{-\pi/6}^{\pi/6} (x + \sin 5x)\, dx = 0.$

$\boxed{37}$ (a) $u = x^2 + 9 \Rightarrow \frac{1}{2} du = x\, dx.\quad x = 0, 4 \Rightarrow u = 9, 25.$

Thus, $\displaystyle\int_0^4 \frac{x}{\sqrt{x^2 + 9}}\, dx = \frac{1}{2}\int_9^{25} u^{-1/2}\, du = \frac{1}{2}\Big[2u^{1/2}\Big]_9^{25} = 5 - 3 = 2.$

Let $f(x) = \dfrac{x}{\sqrt{x^2 + 9}}.\quad \displaystyle\int_0^4 \frac{x}{\sqrt{x^2 + 9}}\, dx = f(z)(4 - 0) \Rightarrow 2 = \dfrac{z}{\sqrt{z^2 + 9}}(4) \Rightarrow$

$2z = \sqrt{z^2 + 9} \Rightarrow 4z^2 = z^2 + 9 \Rightarrow 3z^2 = 9 \Rightarrow z = \sqrt{3}$, since $-\sqrt{3} \notin (0, 4).$

(b) $f_{av} = \dfrac{1}{4 - 0}\displaystyle\int_0^4 f(x)\, dx = \dfrac{1}{4 - 0}(2) = \frac{1}{2}.$

$\boxed{41}$ To evaluate this expression, we would find an antiderivative of $\sqrt{x^2 + 16}$, and then evaluate this antiderivative using the values $x = 0$ and $x = 3$. This would yield a constant. Differentiating the constant would give us zero.

$$\text{Hence, } D_x \int_0^3 \sqrt{x^2 + 16}\, dx = 0.$$

$\boxed{43}$ Let $f(t) = \dfrac{1}{t + 1}$ and $c = 0$ in (5.35). Then, $D_x \displaystyle\int_0^x f(t)\, dt = f(x) = \dfrac{1}{x + 1}.$

$\boxed{45}$ By (5.29), the average value of a on $[t_1, t_2] = a_{av} = \dfrac{1}{t_2 - t_1}\displaystyle\int_{t_1}^{t_2} a(t)\, dt =$

$$\frac{1}{t_2 - t_1}\Big[v(t)\Big]_{t_1}^{t_2} = \frac{v(t_2) - v(t_1)}{t_2 - t_1}, \text{ which is the average acceleration.}$$

$\boxed{47}$ (a) $\quad v_{av} = \dfrac{1}{d - 0}\displaystyle\int_0^d v\, dy \qquad\qquad \{\text{by (5.29)}\}$

$\qquad\qquad = \dfrac{1}{d}\displaystyle\int_0^d c(d - y)^{1/6}\, dy \qquad \{\text{definition of } v\}$

$\qquad\qquad = \dfrac{c}{d}\Big[-\frac{6}{7}(u)^{7/6}\Big]_d^0 \qquad\quad \{\text{substitute } u = d - y, -du = dy\}$

$\qquad\qquad = -\dfrac{6c}{7d}(0 - d^{7/6}) \qquad\quad \{\text{evaluate at } u = 0 \text{ and } u = d\}$

$\qquad\qquad = \frac{6}{7}cd^{1/6} \qquad\qquad\qquad \{\text{simplify}\}$

(b) At the surface, $y = 0$ and $v = c(d - y)^{1/6} = c(d - 0)^{1/6} = cd^{1/6} = v_0.$

$$\text{From part (a), } v_{av} = \tfrac{6}{7}cd^{1/6} = \tfrac{6}{7}(cd^{1/6}) = \tfrac{6}{7}v_0.$$

$\boxed{49}$ If $s(t)$ denotes the height of the ball, then $s(t) = s_0 - 16t^2.$

The ball hits the ground when $s(t) = 0 \Rightarrow s_0 - 16t^2 = 0 \Rightarrow t^2 = \dfrac{s_0}{16} \Rightarrow t = \frac{1}{4}\sqrt{s_0}.$

Call this time t_1. The ball is in the air from $t_0 = 0\ \{s(t) = s_0\}$ to $t_1 = \frac{1}{4}\sqrt{s_0}.$

Thus, $v_{av} = \dfrac{1}{t_1 - t_0}\displaystyle\int_{t_0}^{t_1} v(t)\, dt = \dfrac{1}{\frac{1}{4}\sqrt{s_0} - 0}\displaystyle\int_{t_0}^{t_1} s'(t)\, dt = \dfrac{4}{\sqrt{s_0}}\Big[s(t)\Big]_{t_0}^{t_1} =$

$\dfrac{4}{\sqrt{s_0}}\Big[s(t_1) - s(t_0)\Big] = \dfrac{4}{\sqrt{s_0}}\Big[0 - s_0\Big] = -4\sqrt{s_0}.$ (The answer is negative since the ball is moving toward the ground.)

$\boxed{51}$ By Part I of Theorem (5.30), the function $G(u) = \displaystyle\int_a^u f(t)\,dt$ is an antiderivative of f.

Hence, we may conclude that $G'(u) = f(u)$.

By the chain rule, we have $D_x\,G(u) = D_u\,G(u)\,D_x\,u = f(u)\,D_x\,u$. Let $u = g(x)$.

Then, $D_x\displaystyle\int_a^{g(x)} f(t)\,dt = D_x\displaystyle\int_a^u f(t)\,dt =$

$$D_x\,G(u) = f(u)\,D_x\,u = f(g(x))\,D_x\big[g(x)\big] = f(g(x))\,g'(x).$$

$\boxed{53}$ By Exercise 51 with $g(x) = x^4$ and $f(t) = \dfrac{t}{\sqrt{t^3 + 2}}$, we have $D_x\displaystyle\int_2^{x^4} \dfrac{t}{\sqrt{t^3 + 2}}\,dt =$

$$f(g(x))\,g'(x) = \dfrac{x^4}{\sqrt{(x^4)^3 + 2}}(4x^3) = \dfrac{4x^7}{\sqrt{x^{12} + 2}}.$$

$\boxed{55}$ By Exercise 52 with $g(x) = x^3$, $k(x) = 3x$ and $f(t) = (t^3 + 1)^{10}$,

we have $D_x\displaystyle\int_{3x}^{x^3} (t^3 + 1)^{10}\,dt = f(g(x))\,g'(x) - f(k(x))\,k'(x) =$

$$\big[(x^3)^3 + 1\big]^{10}(3x^2) - \big[(3x)^3 + 1\big]^{10}(3) = 3x^2(x^9 + 1)^{10} - 3(27x^3 + 1)^{10}.$$

$\boxed{\text{ Exercises 5.7 }}$

Note: Let T denote the trapezoidal rule and S, Simpson's rule, in this section and all subsequent sections using numerical integration. The first step listed is

$$\text{T} = \frac{b - a}{2n}\{f(x_0) + 2\sum_{k=1}^{n-1} f(x_k) + f(x_n)\} \text{ or}$$

$$\text{S} = \frac{b - a}{3n}\Big[f(x_0) + 4f(x_1) + 2f(x_2) + \cdots + 4f(x_{n-1}) + f(x_n)\Big].$$

$\boxed{1}$ (a) Let $f(x) = x^2 + 1$, $a = 1$, $b = 3$, and $n = 4$. Then,

$\Delta x = \frac{b - a}{n} = \frac{3 - 1}{4} = \frac{1}{2}$, and $x_k = a + k\Delta x = 1 + \frac{k}{2}$ for $k = 0, 1, \ldots, 4$.

$\text{T} = \frac{3-1}{2(4)}\{f(1) + 2\big[f(1.5) + f(2) + f(2.5)\big] + f(3)\}$

$= \frac{1}{4}\big[2 + 2(3.25 + 5 + 7.25) + 10\big] = \frac{1}{4}(43) = 10.75$

Note: You do not need numerical integration for the first four exercises since you can evaluate the integrals using previously learned methods. They are intended to be a simple introduction to numerical integration. You may want to evaluate the integral using other methods just to see how accurate your answers in parts (a) and (b) are.

(b) $\text{S} = \frac{3-1}{3(4)}\big[f(1) + 4f(1.5) + 2f(2) + 4f(2.5) + 2f(3)\big]$

$= \frac{1}{6}\big[2 + 4(3.25) + 2(5) + 4(7.25) + 10\big] = \frac{1}{6}(64) = 10\frac{2}{3} \approx 10.67$

$\boxed{5}$ Let $f(x) = 1/x$, $a = 1$, $b = 4$, and $n = 6$.

Then, $\Delta x = \frac{b-a}{n} = \frac{4-1}{6} = \frac{1}{2}$, and $x_k = a + k\Delta x = 1 + \frac{k}{2}$ for $k = 0, 1, \ldots, 6$.

(a) $T = \frac{4-1}{2(6)}\{f(1) + 2[f(1.5) + f(2) + f(2.5) + f(3) + f(3.5)] + f(4)\}$

$\approx \frac{1}{4}[1 + 2(0.6667 + 0.5 + 0.4 + 0.3333 + 0.2857) + 0.25]$

$= \frac{1}{4}(5.6214) = 1.40535 \approx 1.41$

(b) $S = \frac{4-1}{3(6)}[f(1) + 4f(1.5) + 2f(2) + 4f(2.5) + 2f(3) + 4f(3.5) + f(4)]$

$\approx \frac{1}{6}[1 + 4(0.6667) + 2(0.5) + 4(0.4) + 2(0.3333) + 4(0.2857) + 0.25]$

$= \frac{1}{6}(8.3262) = 1.3877 \approx 1.39$

$\boxed{9}$ $f(x) = 1/(4 + x^2)$, $a = 0$, $b = 2$, and $n = 6$.

(a) $T = \frac{2-0}{2(6)}\{f(0) + 2[f(\frac{1}{3}) + f(\frac{2}{3}) + f(1) + f(\frac{4}{3}) + f(\frac{5}{3})] + f(2)\}$

$\approx \frac{1}{6}[0.25 + 2(0.2432 + 0.225 + 0.2 + 0.1731 + 0.1475) + 0.125]$

$= \frac{1}{6}(2.3526) = 0.3921 \approx 0.39$

(b) $S = \frac{2-0}{3(6)}[f(0) + 4f(\frac{1}{3}) + 2f(\frac{2}{3}) + 4f(1) + 2f(\frac{4}{3}) + 4f(\frac{5}{3}) + f(2)]$

$\approx \frac{1}{9}[0.25 + 4(0.2432) + 2(0.225) + 4(0.2) + 2(0.1731) + 4(0.1475) + 0.125]$

$= \frac{1}{9}(3.534) \approx 0.3927 \approx 0.39$

$\boxed{11}$ $f(x) = \sqrt{\sin x}$, $a = 0$, $b = \pi$, and $n = 6$.

(a) $T = \frac{\pi-0}{2(6)}\{f(0) + 2[f(\frac{\pi}{6}) + f(\frac{\pi}{3}) + f(\frac{\pi}{2}) + f(\frac{2\pi}{3}) + f(\frac{5\pi}{6})] + f(\pi)\}$

$\approx \frac{\pi}{12}[0 + 2(0.7071 + 0.9306 + 1 + 0.9306 + 0.7071) + 0]$

$= \frac{\pi}{12}(8.5508) \approx 2.2386 \approx 2.24$

(b) $S = \frac{\pi-0}{3(6)}[f(0) + 4f(\frac{\pi}{6}) + 2f(\frac{\pi}{3}) + 4f(\frac{\pi}{2}) + 2f(\frac{2\pi}{3}) + 4f(\frac{5\pi}{6}) + f(\pi)]$

$\approx \frac{\pi}{18}[0 + 4(0.7071) + 2(0.9306) + 4(1) + 2(0.9306) + 4(0.7071) + 0]$

$= \frac{\pi}{18}(13.3792) \approx 2.3351 \approx 2.34$

$\boxed{13}$ (a) To estimate the maximum error, use (5.37). First, we must find the maximum value M of $|f''(x)|$ on $[-2, 3]$. To do this, we will find the *CN* of f'' and then evaluate the *CN* and endpoints in $|f''(x)|$. The largest of these values is the value of M. Remember, f'' is just a function—so finding its maximum value is no more than an application of Guidelines (4.9). The only difference is that we examine the absolute value of the function results. If $f(x) = \frac{1}{360}x^6 + \frac{1}{60}x^5$, then $f''(x) = \frac{1}{12}x^4 + \frac{1}{3}x^3$ and $f'''(x) = \frac{1}{3}x^3 + x^2$. (The *CN* for f'' occur when f''' is equal to zero or undefined.) $f'''(x) = 0 \Rightarrow x = -3, 0$; $-3 \notin [-2, 3]$. $|f''(-2)| = \frac{4}{3}$, $|f''(0)| = 0$, and $|f''(3)| = \frac{63}{4} \Rightarrow M = \frac{63}{4}$.

$$|\text{error}| \le \frac{M(b-a)^3}{12n^2} = \frac{63(5)^3}{4 \cdot 12 \cdot 4^2} = \frac{2625}{256} \approx 10.25.$$

(b) To estimate the maximum error, use (5.39). First, we must find the maximum value M of $\left|f^{(4)}(x)\right|$ on $[-2, 3]$. To do this, we will find the *CN* of $f^{(4)}$ and then evaluate the *CN* and endpoints in $\left|f^{(4)}(x)\right|$. The largest of these values is the value of M. $f^{(4)}(x) = x^2 + 2x$ and $f^{(5)}(x) = 2x + 2$. $f^{(5)}(x) = 0 \Rightarrow x = -1$. $\left|f^{(4)}(-2)\right| = 0$, $\left|f^{(4)}(-1)\right| = 1$, and $\left|f^{(4)}(3)\right| = 15 \Rightarrow M = 15$.

$$|\text{error}| \leq \frac{M(b-a)^5}{180n^4} = \frac{15(5)^5}{180(4)^4} = \frac{3125}{3072} \approx 1.02.$$

[15] (a) Again, we use (5.37). $f(x) = 1/x^2 \Rightarrow f''(x) = 6/x^4$ and $f'''(x) = -24/x^5$.

Since there are no critical numbers for f'' in $[1, 5]$, we need only examine the endpoints. $\left|f''(1)\right| = 6$ and $\left|f''(5)\right| = \frac{6}{625} \Rightarrow M = 6$.

$$|\text{error}| \leq \frac{M(b-a)^3}{12n^2} = \frac{6(4)^3}{12(8)^2} = \frac{1}{2} = 0.5.$$

(b) Using (5.39), $f^{(4)}(x) = 120/x^6$ and $f^{(5)}(x) = -720/x^7$.

As in part (a), $\left|f^{(4)}(1)\right| = 120$ and $\left|f^{(4)}(5)\right| = \frac{120}{15,625} \Rightarrow M = 120$.

$$|\text{error}| \leq \frac{M(b-a)^5}{180n^4} = \frac{120(4)^5}{180(8)^4} = \frac{1}{6} \approx 0.17.$$

[17] (a) See Exercises 13 and 15 for help with finding M. $f(x) = 81x^{8/3} \Rightarrow$
$f''(x) = 360x^{2/3}$. $f'''(x) = 240x^{-1/3}$ has no *CN* in $[1, 8]$. Evaluating the endpoints, we find that $\left|f''(1)\right| = 360$ and $\left|f''(8)\right| = 1440 \Rightarrow M = 1440$.

Using (5.37), $|\text{error}| \leq \frac{1440(7)^3}{12n^2} \leq 0.001 \Rightarrow n^2 \geq \frac{1440(7)^3(1000)}{12} = 41,160,000$

$$\Rightarrow n \geq 6415.6\ldots, \text{ or } n \geq 6416.$$

(b) $f^{(4)}(x) = -80x^{-4/3}$. $f^{(5)}(x) = \frac{320}{3}x^{-7/3}$ has no *CN* in $[1, 8]$.

Evaluating the endpoints, we find that $\left|f^{(4)}(1)\right| = 80$ and $\left|f^{(4)}(8)\right| = 5 \Rightarrow$

$M = 80$. Using (5.39), $|\text{error}| \leq \frac{80(7)^5}{180n^4} \leq 0.001 \Rightarrow n^4 \geq \frac{80(7)^5(1000)}{180} \Rightarrow$

$$n \geq 52.2\ldots, \text{ or } n \geq 54, \text{ since } n \text{ must be even.}$$

[19] (a) Since $f''(x) = 2/x^3$ is $\downarrow$ on $[\frac{1}{2}, 1]$, it assumes its maximum at $x = \frac{1}{2}$.

$\left|f''(x)\right| \leq 2(2)^3 = 16 = M$ on $[\frac{1}{2}, 1]$. $|\text{error}| \leq \frac{M(b-a)^3}{12n^2} = \frac{16(\frac{1}{2})^3}{12n^2} \leq 0.0001 \Rightarrow$

$$n^2 \geq \frac{16(\frac{1}{2})^3(10,000)}{12} \Rightarrow n \geq \frac{100}{\sqrt{6}} \approx 40.82 \Rightarrow n = 41.$$

(b) Since $f^{(4)}(x) = 24/x^5$ is $\downarrow$ on $[\frac{1}{2}, 1]$, it assumes its maximum at $x = \frac{1}{2}$.

$$\left| f^{(4)}(x) \right| \le 24(2^5) = 768 = M \text{ on } [\frac{1}{2}, 1].$$

$$|error| \le \frac{M(b-a)^5}{180n^4} \le \frac{768(\frac{1}{2})^5}{180n^4} \le 0.0001 \Rightarrow n^4 \ge \frac{768(\frac{1}{2})^5(10,000)}{180} \Rightarrow$$

$$n \ge \sqrt[4]{\frac{4000}{3}} \approx 6.04 \Rightarrow n = 8, \text{ since } n \text{ must be even.}$$

$\boxed{21}$ In this exercise, $a = 2$ and $b = 4$. Since there are 4 subintervals in the table, $n = 4$.

(a) $T = \frac{4-2}{2(4)}\left[3 + 2(2 + 4 + 3) + 5\right] = \frac{1}{4}(26) = 6.5$

(b) $S = \frac{4-2}{3(4)}\left[3 + 4(2) + 2(4) + 4(3) + 5\right] = \frac{1}{6}(36) = 6$

$\boxed{27}$ (5.39) is valid for any $M \ge \left| f^{(4)}(x) \right|$. Let $f(x) = ax^3 + bx^2 + cx + d$ be a polynomial of degree less than four, where a, b, c, and d are constants. Then, $f'(x) = 3ax^2 + 2bx + c$, $f''(x) = 6ax + 2b$, $f'''(x) = 6a$, and $f^{(4)}(x) = 0$. Since $f^{(4)}(x) = 0$ for a polynomial of degree 3 or less, it follows that we can let $M = 0$ and the error bound can be set at zero. (It is interesting to note that Simpson's rule, with equal subintervals, is exact for a cubic polynomial even though it is using only a quadratic polynomial to interpolate the function.)

$\boxed{31}$ In (5.36), let $f = v$, $a = 0$, $b = k$, and $n = 5$. Then, $\bar{v}_x = \frac{1}{k}\int_0^k v(y)\, dy$

$$\approx \frac{1}{k} \cdot \frac{k-0}{2(5)}\left\{ v(0) + 2\left[v(0.2k) + v(0.4k) + v(0.6k) + v(0.8k)\right] + v(k) \right\}$$

$$= \frac{1}{10}\left[0.28 + 2(0.23 + 0.19 + 0.17 + 0.13) + 0.02\right] = 0.174 \text{ m/sec}$$

$\boxed{35}$ By the fundamental theorem of calculus (5.30),

$$\int_0^1 f'(x)\, dx = f(1) - f(0) \Rightarrow f(1) = f(0) + \int_0^1 \frac{\sqrt{x}}{x^2 + 1}\, dx.$$

To estimate $\int_0^1 \frac{\sqrt{x}}{x^2 + 1}\, dx$, let $a = 0$, $b = 1$, and $n = 10$ in (5.36).

$$T = \frac{1-0}{2(10)}\{f'(0) + 2\left[f'(0.1) + f'(0.2) + \cdots + f'(0.9)\right] + f'(1)\}$$

$$= \frac{1}{20}[0 + 2(0.3131 + 0.4300 + 0.5025 + 0.5452 + 0.5657 +$$

$$0.5696 + 0.5615 + 0.5454 + 0.5241) + 0.5]$$

$$\approx \frac{1}{20}(9.6142) \approx 0.4807. \text{ Thus, } f(1) \approx f(0) + 0.4807 = 1 + 0.4807 = 1.4807.$$

5.8 Review Exercises

$\boxed{1}$ $\displaystyle\int \frac{8x^2 - 4x + 5}{x^4}\, dx = \int\left(\frac{8}{x^2} - \frac{4}{x^3} + \frac{5}{x^4}\right) dx = \int (8x^{-2} - 4x^{-3} + 5x^{-4})\, dx =$

$$-\frac{8}{x} + \frac{2}{x^2} - \frac{5}{3x^3} + C$$

$\boxed{5}$ $u = 2x + 1$, $\frac{1}{2}\, du = dx \Rightarrow$

$$\int (2x + 1)^7\, dx = \frac{1}{2}\int u^7\, du = \frac{1}{16}u^8 + C = \frac{1}{16}(2x + 1)^8 + C$$

⑨ $u = 1 + \sqrt{x}$, $2\,du = \frac{1}{\sqrt{x}}\,dx \Rightarrow \int \frac{1}{\sqrt{x}\,(1+\sqrt{x})^2}\,dx = \int \frac{1}{u^2}(2)\,du = 2\int u^{-2}\,du =$

$$2 \cdot \frac{u^{-1}}{-1} + C = -\frac{2}{u} + C = -\frac{2}{1+\sqrt{x}} + C$$

⑬ $u = 4x^2 + 2x - 7$, $du = (8x+2)\,dx$ or $\frac{1}{2}\,du = (4x+1)\,dx \Rightarrow$

$$\int (4x+1)(4x^2+2x-7)^2\,dx = \frac{1}{2}\int u^2\,du = \frac{1}{6}u^3 + C = \frac{1}{6}(4x^2+2x-7)^3 + C$$

⑲ $u = 1 + x^3 \Rightarrow \frac{1}{3}\,du = x^2\,dx$. $x = 0, 1 \Rightarrow u = 1, 2$.

$$\int_0^1 \frac{x^2}{(1+x^3)^2}\,dx = \frac{1}{3}\int_1^2 u^{-2}\,du = \frac{1}{3}\left[-\frac{1}{u}\right]_1^2 = -\frac{1}{3}\left(\frac{1}{2} - 1\right) = \frac{1}{6}.$$

㉑ $u = x^2 + 2x \Rightarrow \frac{1}{2}\,du = (x+1)\,dx$. $x = 1, 2 \Rightarrow u = 3, 8$.

$$\int_1^2 \frac{x+1}{\sqrt{x^2+2x}}\,dx = \frac{1}{2}\int_3^8 u^{-1/2}\,du = \frac{1}{2}\left[2u^{1/2}\right]_3^8 = \sqrt{8} - \sqrt{3} \approx 1.10.$$

㉕ $\int_0^1 (2x-3)(5x+1)\,dx = \int_0^1 (10x^2 - 13x - 3)\,dx = \left[\frac{10}{3}x^3 - \frac{13}{2}x^2 - 3x\right]_0^1 =$

$$\left(\frac{10}{3} - \frac{13}{2} - 3\right) = -\frac{37}{6}$$

㉗ $\int_0^4 \sqrt{3x}\,(\sqrt{x} + \sqrt{3})\,dx = \int_0^4 (\sqrt{3}\,x + 3x^{1/2})\,dx = \left[\frac{1}{2}\sqrt{3}\,x^2 + 2x^{3/2}\right]_0^4 =$

$$8\sqrt{3} + 16 \approx 29.86$$

㉛ $u = \sin 3x$, $\frac{1}{3}\,du = \cos 3x\,dx \Rightarrow$

$$\int \cos 3x \sin^4 3x\,dx = \frac{1}{3}\int u^4\,du = \frac{1}{15}u^5 + C = \frac{1}{15}\sin^5 3x + C$$

㉟ $u = 3 + 5\sin x \Rightarrow \frac{1}{5}\,du = \cos x\,dx$. $x = 0, \frac{\pi}{2} \Rightarrow u = 3, 8$.

$$\int_0^{\pi/2} \cos x\sqrt{3 + 5\sin x}\,dx = \frac{1}{5}\int_3^8 u^{1/2}\,du = \frac{1}{5}\left[\frac{2}{3}u^{3/2}\right]_3^8 = \frac{2}{15}(16\sqrt{2} - 3\sqrt{3}) \approx 2.32.$$

㊳ By (5.5)(i), $\int D_x \sqrt[5]{x^4 + 2x^2 + 1}\,dx = \sqrt[5]{x^4 + 2x^2 + 1} + C.$

㊶ Since the definite integral is a constant, its derivative is 0.

㊸

$$\frac{d^2y}{dx^2} = 6x - 4 \qquad \{\text{given differential equation}\}$$

$$\int \frac{d^2y}{dx^2}\,dx = \int (6x - 4)\,dx \qquad \{\text{integrate both sides}\}$$

$$\frac{dy}{dx} = 3x^2 - 4x + C \qquad \{\text{find antiderivatives}\}$$

$x = 2 \Rightarrow y' = 3(2)^2 - 4(2) + C = 4 + C.$ $y' = 5 \Rightarrow C = 1.$

$$\frac{dy}{dx} = 3x^2 - 4x + 1 \qquad \{\text{replace } C \text{ with } 1\}$$

$$\int \frac{dy}{dx}\,dx = \int (3x^2 - 4x + 1)\,dx \qquad \{\text{integrate both sides}\}$$

$$y = x^3 - 2x^2 + x + D \qquad \{\text{find antiderivatives}\}$$

$x = 2 \Rightarrow y = (2)^3 - 2(2)^2 + 2 + D = 2 + D.$ $y = 4 \Rightarrow D = 2.$

$$y = x^3 - 2x^2 + x + 2.$$

$\boxed{45}$ $\forall k$, $\Delta x_k = 1 \Rightarrow R_P = \sum\limits_{k=1}^{5} f(w_k)(1) =$

$$f(-\tfrac{3}{2}) + f(-\tfrac{1}{2}) + f(\tfrac{1}{2}) + f(\tfrac{3}{2}) + f(\tfrac{5}{2}) = \tfrac{27}{4} + \tfrac{35}{4} + \tfrac{35}{4} + \tfrac{27}{4} + \tfrac{11}{4} = \tfrac{135}{4}.$$

$\boxed{47}$ Using (5.27) with $f(x) = x^2$ and $g(x) = x^3$, and the fact that $x^2 \geq x^3$ on $[0, 1]$,

$$\text{we have } \int_0^1 x^2 \, dx \geq \int_0^1 x^3 \, dx.$$

$\boxed{49}$ $\displaystyle\int_c^e f(x)\,dx + \int_a^b f(x)\,dx - \int_c^b f(x)\,dx - \int_d^b f(x)\,dx$

$$= \int_c^e f(x)\,dx + \int_a^b f(x)\,dx + \int_b^c f(x)\,dx - 0 \ \{(5.17),\,(5.18)\}$$

$$= \int_c^e f(x)\,dx + \int_a^c f(x)\,dx \ \{(5.25)\}$$

$$= \int_a^c f(x)\,dx + \int_c^e f(x)\,dx$$

$$= \int_a^e f(x)\,dx \ \{(5.25)\}$$

$\boxed{51}$ (a) $a(t) = v'(t) = -32 \Rightarrow \int v'(t)\,dt = \int -32\,dt \Rightarrow v(t) = -32t + C$.

$v(0) = -30 \Rightarrow C = -30$. $v(t) = s'(t) = -32t - 30 \Rightarrow$

$\int s'(t)\,dt = \int(-32t - 30)\,dt \Rightarrow s(t) = -16t^2 - 30t + D$.

$$s(0) = 900 \Rightarrow D = 900. \text{ Thus, } s(t) = -16t^2 - 30t + 900.$$

(b) $v(5) = -32(5) - 30 = -190$ ft/sec.

(c) $s(t) = -16t^2 - 30t + 900 = 0$ $(t > 0)$ when $t = \tfrac{1}{16}(-15 + \sqrt{14{,}625})$

$$\{\text{by the quadratic formula}\} = \tfrac{15}{16}(-1 + \sqrt{65}) \approx 6.6 \text{ sec.}$$

$\boxed{53}$ (a) $f(x) = \sqrt{1 + x^4}$, $a = 0$, $b = 10$, and $n = 5$.

$\text{T} = \tfrac{10-0}{2(5)}\{f(0) + 2[f(2) + f(4) + f(6) + f(8)] + f(10)\}$

$= (1)\big[1 + 2(4.1231 + 16.0312 + 36.0139 + 64.0078) + 100.0050\big]$

$= 341.3570 \approx 341.36$

(b) $\text{S} = \tfrac{10-0}{3(8)}[f(0) + 4f(1.25) + 2f(2.5) + 4f(3.75) + 2f(5) + 4f(6.25) +$

$$2f(7.5) + 4f(8.75) + f(10)]$$

$= \tfrac{5}{12}[1 + 4(1.8551) + 2(6.3295) + 4(14.0980) + 2(25.0200) + 4(39.0753) +$

$$2(56.2589) + 4(76.5690) + 100.0050]$$

$= \tfrac{5}{12}(802.6114) \approx 334.42$

Chapter 6: Applications of the Definite Integral

Exercises 6.1

1 The shaded region is an R_x region since it lies between

$y = f(x) = x^2 + 1$ and $y = g(x) = x - 2$ on the interval $-2 \leq x \leq 2$.

Using (6.1) with $a = -2$, $b = 2$, and $f \geq g$, $A = \int_{-2}^{2} \left[(x^2 + 1) - (x - 2) \right] dx$.

Note: For an R_x region, always use "upper boundary minus lower boundary."

3 The shaded region is an R_y region since it lies between $x = f(y) = -3y^2 + 4$ and

$x = g(y) = y^3$ on the interval $-2 \leq y \leq 1$. From the discussion on page 309 with

$c = -2$, $d = 1$, and $f \geq g$, $A = \int_{-2}^{1} \left[(-3y^2 + 4) - y^3 \right] dy$.

Note: For an R_y region, always use "right boundary minus left boundary."

If you use left boundary minus right boundary,

you will obtain the negative of the value of the area.

5 To sketch the region, we must first find the points of intersection of the graphs of the

two equations. $x^2 = 4x \Rightarrow x^2 - 4x = 0 \Rightarrow x(x - 4) = 0 \Rightarrow x = 0, 4$ and $y = 0$,

16. $4x \geq x^2$ on $[0, 4] \Rightarrow A = \int_{0}^{4} (4x - x^2) \, dx = \left[2x^2 - \frac{1}{3}x^3 \right]_{0}^{4} = 32 - \frac{64}{3} = \frac{32}{3}$.

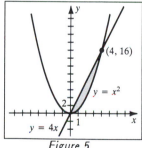

Figure 5

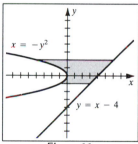

Figure 11

11 Equivalent equations are $x = -y^2$ and $x = y + 4$. $y + 4 > -y^2$ on $[-1, 2] \Rightarrow$

$A = \int_{-1}^{2} \left[(4 + y) - (-y^2) \right] dy = \left[4y + \frac{1}{2}y^2 + \frac{1}{3}y^3 \right]_{-1}^{2} = \frac{38}{3} - (-\frac{23}{6}) = \frac{33}{2}$.

17 $x^3 - x = x(x^2 - 1) = x(x + 1)(x - 1) = 0 \Rightarrow x = 0, \pm 1$.

$x^3 - x \geq 0$ on $[-1, 0]$ and $0 \geq x^3 - x$ on $[0, 1]$.

By symmetry, $A = 2 \int_0^1 \left[0 - (x^3 - x)\right] dx = 2\left[-\frac{1}{4}x^4 + \frac{1}{2}x^2\right]_0^1 = 2(\frac{1}{4}) = \frac{1}{2}$.

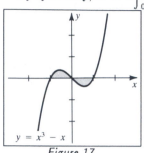

$y = x^3 - x$

Figure 17

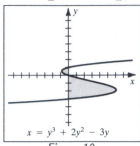

$x = y^3 + 2y^2 - 3y$

Figure 19

19 $x = y^3 + 2y^2 - 3y = y(y^2 + 2y - 3) = y(y + 3)(y - 1) \geq 0$ on $[-3, 0]$ and

$x = y(y + 3)(y - 1) \leq 0$ on $[0, 1]$.

$$A = \int_{-3}^0 \left[(y^3 + 2y^2 - 3y) - 0\right] dy + \int_0^1 \left[0 - (y^3 + 2y^2 - 3y)\right] dy \quad \oplus$$

$$= \left[\frac{1}{4}y^4 + \frac{2}{3}y^3 - \frac{3}{2}y^2\right]_{-3}^0 + \left[-(\frac{1}{4}y^4 + \frac{2}{3}y^3 - \frac{3}{2}y^2)\right]_0^1$$

$$= \left[0 - (-\frac{45}{4})\right] + \left[-(-\frac{7}{12}) - 0\right] = \frac{45}{4} + \frac{7}{12} = \frac{71}{6}.$$

21 Since $\sqrt{4 - x^2} \geq 0$ on $[-2, 2]$, $y = x\sqrt{4 - x^2} \leq 0$ on $[-2, 0]$ and

$y = x\sqrt{4 - x^2} \geq 0$ on $[0, 2]$. $y = x\sqrt{4 - x^2} = 0$ when $x = 0, \pm 2$.

By symmetry, $A = 2 \int_0^2 x\sqrt{4 - x^2}\, dx$. $u = 4 - x^2 \Rightarrow -du = 2x\, dx$.

$x = 0, 2 \Rightarrow u = 4, 0$. $A = -\int_4^0 u^{1/2}\, du = -\left[\frac{2}{3}u^{3/2}\right]_4^0 = -\frac{2}{3}(0 - 8) = \frac{16}{3}$.

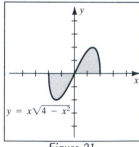

$y = x\sqrt{4 - x^2}$

Figure 21

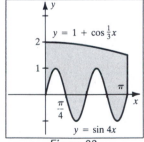

$y = 1 + \cos\frac{1}{3}x$

2

1

π

$\frac{\pi}{4}$

$y = \sin 4x$

Figure 23

23 $\sin 4x \leq 1$ $\forall x$ and $1 + \cos\frac{1}{3}x \geq 1$ for $x \in [0, \pi]$. $\sin 4x \leq 1 + \cos\frac{1}{3}x \Rightarrow$

$$A = \int_0^\pi \left[(1 + \cos\frac{1}{3}x) - \sin 4x\right] dx$$

$$= \left[x + 3\sin\frac{1}{3}x + \frac{1}{4}\cos 4x\right]_0^\pi$$

$$= (\pi + \frac{3}{2}\sqrt{3} + \frac{1}{4}) - (\frac{1}{4}) = \pi + \frac{3}{2}\sqrt{3} \approx 5.74.$$

25 (a) If you draw any vertical line $x = k$, where $0 < k < 1$, it will pass through the region with upper boundary $y = 3x$ and lower boundary $y = x$. Similarly, if $1 < k < 2$, then the upper boundary is $y = 4 - x$ and the lower boundary is $y = x$. Since the upper boundary changes at $x = 1$, we must represent the area of the region with a sum of two integrals. On $[0, 1]$, $3x \geq x$ and on $[1, 2]$, $(4 - x) \geq x$. $A = \int_0^1 (3x - x)\, dx + \int_1^2 \left[(4 - x) - x\right] dx$.

(b) If you draw any *horizontal* line $y = k$, where $0 < k < 2$, it will pass through the region with right boundary $x = y$ and left boundary $x = \frac{1}{3}y$. Similarly, if $2 < k < 3$, then the right boundary is $x = 4 - y$ and the left boundary is $x = \frac{1}{3}y$. Since the right boundary changes at $y = 2$, we must represent the area of the region with a sum of two integrals. On $[0, 2]$, $y \geq \frac{1}{3}y$ and on $[2, 3]$, $(4 - y) \geq \frac{1}{3}y$. $A = \int_0^2 (y - \frac{1}{3}y)\, dy + \int_2^3 \left[(4 - y) - \frac{1}{3}y\right] dy$.

27 (a) On $[1, 4]$, $\sqrt{x} > -x$. $A = \int_1^4 \left[\sqrt{x} - (-x)\right] dx$.

(b) There are three regions to consider.

On all three regions, $x = 4$ is the right boundary. The left boundary is $x = -y$ on $[-4, -1]$, $x = 1$ on $[-1, 1]$, and $x = y^2$ on $[1, 2]$.

$$\text{Hence, } A = \int_{-4}^{-1} \left[4 - (-y)\right] dy + \int_{-1}^1 (4 - 1)\, dy + \int_1^2 (4 - y^2)\, dy.$$

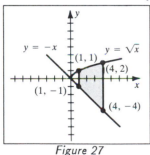

Figure 27

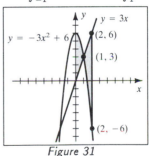

Figure 31

31 $6 - 3x^2 = 3x \Rightarrow 3x^2 + 3x - 6 = 0 \Rightarrow 3(x + 2)(x - 1) = 0 \Rightarrow x = 1$ on $[0, 2]$.

$f(x) = 6 - 3x^2 \geq g(x) = 3x$ on $[0, 1]$ and $f(x) \leq g(x)$ on $[1, 2]$.

$$A = \int_0^1 \left[(6 - 3x^2) - 3x\right] dx + \int_1^2 \left[3x - (6 - 3x^2)\right] dx$$

$$= \left[6x - x^3 - \tfrac{3}{2}x^2\right]_0^1 + \left[\tfrac{3}{2}x^2 - 6x + x^3\right]_1^2$$

$$= \left[\tfrac{7}{2} - 0\right] + \left[2 - (-\tfrac{7}{2})\right] = \tfrac{7}{2} + \tfrac{11}{2} = 9.$$

$\boxed{35}$ $\sin x = \cos x \Rightarrow \frac{\sin x}{\cos x} = 1 \Rightarrow \tan x = 1 \Rightarrow x = \frac{\pi}{4}, \frac{5\pi}{4}$ on $[0, 2\pi]$.

$f(x) = \sin x \geq g(x) = \cos x$ on $[\frac{\pi}{4}, \frac{5\pi}{4}]$, $f(x) \leq g(x)$ on $[0, \frac{\pi}{4}] \cup [\frac{5\pi}{4}, 2\pi]$.

$$A = \int_0^{\pi/4} (\cos x - \sin x)\, dx + \int_{\pi/4}^{5\pi/4} (\sin x - \cos x)\, dx + \int_{5\pi/4}^{2\pi} (\cos x - \sin x)\, dx$$

$$= \Big[\sin x + \cos x\Big]_0^{\pi/4} + \Big[-\cos x - \sin x\Big]_{\pi/4}^{5\pi/4} + \Big[\sin x + \cos x\Big]_{5\pi/4}^{2\pi}$$

$$= \Big[(\tfrac{\sqrt{2}}{2} + \tfrac{\sqrt{2}}{2}) - (0 + 1)\Big] + \Big[(\tfrac{\sqrt{2}}{2} + \tfrac{\sqrt{2}}{2}) - (-\tfrac{\sqrt{2}}{2} - \tfrac{\sqrt{2}}{2})\Big] +$$

$$\Big[(0 + 1) - (-\tfrac{\sqrt{2}}{2} - \tfrac{\sqrt{2}}{2})\Big]$$

$$= (\sqrt{2} - 1) + 2\sqrt{2} + (1 + \sqrt{2}) = 4\sqrt{2}.$$

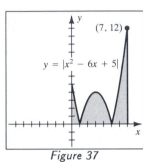

Figure 35   Figure 37

$\boxed{37}$ Let $g(x) = x^2 - 6x + 5 = (x - 1)(x - 5)$. To graph $f(x) = |x^2 - 6x + 5|$,

refer to Example 5 in §4.5. For the interval $[0, 7]$, $g(x) \leq 0$ on $[1, 5]$, and $g(x) \geq 0$

on $[0, 1] \cup [5, 7]$. Recalling the definition of absolute value, $f(x) = g(x)$ if $g(x) \geq 0$

and $f(x) = -g(x)$ if $g(x) < 0$.

Hence, $A = \int_0^1 (x^2 - 6x + 5)\, dx + \int_1^5 -(x^2 - 6x + 5)\, dx + \int_5^7 (x^2 - 6x + 5)\, dx.$

$\boxed{39}$ *Note:* Answers may vary depending upon the estimates of the function values.

Let $f(x)$ be the difference between the ordinates of the top and the bottom curves,

that is, $f(x_k) = y_k(\text{top}) - y_k(\text{bottom})$. Then, $\int_0^6 f(x)\, dx \approx$

(a) $T = \frac{6-0}{2(6)}\{f(0) + 2[f(1) + f(2) + f(3) + f(4) + f(5)] + f(6)\}$

$$= \tfrac{1}{2}\{0 + 2[(1.5 - 1) + (2 - 1.25) + (2.5 - 1.5) + (3 - 2) + (3.5 - 2.5)]$$
$$+ 0\}$$

$$= \tfrac{1}{2}[0 + 2(0.5 + 0.75 + 1 + 1 + 1) + 0] = \tfrac{1}{2}(8.5) = 4.25$$

(b) $S = \frac{6-0}{3(6)}[f(0) + 4f(1) + 2f(2) + 4f(3) + 2f(4) + 4f(5) + f(6)]$

$$= \tfrac{1}{3}[0 + 4(0.5) + 2(0.75) + 4(1) + 2(1) + 4(1) + 0] = \tfrac{1}{3}(13.5) = 4.50$$

Exercises 6.2

1 Each disk is perpendicular to the x-axis. We are summing up the volume of the disks from $x = -1$ to $x = 2$. The radius of each disk generated is $r = \frac{1}{2}x^2 + 2$. In (6.5), let $f(x) = \frac{1}{2}x^2 + 2$, $a = -1$, and $b = 2$. $V = \pi \int_a^b [f(x)]^2 \, dx = \pi \int_{-1}^2 (\frac{1}{2}x^2 + 2)^2 \, dx$.

3 For each washer generated, the outer radius is $R = \sqrt{25 - y^2}$ and the inner radius is $r = 3$. Using (6.9) on the interval $[-4, 4]$, $V = \pi \int_{-4}^4 \left[(\sqrt{25 - y^2})^2 - 3^2 \right] dy$.

By symmetry, $V = 2 \cdot \pi \int_0^4 \left[(\sqrt{25 - y^2})^2 - 3^2 \right] dy$.

7 $x^2 - 4x = 0 \Rightarrow x = 0, 4.$ $y = x^2 - 4x \le 0$ on $[0, 4]$. The radius of each disk generated is the distance from the x-axis to the function. As in §6.1, to find this positive distance, we subtract the equation of the lower boundary ($y = x^2 - 4x$) from the equation of the upper boundary ($y = 0$), that is, $(0) - (x^2 - 4x)$, or, equivalently, $4x - x^2$. The values of $4x - x^2$ are positive on $[0, 4]$. Since we are squaring the radius in the integrand, it doesn't matter whether we use $4x - x^2$ or $x^2 - 4x$, because their squares are equal. Therefore, we will refer to the radius as $r = x^2 - 4x$, even though the radius is normally considered a positive quantity. This is a fairly subtle point, and may be better understood in the next section. Using (6.5),

$$V = \pi \int_0^4 (x^2 - 4x)^2 \, dx = \pi \left[\frac{1}{5}x^5 - 2x^4 + \frac{16}{3}x^3 \right]_0^4 = \pi(\frac{1024}{5} - 512 + \frac{1024}{3}) = \frac{512\pi}{15}.$$

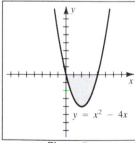

Figure 7

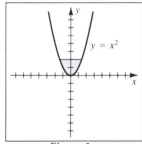

Figure 9

9 We will be summing up the volumes of the disks along the y-axis from $y = 0$ to $y = 2$. Thus, we need an equation of the form $x = g(y)$. Solving $y = x^2$ for x yields $x = \pm \sqrt{y}$. $x = \sqrt{y}$ represents the right side of the parabola and $x = -\sqrt{y}$ represents the left side. The volume will be generated by revolving *either* curve about the y-axis. By (6.6) with $g(y) = \sqrt{y}$, $c = 0$, and $d = 2$,

$$V = \pi \int_0^2 (\sqrt{y})^2 \, dy = \pi \left[\frac{1}{2}y^2 \right]_0^2 = \frac{\pi}{2}(4 - 0) = 2\pi.$$

$\boxed{13}$ $x^2 = 4 - x^2 \Rightarrow x = \pm\sqrt{2}$. $4 - x^2 \geq x^2$ on $[-\sqrt{2}, \sqrt{2}]$. The outer radius is $4 - x^2$

and the inner radius is x^2. Since the volume generated on the right side of the y-axis

is equal to the volume generated on the left side of the y-axis, there is symmetry in

the region with respect to the y-axis. By (6.9),

$$V = \pi\int_{-\sqrt{2}}^{\sqrt{2}}\left[(4 - x^2)^2 - (x^2)^2\right]dx = 2\cdot\pi\int_0^{\sqrt{2}}\left[(4 - x^2)^2 - (x^2)^2\right]dx$$

$$= 2\pi\left[16x - \tfrac{8}{3}x^3\right]_0^{\sqrt{2}} = 2\sqrt{2}\pi(16 - \tfrac{16}{3}) = \frac{64\pi\sqrt{2}}{3}.$$

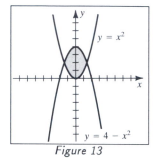

Figure 13

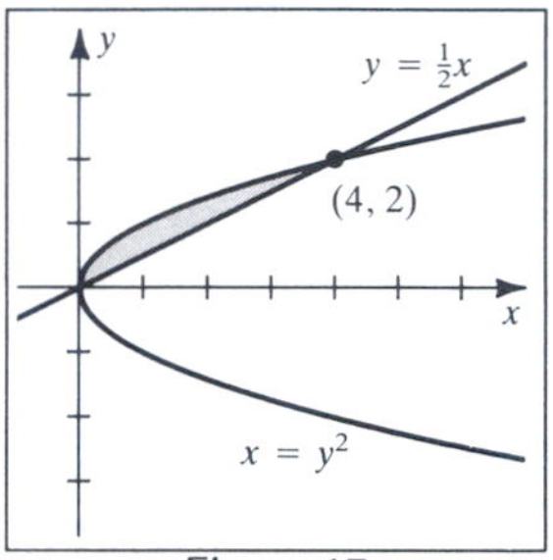

Figure 17

$\boxed{17}$ $y^2 = 2y \Rightarrow y = 0, 2$. $2y \geq y^2$ on $[0, 2]$. The outer radius is $2y$ and the inner radius

is y^2. By (6.9), $V = \pi\int_0^2\left[(2y)^2 - (y^2)^2\right]dy = \pi\left[\tfrac{4}{3}y^3 - \tfrac{1}{5}y^5\right]_0^2 = \pi(\tfrac{32}{3} - \tfrac{32}{5}) = \frac{64\pi}{15}$.

$\boxed{21}$ The graph of $y = \sin 2x$ is a horizontal compression of

the graph of $y = \sin x$ by a factor of 2. By (6.5),

$$V = \pi\int_0^\pi (\sin 2x)^2\, dx$$

$$= \tfrac{\pi}{2}\int_0^\pi (1 - \cos 4x)\, dx \left\{\sin^2 2x = \frac{1 - \cos(2\cdot 2x)}{2}\right\}$$

$$= \tfrac{\pi}{2}\left[x - \tfrac{1}{4}\sin 4x\right]_0^\pi = \tfrac{1}{2}\pi^2.$$

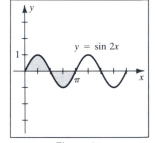

Figure 21

$\boxed{25}$ (a) The radius of a typical disk is $4 - x^2$ (upper minus lower).

$$V = \pi\int_{-2}^2 (4 - x^2)^2\, dx = 2\cdot\pi\int_0^2 (4 - x^2)^2\, dx \{\text{symmetry about } x = 0\}$$

$$= 2\pi\left[16x - \tfrac{8}{3}x^3 + \tfrac{1}{5}x^5\right]_0^2 = \tfrac{512\pi}{15}. \text{ See } \textit{Figure 25a \& b.}$$

(b) The outer radius is $5 - x^2$ (upper minus lower) and the

inner radius is $5 - 4$ (upper minus lower).

$$V = 2\cdot\pi\int_0^2\left[(5 - x^2)^2 - (5 - 4)^2\right]dx \{\text{by symmetry}\}$$

$$= 2\pi\left[24x - \tfrac{10}{3}x^3 + \tfrac{1}{5}x^5\right]_0^2 = \tfrac{832\pi}{15}.$$

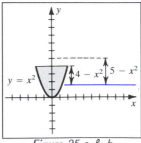

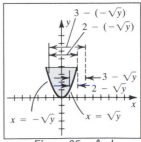

Figure 25 a & b   Figure 25 c & d

(c) For the revolution about the vertical line $x = 2$, the y-interval is $0 \le y \le 4$.

From *Figure 25 c & d*, $x = \pm \sqrt{y}$ and the outer radius is

$2 - (-\sqrt{y})$ (right minus left) and the inner radius is $2 - \sqrt{y}$ (right minus left).

$$V = \pi \int_0^4 \{ [2 - (-\sqrt{y})]^2 - [2 - \sqrt{y}]^2 \} \, dy = \pi \int_0^4 8\sqrt{y} \, dy = 8\pi \left[\tfrac{2}{3} y^{3/2} \right]_0^4 = \tfrac{128\pi}{3}.$$

(d) The outer radius is $3 - (-\sqrt{y})$ and the inner radius is $3 - \sqrt{y}$. $V =$

$$\pi \int_0^4 \{ [3 - (-\sqrt{y})]^2 - [3 - \sqrt{y}]^2 \} \, dy = \pi \int_0^4 12\sqrt{y} \, dy = 12\pi \left[\tfrac{2}{3} y^{3/2} \right]_0^4 = 64\pi.$$

27 (a) $x + 2y = 4 \Leftrightarrow y = -\tfrac{1}{2}x + 2$. On the x-interval $[0, 4]$, the outer radius is

$(-\tfrac{1}{2}x + 2) - (-2)$ (upper minus lower) and the inner radius is $0 - (-2)$

(upper minus lower). $V = \pi \int_0^4 \left\{ \left[(-\tfrac{1}{2}x + 2) - (-2) \right]^2 - \left[0 - (-2) \right]^2 \right\} dx.$

(b) The outer radius is $5 - 0$ (upper minus lower) and

the inner radius is $5 - (-\tfrac{1}{2}x + 2)$ (upper minus lower).

$$V = \pi \int_0^4 \left\{ (5 - 0)^2 - \left[5 - (-\tfrac{1}{2}x + 2) \right]^2 \right\} dx.$$

(c) $x + 2y = 4 \Leftrightarrow x = -2y + 4$. On the y-interval $[0, 2]$, the outer radius is

$7 - 0$ (right minus left) and the inner radius is $7 - (-2y + 4)$ (right minus

left). $V = \pi \int_0^2 \left\{ (7 - 0)^2 - \left[7 - (-2y + 4) \right]^2 \right\} dy.$

(d) The outer radius is $(-2y + 4) - (-4)$ (right minus left) and

the inner radius is $0 - (-4)$ (right minus left).

$$V = \pi \int_0^2 \left\{ \left[(-2y + 4) - (-4) \right]^2 - \left[0 - (-4) \right]^2 \right\} dy.$$

29 $x^3 = 4x \Rightarrow x^3 - 4x = x(x^2 - 4) = x(x - 2)(x + 2) = 0 \Rightarrow x = 0, \pm 2.$

On $[-2, 0]$, $4x \le x^3$, and on $[0, 2]$, $4x \ge x^3$. For $x \le 0$, the radius of the outer disk is $(8 - 4x)$ (upper minus lower) and the radius of the inner disk is $(8 - x^3)$ (upper minus lower). For $x \ge 0$, the outer disk has radius $(8 - x^3)$ (upper minus lower) and the inner disk has radius $(8 - 4x)$ (upper minus lower).

$$V = \pi \int_{-2}^{0} \left[(8 - 4x)^2 - (8 - x^3)^2 \right] dx + \pi \int_{0}^{2} \left[(8 - x^3)^2 - (8 - 4x)^2 \right] dx.$$

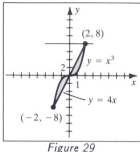

Figure 29

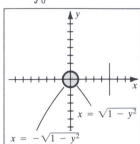

Figure 33

33 For $-1 \le y \le 1$, the outer radius is the distance between the line and the left side of the semicircle, $\left[5 - (-\sqrt{1 - y^2}) \right]$ (right minus left). The inner radius is the distance between the line and the right side of the semicircle, $(5 - \sqrt{1 - y^2})$. Since there is symmetry about the x-axis, $V = 2 \cdot \pi \int_{0}^{1} \left\{ [5 - (-\sqrt{1 - y^2})]^2 - [5 - \sqrt{1 - y^2}]^2 \right\} dy.$

35 Revolving a rectangle in the first quadrant, with coordinate axes for two sides, about the y-axis will generate a cylinder. Since we want altitude h and radius r, we'll use the lines $y = h$ and $x = r$ to form the rectangle. Summing disks along the y-axis from $y = 0$ to $y = h$ with $g(y) = r$, we have $V = \pi \int_{0}^{h} r^2 \, dy = \pi r^2 \left[y \right]_{0}^{h} = \pi r^2 h.$

37 Revolving a triangle with vertices $(0, 0)$, $(h, 0)$ and (h, r) about the x-axis will generate a right circular cone. The equation of the line containing the side of the triangle with endpoints $(0, 0)$ and (h, r) is $y = \frac{\Delta y}{\Delta x} x = \frac{r}{h} x$. The last expression is the radius of each disk. Summing the volume of the disks from $x = 0$ to $x = h$ yields $V = \pi \int_{0}^{h} \left(\frac{r}{h} x \right)^2 dx = \pi \left(\frac{r}{h} \right)^2 \left[\frac{1}{3} x^3 \right]_{0}^{h} = \frac{1}{3} \pi r^2 h.$

39 Revolve the trapezoid with vertices $(0, 0)$, $(0, r)$, $(h, 0)$ and (h, R) {see *Figure 39*}
about the x-axis. The equation of the line containing $(0, r)$ and (h, R) is of the form
$y = mx + b$, specifically, $y = \left(\dfrac{R - r}{h}\right)x + r$. If we sum the volume of the disks
along the x-axis from $x = 0$ to $x = h$, we have

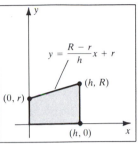

$$V = \pi \int_0^h \left(\frac{R - r}{h} \cdot x + r\right)^2 dx$$

$$= \pi \int_0^h \left[\left(\frac{R - r}{h}\right)^2 x^2 + 2r\left(\frac{R - r}{h}\right)x + r^2\right] dx$$

$$= \pi \left[\left(\frac{R - r}{h}\right)^2 \frac{x^3}{3} + 2r\left(\frac{R - r}{h}\right)\frac{x^2}{2} + r^2 x\right]_0^h$$

$$= \pi \left[\frac{(R - r)^2}{h^2}\frac{h^3}{3} + 2r\frac{(R - r)}{h}\frac{h^2}{2} + r^2 h\right]$$

$$= \tfrac{1}{3}\pi h\left[(R - r)^2 + 3r(R - r) + 3r^2\right]$$

$$= \tfrac{1}{3}\pi h\left[(R^2 - 2Rr + r^2) + 3rR - 3r^2 + 3r^2\right] = \tfrac{1}{3}\pi h(R^2 + Rr + r^2).$$

Figure 39

41 The volume $V = \pi \displaystyle\int_0^6 [f(x)]^2\, dx$. Let $g(x) = [f(x)]^2$ to estimate V.

$$T = \pi \cdot \tfrac{6 - 0}{2(6)}\{g(0) + 2\left[g(1) + g(2) + g(3) + g(4) + g(5)\right] + g(6)\}$$

$$= \tfrac{\pi}{2}\left[2^2 + 2(1^2 + 2^2 + 4^2 + 2^2 + 2^2) + 1^2\right] = \tfrac{63\pi}{2} \approx 98.96.$$

Exercises 6.3

Note: Students sometimes confuse the disk method of §6.2 with the shell method of this
section, especially in terms of the variable of integration. The following may help
avoid this confusion. When using the disk method, revolving about the x-axis
leads to an integration with respect to x, and revolving about the y-axis leads to
an integration with respect to y. When using the shell method, revolving about
the x-axis leads to an integration with respect to y, and revolving about the y-axis
leads to an integration with respect to x.

1 To find the radius of a typical shell, think of it as the distance between the shell and
the line that it is being revolved about. In this case, a shell is x units from the y-axis,
or $x - 0 = x$. The altitude of the shell is computed exactly the same way as the
height of a rectangle was in §6.1 for finding the area between two curves. In this
case, the altitude is $\sqrt{x - 2}$. Using (6.11) with $f(x) = \sqrt{x - 2}$, $a = 2$, and $b = 11$,
we have $V = 2\pi \displaystyle\int_2^{11} x\sqrt{x - 2}\, dx$.

3 $2x + y = 6 \Rightarrow x = -\tfrac{1}{2}y + 3$. In this exercise, $g(y) = -\tfrac{1}{2}y + 3$ is revolved about
the x-axis. The radius of each shell is $y - 0 = y$. The altitude of each shell is
$(-\tfrac{1}{2}y + 3)$. $V = 2\pi \displaystyle\int_c^d y\, g(y)\, dy = 2\pi \displaystyle\int_0^6 y(-\tfrac{1}{2}y + 3)\, dy.$

$\boxed{5}$ Using (6.11) with $f(x) = \sqrt{x}$, $a = 0$, and $b = 4$ yields

$$V = 2\pi \int_0^4 x\sqrt{x}\,dx = 2\pi\left[\tfrac{2}{5}x^{5/2}\right]_0^4 = \tfrac{4\pi}{5}(32) = \tfrac{128\pi}{5}.$$

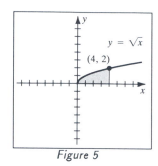

Figure 5

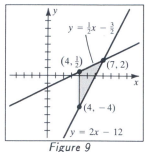

Figure 9

$\boxed{9}$ $y = 2x - 12$ and $y = \tfrac{1}{2}x - \tfrac{3}{2}$ intersect at $(7, 2)$. $2x - 12 \le \tfrac{1}{2}x - \tfrac{3}{2}$ on $[4, 7]$.
From *Figure 9*, the radius of a typical shell is x and the altitude is

$$(\tfrac{1}{2}x - \tfrac{3}{2}) - (2x - 12). \quad V = 2\pi \int_4^7 x\left[(\tfrac{1}{2}x - \tfrac{3}{2}) - (2x - 12)\right] dx =$$

$$2\pi \int_4^7 (\tfrac{21}{2}x - \tfrac{3}{2}x^2)\,dx = \pi\left[\tfrac{21}{2}x^2 - x^3\right]_4^7 = \pi\left(\tfrac{343}{2} - 104\right) = \tfrac{135\pi}{2}.$$

$\boxed{11}$ From *Figure 11*, we see that $2x - 4 \le 0$ on $[0, 2]$. The radius of a typical shell is x.
The altitude of a typical shell is $0 - (2x - 4) = 4 - 2x$. Note that $4 - 2x$ is
positive on the interval $[0, 2]$. If we had simply used $2x - 4$ as the height, the value
of the integral would be the negative of the true value.

$$V = 2\pi \int_0^2 x\left[0 - (2x - 4)\right] dx = 2\pi \int_0^2 (4x - 2x^2)\,dx$$

$$= 2\pi\left[2x^2 - \tfrac{2}{3}x^3\right]_0^2 = 2\pi(8 - \tfrac{16}{3}) = \tfrac{16\pi}{3}.$$

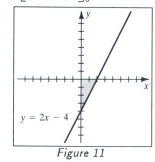

Figure 11

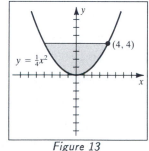

Figure 13

$\boxed{13}$ $x^2 = 4y$ and $y = 4$ intersect when $x^2 = 4(4) \Rightarrow x = \pm 4$.
Using symmetry about $x = 0$ with $g(y) = \sqrt{4y}$,

$$V = 2 \cdot 2\pi \int_0^4 y\,g(y)\,dy = 2 \cdot 2\pi \int_0^4 y(\sqrt{4y})\,dy = 8\pi \int_0^4 y^{3/2}\,dy = 8\pi\left[\tfrac{2}{5}y^{5/2}\right]_0^4 = \tfrac{512\pi}{5}.$$

$\boxed{17}$ $y = \sqrt{x + 4} \Rightarrow x = g(y) = y^2 - 4$. $y^2 - 4 \le 0$ on $0 \le y \le 2$.

Radius $= y$ and altitude $= 0 - (y^2 - 4)$ (right minus left).

$$V = 2\pi \int_0^2 y\, g(y)\, dy = 2\pi \int_0^2 y\Big[0 - (y^2 - 4)\Big]\, dy = 2\pi \int_0^2 (4y - y^3)\, dy$$

$$= 2\pi \Big[2y^2 - \tfrac{1}{4}y^4\Big]_0^2 = 2\pi(8 - 4) = 8\pi.$$

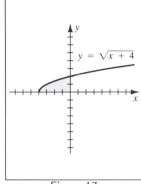

Figure 17

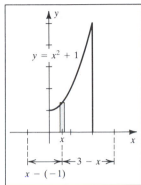

Figure 19

$\boxed{19}$ (a) On the interval $0 \le x \le 2$, a typical rectangle has radius $3 - x$ (right minus left)

and altitude $x^2 + 1$. $V = 2\pi \int_0^2 (3 - x)(x^2 + 1)\, dx$.

(b) The radius is $\Big[x - (-1)\Big]$ (right minus left) and the altitude is $x^2 + 1$.

$$V = 2\pi \int_0^2 [x - (-1)](x^2 + 1)\, dx.$$

Note: In Exercises 23–26, refer to the figures of Exercises 31–34 in §6.2.

$\boxed{25}$ Let $y = \pm\sqrt{1 - x^2}$ be the two functions on $-1 \le x \le 1$. The radius of a typical

shell is $(5 - x)$ (right minus left) and the altitude is $\Big[\sqrt{1 - x^2} - (-\sqrt{1 - x^2})\Big]$

(upper minus lower) $= 2\sqrt{1 - x^2}$. Do not make the mistake of trying to use

symmetry for this exercise. The volume generated by the region on $-1 \le x \le 0$ is

greater than the volume generated by the region on $0 \le x \le 1$.

$$V = 2 \cdot 2\pi \int_{-1}^1 (5 - x)\sqrt{1 - x^2}\, dx.$$

$\boxed{27}$ (a) $y = 1/\sqrt{x} \Rightarrow x = 1/y^2$. When integrating with
respect to y, the region must be broken into two
parts. For both parts, the radius is y. We will be
summing up the shells from $y = 0$ to $y = 1$. On
$0 \le y \le \tfrac{1}{2}$, the altitude of a cylinder is $(4 - 1)$
(right minus left). On $\tfrac{1}{2} \le y \le 1$, the altitude is
$\Big[(1/y^2) - 1\Big]$ (right minus left).

$$V = 2\pi \int_0^{1/2} y(4 - 1)\, dy + 2\pi \int_{1/2}^1 y\Big[(1/y^2) - 1\Big]\, dy.$$

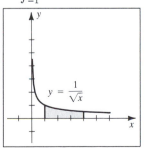

Figure 27

(b) On $1 \leq x \leq 4$, the radius of a disk is $1/\sqrt{x}$. $V = \pi \int_1^4 \left(\frac{1}{\sqrt{x}}\right)^2 dx$.

31 The volume $V = 2\pi \int_0^6 xf(x)\, dx$. Let $g(x) = xf(x)$ to estimate V.

$$T = 2\pi \cdot \frac{6-0}{2(6)} \{ g(0) + 2\big[g(1) + g(2) + g(3) + g(4) + g(5)\big] + g(6) \}$$

$$= \pi \big[0 \cdot 2 + 2(1 \cdot 1 + 2 \cdot 2 + 3 \cdot 4 + 4 \cdot 2 + 5 \cdot 2) + 6 \cdot 1 \big] = 76\pi \approx 238.76.$$

Exercises 6.4

Note: Exercises 1–26: The first integral represents a general formula for the volume. In
Exercises 1–8, the vertical distance between the graphs of $y = \sqrt{x}$ and $y = -\sqrt{x}$ is
$\big[\sqrt{x} - (-\sqrt{x})\big]$, denoted by $2\sqrt{x}$.

1 Let s denote the length of a side of the square.

Since s is the distance from $y = \sqrt{x}$ to $y = -\sqrt{x}$, $s = 2\sqrt{x}$.

$$V = \int_c^d s^2\, dx = \int_0^9 A(x)\, dx = \int_0^9 (2\sqrt{x})^2\, dx = \int_0^9 4x\, dx = 4\big[\tfrac{1}{2}x^2\big]_0^9 = 162.$$

7 Let B denote the length of the lower base, b the length of the upper base,
and h the height. Then $B = 2\sqrt{x}$, $b = \tfrac{1}{2}(2\sqrt{x})$, and $h = \tfrac{1}{4}(2\sqrt{x})$.

$$V = \int_c^d \tfrac{1}{2}(B + b)h\, dx = \int_0^9 A(x)\, dx = \int_0^9 \tfrac{1}{2}\big[2\sqrt{x} + \tfrac{1}{2}(2\sqrt{x})\big]\big[\tfrac{1}{4}(2\sqrt{x})\big]\, dx =$$

$$\int_0^9 \tfrac{1}{2}(3\sqrt{x})(\tfrac{1}{2}\sqrt{x})\, dx = \tfrac{3}{4}\int_0^9 x\, dx = \tfrac{3}{4}\big[\tfrac{1}{2}x^2\big]_0^9 = \tfrac{243}{8}.$$

9 The boundary of the base, $x^2 + y^2 = a^2$, can be
thought of as two semicircles; an upper semicircle
with equation $y = \sqrt{a^2 - x^2}$, and a lower semicircle
with equation $y = -\sqrt{a^2 - x^2}$.   The side s of the
square in the base runs from the lower semicircle to
the upper semicircle.   Hence,

$s = \sqrt{a^2 - x^2} - (-\sqrt{a^2 - x^2}) = 2\sqrt{a^2 - x^2}$.

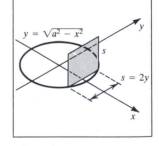

Figure 9

Note: It is sometimes helpful to examine the value of s
(or whatever expression you are considering) by checking its value at the extreme
points. In this case, if $x = 0$, then $s = 2a$, and if $x = a$, then $s = 0$ (no square is
formed). Both calculations make sense in terms of Figure 9. We next wish to add up
the area of all of the squares from $x = -a$ to $x = a$. Thus,

$$V = \int_c^d s^2\, dx = \int_{-a}^a A(x)\, dx = \int_{-a}^a (2\sqrt{a^2 - x^2})^2\, dx = 4\int_{-a}^a (a^2 - x^2)\, dx$$

$$= 8\int_0^a (a^2 - x^2)\, dx \; \{\text{by symmetry}\} = 8\big[a^2 x - \tfrac{1}{3}x^3\big]_0^a = \tfrac{16}{3}a^3.$$

$\boxed{13}$ In Figure 6.37(i), the base has dimensions a and a, which may be represented by $2y$ and $2y$ as in 6.37(ii). In this exercise, we have a base with dimensions a and $2a$. Thus, the cross-sectional rectangle x units from O will have dimensions $2y$ by $4y$. By Example 1, $y = \frac{ax}{2h}$. $V = \int_c^d lw \, dx = \int_0^h A(x) \, dx = \int_0^h (4y)(2y) \, dx = \int_0^h \left(\frac{2ax}{h}\right)\left(\frac{ax}{h}\right) dx$

$$= \frac{2a^2}{h^2}\left[\frac{1}{3}x^3\right]_0^h = \frac{2}{3}a^2 h.$$

$\boxed{15}$ From *Figure 15*, we see that the diameter d of each semicircular cross section has length $4 - \frac{1}{4}y^2$. We need a formula for the radius to compute the area of a cross section. Hence, $r = \frac{1}{2}d = \frac{1}{2}(4 - \frac{1}{4}y^2)$. $V = \int_c^d \frac{1}{2}\pi r^2 \, dy = \int_{-4}^4 A(y) \, dy =$

$2 \int_0^4 \frac{1}{2}\pi \left[\frac{1}{2}(4 - \frac{1}{4}y^2)\right]^2 dy \, \{\text{by symmetry}\} = \frac{\pi}{4} \int_0^4 (16 - 2y^2 + \frac{1}{16}y^4) \, dy =$

$$\frac{\pi}{4}\left[16y - \frac{2}{3}y^3 + \frac{1}{80}y^5\right]_0^4 = \frac{\pi}{4}\left[16(4) - \frac{2}{3}(64) + \frac{1}{80}(1024)\right] = \frac{\pi}{4}\left(\frac{512}{15}\right) = \frac{128\pi}{15}.$$

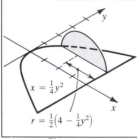

Figure 15

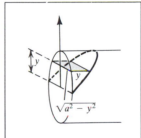

Figure 17

$\boxed{17}$ The wedge can be considered as a solid whose base is a semicircle of radius a and every cross section of the wedge, by a plane perpendicular to the vertical radius, is a rectangle. Using $x^2 + y^2 = a^2$ for the equation of the circle, then if $0 \le y \le a$, the length of this rectangle is the distance from the right side of the semicircle, $x = \sqrt{a^2 - y^2}$, to the left side of the semicircle, $x = -\sqrt{a^2 - y^2}$. Thus, the length of the rectangle is $2\sqrt{a^2 - y^2}$. Because of the 45° cut, the width of this rectangle is y.

$V = \int_c^d lw \, dy = \int_0^a A(y) \, dy = 2\int_0^a (\sqrt{a^2 - y^2}) \, y \, dy$

$= -\int_{a^2}^0 \sqrt{u} \, du \, \{ u = a^2 - y^2, \, -du = 2y \, dy; \, y = 0, \, a \Rightarrow u = a^2, \, 0 \}$

$= -\left[\frac{2}{3}u^{3/2}\right]_{a^2}^0 = -\frac{2}{3}(0 - a^3) = \frac{2}{3}a^3.$

Alternate solution: If the x-axis is the intersection of the two cuts, we can sum the areas of the isosceles right triangles from $x = -a$ to $x = a$. The height of each triangle is $y = \sqrt{a^2 - x^2}$, which is also the length of the base.   (cont.)

$$V = \int_c^d \tfrac{1}{2}bh\,dx = \int_{-a}^a A(x)\,dx = 2\int_0^a \tfrac{1}{2}\sqrt{a^2 - x^2}\,\sqrt{a^2 - x^2}\,dx$$

$$= \int_0^a (a^2 - x^2)\,dx = \left[a^2x - \tfrac{1}{3}x^3\right]_0^a = \tfrac{2}{3}a^3.$$

19 From *Figure 19*, we see that the base of each cross section is $2y = 2\sqrt{a^2 - x^2}$.
The height h is constant.

$$V = \int_c^d \tfrac{1}{2}bh\,dx = \int_{-a}^a A(x)\,dx = \int_{-a}^a \tfrac{1}{2}\cdot 2\left(\sqrt{a^2 - x^2}\right)h\,dx = h\int_{-a}^a \sqrt{a^2 - x^2}\,dx.$$

To interpret $\displaystyle\int_{-a}^a \sqrt{a^2 - x^2}\,dx$ as an area (as stated in the *Hint*), we recognize it as the

area under the curve $y = \sqrt{a^2 - x^2}$ from $x = -a$ to $x = a$. But this is just a

semicircle of radius a, and its area is $\tfrac{1}{2}\pi a^2$. Thus, $V = h(\tfrac{1}{2}\pi a^2) = \tfrac{1}{2}\pi a^2 h.$

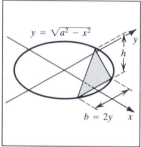

Figure 19

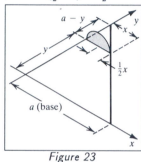

Figure 23

21 We place a coordinate line along the 4-cm side with the origin at the vertex as shown
in *Figure 21a*. Consider a point on this line x units from the vertex. A plane through
this point, perpendicular to the coordinate line, intersects the solid in a triangular
cross section. Let b denote the base and h the height of this triangle. We can express
b and h in terms of x using similar triangles. From *Figure 21b*, $b = \tfrac{2}{4}x$. From *Figure
21c*, $h = \tfrac{3}{4}x.$

$$V = \int_c^d \tfrac{1}{2}bh\,dx = \int_0^4 A(x)\,dx = \int_0^4 \tfrac{1}{2}(\tfrac{2}{4}x)(\tfrac{3}{4}x)\,dx = \tfrac{3}{16}\int_0^4 x^2\,dx = \tfrac{3}{16}\left[\tfrac{1}{3}x^3\right]_0^4 = 4\text{ cm}^3.$$

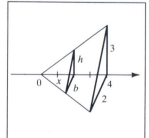

Figure 21a

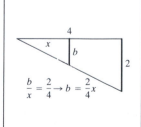

Figure 21b

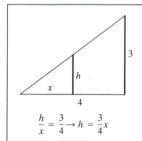

Figure 21c

$\boxed{23}$ The diameter of a cross section is x, hence, its radius is $\frac{1}{2}x$. The area of a cross section of radius $\frac{1}{2}x$ is $\frac{1}{2}\pi(\frac{1}{2}x)^2$. From *Figure 23* (p. 162), we see that the equation of the line containing the hypotenuse is $y = a - x$, or, equivalently, $x = a - y$.

$$V = \int_c^d \tfrac{1}{2}\pi r^2 \, dy = \int_0^a A(y)\, dy = \int_0^a \tfrac{1}{2}\pi \left[\tfrac{1}{2}(a - y)\right]^2 dy$$

$$= \tfrac{\pi}{2} \int_0^a \tfrac{1}{4}(a^2 - 2ay + y^2)\, dy = \tfrac{\pi}{8}\left[a^2 y - ay^2 + \tfrac{1}{3}y^3\right]_0^a = \tfrac{\pi}{24}a^3.$$

Exercises 6.5

$\boxed{1}$ (a) On $1 \le x \le 3$, $y = x^3 + 1 \Rightarrow y' = 3x^2$ and $L = \int_1^3 \sqrt{1 + (3x^2)^2}\, dx$ by (6.14).

(b) To integrate with respect to y, we must have a function of the form $x = g(y)$.

Solving $y = x^3 + 1$ for x yields $y = x^3 + 1 \Rightarrow y - 1 = x^3 \Rightarrow x = \sqrt[3]{y - 1}$.

On $2 \le y \le 28$, $x = \sqrt[3]{y - 1} \Rightarrow x' = \tfrac{1}{3}(y - 1)^{-2/3}$ and

$$L = \int_2^{28} \sqrt{1 + \left[\tfrac{1}{3}(y - 1)^{-2/3}\right]^2}\, dy \text{ by (6.15)}.$$

$\boxed{5}$ $y = f(x) = \tfrac{2}{3}x^{2/3} \Rightarrow f'(x) = \tfrac{4}{9}x^{-1/3}$. Thus, $\sqrt{1 + [f'(x)]^2} = \sqrt{1 + \tfrac{16}{81}x^{-2/3}} = $

$\sqrt{1 + \dfrac{\tfrac{16}{81}}{x^{2/3}}} = \sqrt{\dfrac{x^{2/3} + \tfrac{16}{81}}{x^{2/3}}} = \dfrac{\sqrt{x^{2/3} + \tfrac{16}{81}}}{\sqrt{(x^{1/3})^2}} = \dfrac{\sqrt{x^{2/3} + \tfrac{16}{81}}}{|x^{1/3}|}$. To determine if we should

use $x^{1/3}$ or $-(x^{1/3})$ for $|x^{1/3}|$, we must examine the range of values that $x^{1/3}$ takes on over the interval of integration, which is $1 \le x \le 8$. On $[1, 8]$, $1 \le x^{1/3} \le 2$, and these are positive values. Thus, let $|x^{1/3}| = x^{1/3}$ and the expression becomes

$x^{-1/3}\sqrt{x^{2/3} + \tfrac{16}{81}}$. By (6.14), $L = \int_1^8 x^{-1/3}\sqrt{x^{2/3} + \tfrac{16}{81}}\, dx$. To evaluate this integral,

let $u = x^{2/3} + \tfrac{16}{81}$ and then $\tfrac{3}{2}\, du = x^{-1/3}\, dx$. $x = 1 \Rightarrow u = 1 + \tfrac{16}{81} = c$ and $x = 8$

$\Rightarrow u = 4 + \tfrac{16}{81} = d$. $L = \tfrac{3}{2}\int_c^d u^{1/2}\, du = \tfrac{3}{2}\left[\tfrac{2}{3}u^{2/3}\right]_c^d = $

$$d^{3/2} - c^{3/2} = (4 + \tfrac{16}{81})^{3/2} - (1 + \tfrac{16}{81})^{3/2} \approx 7.29.$$

$\boxed{9}$ $y = f(x) = \tfrac{1}{12}x^3 + 1/x \Rightarrow y' = f'(x) = \tfrac{1}{4}x^2 - 1/x^2 \Rightarrow 1 + [f'(x)]^2 = $

$$1 + (\tfrac{1}{16}x^4 - \tfrac{1}{2} + 1/x^4) = \tfrac{1}{16}x^4 + \tfrac{1}{2} + 1/x^4 = (\tfrac{1}{4}x^2 + 1/x^2)^2.$$

The last factorization may not be obvious. Several solutions in this section use the following steps (where p is an algebraic expression).

$$1 + \left(\tfrac{p}{2} - \tfrac{1}{2p}\right)^2 = 1 + \left(\tfrac{p^2}{4} - \tfrac{1}{2} + \tfrac{1}{4p^2}\right) = \tfrac{p^2}{4} + \tfrac{1}{2} + \tfrac{1}{4p^2} = \left(\tfrac{p}{2} + \tfrac{1}{2p}\right)^2.$$

There could also be a constant associated with the expression p. Hence,   (cont.)

$$L = \int_1^2 \sqrt{(\tfrac{1}{4}x^2 + 1/x^2)^2}\, dx \qquad \{\text{by (6.14)}\}$$

$$= \int_1^2 \left|\tfrac{1}{4}x^2 + 1/x^2\right| dx \qquad \{\sqrt{x^2} = |x|\}$$

$$= \int_1^2 (\tfrac{1}{4}x^2 + 1/x^2)\, dx \qquad \{\text{since } (\tfrac{1}{4}x^2 + 1/x^2) > 0 \text{ on } [1, 2]\}$$

$$= \left[\tfrac{1}{12}x^3 - \tfrac{1}{x}\right]_1^2 \qquad \{\text{antiderivative}\}$$

$$= \left[(\tfrac{2}{3} - \tfrac{1}{2}) - (\tfrac{1}{12} - 1)\right] = \tfrac{13}{12} \qquad \{\text{simplify}\}$$

[11] Since it would be difficult to solve this equation for y, we will solve it for x and use

(6.15). $30xy^3 - y^8 = 15 \Rightarrow x = g(y) = \dfrac{15 + y^8}{30y^3} = \tfrac{1}{2}y^{-3} + \tfrac{1}{30}y^5 \Rightarrow$

$g'(y) = -\tfrac{3}{2}y^{-4} + \tfrac{1}{6}y^4 \Rightarrow 1 + [g'(y)]^2 = 1 + (\tfrac{9}{4}y^{-8} - \tfrac{1}{2} + \tfrac{1}{36}y^8) =$

$\tfrac{9}{4}y^{-8} + \tfrac{1}{2} + \tfrac{1}{36}y^8 = (\tfrac{3}{2}y^{-4} + \tfrac{1}{6}y^4)^2$. Hence, $L = \displaystyle\int_1^2 \sqrt{(\tfrac{3}{2}y^{-4} + \tfrac{1}{6}y^4)^2}\, dy =$

$\displaystyle\int_1^2 \left|\tfrac{3}{2}y^{-4} + \tfrac{1}{6}y^4\right| dy = \int_1^2 (\tfrac{3}{2}y^{-4} + \tfrac{1}{6}y^4)\, dy = \left[-\tfrac{1}{2}y^{-3} + \tfrac{1}{30}y^5\right]_1^2 =$

$$\left[(-\tfrac{1}{16} + \tfrac{32}{30}) - (-\tfrac{1}{2} + \tfrac{1}{30})\right] = \tfrac{353}{240}.$$

[13] Since it would be difficult to solve this equation for y, we will solve it for x and use

(6.15). $2y^3 - 7y + 2x = 8 \Rightarrow x = g(y) = \tfrac{1}{2}(8 + 7y - 2y^3) = 4 + \tfrac{7}{2}y - y^3 \Rightarrow$

$$1 + [g'(y)]^2 = 1 + (\tfrac{7}{2} - 3y^2)^2. \text{ Hence, } L = \int_0^2 \sqrt{1 + (\tfrac{7}{2} - 3y^2)^2}\, dy.$$

[15] Let $y = x$ intersect the graph at $x = y = a$, $a > 0$.

See *Figure 15.* The point of intersection occurs when

$a^{2/3} + a^{2/3} = 1 \Rightarrow 2a^{2/3} = 1 \Rightarrow a^{2/3} = \tfrac{1}{2} \Rightarrow$

$a = (\tfrac{1}{2})^{3/2}$. Solving $x^{2/3} + y^{2/3} = 1$ for y yields

$y^{2/3} = 1 - x^{2/3} \Rightarrow y = \pm(1 - x^{2/3})^{3/2}$.

Since $y > 0$, $y = +(1 - x^{2/3})^{3/2}$ and

$y' = \tfrac{3}{2}(1 - x^{2/3})^{1/2}(-\tfrac{2}{3}x^{-1/3}) = -\dfrac{(1 - x^{2/3})^{1/2}}{x^{1/3}}.$

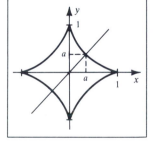

Figure 15

Thus, $\sqrt{1 + (y')^2} = \left(1 + \dfrac{1 - x^{2/3}}{x^{2/3}}\right)^{1/2} =$

$\left(\dfrac{x^{2/3}}{x^{2/3}} + \dfrac{1 - x^{2/3}}{x^{2/3}}\right)^{1/2} = \left(\dfrac{1}{x^{2/3}}\right)^{1/2} = \dfrac{1}{|x^{1/3}|} = \dfrac{1}{x^{1/3}} = x^{-1/3}.$

By symmetry, $L = 8\displaystyle\int_a^1 x^{-1/3}\, dx = 8\left[\tfrac{3}{2}x^{2/3}\right]_a^1 = 12(1 - a^{2/3}) =$

$$12 \cdot \left\{1 - \left[(\tfrac{1}{2})^{3/2}\right]^{2/3}\right\} = 12(1 - \tfrac{1}{2}) = 6. \text{ See } \textit{Note} \text{ on page 165.}$$

Note: If $L = 4\int_0^1 x^{-1/3}\,dx = 6$ is used, the integrand is discontinuous at $x = 0$ and

the fundamental theorem of calculus (5.30) does not apply. This type of

integral is called *improper* and will be studied in Chapter 10.

17 (a) $f(x) = x^{2/3} \Rightarrow f'(x) = \frac{2}{3}x^{-1/3} \Rightarrow 1 + [f'(x)]^2 = 1 + \frac{4}{9}x^{-2/3} =$

$(x^{2/3} + \frac{4}{9})x^{-2/3}$. An arc length function is

$$
\begin{aligned}
s(x) &= \int_a^x \sqrt{1 + \left[f'(t)\right]^2}\,dt && \{\,\text{by (6.16)}\,\} \\
&= \int_1^x \sqrt{(t^{2/3} + \frac{4}{9})t^{-2/3}}\,dt && \{\,\text{from above}\,\} \\
&= \int_1^x \sqrt{t^{2/3} + \frac{4}{9}}\,(t^{-1/3})\,dt && \{\,\sqrt{t^{-2/3}} = t^{-1/3} \text{ on } [1, x]\,\} \\
&= \left[(t^{2/3} + \frac{4}{9})^{3/2}\right]_1^x && \{\,u = t^{2/3} + \frac{4}{9} \text{ and } \frac{3}{2}\,du = t^{-1/3}\,dt\,\} \\
&= (x^{2/3} + \frac{4}{9})^{3/2} - (1 + \frac{4}{9})^{3/2} && \{\,\text{evaluate at } t = x \text{ and at } t = 1\,\} \\
&= \left(\frac{9x^{2/3} + 4}{9}\right)^{3/2} - \left(\frac{13}{9}\right)^{3/2} && \{\,\text{simplify}\,\} \\
&= \frac{1}{27}\left[(9x^{2/3} + 4)^{3/2} - 13^{3/2}\right] && \{\,(\frac{1}{9})^{3/2} = \frac{1}{27}\,\}
\end{aligned}
$$

The function $s(x)$ gives the exact length of the curve from the fixed point with

x-coordinate 1 to the variable point with x-coordinate x.

(b) Using (3.26), $\Delta s = s(x + \Delta x) - s(x) = s(1.1) - s(1) =$

$$\frac{1}{27}\left[[9(1.1)^{2/3} + 4]^{3/2} - 13^{3/2}\right] - 0 \approx 0.1196.$$

To find ds, we use (6.17)(i).

$$
\begin{aligned}
ds &= \sqrt{1 + [f'(x)]^2}\,dx && \{\,(6.17)(i)\,\} \\
&= \sqrt{x^{2/3} + \frac{4}{9}}\,(x^{-1/3})\,dx && \{\,\text{from part (a)}\,\} \\
&= \sqrt{(1)^{2/3} + \frac{4}{9}}\,(1)^{-1/3}\,(0.1) && \{\,x = 1 \text{ and } dx = 0.1\,\} \\
&= \frac{\sqrt{13}}{30} \approx 0.1202 && \{\,\text{simplify and approximate}\,\}
\end{aligned}
$$

Alternatively, we could use $ds = s'(x)\,dx$ and employ the fundamental theorem of

calculus to find $s'(x)$.

19 Let $f(x) = x^2$. Then, $\Delta s \approx ds = \sqrt{1 + [f'(x)]^2}\,dx = \sqrt{1 + (2x)^2}\,dx$.

With $x = 2$, $dx = 0.1$, $ds = \sqrt{17}(0.1) \approx 0.4123$.

21 $f(x) = \cos x$, $ds = \sqrt{1 + [f'(x)]^2}\,dx = \sqrt{1 + (-\sin x)^2}\,dx$. With $x = a = \frac{\pi}{6}$ and

$$dx = b - a = \frac{31\pi}{180} - \frac{\pi}{6} = \frac{\pi}{180},\ ds = \sqrt{1 + (-\frac{1}{2})^2}\,(\frac{\pi}{180}) = \frac{\pi\sqrt{5}}{360} \approx 0.0195.$$

[23] Let $f(x) = x^2 + x + 3$. $L = \int_{-2}^{2} \sqrt{1 + [f'(x)]^2}\, dx =$

$\int_{-2}^{2} \sqrt{1 + (2x + 1)^2}\, dx = \int_{-2}^{2} g(x)\, dx$, where $g(x) = \sqrt{1 + (2x + 1)^2}$.

We now use Simpson's rule with $n = 4$ to approximate L.

$S = \frac{2 - (-2)}{3(4)}\Big[g(-2) + 4\,g(-1) + 2\,g(0) + 4\,g(1) + g(2)\Big]$

$= \frac{1}{3}(\sqrt{10} + 4\sqrt{2} + 2\sqrt{2} + 4\sqrt{10} + \sqrt{26}) \approx 9.80$.

[29] To use (6.19), we need a function $f(x) \geq 0$ on $[0, 1]$. $4x = y^2 \Rightarrow y = \pm\sqrt{4x}$.

Since y is positive (0 to 2), $y = f(x) = \sqrt{4x} = 2\sqrt{x}$ and $f'(x) = \frac{1}{\sqrt{x}}$.

$S = \int_{a}^{b} 2\pi f(x)\sqrt{1 + [f'(x)]^2}\, dx = 2\pi \int_{0}^{1} 2\sqrt{x}\,\sqrt{1 + \frac{1}{x}}\, dx = 4\pi \int_{0}^{1} \sqrt{x \cdot 1 + x \cdot \frac{1}{x}}\, dx =$

$4\pi \int_{0}^{1} \sqrt{x + 1}\, dx = 4\pi \Big[\frac{2}{3}(x + 1)^{3/2}\Big]_{0}^{1} = \frac{8\pi}{3}(2^{3/2} - 1) \approx 15.32$.

[33] To use the formula at the bottom of page 340, we need a function $g(y) \geq 0$ on $[2, 4]$.

$y = 2x^{1/3} \Rightarrow x = (\frac{1}{2}y)^3 \Rightarrow x = g(y) = \frac{1}{8}y^3$ and $x' = \frac{3}{8}y^2$.

$1 + [g'(y)]^2 = 1 + \frac{9}{64}y^4 = \frac{1}{64}(64 + 9y^4)$.

$S = 2\pi \int_{2}^{4} g(y)\sqrt{1 + [g'(y)]^2}\, dy = 2\pi \int_{2}^{4} \frac{1}{8}y^3 \Big(\frac{1}{8}\sqrt{64 + 9y^4}\Big)\, dy = \frac{\pi}{32}\int_{2}^{4} y^3 \sqrt{64 + 9y^4}\, dy$

$= \frac{\pi}{32}\Big[\frac{1}{36} \cdot \frac{2}{3}(64 + 9y^4)^{3/2}\Big]_{2}^{4} \ \{u = 64 + 9y^4,\ \frac{1}{36}\,du = y^3\,dy\}$

$= \frac{\pi}{1728}(2368^{3/2} - 208^{3/2}) = \frac{\pi}{1728}\Big[(64 \cdot 37)^{3/2} - (16 \cdot 13)^{3/2}\Big]$

$= \frac{\pi}{1728}\Big[512(37)^{3/2} - 64(13)^{3/2}\Big] = \frac{\pi}{27}\Big[8(37)^{3/2} - 13^{3/2}\Big] \approx 204.04$.

[35] The entire surface is generated when the arc between the points $(3, 4)$ and $(0, 5)$ is

revolved about the y-axis. $x^2 + y^2 = 25 \Rightarrow x = \pm\sqrt{25 - y^2}$.

$x > 0 \Rightarrow g(y) = \sqrt{25 - y^2} \Rightarrow 1 + [g'(y)]^2 = 1 + \Big(\dfrac{-y}{\sqrt{25 - y^2}}\Big)^2 = \dfrac{25}{25 - y^2}$.

$S = 2\pi \int_{4}^{5} g(y)\sqrt{1 + [g'(y)]^2}\, dy = 2\pi \int_{4}^{5} \sqrt{25 - y^2}\,\sqrt{\dfrac{25}{25 - y^2}}\, dy = 2\pi \int_{4}^{5} 5\, dy =$

$10\pi \int_{4}^{5} dy = 10\pi(5 - 4) = 10\pi$.

[37] The cone can be obtained by revolving the line segment with endpoints $(0, 0)$ and

(h, r) about the x-axis. An equation of the line containing those endpoints is

$y = \frac{r}{h}x$. Then, $y' = \frac{r}{h}$ and $1 + [y']^2 = 1 + \dfrac{r^2}{h^2} = \dfrac{h^2 + r^2}{h^2}$.

$S = 2\pi \int_{0}^{a} \frac{r}{h}x\Big(\dfrac{h^2 + r^2}{h^2}\Big)^{1/2}\, dx$ {note here that the integration variable is x and that

all terms except the single x are constant} $=$

$$2\pi \cdot \frac{r}{h} \cdot \frac{\sqrt{h^2 + r^2}}{h}\Big[\frac{1}{2}x^2\Big]_{0}^{h} = \frac{2\pi r\sqrt{h^2 + r^2}}{h^2} \cdot \frac{h^2}{2} = \pi r\sqrt{h^2 + r^2}.$$

41 Regard ds as the slant height of the frustum of a cone that has average radius x.

Using (6.18), the surface area of this frustum is $2\pi x\,ds = 2\pi x\sqrt{1 + \left[f'(x)\right]^2}\,dx$.

Apply the limit of sums operator $\int_a^b$ to obtain the formula

$$S = \int_a^b 2\pi x\sqrt{1 + \left[f'(x)\right]^2}\,dx.$$

43 For $k = 1$ and $f(x) = 1 - x^3$, $\frac{1}{4}k = \frac{1}{4}$, $Q_k = Q_1 = (\frac{1}{4}, f(\frac{1}{4})) = (\frac{1}{4}, \frac{63}{64})$,

and $Q_{k-1} = Q_0 = (0, f(0)) = (0, 1)$. The expression in the summation is

$\pi\left[f(x_0) + f(x_1)\right] d(Q_0, Q_1) = \pi\left[f(0) + f(\frac{1}{4})\right] d\left[(0, 1), (\frac{1}{4}, \frac{63}{64})\right] =$

$\pi\left(1 + \frac{63}{64}\right)\sqrt{(\frac{1}{4} - 0)^2 + (\frac{63}{64} - 1)^2} \approx 1.561566$. The other values are found in a

similar manner. $S = \sum_{k=1}^{4} 2\pi\dfrac{f(x_{k-1}) + f(x_k)}{2} d(Q_{k-1}, Q_k) \approx 1.561566 +$

1.593994 + 1.771803 + 1.143980 \approx 6.07. The true value is approximately 6.16.

Exercises 6.6

1 (a) and (b) The elapsed time is not relevant to the amount of work done.

It is relevant to Power = Work/Time. Thus, $F = (400)(15) = 6000$ ft-lb.

3 By Hooke's law, $f(x) = kx$. $f(1.5) = 1.5k = 8 \Rightarrow k = \frac{16}{3}$.

(a) $W = \displaystyle\int_0^{14-10} \frac{16}{3}x\,dx = \left[\frac{8}{3}x^2\right]_0^4 = \frac{128}{3}$ in.-lb.

(b) $W = \displaystyle\int_{11-10}^{13-10} \frac{16}{3}x\,dx = \left[\frac{8}{3}x^2\right]_1^3 = \frac{64}{3}$ in.-lb.

7 The work done in lifting only the elevator is $(3000)(9) = Fd = 27{,}000$ ft-lb. We must add the work of lifting the cable to this value. Let the cable be placed along the y-axis with $y = 0$ corresponding to the initial position of the bottom of the cable. On $0 \le y \le 9$, if the bottom of the cable is at y, there is $(12 - y)$ ft of cable still suspended with a weight of $14(12 - y)$. Thus, the work done in lifting the cable is

$$W = \int_0^9 14(12 - y)\,dy = 14\left[12y - \frac{1}{2}y^2\right]_0^9 = 945.$$

Total work = work$_{elevator}$ + work$_{cable}$ = 27,000 + 945 = 27,945 ft-lb.

9 Since the rates of vertical lift and water loss are constant with respect to time, the rate of water loss with respect to vertical lift is $\dfrac{-0.25 \text{ lb/sec}}{1.5 \text{ ft/sec}} = -\frac{1}{6}$ lb/ft. The weight of the water in the bucket at x feet is $20 - \frac{1}{6}x$ for $0 \le x \le 12$. The bucket will still contain some water at 12 feet (18 pounds) and the work required to lift just the water is $\displaystyle\int_0^{12} (20 - \frac{1}{6}x)\,dx = \left[20x - \frac{1}{12}x^2\right]_0^{12} = 228$ ft-lb. The work required to lift the empty bucket is $(4)(12) = 48$ ft-lb.

Thus, the total work is work$_{bucket}$ + work$_{water}$ = 48 + 228 = 276 ft-lb.

13 (a) Place the cylinder as shown in *Figure 13*. On $0 \le y \le 6$, a slice of water at height y has a volume of $\pi r^2\, dy = \pi(\frac{3}{2})^2\, dy$, a weight of $62.5\pi(\frac{3}{2})^2\, dy$, and is lifted a distance of $(6 - y)$. $W = 62.5\pi(\frac{9}{4})\int_0^6 (6 - y)\, dy = 62.5\pi(\frac{9}{4})\Big[6y - \frac{1}{2}y^2\Big]_0^6 =$

$$\frac{81\pi}{2}(62.5) \approx 7952 \text{ ft-lb.}$$

(b) A slice of water at height y is now lifted a distance of $(10 - y)$. The volume and weight of the slice are the same as in part (a). $W = 62.5\pi(\frac{9}{4})\int_0^6 (10 - y)\, dy =$

$$62.5\pi(\frac{9}{4})\Big[10y - \frac{1}{2}y^2\Big]_0^6 = \frac{189\pi}{2}(62.5) \approx 18{,}555 \text{ ft-lb.}$$

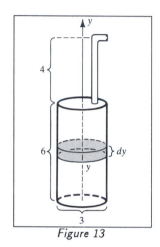

Figure 13

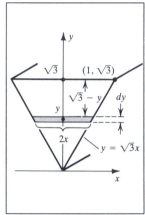

Figure 15

15 Place the trough as shown in *Figure 15*. Partition $0 \le y \le \sqrt{3}$. At height y, the width of the trough is $2x$. The end of the trough and the lower triangle are similar triangles. Thus, $\frac{2}{\sqrt{3}} = \frac{2x}{y} \Rightarrow x = \frac{y}{\sqrt{3}}$. A slice of water at height y is nearly a rectangle with a volume of (length × width × height) $= 8(2x)\, dy = 8\Big(2\frac{y}{\sqrt{3}}\Big) dy = \frac{16y}{\sqrt{3}}\, dy$, weight $(62.5)\frac{16y}{\sqrt{3}}\, dy$, and must be lifted a distance of $(\sqrt{3} - y)$.

$W = \int_0^{\sqrt{3}} (62.5)\frac{16y}{\sqrt{3}}(\sqrt{3} - y)\, dy = \frac{16(62.5)}{\sqrt{3}}\int_0^{\sqrt{3}} y(\sqrt{3} - y)\, dy = \frac{1000}{\sqrt{3}}\Big[\frac{\sqrt{3}}{2}y^2 - \frac{1}{3}y^3\Big]_0^{\sqrt{3}}$

$$= \frac{1000}{\sqrt{3}}\Big[\frac{3\sqrt{3}}{2} - \sqrt{3}\Big] = \frac{1000}{\sqrt{3}}\cdot\frac{\sqrt{3}}{2} = \boxed{500 \text{ ft-lb.}}$$

17 $pv^{1.2} = 115 \Rightarrow p = 115v^{-1.2}$. The work done is

$W = \int_{32}^{40} p\, dv = 115\int_{32}^{40} v^{-1.2}\, dv = 115\int_{32}^{40} v^{-6/5}\, dv = 115\Big[\frac{v^{-1/5}}{-1/5}\Big]_{32}^{40} =$

$$-575(40^{-1/5} - 32^{-1/5}) = 575(\tfrac{1}{2} - 40^{-1/5}) \approx 12.55 \text{ in.-lb.}$$

$\boxed{19}$ $F(s) = \dfrac{Gm_1 m_2}{s^2}$. Hence,

$$W = \int_{4000}^{4000+h} \frac{Gm_1 m_2}{s^2}\, ds = Gm_1 m_2 \int_{4000}^{4000+h} s^{-2}\, ds$$

$$= Gm_1 m_2 \left[-\frac{1}{s}\right]_{4000}^{4000+h} = -Gm_1 m_2 \left[\frac{1}{4000+h} - \frac{1}{4000}\right]$$

$$= -Gm_1 m_2 \left[\frac{4000-(4000+h)}{(4000+h)(4000)}\right] = \frac{Gm_1 m_2 h}{(4000)(4000+h)}.$$

$\boxed{23}$ (a) Let $f(x)$ be the force between two electrons. Then $f(x) = k/d^2$, where k is a

constant and d is the distance between the electrons. The distance between the

two electrons is given by $5 - x$, where x is the x-coordinate of the moving

electron. Hence, $f(x) = \dfrac{k}{(5-x)^2}$ and the work done in moving the electron from

$x = 0$ to $x = 3$ is $W = \displaystyle\int_0^3 \frac{k}{(5-x)^2}\, dx = k\left[\frac{1}{5-x}\right]_0^3 = \frac{3}{10}k$ J.

(b) The distance between the moving electron and the electron at $(-5,\ 0)$ is

$\left[x - (-5)\right]$. Since the forces are in opposite directions they have opposite signs.

Thus, the force is given by $-\dfrac{k}{(5+x)^2}$. The net force is the sum of these two

forces, and the work is

$$W = \int_0^3 \frac{k}{(5-x)^2}\, dx + \int_0^3 -\frac{k}{(5+x)^2}\, dx$$

$\{\, u = 5 - x,\ -du = dx$ in the 1st integral, and $u = 5 + x,\ du = dx$ in the 2nd $\}$

$$= -k\int_5^2 \frac{1}{u^2}\, du - k\int_5^8 \frac{1}{u^2}\, du$$

$$= -k\left[-\frac{1}{u}\right]_5^2 - k\left[-\frac{1}{u}\right]_5^8 = k(\tfrac{1}{2} - \tfrac{1}{5}) + k(\tfrac{1}{8} - \tfrac{1}{5}) = \frac{9}{40}k \text{ J.}$$

Exercises 6.7

$\boxed{1}$ m is the total mass. $m = \displaystyle\sum_{k=1}^{3} m_k = 100 + 80 + 70 = 250$. M_0 is the moment

about the origin. $M_0 = \displaystyle\sum_{k=1}^{3} m_k x_k = 100(-3) + 80(2) + 70(4) = 140$. The center

of mass is $\bar{x} = M_0/m = \frac{140}{250} = 0.56$.

$\boxed{3}$ m is the total mass. $m = \displaystyle\sum_{k=1}^{3} m_k = 2 + 7 + 5 = 14$. M_x is the moment about the

x-axis. $M_x = \displaystyle\sum_{k=1}^{3} m_k y_k = 2(-1) + 7(0) + 5(-5) = -27$. M_y is the moment

about the y-axis. $M_y = \displaystyle\sum_{k=1}^{3} m_k x_k = 2(4) + 7(-2) + 5(-8) = -46$.

$$\bar{x} = M_y/m = \frac{-46}{14} = -\frac{23}{7} \text{ and } \bar{y} = M_x/m = -\frac{27}{14}.$$

$\boxed{5}$ Let $f(x) = x^3$ and $g(x) = 0$ in (6.25).

The mass of the region is $m = \displaystyle\int_0^1 \rho(x^3 - 0)\,dx = \rho\left[\tfrac{1}{4}x^4\right]_0^1 = \tfrac{1}{4}\rho.$

The moment about the x-axis is $M_x = \displaystyle\int_0^1 \left[\tfrac{1}{2}(x^3 + 0)\cdot\rho(x^3 - 0)\right]dx = \tfrac{1}{2}\rho\left[\tfrac{1}{7}x^7\right]_0^1 =$

$\tfrac{1}{14}\rho.$ The moment about the y-axis is $M_y = \displaystyle\int_0^1 \left[x\cdot\rho(x^3 - 0)\right]dx = \rho\left[\tfrac{1}{5}x^5\right]_0^1 = \tfrac{1}{5}\rho.$

$$\bar{x} = \frac{M_y}{m} = \tfrac{4}{5} \text{ and } \bar{y} = \frac{M_x}{m} = \tfrac{2}{7}.$$

Note: Without loss of generality, we assume $\rho = 1$ for the remainder of this section.

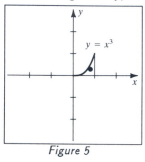

Figure 5

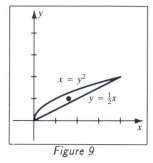

Figure 9

$\boxed{9}$ Let $f(x) = \sqrt{x}$ and $g(x) = \tfrac{1}{2}x$ in (6.25). $\sqrt{x} = \tfrac{1}{2}x \Rightarrow 4x = x^2 \Rightarrow x(4 - x) = 0 \Rightarrow$

$x = 0, 4$. For $0 \le x \le 4$, $\sqrt{x} \ge \tfrac{1}{2}x$. $m = \displaystyle\int_0^4 (\sqrt{x} - \tfrac{1}{2}x)\,dx = \left[\tfrac{2}{3}x^{3/2} - \tfrac{1}{4}x^2\right]_0^4 = \tfrac{4}{3}.$

$M_x = \tfrac{1}{2}\displaystyle\int_0^4 \left[(\sqrt{x})^2 - (\tfrac{1}{2}x)^2\right]dx = \tfrac{1}{2}\left[\tfrac{1}{2}x^2 - \tfrac{1}{12}x^3\right]_0^4 = \tfrac{4}{3}.$

$M_y = \displaystyle\int_0^4 x(\sqrt{x} - \tfrac{1}{2}x)\,dx = \left[\tfrac{2}{5}x^{5/2} - \tfrac{1}{6}x^3\right]_0^4 = \tfrac{32}{15}.$ $\bar{x} = \dfrac{M_y}{m} = \tfrac{8}{5}$ and $\bar{y} = \dfrac{M_x}{m} = 1.$

$\boxed{13}$ This is an R_y region. $y^2 = y + 2 \Rightarrow$

$y^2 - y - 2 = (y - 2)(y + 1) = 0 \Rightarrow y = -1, 2.$

For $-1 \le y \le 2$, $y + 2 \ge y^2$.

$m = \displaystyle\int_{-1}^2 \left[(y + 2) - y^2\right]dy$

$= \left[\tfrac{1}{2}y^2 + 2y - \tfrac{1}{3}y^3\right]_{-1}^2 = \tfrac{10}{3} - (-\tfrac{7}{6}) = \tfrac{9}{2}.$

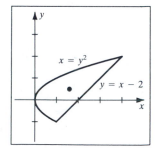

Figure 13

$M_x = \displaystyle\int_{-1}^2 y\left[(y + 2) - y^2\right]dy$

$= \left[\tfrac{1}{3}y^3 + y^2 - \tfrac{1}{4}y^4\right]_{-1}^2 = \tfrac{8}{3} - \tfrac{5}{12} = \tfrac{9}{4}.$

$M_y = \tfrac{1}{2}\displaystyle\int_{-1}^2 \left[(y + 2)^2 - (y^2)^2\right]dy = \tfrac{1}{2}\displaystyle\int_{-1}^2 \left[y^2 + 4y + 4 - y^4\right]dy$

$= \tfrac{1}{2}\left[\tfrac{1}{3}y^3 + 2y^2 + 4y - \tfrac{1}{5}y^5\right]_{-1}^2 = \tfrac{92}{15} - (-\tfrac{16}{15}) = \tfrac{36}{5}.$

$$\bar{x} = \frac{M_y}{m} = \tfrac{8}{5} \text{ and } \bar{y} = \frac{M_x}{m} = \tfrac{1}{2}.$$

17 In *Figure 17*, $\bar{x} = 0$ by symmetry. By the additivity of moments,

$M_x = M_x(\text{semicircle}) + M_x(\text{square})$. From Example 5, $M_x(\text{semicircle}) = \frac{2}{3}a^3$.

To find $M_x(\text{square})$, use (6.25) with $f(x) = 0$ and $g(x) = -2a$ for $-a \leq x \leq a$.

$M_x(\text{square}) = \int_{-a}^{a} \frac{1}{2}[0 + (-2a)] \cdot [0 - (-2a)] \, dx = 2 \cdot \frac{1}{2}(-4a^2) \int_{0}^{a} dx$

$\{\text{by symmetry}\} = -4a^3$. Thus, $M_x = \frac{2}{3}a^3 + (-4a^3) = -\frac{10}{3}a^3$. $m = \text{area} =$

$4a^2 + \frac{1}{2}\pi a^2 = \frac{1}{2}a^2(8 + \pi)$ and so $\bar{y} = \dfrac{M_x}{m} = \dfrac{-\frac{10}{3}a^3}{\frac{1}{2}a^2(8 + \pi)} = \dfrac{-20a}{3(8 + \pi)} \approx -0.60a$.

This is about $\dfrac{-0.60a}{-2a} = 0.30 = 30\%$ of the distance from the origin to the bottom of

the square.

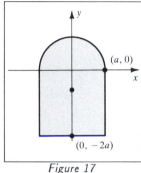

Figure 17

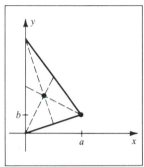

Figure 19

19 The midpoints of the three sides are $(\frac{a}{2}, \frac{b}{2})$, $(\frac{a}{2}, \frac{b + c}{2})$, and $(0, \frac{c}{2})$.

The three medians have equations $y = (\frac{b + c}{a})x$, $y = (\frac{2b - c}{2a})x + \frac{1}{2}c$, and

$y = (\frac{b - 2c}{a})x + c$. To find the intersection of the medians, let

$(\frac{b + c}{a})x = (\frac{2b - c}{2a})x + \frac{1}{2}c$. Solving for x, $\frac{b}{a}x + \frac{c}{a}x = \frac{b}{a}x - \frac{1}{2}\frac{c}{a}x + \frac{1}{2}c \Rightarrow$

$\frac{3}{2}(\frac{c}{a})x = \frac{1}{2}c \Rightarrow x = \frac{1}{3}a$ and $y = (\frac{b + c}{a})\frac{1}{3}a = \frac{1}{3}(b + c)$. This point also satisfies the

equation of the third median, $y = (\frac{b - 2c}{a})x + c$. Hence, all three medians intersect

at $(\frac{1}{3}a, \frac{1}{3}(b + c))$. The triangle has area $\frac{1}{2}(\text{base})(\text{height}) = \frac{1}{2}ca = \frac{1}{2}ac$.

The upper boundary is $f(x) = (\frac{b - c}{a})x + c$ and the lower boundary is $g(x) = \frac{b}{a}x$.

$M_y = \int_{0}^{a} x[f(x) - g(x)] \, dx = \int_{0}^{a} x[(\frac{b - c}{a})x + c - \frac{b}{a}x] \, dx$

$= \int_{0}^{a} x[c - \frac{c}{a}x] \, dx = c \int_{0}^{a} (x - x^2/a) \, dx = c[\frac{1}{2}x^2 - \frac{x^3}{3a}]_{0}^{a} = \frac{1}{6}ca^2$.   (cont.)

$$M_x = \frac{1}{2}\int_0^a \{[f(x)]^2 - [g(x)]^2\}\,dx = \frac{1}{2}\int_0^a \left\{\left[\left(\frac{b-c}{a}\right)x + c\right]^2 - \left[\frac{b}{a}x\right]^2\right\}\,dx$$

$$= \frac{1}{2}\int_0^a \left[\left(\frac{b-c}{a}\right)^2 x^2 + 2c\left(\frac{b-c}{a}\right)x + c^2 - \frac{b^2}{a^2}x^2\right]\,dx$$

$$= \frac{1}{2}\int_0^a \left[\left(\frac{b^2 - 2bc + c^2}{a^2}\right)x^2 + 2c\left(\frac{b-c}{a}\right)x + c^2 - \frac{b^2}{a^2}x^2\right]\,dx$$

$$= \frac{1}{2}\int_0^a \left[\left(\frac{c^2 - 2bc}{a^2}\right)x^2 + 2c\left(\frac{b-c}{a}\right)x + c^2\right]\,dx$$

$$= \frac{c}{2}\int_0^a \left[\frac{c - 2b}{a^2}x^2 + \frac{2(b-c)}{a}x + c\right]\,dx = \frac{c}{2}\left[\frac{c-2b}{3a^2}x^3 + \frac{b-c}{a}x^2 + cx\right]_0^a$$

$$= \frac{1}{6}ac(b + c). \qquad \bar{x} = \frac{M_y}{m} = \frac{\frac{1}{6}ca^2}{\frac{1}{2}ac} = \frac{1}{3}a \text{ and } \bar{y} = \frac{M_x}{m} = \frac{\frac{1}{6}ac(b+c)}{\frac{1}{2}ac} = \frac{1}{3}(b+c).$$

[21] Sketch the region R with vertices $B(1, 2)$, $C(2, 1)$, $D(5, 4)$, and $E(4, 5)$. To apply the Theorem of Pappus, we need to find the area of R and its centroid. The sides of the rectangle have length $d(B, C) = \sqrt{2}$ and $d(C, D) = \sqrt{18}$. From the symmetry of R, $(\bar{x}, \bar{y}) = (3, 3)$. Alternatively, we could have shown that $M(B, D) = (3, 3)$. The area A of R is $\sqrt{2} \cdot \sqrt{18} = 6$. The distance traveled by the centroid is $2\pi\bar{x}$ is $2\pi(3) = 6\pi$. By (6.26), $V = (2\pi\bar{x})A = 6\pi(6) = 36\pi$.

[23] The graph of $y = \sqrt{a^2 - x^2}$ in the first quadrant is a quarter of a circle that is centered at the origin with radius $r = a$. The area of the quarter circle is $\frac{1}{4}\pi r^2 = \frac{\pi}{4}a^2$ and its volume of revolution (a hemisphere) about the y-axis is $\frac{1}{2}(\frac{4}{3}\pi r^3) = \frac{2}{3}\pi a^3$. Thus, $\bar{x} = \frac{V}{2\pi A} = \frac{\frac{2}{3}\pi a^3}{2\pi(\frac{\pi}{4}a^2)} = \frac{4a}{3\pi}$. Since the centroid lies on the line $y = x$,

$$\bar{y} = \bar{x} = \frac{4a}{3\pi}.$$

Exercises 6.8

[1] (a) A horizontal rectangle along one of the ends is a distance of y units from the bottom of the tank. It has depth $1 - y$, width 1, and area = width × height = $1\,dy$. $F = \int_c^d \rho\, h(y)\, L(y)\, dy = \rho\int_0^1 (1 - y)(1)\, dy = \rho\left[y - \frac{1}{2}y^2\right]_0^1 = \frac{1}{2}\rho =$

$$\frac{1}{2}(62.5) = 31.25 \text{ lb.}$$

(b) A horizontal rectangle along one of the sides has width 3. $F = \int_c^d \rho\, h(y)\, L(y)\, dy$

$$= \rho\int_0^1 (1 - y)(3)\, dy = 3\rho\left[y - \frac{1}{2}y^2\right]_0^1 = \frac{3}{2}\rho = \frac{3}{2}(62.5) = 93.75 \text{ lb.}$$

[5] Let the center of one of the circular ends of the tank be placed at the origin. This circle has equation $x^2 + y^2 = 4$. The right and left boundaries of the circle are $x = \pm\sqrt{4 - y^2}$. The surface of the oil coincides with $y = 0$ and the bottom of the tank coincides with $y = -2$. A horizontal rectangle a distance y units below the origin has depth $(0 - y)$, width $= \sqrt{4 - y^2} - (-\sqrt{4 - y^2}) = 2\sqrt{4 - y^2}$, and area $2\sqrt{4 - y^2}\, dy$.

$$F = \rho\int_{-2}^{0}(0 - y)(2\sqrt{4 - y^2})\, dy \ \{u = 4 - y^2,\ -\tfrac{1}{2}\, du = y\, dy\}$$

$$= -2\rho\int_{0}^{4}\sqrt{u}\,(-\tfrac{1}{2}\, du) = \rho\left[\tfrac{2}{3}u^{3/2}\right]_{0}^{4} = \tfrac{2}{3}\rho(8 - 0) = \tfrac{16}{3}\rho = \tfrac{16}{3}(60) = 320 \text{ lb.}$$

[9] (a) Place the left edge of the plate along the positive y-axis and the bottom edge on the positive x-axis. Then, the right edge has equation $x = 3$ and the left, $x = 0$. The surface of the oil is located at $y = 8$, the top of the plate at $y = 6$, and the bottom at $y = 0$. Then, $L(y) = 3$ and $h(y) = 8 - y \Rightarrow F = \rho\int_{0}^{6}(8 - y)(3)\, dy$

$$= 3\rho\left[8y - \tfrac{1}{2}y^2\right]_{0}^{6} = 3\rho(48 - 18) = 90\rho = 90(50) = 4500 \text{ lb.}$$

(b) An equation representing the diagonal can be found by using the points $(0, 0)$ and $(3, 6)$, and then finding the equation of the line containing them.

Thus, an equation of the diagonal is $y = 2x$, or $x = \tfrac{1}{2}y$.

On the lower triangle, $h(y) = 8 - y$ and $L(y) = 3 - \tfrac{1}{2}y$ (right minus left) $\Rightarrow$

$$F = \rho\int_{0}^{6}(8 - y)(3 - \tfrac{1}{2}y)\, dy = \rho\int_{0}^{6}(24 - 7y + \tfrac{1}{2}y^2)\, dy =$$

$$\rho\left[24y - \tfrac{7}{2}y^2 + \tfrac{1}{6}y^3\right]_{0}^{6} = \rho(54 - 0) = 54(50) = 2700 \text{ lb.}$$

On the upper triangle, $L(y) = \tfrac{1}{2}y - 0 \Rightarrow F = \rho\int_{0}^{6}(8 - y)(\tfrac{1}{2}y)\, dy =$

$$\tfrac{1}{2}\rho\int_{0}^{6}(8y - y^2)\, dy = \tfrac{1}{2}\rho\left[4y^2 - \tfrac{1}{3}y^3\right]_{0}^{6} = \tfrac{1}{2}\rho(72 - 0) = 36(50) = 1800 \text{ lb.}$$

Notice that 2700 lb + 1800 lb = 4500 lb, which is the answer in part (a).

[11] Note that both $c(t)$ and Q_0 are given using milligrams. (6.28) will be valid if the units of mass are either grams or milligrams. (The derivation of (6.28) is exactly the same if we use milligrams.) You could also change all units in milligrams to grams if you wish. Estimating $\int_{0}^{12} c(t)\, dt$ with $n = 12$,

$$S = \tfrac{12 - 0}{3(12)}[0 + 4(0) + 2(0.15) + 4(0.48) + 2(0.86) + 4(0.72) + 2(0.48) +$$

$$4(0.26) + 2(0.15) + 4(0.09) + 2(0.05) + 4(0.01) + 0] = \tfrac{1}{3}(9.62).$$

Solving (6.28) for F, we have $F = \dfrac{Q_0}{\displaystyle\int_{0}^{T} c(t)\, dt} = \dfrac{5}{9.62/3} \approx 1.56 \text{ liters/min.}$

13 (a) Approximating the time required to assemble 1 item with the given function
yields $f(1) = 20(1 + 1)^{-0.4} + 3 \approx 18.16$ min. Using a definite integral to
approximate the time required to assemble 1 item yields

$$f(1) \approx \int_0^1 \left[20(x + 1)^{-0.4} + 3 \right] dx = \left[\tfrac{20}{0.6}(x + 1)^{0.6} + 3x \right]_0^1 \approx$$

$$53.52 - 33.33 \approx 20.2 \text{ min.}$$

(b) $\int_0^4 \left[20(x + 1)^{-0.4} + 3 \right] dx = \left[\tfrac{20}{0.6}(x + 1)^{0.6} + 3x \right]_0^4 \approx$

$$99.55 - 33.33 \approx 66.2 \text{ min.}$$

(c) $\int_0^8 \left[20(x + 1)^{-0.4} + 3 \right] dx = \left[\tfrac{20}{0.6}(x + 1)^{0.6} + 3x \right]_0^8 \approx$

$$148.57 - 33.33 \approx 115.2 \text{ min.}$$

(d) $\int_0^{16} \left[20(x + 1)^{-0.4} + 3 \right] dx = \left[\tfrac{20}{0.6}(x + 1)^{0.6} + 3x \right]_0^{16} \approx$

$$230.45 - 33.33 \approx 197.1 \text{ min.}$$

17 $\displaystyle\sum_{k=1}^{100} k(k^2 + 1)^{-1/4} \approx \int_0^{100} x(x^2 + 1)^{-1/4} dx = \left[\tfrac{2}{3}(x^2 + 1)^{3/4} \right]_0^{100} =$

$$\tfrac{2}{3}\left[(10{,}001)^{3/4} - 1 \right] \approx 666.$$

19 Note that the graph has units measured in minutes and mi/hr. We will change all
the time measurements to hours by dividing by 60. Using the trapezoidal rule with
$n = 12$, the distance traveled (in miles) is $\int_0^{12/60} v(t)\, dt$, where time t is in <u>hours</u>, not
minutes.

$$T = \tfrac{12/60 - 0}{2(12)} \left\{ v(0) + 2\left[v(\tfrac{1}{60}) + v(\tfrac{2}{60}) + v(\tfrac{3}{60}) + \cdots + v(\tfrac{11}{60}) \right] + v(\tfrac{12}{60}) \right\}$$

$$= \tfrac{1}{120}[40 + 2(45 + 40 + 50 + 55 + 65 + 60 + 55 + 55 + 60 + 65 + 65) + 55]$$

$$= \tfrac{1}{120}(1325) \approx 11 \text{ mi.} \quad \{ \text{Answers may vary depending on graph interpretations.} \}$$

23 If we integrate the rate at which gasoline is used (measured in gal/hr),
we will find the total amount of gasoline used (measured in gal).

$$\text{Gasoline used} = \int_0^2 t\sqrt{9 - t^2}\, dt \ \{ u = 9 - t^2,\ -\tfrac{1}{2}du = t\, dt \} = -\tfrac{1}{2}\int_9^5 \sqrt{u}\, du$$

$$= -\tfrac{1}{2} \cdot \tfrac{2}{3}\left[u^{3/2} \right]_9^5 = -\tfrac{1}{3}(5^{3/2} - 27) = 9 - \tfrac{5\sqrt{5}}{3} \approx 5.27 \text{ gal.}$$

27 (a) Inhaling occurs when $V'(t) > 0$. Since the sine function is positive on $(0, \pi)$, we
need to solve $0 < 30\pi t < \pi$ for t. Thus, t must be in $(0, \tfrac{1}{30})$ for $V'(t)$ to be
positive. Hence, the volume of air inhaled in one breath is given by

$$\int_0^{1/30} V'(t)\, dt = \int_0^{1/30} 12{,}450\pi \sin(30\pi t)\, dt = 12{,}450\pi \left[-\tfrac{1}{30\pi}\cos(30\pi t) \right]_0^{1/30} =$$

$$-415(-1 - 1) = 830 \text{ cm}^3.$$

(b) Since an adult inhales 830 cm^3/breath and there is 4.1×10^{-12} joule/cm^3, this amounts to $830 \times 4.1 \times 10^{-12}$ joules/breath. Next, we must determine how many times an adult breaths in one year. One cycle of inhaling and exhaling corresponds to a period of $V'(t)$. This takes $\frac{1}{30} + \frac{1}{30} = \frac{1}{15}$ min. Thus, there are $15 \frac{\text{breaths}}{\text{min}}$, or $15 \frac{\text{breaths}}{\text{min}} \times \frac{60 \text{ min}}{1 \text{ hr}} \times \frac{24 \text{ hr}}{1 \text{ day}} \times \frac{365 \text{ day}}{1 \text{ yr}} = 7.884 \times 10^6 \frac{\text{breaths}}{\text{yr}}$.

In one year, the number of joules that an adult will inhale is (7.884×10^6) $\frac{\text{breaths}}{\text{yr}} \times (830 \times 4.1 \times 10^{-12}) \frac{\text{joule}}{\text{breath}} \approx 26.8 \times 10^{-3}$ joule ≈ 0.026 joule > 0.02 joule. No, it is not safe.

[29] If R is the rate of growth in cm/yr, then $\int_{10}^{15} R(t)\, dt$ will give the total number of centimeters of growth between the ages of 10 and 15.

$$T = \tfrac{15-10}{2(5)} \{ R(10) + 2\big[R(11) + R(12) + R(12) + R(13) + R(14) \big] + R(15) \}$$

$$= \tfrac{1}{2}\big[5.3 + 2(5.2 + 4.9 + 6.5 + 9.3) + 7.0 \big] = 32.05 \approx 32 \text{ cm}.$$

6.9 Review Exercises

[1] (a) $-x^2 = x^2 - 8 \Rightarrow 2x^2 - 8 = 0 \Rightarrow x^2 = 4 \Rightarrow x = \pm 2$. Since $-x^2 \geq x^2 - 8$ on $[-2, 2]$ and the region is symmetric with respect to the y-axis,

$$A = 2\int_0^2 \big[(-x^2) - (x^2 - 8)\big]\, dx = 2\int_0^2 (-2x^2 + 8)\, dx = 4\int_0^2 (4 - x^2)\, dx =$$

$$4\Big[4x - \tfrac{1}{3}x^3\Big]_0^2 = 4(8 - \tfrac{8}{3}) = \tfrac{64}{3}.$$

(b) The region is also symmetric about the line $y = -4$. We will find the area of the region in the fourth quadrant with $-4 \leq y \leq 0$, and multiply it by 4.

Thus, $A = 4\int_{-4}^0 \sqrt{-y}\, dy = 4\Big[-\tfrac{2}{3}(-y)^{3/2}\Big]_{-4}^0 = -\tfrac{8}{3}(0 - 8) = \tfrac{64}{3}.$

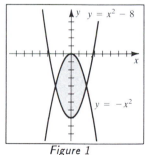

Figure 1

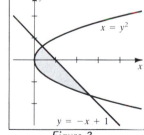

Figure 3

[3] $y^2 = 1 - y \Rightarrow y^2 + y - 1 = 0$. Using the quadratic formula, $y = a, b$, where $a = \tfrac{1}{2}(-1 - \sqrt{5})$ and $b = \tfrac{1}{2}(-1 + \sqrt{5})$. If we integrate with respect to y, we can find the area using one integral. Since $(1 - y) \geq y^2$ on $a \leq y \leq b$,   (cont.)

$$A = \int_a^b \left[(1 - y) - y^2\right] dy = \left[y - \tfrac{1}{2}y^2 - \tfrac{1}{3}y^3\right]_a^b =$$

$$\left(\tfrac{5\sqrt{5}}{12} - \tfrac{7}{12}\right) - \left(-\tfrac{5\sqrt{5}}{12} - \tfrac{7}{12}\right) = \tfrac{5\sqrt{5}}{6}.$$

$\boxed{7}$ Note that the x-intercept of $y = \sqrt{4x + 1}$ is $(-\tfrac{1}{4}, 0)$, but R is bounded by $x = 0$.

Using disks with $f(x) = \sqrt{4x + 1}$, $V = \pi \int_a^b [f(x)]^2 \, dx = \pi \int_0^2 (\sqrt{4x + 1})^2 \, dx =$

$$\pi \int_0^2 (4x + 1) \, dx = \pi \left[2x^2 + x\right]_0^2 = \pi(10 - 0) = 10\pi.$$

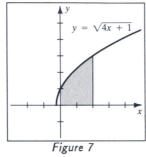

Figure 7

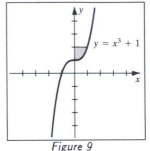

Figure 9

$\boxed{9}$ Using shells with $f(x) = 2 - (x^3 + 1)$,

$$V = 2\pi \int_a^b x f(x) \, dx = 2\pi \int_0^1 x \left[2 - (x^3 + 1)\right] dx = 2\pi \int_0^1 (-x^4 + x) \, dx =$$

$$2\pi \left[-\tfrac{1}{5}x^5 + \tfrac{1}{2}x^2\right]_0^1 = 2\pi(\tfrac{3}{10} - 0) = \tfrac{3\pi}{5}.$$

$\boxed{11}$ Using shells with $f(x) = \cos x^2$,

$$V = 2\pi \int_a^b x f(x) \, dx$$

$$= 2\pi \int_0^{\sqrt{\pi/2}} x(\cos x^2) \, dx$$

$$= \pi \int_0^{\pi/2} \cos u \, du \ \{ u = x^2, \tfrac{1}{2} du = x \, dx; \ x = 0, \sqrt{\pi/2} \Rightarrow u = 0, \pi/2 \}$$

$$= \pi \left[\sin u\right]_0^{\pi/2} = \pi(1 - 0) = \pi.$$

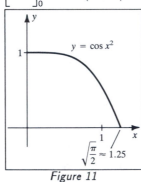

Figure 11

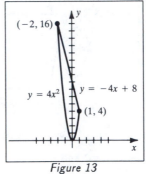

Figure 13

13 (a) See *Figure 13.* $4x^2 = -4x + 8 \Rightarrow 4x^2 + 4x - 8 = 0 \Rightarrow x^2 + x - 2 = 0 \Rightarrow$

$(x + 2)(x - 1) = 0 \Rightarrow x = -2, 1;\ y = 16, 4.$

Using washers with $f(x) = -4x + 8$ and $g(x) = 4x^2$,

$$V = \pi \int_a^b \left\{ \left[f(x)\right]^2 - \left[g(x)\right]^2 \right\} dx = \pi \int_{-2}^1 \left[(-4x + 8)^2 - (4x^2)^2 \right] dx$$

$$= \pi \int_{-2}^1 (-16x^4 + 16x^2 - 64x + 64)\, dx$$

$$= -16\pi \int_{-2}^1 (x^4 - x^2 + 4x - 4)\, dx$$

$$= -16\pi \left[\tfrac{1}{5}x^5 - \tfrac{1}{3}x^3 + 2x^2 - 4x \right]_{-2}^1$$

$$= -16\pi(-\tfrac{32}{15} - \tfrac{184}{15}) = -16\pi(-\tfrac{72}{5}) = \tfrac{1152\pi}{5}.$$

(b) A typical shell has radius $(1 - x)$ and altitude $\left[(-4x + 8) - 4x^2 \right].$

$$V = 2\pi \int_{-2}^1 (1 - x)\left[(-4x + 8) - 4x^2 \right] dx$$

$$= 2\pi \int_{-2}^1 (4x^3 - 12x + 8)\, dx$$

$$= 2\pi \left[x^4 - 6x^2 + 8x \right]_{-2}^1 = 2\pi \left[3 - (-24) \right] = 54\pi.$$

(c) A typical washer has outer radius $(16 - 4x^2)$ and

inner radius $\left[16 - (-4x + 8) \right].$

$$V = \pi \int_{-2}^1 \left\{ (16 - 4x^2)^2 - \left[16 - (-4x + 8) \right]^2 \right\} dx$$

$$= \pi \int_{-2}^1 \left[4^2(4 - x^2)^2 - 4^2(x + 2)^2 \right] dx$$

$$= 16\pi \int_{-2}^1 (x^4 - 9x^2 - 4x + 12)\, dx$$

$$= 16\pi \left[\tfrac{1}{5}x^5 - 3x^3 - 2x^2 + 12x \right]_{-2}^1$$

$$= 16\pi \left[\tfrac{36}{5} - (-\tfrac{72}{5}) \right] = \tfrac{1728\pi}{5}.$$

15 $(x + 3)^2 = 8(y - 1)^3 \Rightarrow (y - 1)^3 = \tfrac{1}{8}(x + 3)^2 \Rightarrow y - 1 = \tfrac{1}{2}(x + 3)^{2/3} \Rightarrow$

$y = f(x) = 1 + \tfrac{1}{2}(x + 3)^{2/3} \Rightarrow f'(x) = \tfrac{1}{3}(x + 3)^{-1/3}.$

$1 + [f'(x)]^2 = 1 + \tfrac{1}{9}(x + 3)^{-2/3} = \tfrac{1}{9}\left[9(x + 3)^{2/3} + 1 \right](x + 3)^{-2/3}.$ (cont.)

$$L = \int_a^b \sqrt{1 + \left[f'(x)\right]^2} \, dx = \int_{-2}^{5} \sqrt{\tfrac{1}{9}\left[9(x+3)^{2/3} + 1\right](x+3)^{-2/3}} \, dx$$

$$= \tfrac{1}{3}\int_{-2}^{5} \left|(x+3)^{-1/3}\right| \sqrt{9(x+3)^{2/3} + 1} \, dx$$

$$= \tfrac{1}{3}\int_{-2}^{5} (x+3)^{-1/3} \sqrt{9(x+3)^{2/3} + 1} \, dx \; \{(x+3)^{-1/3} > 0 \text{ on } [-2, 5]\}$$

$$= \tfrac{1}{3}\int_{10}^{37} \tfrac{1}{6}\sqrt{u} \, du \; \{u = 9(x+3)^{2/3} + 1, \; \tfrac{1}{6}\,du = (x+3)^{-1/3}\,dx\}$$

$$= \tfrac{1}{18}\left[\tfrac{2}{3}u^{3/2}\right]_{10}^{37} = \tfrac{1}{27}(37^{3/2} - 10^{3/2}) \approx 7.16.$$

19 See *Figure 19*. The diagonals of the square have length $L^2 = 4^2 + 4^2 = 32 \Rightarrow$
$L = 2\sqrt{8}$. Thus, the corners of the plate are at $(\pm\sqrt{8},\, 0)$ and $(0,\, \pm\sqrt{8})$,
and the surface is at $y = 6$. Consider the upper half of the plate $(y > 0)$.
The right edge is $x = f(y) = \sqrt{8} - y$. By symmetry, $L(y) = 2(\sqrt{8} - y)$ and

$$F_1 = \rho \int_0^{\sqrt{8}} (6 - y)\,2(\sqrt{8} - y) \, dy$$

$$= 2\rho \int_0^{\sqrt{8}} (y^2 - 2\sqrt{2}\,y - 6y + 12\sqrt{2}) \, dy$$

$$= 2\rho\left[\tfrac{1}{3}y^3 - (3 + \sqrt{2})y^2 + 12\sqrt{2}\,y\right]_0^{\sqrt{8}}$$

$$= 2\rho(\tfrac{16}{3}\sqrt{2} - 24 - 8\sqrt{2} + 48) = 2\rho(24 - \tfrac{8}{3}\sqrt{2}) = \rho(48 - \tfrac{16}{3}\sqrt{2}).$$

Now consider the lower half of the plate $(y < 0)$. $L(y) = 2(y + \sqrt{8})$ and

$$F_2 = \rho \int_{-\sqrt{8}}^{0} (6 - y)\,2(y + \sqrt{8}) \, dy$$

$$= 2\rho \int_{-\sqrt{8}}^{0} (-y^2 - 2\sqrt{2}\,y + 6y + 12\sqrt{2}) \, dy$$

$$= 2\rho\left[-\tfrac{1}{3}y^3 + (3 - \sqrt{2})y^2 + 12\sqrt{2}\,y\right]_{-\sqrt{8}}^{0}$$

$$= -2\rho(\tfrac{16}{3}\sqrt{2} + 24 - 8\sqrt{2} - 48) = -2\rho(-24 - \tfrac{8}{3}\sqrt{2}) = \rho(48 + \tfrac{16}{3}\sqrt{2}).$$

Total force $= F_1 + F_2 = 96\rho = 96(62.5) = 6000$ lb.

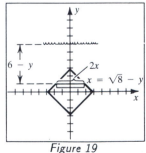

Figure 19

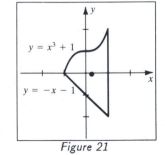

Figure 21

Note: For Exercises 21–22, without loss of generality, assume $\rho = 1$.

$\boxed{21}$ From *Figure 21*, we see that

$$m = \text{area} = \int_{-1}^{1}\left[(x^3 + 1) - (-x - 1)\right]dx = \int_{-1}^{1}(x^3 + x + 2)\,dx$$

$$= 2\int_{0}^{1}2\,dx \;\{\,x^3 \text{ and } x \text{ are odd, 2 is even; symmetric interval}\,\} = 4(1 - 0) = 4.$$

$$M_x = \tfrac{1}{2}\int_{-1}^{1}\left[(x^3 + 1)^2 - (-x - 1)^2\right]dx$$

$$= \tfrac{1}{2}\int_{-1}^{1}(x^6 + 2x^3 - x^2 - 2x)\,dx$$

$$= 2\cdot\tfrac{1}{2}\int_{0}^{1}(x^6 - x^2)\,dx \;\{\,2x^3 \text{ and } -2x \text{ are odd, } x^6 \text{ and } -x^2 \text{ are even}\,\}$$

$$= \left[\tfrac{1}{7}x^7 - \tfrac{1}{3}x^3\right]_0^1 = \tfrac{1}{7} - \tfrac{1}{3} = -\tfrac{4}{21}.$$

$$M_y = \int_{-1}^{1}x\left[(x^3 + 1) - (-x - 1)\right]dx$$

$$= \int_{-1}^{1}(x^4 + x^2 + 2x)\,dx$$

$$= 2\int_{0}^{1}(x^4 + x^2)\,dx \;\{\,2x \text{ is odd, } x^4 \text{ and } x^2 \text{ are even}\,\}$$

$$= 2\left[\tfrac{1}{5}x^5 + \tfrac{1}{3}x^3\right]_0^1 = 2(\tfrac{1}{5} + \tfrac{1}{3}) = \tfrac{16}{15}. \qquad \bar{x} = \frac{M_y}{m} = \tfrac{4}{15} \text{ and } \bar{y} = \frac{M_x}{m} = -\tfrac{1}{21}.$$

$\boxed{23}$ $12y = 4x^3 + (3/x) \Rightarrow y = f(x) = \tfrac{1}{3}x^3 + \tfrac{1}{4}x^{-1} \Rightarrow$

$$1 + [f'(x)]^2 = 1 + (x^4 - \tfrac{1}{2} + x^{-4}) = (x^4 + \tfrac{1}{2} + x^{-4}) = (x^2 + \tfrac{1}{4}x^{-2})^2.$$

$$S = 2\pi\int_{a}^{b}f(x)\sqrt{1 + [f'(x)]^2}\,dx = 2\pi\int_{1}^{2}(\tfrac{1}{3}x^3 + \tfrac{1}{4}x^{-1})(x^2 + \tfrac{1}{4}x^{-2})\,dx$$

$$= 2\pi\int_{1}^{2}(\tfrac{1}{3}x^5 + \tfrac{1}{3}x + \tfrac{1}{16}x^{-3})\,dx$$

$$= 2\pi\left[\frac{x^6}{18} + \frac{x^2}{6} - \frac{1}{32x^2}\right]_1^2$$

$$= \pi\left[\frac{x^6}{9} + \frac{x^2}{3} - \frac{1}{16x^2}\right]_1^2 \;\{\text{factor 2 out of each denominator}\}$$

$$= \pi\left[(\tfrac{64}{9} + \tfrac{4}{3} - \tfrac{1}{64}) - (\tfrac{1}{9} + \tfrac{1}{3} - \tfrac{1}{16})\right]$$

$$= \pi(7 + 1 + \tfrac{3}{64}) = \tfrac{515\pi}{64} \approx 25.3.$$

Chapter 7: Logarithmic and Exponential Functions

$\boxed{1}$ We will follow the guidelines listed in (7.5).

1) $f(x) = 3x + 5 \Rightarrow f'(x) = 3$. Since $f' > 0$ for all x, f is an increasing function and hence is a one-to-one function. This tells us that an inverse function exists.

2) $y = 3x + 5 \Rightarrow y - 5 = 3x \Rightarrow x = \dfrac{y - 5}{3} \Rightarrow f^{-1}(y) = \dfrac{y - 5}{3}$.

$$\text{Substituting } x \text{ for } y \text{ gives us } f^{-1}(x) = \frac{x - 5}{3}.$$

3) We now verify that $f^{-1}(x)$ is the inverse function for $f(x)$.

$$f^{-1}(f(x)) = f^{-1}(3x + 5) = \frac{(3x + 5) - 5}{3} = \frac{3x}{3} = x.$$

$$f(f^{-1}(x)) = f\left(\frac{x - 5}{3}\right) = 3\left(\frac{x - 5}{3}\right) + 5 = (x - 5) + 5 = x.$$

$\boxed{3}$ 1) $f(x) = \dfrac{1}{3x - 2} \Rightarrow f'(x) = -\dfrac{3}{(3x - 2)^2}$. $f' < 0$ for $x \in \mathbb{R} - \{\frac{2}{3}\} \Rightarrow f$ is decreasing throughout its domain. Thus, f is one-to-one and an inverse exists.

2) $y = \dfrac{1}{3x - 2} \Rightarrow 3xy - 2y = 1 \Rightarrow 3xy = 2y + 1 \Rightarrow x = \dfrac{2y + 1}{3y} \Rightarrow$

$$f^{-1}(y) = \frac{2y + 1}{3y} \Rightarrow f^{-1}(x) = \frac{2x + 1}{3x}$$

3) $f^{-1}(f(x)) = f^{-1}\left(\dfrac{1}{3x - 2}\right) = \dfrac{2\left(\dfrac{1}{3x - 2}\right) + 1}{3\left(\dfrac{1}{3x - 2}\right)}$ {multiply each term by $3x - 2$} $=$

$$\frac{2 + (3x - 2)}{3} = x.$$

$f(f^{-1}(x)) = f\left(\dfrac{2x + 1}{3x}\right) = \dfrac{1}{3\left(\dfrac{2x + 1}{3x}\right) - 2}$ {multiply each term by x} $=$

$$\frac{x}{(2x + 1) - 2x} = x.$$

$\boxed{5}$ 1) $f(x) = \dfrac{3x + 2}{2x - 5} \Rightarrow f'(x) = \dfrac{(2x - 5)(3) - (3x + 2)(2)}{(2x - 5)^2} = -\dfrac{19}{(2x - 5)^2}$.

$f' < 0$ for $x \in \mathbb{R} - \{\frac{5}{2}\} \Rightarrow f$ is decreasing throughout its domain.

$$\text{Thus, } f \text{ is one-to-one and an inverse exists.}$$

2) $y = \dfrac{3x + 2}{2x - 5} \Rightarrow 2xy - 5y = 3x + 2 \Rightarrow 2xy - 3x = 5y + 2 \Rightarrow$

$$x = \frac{5y + 2}{2y - 3} \Rightarrow f^{-1}(y) = \frac{5y + 2}{2y - 3} \Rightarrow f^{-1}(x) = \frac{5x + 2}{2x - 3}$$

3) $f^{-1}(f(x)) = f^{-1}\left(\dfrac{3x+2}{2x-5}\right) = \dfrac{5\left(\dfrac{3x+2}{2x-5}\right)+2}{2\left(\dfrac{3x+2}{2x-5}\right)-3}$ { multiply each term by $2x-5$ } $=$

$$\dfrac{5(3x+2)+2(2x-5)}{2(3x+2)-3(2x-5)} = \dfrac{19x}{19} = x.$$

$f(f^{-1}(x)) = f\left(\dfrac{5x+2}{2x-3}\right) = \dfrac{3\left(\dfrac{5x+2}{2x-3}\right)+2}{2\left(\dfrac{5x+2}{2x-3}\right)-5}$ { multiply each term by $2x-3$ } $=$

$$\dfrac{3(5x+2)+2(2x-3)}{2(5x+2)-5(2x-3)} = \dfrac{19x}{19} = x.$$

$\boxed{7}$ 1) $f(x) = 2 - 3x^2 \Rightarrow f'(x) = -6x.$ $f' < 0$ for $x < 0$ and $f' = 0$ for $x = 0 \Rightarrow f$ is decreasing throughout its domain. Thus, f is one-to-one and an inverse exists.

2) $y = 2 - 3x^2,\ x \le 0 \Rightarrow y + 3x^2 = 2 \Rightarrow x^2 = \dfrac{2-y}{3} \Rightarrow$

$x = \pm\frac{1}{3}\sqrt{6-3y}$ { Choose the "$-$" since $x \le 0.$ } $\Rightarrow$

$$f^{-1}(y) = -\tfrac{1}{3}\sqrt{6-3y} \Rightarrow f^{-1}(x) = -\tfrac{1}{3}\sqrt{6-3x}$$

3) $f^{-1}(f(x)) = f^{-1}(2 - 3x^2) = -\frac{1}{3}\sqrt{6 - 3(2-3x^2)} = -\frac{1}{3}\sqrt{9x^2} = -\frac{1}{3}(3\,|x|) =$

$$-|x| = -(-x)\ \{\text{since } x \le 0 \text{ for } f\} = x.$$

$f(f^{-1}(x)) = f(-\tfrac{1}{3}\sqrt{6-3x}) = 2 - 3(-\tfrac{1}{3}\sqrt{6-3x})^2 = 2 - 3\left[\tfrac{1}{9}(6-3x)\right] =$

$$2 - \tfrac{1}{3}(6-3x) = x.$$

$\boxed{9}$ 1) $f(x) = \sqrt{3-x} \Rightarrow f'(x) = -\dfrac{1}{2\sqrt{3-x}}.$ The domain of f is $(-\infty, 3]$.

Since $f' < 0$ on $(-\infty, 3)$, f is a decreasing function on $(-\infty, 3]$.

Thus, f is one-to-one and an inverse exists.

2) $y = \sqrt{3-x} \Rightarrow y^2 = 3 - x \Rightarrow x = 3 - y^2$ { Since $y \ge 0$ for f, $x \ge 0$ for f^{-1}.

See (7.4). } $\Rightarrow f^{-1}(y) = 3 - y^2 \Rightarrow f^{-1}(x) = 3 - x^2,\ x \ge 0$

3) $f^{-1}(f(x)) = f^{-1}(\sqrt{3-x}) = 3 - (\sqrt{3-x})^2 = 3 - (3-x) = x.$

$f(f^{-1}(x)) = f(3 - x^2) = \sqrt{3 - (3 - x^2)} = \sqrt{x^2} =$

$$|x| = x\ \{\text{since } x \ge 0 \text{ for } f^{-1}\}.$$

$\boxed{13}$ (a) Since $f'(x) = a \ne 0$, f is increasing or f is decreasing. Thus, f is one-to-one.

Since f is one-to-one, an inverse exists. If $f(x) = ax + b$, then $y = ax + b \Rightarrow$

$$y - b = ax \Rightarrow x = \frac{y-b}{a} \Rightarrow f^{-1}(x) = \frac{x-b}{a} \text{ for } a \ne 0.$$

(b) No, because a constant function has a horizontal line for its graph and is not one-to-one.

15 (a) Reflecting the graph of f through the line $y = x$ gives us *Figure 15*. Note that $(-1, \frac{1}{2})$ and $(2, 4)$ are on f, and that $(\frac{1}{2}, -1)$ and $(4, 2)$ are on f^{-1}.

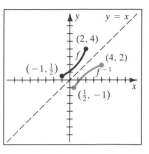

(b) The domain of f is the set of x values on the graph of f, $[-1, 2]$. The range of f is the set of y values on the graph of f, $[\frac{1}{2}, 4]$.

(c) From (7.4), the domain and range of f^{-1} are the range and domain of f, respectively. $D = [\frac{1}{2}, 4]$; $R = [-1, 2]$.

Figure 15

21 (a) $f(x) = \sqrt{2x + 3} \Rightarrow f'(x) = \frac{1}{2}(2x + 3)^{-1/2}(2) = \dfrac{1}{\sqrt{2x + 3}} > 0 \Rightarrow f$ is increasing on $[-\frac{3}{2}, \infty)$ and hence is one-to-one. Thus, f has an inverse function.

Finding this inverse, we have

$$y = \sqrt{2x + 3} \Rightarrow y^2 = 2x + 3 \Rightarrow x = \frac{y^2 - 3}{2} \Rightarrow f^{-1}(x) = \frac{x^2 - 3}{2}.$$

(b) Since the result of a square root is nonnegative, the range of f is $[0, \infty)$.

This is also the domain of f^{-1}.

(c) Using (7.8) with $g = f^{-1}$, $D_x f^{-1}(x) = \dfrac{1}{f'(f^{-1}(x))} = \dfrac{1}{\dfrac{1}{\sqrt{2f^{-1}(x) + 3}}} =$

$$\sqrt{2f^{-1}(x) + 3} = \sqrt{(x^2 - 3) + 3} = \sqrt{x^2} = |x| = x \ \{\text{since } x > 0 \text{ for } f^{-1}\}.$$

23 (a) $f(x) = 4 - x^2 \Rightarrow f'(x) = -2x < 0$ if $x > 0 \Rightarrow f$ is decreasing on $[0, \infty)$ and hence is one-to-one. Thus, f has an inverse function.

$$y = 4 - x^2 \Rightarrow x^2 = 4 - y \Rightarrow x = \sqrt{4 - y} \ \{x \geq 0\} \Rightarrow f^{-1}(x) = \sqrt{4 - x}.$$

(b) The range of f is $(-\infty, 4]$. This is also the domain of f^{-1}.

(c) Since $f'(x) = -2x$, $D_x f^{-1}(x) = \dfrac{1}{f'(f^{-1}(x))} = \dfrac{1}{-2f^{-1}(x)} = \dfrac{-1}{2\sqrt{4 - x}}.$

27 (a) $f(x) = x^5 + 3x^3 + 2x - 1 \Rightarrow f'(x) = 5x^4 + 9x^2 + 2 > 0 \Rightarrow f$ is ↑ ⇒

f is one-to-one $\Rightarrow f^{-1}$ exists.

(b) The point $(1, 5)$ is on the graph of $f \Rightarrow P(5, 1)$ is on the graph of f^{-1}. Using

(7.7) with $g = f^{-1}$, $D_x f^{-1}(5) = \dfrac{1}{f'(f^{-1}(5))} = \dfrac{1}{f'(1)} = \dfrac{1}{5(1)^4 + 9(1)^2 + 2} = \dfrac{1}{16}.$

We used (7.7) instead of (7.8) since we are evaluating at a particular point.

Alternate solution: Since $f'(1) = 16$, we know that $(f^{-1})'(5) = \frac{1}{16}$.

See the comments on page 380.

29 (a) $f(x) = -2x + (8/x^3) \Rightarrow f'(x) = -2 - 24/x^4 = -2(1 + 12/x^4) < 0$ for $x > 0$

$$\Rightarrow f \text{ is } \downarrow \Rightarrow f \text{ is one-to-one} \Rightarrow f^{-1} \text{ exists.}$$

(b) $D_x f^{-1}(-3) = \dfrac{1}{f'(f^{-1}(-3))} = \dfrac{1}{f'(2)} = \dfrac{1}{-2(1 + 12/2^4)} = \dfrac{1}{-7/2} = -\dfrac{2}{7}.$

Exercises 7.2

1 Using (7.11)(i), $f(x) = \ln(9x + 4) \Rightarrow$

$$f'(x) = \frac{1}{9x + 4} \cdot D_x(9x + 4) = \frac{1}{9x + 4} \cdot 9 = \frac{9}{9x + 4}.$$

3 Using (7.11)(i), $f(x) = \ln(3x^2 - 2x + 1) \Rightarrow f'(x) =$

$$\frac{1}{3x^2 - 2x + 1} D_x(3x^2 - 2x + 1) = \frac{1}{3x^2 - 2x + 1} \cdot (6x - 2) = \frac{2(3x - 1)}{3x^2 - 2x + 1}.$$

7 Using (7.12)(iii), $f(x) = \ln|2 - 3x|^5 = 5\ln|2 - 3x|$. We now apply (7.11)(ii) to find

$$f'(x). \quad f'(x) = 5 \cdot \frac{1}{2 - 3x} D_x(2 - 3x) = 5 \cdot \frac{1}{2 - 3x} \cdot (-3) = \frac{15}{3x - 2}.$$

9 Using (7.12)(iii), $f(x) = \ln\sqrt{7 - 2x^3} = \ln(7 - 2x^3)^{1/2} = \frac{1}{2}\ln(7 - 2x^3).$

$$\text{Thus, } f'(x) = \frac{1}{2} \cdot \frac{1}{7 - 2x^3} D_x(7 - 2x^3) = \frac{1}{2} \cdot \frac{-6x^2}{7 - 2x^3} = \frac{3x^2}{2x^3 - 7}.$$

13 Using (7.12)(iii), $f(x) = \ln\sqrt{x} + \sqrt{\ln x} = \ln(x^{1/2}) + (\ln x)^{1/2} = \frac{1}{2}\ln x + (\ln x)^{1/2}.$

$$\text{Thus, } f'(x) = \frac{1}{2} \cdot \frac{1}{x} + \frac{1}{2}(\ln x)^{-1/2} \cdot \frac{1}{x} = \frac{1}{2x} + \frac{1}{2x\sqrt{\ln x}} = \frac{1}{2x}\left(1 + \frac{1}{\sqrt{\ln x}}\right).$$

17 $\quad f(x) = \ln\left[(5x - 7)^4(2x + 3)^3\right] \qquad \{\text{given}\}$

$$= \ln(5x - 7)^4 + \ln(2x + 3)^3 \qquad \{\text{apply (7.12)(i)}\}$$

$$= 4\ln(5x - 7) + 3\ln(2x + 3) \qquad \{\text{apply (7.12)(iii)}\}$$

$$f'(x) = 4 \cdot \frac{1}{5x - 7} \cdot 5 + 3 \cdot \frac{1}{2x + 3} \cdot 2 \qquad \{\text{differentiating}\}$$

$$= \frac{20}{5x - 7} + \frac{6}{2x + 3} \qquad \{\text{simplifying}\}$$

19 $\quad f(x) = \ln\dfrac{\sqrt{x^2 + 1}}{(9x - 4)^2} \qquad \{\text{given}\}$

$$= \ln\sqrt{x^2 + 1} - \ln(9x - 4)^2 \qquad \{\text{apply (7.12)(ii)}\}$$

$$= \frac{1}{2}\ln(x^2 + 1) - 2\ln(9x - 4) \qquad \{\text{apply (7.12)(iii)}\}$$

$$f'(x) = \frac{1}{2} \cdot \frac{1}{x^2 + 1} \cdot 2x - 2 \cdot \frac{1}{9x - 4} \cdot 9 \qquad \{\text{differentiating}\}$$

$$= \frac{x}{x^2 + 1} - \frac{18}{9x - 4} \qquad \{\text{simplifying}\}$$

21 $f(x) = \ln\sqrt{\dfrac{x^2 - 1}{x^2 + 1}} = \frac{1}{2} \cdot \ln\dfrac{x^2 - 1}{x^2 + 1} = \frac{1}{2}\left[\ln(x^2 - 1) - \ln(x^2 + 1)\right] \Rightarrow$

$$f'(x) = \frac{1}{2}\left(\frac{2x}{x^2 - 1} - \frac{2x}{x^2 + 1}\right) = \frac{x}{x^2 - 1} - \frac{x}{x^2 + 1}$$

$\boxed{23}$ $f(x) = \ln\left(x + \sqrt{x^2 - 1}\right) \Rightarrow$

$$f'(x) = \frac{1}{x + \sqrt{x^2 - 1}} D_x\left(x + \sqrt{x^2 - 1}\right)$$

$$= \frac{1}{x + \sqrt{x^2 - 1}} \cdot \left[1 + \tfrac{1}{2}(x^2 - 1)^{-1/2}(2x)\right]$$

$$= \frac{1 + x/\sqrt{x^2 - 1}}{x + \sqrt{x^2 - 1}} \cdot \frac{\sqrt{x^2 - 1}}{\sqrt{x^2 - 1}} = \frac{\sqrt{x^2 - 1} + x}{\sqrt{x^2 - 1}\,(x + \sqrt{x^2 - 1})} = \frac{1}{\sqrt{x^2 - 1}}$$

$\boxed{27}$ $f(x) = \ln\tan^3 3x = \ln(\tan 3x)^3 = 3\ln(\tan 3x) \Rightarrow f'(x) = 3 \cdot \dfrac{1}{\tan 3x} D_x(\tan 3x) =$

$$3 \cdot \frac{1}{\tan 3x} \cdot \sec^2 3x \cdot D_x(3x) = 3 \cdot \frac{\sec 3x}{\tan 3x} \cdot \sec 3x \cdot 3 = 9\csc 3x \sec 3x$$

$\boxed{29}$ $f(x) = \ln\ln\sec 2x = \ln\left[\ln(\sec 2x)\right] \Rightarrow$

$$f'(x) = \frac{1}{\ln(\sec 2x)} D_x\left[\ln(\sec 2x)\right] = \frac{1}{\ln(\sec 2x)} \cdot \frac{1}{\sec 2x} \cdot 2\sec 2x \tan 2x = \frac{2\tan 2x}{\ln\sec 2x}$$

$\boxed{33}$ $f(x) = \ln|\sec x + \tan x| \Rightarrow f'(x) = \dfrac{1}{\sec x + \tan x} D_x(\sec x + \tan x) =$

$$\frac{1}{\sec x + \tan x} \cdot (\sec x \tan x + \sec^2 x) = \frac{\sec x(\tan x + \sec x)}{\sec x + \tan x} = \sec x$$

$\boxed{35}$ $3y - x^2 + \ln xy = 2 \Leftrightarrow 3y - x^2 + \ln x + \ln y = 2$ { using (7.12)(i) } $\Rightarrow$

$$3y' - 2x + \frac{1}{x} + \frac{1}{y} D_x y = 0 \Rightarrow 3y' + \frac{y'}{y} = 2x - \frac{1}{x} \Rightarrow$$

$$y'\left(3 + \frac{1}{y}\right) = 2x - \frac{1}{x} \Rightarrow y' = \frac{2x - 1/x}{3 + 1/y} \cdot \frac{xy}{xy} = \frac{y(2x^2 - 1)}{x(3y + 1)}$$

$\boxed{39}$ $y = (5x + 2)^3(6x + 1)^2$ { given }

     $\ln y = \ln\left[(5x + 2)^3(6x + 1)^2\right]$ { take ln of both sides }

         $= \ln(5x + 2)^3 + \ln(6x + 1)^2$ { apply (7.12)(i) }

         $= 3\ln(5x + 2) + 2\ln(6x + 1)$ { apply (7.12)(iii) }

$\dfrac{1}{y} D_x y = 3 \cdot \dfrac{1}{5x + 2} \cdot 5 + 2 \cdot \dfrac{1}{6x + 1} \cdot 6$ { differentiate with respect to x }

   $\dfrac{y'}{y} = \dfrac{15}{5x + 2} + \dfrac{12}{6x + 1}$ { simplifying }

     $= \dfrac{15(6x + 1) + 12(5x + 2)}{(5x + 2)(6x + 1)}$ { simplifying }

  $y' = \dfrac{150x + 39}{(5x + 2)(6x + 1)} \cdot y$ { simplify and multiply by y }

     $= \dfrac{150x + 39}{(5x + 2)(6x + 1)} \cdot (5x + 2)^3(6x + 1)^2$ { definition of y }

     $= (150x + 39)(5x + 2)^2(6x + 1)$ { simplifying }

$\boxed{43}$ $\qquad y = \dfrac{(x^2 + 3)^5}{\sqrt{x + 1}}$ $\qquad\qquad\qquad$ {given}

$\qquad \ln y = \ln\left[\dfrac{(x^2 + 3)^5}{\sqrt{x + 1}}\right]$ $\qquad\qquad$ {take ln of both sides}

$\qquad\qquad = \ln(x^2 + 3)^5 - \ln(x + 1)^{1/2}$ $\qquad$ {apply (7.12)(ii)}

$\qquad\qquad = 5\ln(x^2 + 3) - \tfrac{1}{2}\ln(x + 1)$ $\qquad$ {apply (7.12)(iii)}

$\dfrac{1}{y} D_x y = 5 \cdot \dfrac{1}{x^2 + 3} \cdot 2x - \dfrac{1}{2} \cdot \dfrac{1}{x + 1} \cdot 1$ $\qquad$ {differentiate with respect to x}

$\qquad \dfrac{y'}{y} = \dfrac{10x}{x^2 + 3} - \dfrac{1}{2(x + 1)}$ $\qquad\qquad$ {simplifying}

$\qquad\quad = \dfrac{(10x)(2)(x + 1) - (1)(x^2 + 3)}{2(x^2 + 3)(x + 1)}$ $\qquad$ {simplifying}

$\qquad y' = \dfrac{19x^2 + 20x - 3}{2(x^2 + 3)(x + 1)} \cdot y$ $\qquad\qquad$ {simplify and multiply by y}

$\qquad\quad = \dfrac{19x^2 + 20x - 3}{2(x^2 + 3)(x + 1)} \cdot \dfrac{(x^2 + 3)^5}{(x + 1)^{1/2}}$ $\qquad$ {definition of y}

$\qquad\quad = \dfrac{(19x^2 + 20x - 3)(x^2 + 3)^4}{2(x + 1)^{3/2}}$ $\qquad\qquad$ {simplifying}

$\boxed{45}$ We first find the slope of the tangent line at $P(3,\ 9)$. $\quad y = x^2 + \ln(2x - 5) \Rightarrow$ $y' = 2x + \dfrac{2}{2x - 5}$. $\quad x = 3 \Rightarrow y' = m = 6 + 2 = 8$. Using m and P, the equation of the tangent line is $y - 9 = 8(x - 3)$, or, in slope-intercept form, $y = 8x - 15$.

$\boxed{47}$ To find the highest point, we must find the value of x such that $y' = 0$.

$\qquad y = 5\ln x - \tfrac{1}{2}x \Rightarrow y' = \dfrac{5}{x} - \dfrac{1}{2}$. $y' = 0 \Rightarrow \dfrac{5}{x} = \dfrac{1}{2} \Rightarrow x = 10$. When $x = 10$,

$\qquad y = 5\ln 10 - 5 \approx 6.51$, and these are the coordinates of the highest point.

$\qquad\qquad\qquad$ Since $y'' = -(5/x^2) < 0$ for $x > 0$, the graph is CD on $(0,\ \infty)$.

$\boxed{49}$ $\ T = -2.57\ln\left[(87 - L)/63\right] = -2.57\left[\ln(87 - L) - \ln 63\right] \Rightarrow$

$\qquad dT = T'(L)\,dL = -2.57\left[\dfrac{1}{87 - L}(-1) - 0\right]dL$

$\qquad\quad = \dfrac{2.57}{87 - L}\,dL = \dfrac{2.57}{87 - 80}(\pm 2)\ \{L = 80,\ dL = \pm 2\} \approx \pm\,0.73\ \text{yr}.$

$\boxed{51}$ (a) The initial velocity and acceleration are given by $s'(0)$ and $s''(0)$, respectively.

$$s(t) = ct + \tfrac{c}{b}(m_1 + m_2 - bt)\big[\ln(m_1 + m_2 - bt) - \ln(m_1 + m_2)\big] \Rightarrow$$

$$s'(t) = c + \tfrac{c}{b}(m_1 + m_2 - bt)\cdot\frac{-b}{m_1 + m_2 - bt} +$$

$$\tfrac{c}{b}(-b)\big[\ln(m_1 + m_2 - bt) - \ln(m_1 + m_2)\big]$$

$$= c + (-c) + (-c)\big[\ln(m_1 + m_2 - bt) - \ln(m_1 + m_2)\big] \Rightarrow$$

$$= c\big[\ln(m_1 + m_2) - \ln(m_1 + m_2 - bt)\big].$$

Thus, $s'(0) = c\big[\ln(m_1 + m_2) - \ln(m_1 + m_2)\big] = c(0) = 0$ m/sec.

Also, $s''(t) = c\Big[0 - \dfrac{1}{m_1 + m_2 - bt}\cdot(-b)\Big] = \dfrac{bc}{m_1 + m_2 - bt} \Rightarrow$

$$s''(0) = \frac{bc}{m_1 + m_2} \text{ m/sec}^2.$$

(b) $s'\Big(\dfrac{m_2}{b}\Big) = c\Big\{\ln(m_1 + m_2) - \ln\big[m_1 + m_2 - b(m_2/b)\big]\Big\}$

$$= c\big[\ln(m_1 + m_2) - \ln(m_1 + m_2 - m_2)\big]$$

$$= c\big[\ln(m_1 + m_2) - \ln(m_1)\big] = c\ln\Big(\frac{m_1 + m_2}{m_1}\Big)$$

$$s''\Big(\frac{m_2}{b}\Big) = \frac{bc}{m_1 + m_2 - b(m_2/b)} = \frac{bc}{m_1 + m_2 - m_2} = \frac{bc}{m_1}$$

Exercises 7.3

$\boxed{3}$ Using (7.22), $f(x) = e^{3x^2} \Rightarrow f'(x) = e^{3x^2}\cdot D_x(3x^2) = e^{3x^2}\cdot(6x) = 6xe^{3x^2}.$

$\boxed{5}$ $f(x) = \sqrt{1 + e^{2x}} = (1 + e^{2x})^{1/2} \Rightarrow f'(x) = \tfrac{1}{2}(1 + e^{2x})^{-1/2} D_x(1 + e^{2x}) =$

$$\tfrac{1}{2}(1 + e^{2x})^{-1/2}(e^{2x}) D_x(2x) = \tfrac{1}{2}(1 + e^{2x})^{-1/2}(e^{2x})(2) = \frac{e^{2x}}{\sqrt{1 + e^{2x}}}$$

$\boxed{9}$ Using the product rule, $f(x) = x^2 e^{-2x} \Rightarrow f'(x) = x^2 D_x(e^{-2x}) + e^{-2x} D_x(x^2) =$

$$x^2(e^{-2x}) D_x(-2x) + e^{-2x}(2x) = x^2 e^{-2x}(-2) + 2xe^{-2x} = 2xe^{-2x}(1 - x).$$

$\boxed{11}$ Using the quotient rule,

$$f(x) = \frac{e^x}{x^2 + 1} \Rightarrow f'(x) = \frac{(x^2 + 1)e^x - 2xe^x}{(x^2 + 1)^2} = \frac{e^x(x^2 + 1 - 2x)}{(x^2 + 1)^2} = \frac{e^x(x - 1)^2}{(x^2 + 1)^2}.$$

$\boxed{15}$ $f(x) = e^{1/x} + (1/e^x) = e^{1/x} + e^{-x} \Rightarrow$

$$f'(x) = e^{1/x} D_x(1/x) + e^{-x} D_x(-x) = e^{1/x}\cdot\Big(-\frac{1}{x^2}\Big) + (-e^{-x}) = -\frac{e^{1/x}}{x^2} - e^{-x}$$

$\boxed{17}$ Using the quotient rule, $f(x) = \dfrac{e^x - e^{-x}}{e^x + e^{-x}} \Rightarrow$

$$f'(x) \;=\; \frac{(e^x + e^{-x})(e^x - (-e^{-x})) - (e^x - e^{-x})(e^x - e^{-x})}{(e^x + e^{-x})^2}$$

$$= \frac{(e^x + e^{-x})^2 - (e^x - e^{-x})^2}{(e^x + e^{-x})^2}$$

$$= \frac{(e^{2x} + 2 + e^{-2x}) - (e^{2x} - 2 + e^{-2x})}{(e^x + e^{-x})^2} = \frac{4}{(e^x + e^{-x})^2}.$$

$\boxed{19}$ $f(x) = e^{-2x} \ln x \Rightarrow f'(x) = e^{-2x} D_x(\ln x) + (\ln x) D_x(e^{-2x}) =$

$$e^{-2x} \cdot \frac{1}{x} + (\ln x)(-2e^{-2x}) = e^{-2x}\left(\frac{1}{x} - 2\ln x\right)$$

$\boxed{21}$ $f(x) = \sin e^{5x} \Rightarrow f'(x) = \cos e^{5x} \cdot D_x(e^{5x}) = 5e^{5x} \cos e^{5x}$

$\boxed{23}$ $f(x) = \ln \cos e^{-x} \Rightarrow f'(x) = \dfrac{1}{\cos e^{-x}} D_x(\cos e^{-x}) = \dfrac{1}{\cos e^{-x}}(-\sin e^{-x}) D_x(e^{-x}) =$

$$\frac{1}{\cos e^{-x}}(-\sin e^{-x})(-e^{-x}) = \frac{e^{-x} \sin e^{-x}}{\cos e^{-x}} = e^{-x} \tan e^{-x}$$

$\boxed{25}$ $f(x) = e^{3x} \tan \sqrt{x} \Rightarrow$

$$f'(x) = e^{3x} D_x(\tan \sqrt{x}) + (\tan \sqrt{x}) D_x(e^{3x}) = e^{3x} \sec^2 \sqrt{x}\, D_x(\sqrt{x}) + (\tan \sqrt{x})(3e^{3x})$$

$$= e^{3x} \cdot \frac{\sec^2 \sqrt{x}}{2\sqrt{x}} + 3e^{3x} \tan \sqrt{x} = e^{3x}\left(\frac{\sec^2 \sqrt{x}}{2\sqrt{x}} + 3\tan \sqrt{x}\right)$$

$\boxed{29}$ $f(x) = xe^{\cot x} \Rightarrow f'(x) = x D_x(e^{\cot x}) + e^{\cot x} D_x(x) =$

$$x\left[e^{\cot x}(-\csc^2 x)\right] + e^{\cot x} \cdot 1 = e^{\cot x}(1 - x \csc^2 x)$$

$\boxed{31}$ $e^{xy} - x^3 + 3y^2 = 11 \Rightarrow e^{xy}(xy' + y) - 3x^2 + 6yy' = 0 \Rightarrow$

$$y'(xe^{xy} + 6y) = 3x^2 - ye^{xy} \Rightarrow y' = \frac{3x^2 - ye^{xy}}{xe^{xy} + 6y}$$

$\boxed{35}$ $y = (x - 1)e^x + 3\ln x + 2 \Rightarrow y' = (x - 1)e^x + e^x + \dfrac{3}{x}$. At $x = 1$, $y' = e + 3$.

Tangent line: $y - 2 = (e + 3)(x - 1)$, or $y = (e + 3)x - (e + 1) \approx 5.72x - 3.72$.

$\boxed{37}$ $f(x) = xe^x \Rightarrow f'(x) = xe^x + e^x(1) = e^x(x + 1)$. $f'(x) = 0 \Rightarrow x = -1$.

$f''(x) = e^x D_x(x + 1) + (x + 1) \cdot D_x(e^x) = e^x(1) + (x + 1)e^x = e^x(x + 2)$.

$f''(x) = 0 \Rightarrow x = -2$. We now set up first and second derivative charts as we did in Chapter 4.

Interval	Sign	Conclusion
$(-\infty, -1)$	$-$	$\downarrow$ on $(-\infty, -1]$
$(-1, \infty)$	$+$	$\uparrow$ on $[-1, \infty)$

Interval	Sign	Concavity
$(-\infty, -2)$	$-$	CD
$(-2, \infty)$	$+$	CU

(cont.)

Since there is a change in concavity at $x = -2$, we know there is a *PI* at $(-2, -2e^{-2}) \approx (-2, -0.271)$. Employing the second derivative test, we find that $f''(-1) = e^{-1} > 0$, indicating that $f(-1) = -e^{-1} \approx -0.368$ is a *LMIN*.

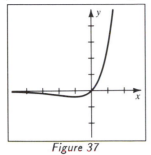

Figure 37

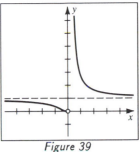

Figure 39

$\boxed{39}$ $f(x) = e^{1/x} \Rightarrow f'(x) = e^{1/x} D_x(1/x) = -\dfrac{e^{1/x}}{x^2}$. Since $e^{1/x} \neq 0$ for any x,

$f'(x) \neq 0$. f' DNE if $x = 0$. There are no local extrema since f is decreasing

throughout its domain. $f''(x) = -\dfrac{x^2\left(\dfrac{-e^{1/x}}{x^2}\right) - e^{1/x}(2x)}{(x^2)^2} = \dfrac{e^{1/x}(2x+1)}{x^4}$.

Interval	Sign	Conclusion
$(-\infty, 0)$	$-$	$\downarrow$ on $(-\infty, 0)$
$(0, \infty)$	$-$	$\downarrow$ on $(0, \infty)$

Interval	Sign	Concavity
$(-\infty, -\frac{1}{2})$	$-$	*CD*
$(-\frac{1}{2}, 0)$	$+$	*CU*
$(0, \infty)$	$+$	*CU*

Since there is a change in concavity at $x = -\frac{1}{2}$,

we know there is a *PI* at $(-\frac{1}{2}, e^{-2}) \approx (-0.5, 0.135)$.

$\boxed{45}$ (a) $C(t) = \dfrac{k}{a-b}(e^{-bt} - e^{-at}) \Rightarrow C'(t) = \dfrac{k}{a-b}(-be^{-bt} + ae^{-at}) = 0 \Rightarrow$

$ae^{-at} = be^{-bt} \Rightarrow \dfrac{a}{b} = \dfrac{e^{-bt}}{e^{-at}} = e^{(a-b)t} \Rightarrow \ln(a/b) = (a-b)t \Rightarrow$

$t = \dfrac{\ln(a/b)}{a-b} = t_0$. Consider the two cases: (i) $a > b$, and (ii) $a < b$.

(i) If $a > b$, then $\dfrac{k}{a-b} > 0$, $e^{at} \geq e^{bt}$, and $\dfrac{a}{e^{at}} - \dfrac{b}{e^{bt}}$ is positive when

$t \in [0, t_0)$, zero when $t = t_0$, and negative when $t > t_0$.

(ii) If $a < b$, then $\dfrac{k}{a-b} < 0$, $e^{at} \leq e^{bt}$, and $\dfrac{a}{e^{at}} - \dfrac{b}{e^{bt}}$ is negative when

$t \in [0, t_0)$, zero when $t = t_0$, and positive when $t > t_0$.

In both cases, $C'(t) > 0$ for $t < t_0$ and $C'(t) < 0$ for $t > t_0 \Rightarrow$

the maximum concentration occurs at $t = t_0$.

(b) $\lim\limits_{t \to \infty} C(t) = \lim\limits_{t \to \infty} \dfrac{k}{a-b}(e^{-bt} - e^{-at}) = \dfrac{k}{a-b}(0 - 0) = 0 \ \{a > 0, b > 0\}$

$\boxed{47}$ (a) The height of the child at age 1 is given by $h(1)$. $h(x) =$
79.041 + 6.39x − $e^{3.261-0.993x}$ $\Rightarrow$ $h(1) = 79.041 + 6.39 − e^{2.268} \approx 75.8$ cm.
The rate of growth of the child at age 1 is given by $h'(1)$. $h'(x) =$
6.39 + 0.993$e^{3.261-0.993x}$ $\Rightarrow$ $h'(1) = 6.39 + 0.993e^{2.268} \approx 15.98$ cm/yr.

(b) To answer the question "When is the rate of growth largest?", we must maximize
the rate of growth function, $h'(x)$.   Refer to the guidelines for finding extrema
(4.9).   To find the critical numbers of $h'(x)$, we begin by finding $h''(x)$. $h''(x) =$
$0.993e^{3.261-0.993x}(-0.993) = -(0.993)^2 e^{3.261-0.993x}$.   Since e raised to any
power is positive, $h''(x)$ is always negative and there are no critical numbers for
$h'(x)$.   Because $h''(x)$ is negative, $h'(x)$ is decreasing on its domain and must
obtain its largest and smallest values at the endpoints.   Thus, the rate of growth
is largest at $x = \frac{1}{4}$ yr and smallest at $x = 6$ yr.   The values of the rate of growth
at these times are $h'(\frac{1}{4}) \approx 26.59$ cm/yr and $h'(6) \approx 6.46$ cm/yr, indicating that
the *growth rate* is rapidly slowing down.

$\boxed{49}$ (a) $f(x) = cx^n e^{-ax} \Rightarrow f'(x) = cx^n e^{-ax}(-a) + cnx^{n-1}e^{-ax} = cx^{n-1}e^{-ax}(n - ax)$.
$f'(x) = 0 \Rightarrow x = \frac{n}{a}$ $\{x = 0$ is an endpoint.$\}$ $f'(x) > 0$ if $0 < x < \frac{n}{a}$ and
$$f'(x) < 0 \text{ for } x > \tfrac{n}{a}. \text{ Thus, } f \text{ has exactly one } LMAX. \text{ It occurs at } x = \tfrac{n}{a}.$$

(b) To determine where $f(x)$ is increasing most rapidly, we must determine when f'
assumes its maximum value.   To do this, we will examine the critical numbers of
f'.   From part (a), with $n = 4$, $f'(x) = cx^3 e^{-ax}(4 - ax)$.   Differentiating f' as a
product of three expressions $\{$ see §3.3, #71 $\}$, we have
$$f''(x) = c\Big[3x^2 e^{-ax}(4 - ax) + (x^3)(-ae^{-ax})(4 - ax) + x^3 e^{-ax}(-a)\Big]$$
$$= cx^2 e^{-ax}\Big[3(4 - ax) - ax(4 - ax) - ax\Big]$$
$$= cx^2 e^{-ax}(a^2 x^2 - 8ax + 12)$$
$$= cx^2 e^{-ax}(ax - 2)(ax - 6).$$
$f''(x) = 0 \Rightarrow x = \frac{2}{a}, \frac{6}{a}$. $f''(x) > 0$ on $(0, \frac{2}{a})$, $f''(x) < 0$ on $(\frac{2}{a}, \frac{6}{a})$, and $f''(x) > 0$
on $(\frac{6}{a}, \infty)$.   This implies a $LMAX$ for f' at $x = \frac{2}{a}$.   Since $f'(0) = 0$ and
$\lim\limits_{x \to \infty} f'(x) = 0$, a greater value than $f'(\frac{2}{a}) > 0$ will not be obtained.   Note that
$f(x)$ is increasing most rapidly exactly halfway between $x = 0$ and the x value of
its $LMAX$.

$\boxed{51}$ $D(r) = ae^{-br + cr^2} \Rightarrow D'(r) = ae^{-br + cr^2}(-b + 2cr)$.

Since $ae^{-br + cr^2} \neq 0$, $D'(r) = 0 \Rightarrow r = \frac{b}{2c}$. D is $\downarrow$ on $[0, \frac{b}{2c}]$ and $\uparrow$ on $[\frac{b}{2c}, \infty)$.

$D''(r) = a\left[e^{-br + cr^2}(2c) + (-b + 2cr)\,e^{-br + cr^2}(-b + 2cr)\right]$

$ = ae^{-br + cr^2}\left[2c + (-b + 2cr)^2\right] > 0 \Rightarrow D$ is always CU.

$D(\frac{b}{2c}) = ae^{-b^2/(4c)}$ is a $LMIN$.

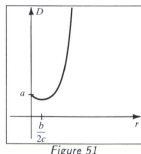

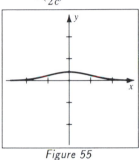

Figure 51   Figure 55

$\boxed{55}$ $y = f(x) = \frac{1}{\sigma\sqrt{2\pi}}e^{-z^2/2}$ with $z = \frac{x - \mu}{\sigma} \Rightarrow$

$f'(x) = \frac{1}{\sigma\sqrt{2\pi}}e^{-z^2/2}(-z)\,D_x(z)$ { by the chain rule, $\frac{dy}{dx} = \frac{dy}{dz} \cdot \frac{dz}{dx}$ }

$ = \frac{1}{\sigma\sqrt{2\pi}}e^{-z^2/2}(-z)(\frac{1}{\sigma})$ { $z = \frac{x - \mu}{\sigma} = \frac{1}{\sigma}x - \frac{\mu}{\sigma} \Rightarrow D_x(z) = \frac{1}{\sigma}$ }

$ = -\frac{1}{\sigma^2\sqrt{2\pi}}e^{-z^2/2}(z)$.

$f'(x) = 0 \Rightarrow z = 0$, or, equivalently, $x = \mu$. f is $\uparrow$ on $(-\infty, \mu]$ and $\downarrow$ on $[\mu, \infty)$.

Thus, $f(\mu) = \frac{1}{\sigma\sqrt{2\pi}}$ is a $LMAX$.

$f''(x) = -\frac{1}{\sigma^2\sqrt{2\pi}}\left[e^{-z^2/2}\,D_x(z) + (z)e^{-z^2/2}(-z)\,D_x(z)\right]$

$ = -\frac{1}{\sigma^2\sqrt{2\pi}}\left[e^{-z^2/2}(\frac{1}{\sigma}) - z^2e^{-z^2/2}(\frac{1}{\sigma})\right]$

$ = -\frac{1}{\sigma^3\sqrt{2\pi}}e^{-z^2/2}(1 - z^2) = 0 \Rightarrow z = \pm 1$, or $x = \mu \pm \sigma$. f is CD on

$(\mu - \sigma, \mu + \sigma)$ and CU on $(-\infty, \mu - \sigma) \cup (\mu + \sigma, \infty)$. PI are $\left(\mu \pm \sigma, \frac{1}{\sigma\sqrt{2\pi e}}\right)$.

As $x \to \pm\infty$, $z \to \infty$, $e^{-z^2/2} \to 0$, and $\lim\limits_{x \to \infty} f(x) = \lim\limits_{x \to -\infty} f(x) = 0$. See *Figure 55*.

$\boxed{57}$ Let $x = r/R > 0$. $v(x) = -kx^2 \ln x \Rightarrow v'(x) = -k\left[x^2 \cdot \frac{1}{x} + (\ln x)(2x)\right] =$

$-kx(1 + 2\ln x)$. $v'(x) = 0 \Rightarrow x = 0$ or $2\ln x = -1$ { $x = r/R = 0$ is discarded } $\Rightarrow$

$\ln x = -\frac{1}{2} \Rightarrow x = e^{-1/2} \approx 0.607$. $v'(x) > 0$ on $(0, e^{-1/2})$ and

$v'(x) < 0$ on $(e^{-1/2}, \infty) \Rightarrow v(e^{-1/2}) = k/(2e)$ is a $LMAX$.

Exercises 7.4

$\boxed{1}$ (a) $u = 2x + 7$, $\frac{1}{2} du = dx \Rightarrow$

$$\int \frac{1}{2x+7}\, dx = \frac{1}{2}\int \frac{1}{u}\, du = \frac{1}{2}\ln|u| + C = \frac{1}{2}\ln|2x+7| + C$$

(b) Using the antiderivative from part (a),

$$\left[\frac{1}{2}\ln|2x+7| \right]_{-2}^{1} = \frac{1}{2}(\ln 9 - \ln 3) = \frac{1}{2}(\ln \tfrac{9}{3}) = \ln(3^{1/2}) = \ln\sqrt{3} \approx 0.55.$$

$\boxed{3}$ (a) $u = x^2 - 9$, $\frac{1}{2} du = x\, dx \Rightarrow$

$$\int \frac{4x}{x^2-9}\, dx = 4 \cdot \frac{1}{2}\int \frac{1}{u}\, du = 2\ln|u| + C = 2\ln|x^2-9| + C$$

(b) $\left[2\ln|x^2-9| \right]_{1}^{2} = 2(\ln|-5| - \ln|-8|) = 2\ln\tfrac{5}{8} = \ln(\tfrac{5}{8})^2 = \ln\tfrac{25}{64} \approx -0.94$

$\boxed{5}$ (a) $u = -4x$, $-\frac{1}{4} du = dx \Rightarrow$

$$\int e^{-4x}\, dx = -\frac{1}{4}\int e^u\, du = -\frac{1}{4}e^u + C = -\frac{1}{4}e^{-4x} + C$$

(b) $\left[-\frac{1}{4}e^{-4x} \right]_{1}^{3} = -\frac{1}{4}(e^{-12} - e^{-4}) \approx 0.0046$

$\boxed{7}$ (a) $u = 2x$, $\frac{1}{2} du = dx \Rightarrow$

$$\int \tan 2x\, dx = \frac{1}{2}\int \tan u\, du = -\frac{1}{2}\ln|\cos u| + C = -\frac{1}{2}\ln|\cos 2x| + C, \text{ by } (7.25)(i).$$

(b) $\left[-\frac{1}{2}\ln|\cos 2x| \right]_{0}^{\pi/8} = -\frac{1}{2}(\ln\cos\tfrac{\pi}{4} - \ln\cos 0) = -\frac{1}{2}(\ln\tfrac{1}{\sqrt{2}} - \ln 1) =$

$$-\frac{1}{2}(\ln 2^{-1/2} - 0) = -\frac{1}{2}(-\tfrac{1}{2}\ln 2) = \tfrac{1}{4}\ln 2 \approx 0.17$$

$\boxed{11}$ $u = x^2 - 4x + 9$, $\frac{1}{2} du = (x - 2)\, dx \Rightarrow$

$$\int \frac{x-2}{x^2-4x+9}\, dx = \frac{1}{2}\int \frac{1}{u}\, du = \frac{1}{2}\ln|u| + C = \frac{1}{2}\ln|x^2-4x+9| + C$$

$\boxed{15}$ $u = \ln x$, $du = \frac{1}{x}\, dx \Rightarrow \int \frac{\ln x}{x}\, dx = \int u\, du = \frac{1}{2}u^2 + C = \frac{1}{2}(\ln x)^2 + C$

$\boxed{19}$ $u = 1 + 2\cos x$, $-\frac{1}{2} du = \sin x\, dx \Rightarrow$

$$\int \frac{3\sin x}{1+2\cos x}\, dx = -\frac{1}{2} \cdot 3\int \frac{1}{u}\, du = -\frac{3}{2}\ln|u| + C = -\frac{3}{2}\ln|1+2\cos x| + C$$

$\boxed{21}$ $\int \frac{(e^x+1)^2}{e^x}\, dx = \int \frac{e^{2x}+2e^x+1}{e^x}\, dx = \int(e^x + 2 + e^{-x})\, dx = e^x + 2x - e^{-x} + C$

$\boxed{25}$ $u = x^{1/3}$, $3\, du = \frac{1}{x^{2/3}}\, dx \Rightarrow$

$$\int \frac{\cot \sqrt[3]{x}}{\sqrt[3]{x^2}}\, dx = 3\int \cot u\, du = 3\ln|\sin u| + C = 3\ln|\sin \sqrt[3]{x}| + C$$

$\boxed{29}$ $u = e^{-3x}$, $-\frac{1}{3} du = e^{-3x}\, dx \Rightarrow \int \frac{\tan e^{-3x}}{e^{3x}}\, dx = \int (\tan e^{-3x})\, e^{-3x}\, dx =$

$$-\frac{1}{3}\int \tan u\, du = -\frac{1}{3}\ln|\sec u| + C = -\frac{1}{3}\ln|\sec e^{-3x}| + C. \text{ \textit{Note}: If we use}$$

$$\int \tan u\, du = -\ln|\cos u| + C, \text{ then the answer would be } \tfrac{1}{3}\ln|\cos e^{-3x}| + C.$$

$\boxed{33}$ $\displaystyle\int \frac{\cos x \sin x}{\cos^2 x - 1}\, dx = \int \frac{\cos x \sin x}{-\sin^2 x}\, dx = -\int \frac{\cos x}{\sin x}\, dx = -\int \cot x\, dx = -\ln|\sin x| + C =$

$\ln|\sin x|^{-1} + C = \ln|\csc x| + C.$ *Note:* $-\ln|\sin x| + C$ is an equivalent answer.

$\boxed{37}$ $A = \displaystyle\int_0^{\ln 3} e^{2x}\, dx = \left[\tfrac{1}{2}e^{2x}\right]_0^{\ln 3} = \tfrac{1}{2}\left[e^{2\ln 3} - e^0\right] = \tfrac{1}{2}\left[(e^{\ln 3})^2 - 1\right] = \tfrac{1}{2}(3^2 - 1) = 4$

$\boxed{39}$ Using shells, $V = 2\pi \displaystyle\int_0^1 xe^{-x^2}\, dx \ \{ u = -x^2,\ -\tfrac{1}{2}\,du = x\,dx;\ x = 0, 1 \Rightarrow u = 0, -1 \}$

$$= 2\pi \cdot (-\tfrac{1}{2}) \int_0^{-1} e^u\, du = -\pi\left[e^u\right]_0^{-1} = -\pi(e^{-1} - e^0) = \pi(1 - e^{-1}) \approx 1.99.$$

$\boxed{41}$ $y' = 4e^{2x} + 3e^{-2x} \Rightarrow y = \int (4e^{2x} + 3e^{-2x})\, dx = 2e^{2x} - \tfrac{3}{2}e^{-2x} + C.$

$$y = 4 \text{ if } x = 0 \Rightarrow 4 = 2 - \tfrac{3}{2} + C \Rightarrow C = \tfrac{7}{2}.$$

$\boxed{43}$ $y'' = 3e^{-x} \Rightarrow y' = \int 3e^{-x}\, dx = -3e^{-x} + C.$ $\ y' = 1 \text{ if } x = 0 \Rightarrow 1 = -3 + C \Rightarrow$

$C = 4.$ $\ y' = -3e^{-x} + 4 \Rightarrow y = \int (-3e^{-x} + 4)\, dx = 3e^{-x} + 4x + C.$

$$y = -1 \text{ if } x = 0 \Rightarrow -1 = 3 + C \Rightarrow C = -4.$$

$\boxed{45}$ $1 = \displaystyle\int_0^3 \frac{cx}{x^2 + 4}\, dx \ \{ u = x^2 + 4,\ \tfrac{1}{2}\,du = x\,dx;\ x = 0, 3 \Rightarrow u = 4, 13 \}$

$$= \tfrac{1}{2}c \int_4^{13} \frac{1}{u}\, du = \tfrac{1}{2}c\left[\ln|u|\right]_4^{13} = \tfrac{1}{2}c(\ln 13 - \ln 4) = c\frac{\ln\frac{13}{4}}{2} \Rightarrow c = \frac{2}{\ln\frac{13}{4}} \approx 1.697$$

$\boxed{47}$ (a) The number of new bacteria at time x is given by $N(x) = \displaystyle\int_0^x 3e^{0.2t}\, dt.$

Thus, $N(5) = \displaystyle\int_0^5 3e^{0.2t}\, dt = 3\left[5e^{0.2t}\right]_0^5 = 15(e - 1) \approx 25.77$, round down to 25.

(b) The sixth through fourteenth hours correspond with $t = 5$ to $t = 14$.

$$N(14) - N(5) = \int_0^{14} 3e^{0.2t}\, dt - \int_0^5 3e^{0.2t}\, dt = \int_5^{14} 3e^{0.2t}\, dt = 3\left[5e^{0.2t}\right]_5^{14} =$$

$$15(e^{2.8} - e) \approx 205.9.$$

(c) $150 = \displaystyle\int_0^x 3e^{0.2t}\, dt \Rightarrow 150 = 15\left[e^{0.2t}\right]_0^x \Rightarrow 10 = e^{0.2x} - 1 \Rightarrow 11 = e^{0.2x} \Rightarrow$

$$\ln 11 = 0.2x \Rightarrow x = 5\ln 11 \approx 11.99, \text{ or } 12, \text{ hours}$$

$\boxed{51}$ (a) $I = \dfrac{dQ}{dt} \Rightarrow I\,dt = dQ \Rightarrow \displaystyle\int I\,dt = \int dQ \Rightarrow Q = \int I\,dt.$

We will use x as the independent variable for I, so as to not confuse the t's.

Since $Q(0) = 0$, $Q(t) = Q(t) - Q(0) =$

$$\int_0^t I(x)\, dx = \int_0^t 10e^{-4x}\, dx = 10\left[-\tfrac{1}{4}e^{-4x}\right]_0^t = \tfrac{5}{2}(1 - e^{-4t}).$$

(b) To compute $Q(t)$ "after a long period of time", we will let t approach "∞".

$$\lim_{t \to \infty} Q(t) = \lim_{t \to \infty} \tfrac{5}{2}(1 - e^{-4t}) = \tfrac{5}{2}(1 - 0) = \tfrac{5}{2} \text{ coulombs.}$$

53 (a) Let $s(t)$ be the distance traveled by the particle in time t. Since $v(t) > 0$ for all t (the particle doesn't "backtrack"), the distance traveled will be given by

$$s(t) = \int_0^t v(x)\,dx = \int_0^t v_0 e^{-x/k}\,dx = v_0\left[-ke^{-x/k}\right]_0^t = kv_0(1 - e^{-t/k}).$$

(b) As in Exercise 51(b), to compute the distance traveled, $s(t)$, after a long period of time, we will let t approach "∞".

$$\lim_{t\to\infty} s(t) = \lim_{t\to\infty}\left[kv_0(1 - e^{-t/k})\right] = kv_0(1 - 0) = kv_0.$$

Exercises 7.5

3   Using (7.28)(ii) with $u = x^2 + 1$, $f(x) = 8^{x^2+1} \Rightarrow$

$$f'(x) = (8^{x^2+1}\ln 8)\,D_x\,(x^2 + 1) = 8^{x^2+1}(2x\ln 8).$$

5   Using (7.31)(ii) with $a = 10$ and $u = x^4 + 3x^2 + 1$,

$$f(x) = \log\,(x^4 + 3x^2 + 1) = \frac{\ln\,(x^4 + 3x^2 + 1)}{\ln 10} \Rightarrow$$

$$f'(x) = \frac{1}{\ln 10}\cdot\frac{1}{x^4 + 3x^2 + 1}\cdot D_x\,(x^4 + 3x^2 + 1) = \frac{4x^3 + 6x}{(x^4 + 3x^2 + 1)\ln 10}.$$

9   Using the product rule, $f(x) = (x^2 + 1)10^{1/x} \Rightarrow$

$$f'(x) = (x^2 + 1)\,D_x\,(10^{1/x}) + (10^{1/x})\,D_x\,(x^2 + 1)$$

$$= (x^2 + 1)(10^{1/x}\ln 10)\,D_x\,(1/x) + (10^{1/x})(2x)$$

$$= (x^2 + 1)(10^{1/x}\ln 10)(-1/x^2) + (2x)\,10^{1/x}$$

$$= -\frac{(x^2 + 1)\,10^{1/x}(\ln 10)}{x^2} + (2x)\,10^{1/x}.$$

11 $f(x) = \log\,(3x^2 + 2)^5 = 5\log\,(3x^2 + 2) = 5\cdot\dfrac{\ln\,(3x^2 + 2)}{\ln 10} \Rightarrow$

$$f'(x) = 5\cdot\frac{1}{\ln 10}\cdot\frac{1}{3x^2 + 2}\cdot D_x\,(3x^2 + 2) = 5\cdot\frac{6x}{(3x^2 + 2)\ln 10} = \frac{30x}{(3x^2 + 2)\ln 10}$$

13 $f(x) = \log_5\left|\dfrac{6x + 4}{2x - 3}\right| = \dfrac{\ln\left|\dfrac{6x + 4}{2x - 3}\right|}{\ln 5} = \dfrac{1}{\ln 5}\Big[\ln|6x + 4| - \ln|2x - 3|\Big] \Rightarrow$

$$f'(x) = \frac{1}{\ln 5}\left(\frac{1}{6x + 4}\cdot 6 - \frac{1}{2x - 3}\cdot 2\right) = \left(\frac{6}{6x + 4} - \frac{2}{2x - 3}\right)\frac{1}{\ln 5}$$

15 $f(x) = \log\ln x = \log_{10}(\ln x) = \dfrac{\ln\,(\ln x)}{\ln 10} \Rightarrow$

$$f'(x) = \frac{1}{\ln 10}\cdot\frac{1}{\ln x}\,D_x\,(\ln x) = \frac{1}{\ln 10}\cdot\frac{1}{\ln x}\cdot\frac{1}{x} = \frac{1}{x\ln x\ln 10}$$

19 Whenever you encounter an exponential expression that has a variable base and a variable exponent, you have two solution methods available—they are both given in Example 5. Using (7.26), $f(x) = (x+1)^x = e^{\ln (x+1)^x} = e^{x \ln (x+1)}$. Thus,

$$f'(x) = D_x \left[e^{x \ln (x+1)} \right] = e^{x \ln (x+1)} D_x \left[x \ln (x+1) \right]$$

$$= e^{x \ln (x+1)} \left[x \cdot \frac{1}{x+1} + 1 \cdot \ln (x+1) \right] = (x+1)^x \left[\frac{x}{x+1} + \ln (x+1) \right].$$

23 Using (7.26), $f(x) = x^{\tan x} = e^{\ln \left[x^{\tan x} \right]} = e^{\tan x \ln x}$. Thus,

$$f'(x) = D_x \left[e^{\tan x \ln x} \right] = e^{\tan x \ln x} D_x (\tan x \ln x)$$

$$= e^{\tan x \ln x} \left(\tan x \cdot \frac{1}{x} + \sec^2 x \ln x \right) = x^{\tan x} \left(\frac{\tan x}{x} + \sec^2 x \ln x \right).$$

25 (a) $f(x) = e^e$ is of the form (constant base)$^{\text{constant exponent}}$.

$\qquad\qquad\qquad$ This expression is a constant and $f'(x) = 0$.

(b) $f(x) = x^5$ is of the form (variable base)$^{\text{constant exponent}}$.

$\qquad\qquad\qquad$ Invoking the power rule, $f'(x) = 5x^{5-1} = 5x^4$.

(c) $f(x) = x^{\sqrt{5}}$ is of the form (variable base)$^{\text{constant exponent}}$.

$\qquad\qquad\qquad$ Invoking the power rule, $f'(x) = \sqrt{5}\, x^{\sqrt{5}-1}$.

(d) $f(x) = (\sqrt{5})^x$ is of the form (constant base)$^{\text{variable exponent}}$.

$\qquad\qquad\qquad$ Invoking (7.28)(i), $f'(x) = (\sqrt{5})^x \ln \sqrt{5}$.

(e) $f(x) = x^{(x^2)}$ is of the form (variable base)$^{\text{variable exponent}}$.

$\qquad$ Changing to a natural exponential form (or using logarithmic differentiation),

$$f(x) = e^{x^2 \ln x} \Rightarrow f'(x) = e^{x^2 \ln x} (x^2 \cdot \tfrac{1}{x} + 2x \ln x) = x^{(x^2)} (x + 2x \ln x) =$$

$$x^{(x^2)} (x^1)(1 + 2\ln x) = x^{1+x^2}(1 + 2\ln x).$$

29 (a) $u = -2x$, $-\frac{1}{2} du = dx \Rightarrow \displaystyle\int 5^{-2x}\, dx = -\frac{1}{2} \int 5^u\, du = -\frac{5^u}{2 \ln 5} + C = -\frac{5^{-2x}}{2 \ln 5} + C$

(b) $\displaystyle\int_1^2 5^{-2x}\, dx = -\frac{1}{2 \ln 5} \left[5^{-2x} \right]_1^2 =$

$$-\frac{1}{2 \ln 5} \left(\frac{1}{625} - \frac{1}{25} \right) = -\frac{1}{2 \ln 5} \left(-\frac{24}{625} \right) = \frac{12}{625 \ln 5} \approx 0.012$$

33 $u = -x^2$, $-\frac{1}{2} du = x\, dx \Rightarrow \displaystyle\int x(3^{-x^2})\, dx = -\frac{1}{2} \int 3^u\, du = -\frac{3^u}{2 \ln 3} + C = -\frac{3^{-x^2}}{2 \ln 3} + C$

35 $u = 2^x + 1$, $du = 2^x \ln 2\, dx$ or $\frac{1}{\ln 2} du = 2^x\, dx \Rightarrow \displaystyle\int \frac{2^x}{2^x + 1}\, dx = \frac{1}{\ln 2} \int \frac{1}{u}\, du =$

$$\frac{1}{\ln 2} \ln u + C \{ \text{since } u > 0, \text{ we do not need to use } \ln |u| \} = \frac{\ln (2^x + 1)}{\ln 2} + C$$

$\boxed{37}$ $u = \log x$, $du = \frac{dx}{x \ln 10}$ or $\ln 10 \, du = \frac{dx}{x}$ $\Rightarrow$

$$\int \frac{1}{x \log x} \, dx = \ln 10 \int \frac{1}{u} \, du = (\ln 10) \ln |u| + C = (\ln 10) \ln |\log x| + C$$

$\boxed{41}$ (a) π^{π} is of the form (constant base)$^{\text{constant exponent}}$.

This expression is just a constant, so $\int \pi^{\pi} \, dx = \pi^{\pi} x + C$.

(b) x^4 is of the form (variable base)$^{\text{constant exponent}}$.

Employing the power rule, $\int x^4 \, dx = \frac{x^{4+1}}{4+1} + C = \frac{1}{5} x^5 + C$.

(c) x^{π} is of the form (variable base)$^{\text{constant exponent}}$.

As in part (b), $\int x^{\pi} \, dx = \frac{x^{\pi+1}}{\pi + 1} + C$.

(d) π^x is of the form (constant base)$^{\text{variable exponent}}$.

Applying (7.29)(i), $\int \pi^x \, dx = \frac{\pi^x}{\ln \pi} + C$.

$\boxed{43}$ The graphs of $y = 2^x$ and $y = 1 - x$ intersect when $x = 0$. On $[0, 1]$, $2^x \geq 1 - x$.

$$A = \int_0^1 \left[2^x - (1 - x) \right] dx$$

$$= \left[\frac{2^x}{\ln 2} - x + \frac{1}{2} x^2 \right]_0^1 = \left(\frac{2}{\ln 2} - 1 + \frac{1}{2} \right) - \left(\frac{1}{\ln 2} \right) = \frac{1}{\ln 2} - \frac{1}{2} \approx 0.94.$$

$\boxed{45}$ (a) The *rate of decrease* is given by $B'(t)$. $B'(t) = (0.95)^t \ln (0.95)$.

$$B'(2) \approx -0.046 \approx \$0.05/\text{yr} \text{ (decreasing)}.$$

(b) Referring to (5.29) for the definition of average value of a function, we have

$$B_{\text{av}} = \frac{1}{2-0} \int_0^2 (0.95)^t \, dt = \frac{1}{2} \left[\frac{(0.95)^t}{\ln (0.95)} \right]_0^2 = \frac{1}{2 \ln (0.95)} \left[(0.95)^2 - 1 \right] \approx \$0.95.$$

$\boxed{47}$ (a) $N(t) = 1000(0.9)^t \Rightarrow dN/dt = N'(t) = 1000(0.9)^t \ln (0.9)$.

$N'(1) = 1000(0.9)^1 \ln (0.9) \approx -95$ trout/yr (a decrease of 95 trout).

$N'(5) = 1000(0.9)^5 \ln (0.9) \approx -62$ trout/yr.   To answer the last part of the question, we first need to find the time t when $N = 500$. In this case, we only find a value for $(0.9)^t$ since it appears in the formula for $N'(t)$. $N = 500 \Rightarrow 1000(0.9)^t = 500 \Rightarrow \frac{1}{2} = (0.9)^t$. Next, we substitute $\frac{1}{2}$ for $(0.9)^t$ in $N'(t)$, which gives us $N'(t) = 1000(\frac{1}{2}) \ln (0.9) \approx -53$ trout/yr.

(b) The total weight of trout in the pond can be found by multiplying the number of trout times their individual weight. If we call this function T, then $T(t) = N(t) \cdot W(t) = 1000(0.9)^t (0.2 + 1.5t)$. To maximize T, we need to find the critical numbers of T.

$$T'(t) = 1000\left[(0.9)^t D_t (0.2 + 1.5t) + (0.2 + 1.5t) D_t (0.9)^t\right]$$

$$= 1000\left[(0.9)^t (1.5) + (0.2 + 1.5t)(0.9)^t (\ln 0.9)\right]$$

$$= 1000(0.9)^t \left[1.5 + \ln(0.9)(0.2 + 1.5t)\right].$$

$$T'(t) = 0 \Rightarrow 1.5 + \ln(0.9)(0.2 + 1.5t) = 0 \{1000(0.9)^t \neq 0\} \Rightarrow$$

$$0.2\ln(0.9) + 1.5\ln(0.9)t = -1.5 \Rightarrow t = t_0 = \frac{-1.5 - 0.2\ln(0.9)}{1.5\ln(0.9)} \approx 9.36 \text{ yr.}$$

Since $T'(t) > 0$ on $[0, t_0)$ and $T'(t) < 0$ on (t_0, ∞),

$$T(t_0) \approx 5311.53 \text{ is a maximum value.}$$

49 Calculating the pH of vinegar, we have

$$pH = -\log\left[H^+\right] = -\log\left[6.3 \times 10^{-3}\right] \approx 2.201.$$

The maximum percentage error in the calculation is $\dfrac{d(pH)}{pH} \times 100\%$ {note that pH is just a single variable}. If we think of pH as a function f of H^+, then

$$pH = f(H^+) = -\log\left[H^+\right] = -\frac{\ln\left[H^+\right]}{\ln 10},$$

and $$d(pH) = f'(H^+) d(H^+) = -\frac{1}{\ln 10} \cdot \frac{1}{H^+} d(H^+).$$

Thus, $\dfrac{d(pH)}{pH} = \dfrac{-\dfrac{d(H^+)}{(\ln 10) H^+}}{-\log\left[H^+\right]} = \dfrac{d(H^+)/H^+}{(\ln 10)\log\left[H^+\right]} = \dfrac{\pm 0.5\%}{\ln 10 \, \log\left(6.3 \times 10^{-3}\right)} \approx \pm 0.1\%.$

51 (a) When $x = x_0$, $R = a\log(x/x_0) = a\log(x_0/x_0) = a\log 1 = a(0) = 0$.

(b) $R = a\log(x/x_0) \Rightarrow S = \dfrac{dR}{dx} = \dfrac{a}{(x/x_0)\ln 10} D_x\left(\dfrac{x}{x_0}\right) = \dfrac{a}{(x/x_0)\ln 10 \,(x_0)} = \dfrac{a}{x\ln 10}.$

Now S is of the form $\dfrac{k}{x}$, where $k = a/\ln 10$ is the constant of proportionality, and this implies that S is inversely proportional to x. $S(x) = \dfrac{k}{x}$.

$S(2x) = \dfrac{k}{2x} \Rightarrow 2 S(2x) = \dfrac{k}{x}$, which is $S(x)$. Hence $S(x) = 2 S(2x)$, indicating that $S(x)$ is twice as sensitive as $S(2x)$. This indicates that a person will be able to detect a small change in the salt content of a dilute salt solution better than in one that is very concentrated.

| Exercises 7.6 |

Note: When working with the formula in (7.33), there are several things to keep in mind.

1) y is the amount present at time t.

2) y_0 is the initial amount corresponding to time $t = 0$. We will usually use either ke^{ct} or $y_0 e^{ct}$.

3) c is the rate of increase or decrease, i.e., if there is a 4% increase per year, then $c = 0.04$. If $c > 0$, the formula describes an increasing function, or law of growth. If $c < 0$, the formula describes a decreasing function, or law of decay.

1 If $q(t)$ is the number of bacteria present at time t, then by (7.33), $q(t) = 5000e^{ct}$ since the initial number of bacteria present is 5000. We could also justify this as follows: $q(t) = ke^{ct} \Rightarrow q(0) = ke^{c(0)} = ke^0 = k(1) = k$. Since $q(0) = 5000$, $k = 5000$. Either way, we now know that the general formula is $q(t) = 5000e^{ct}$.

To determine the value of c, we need to know some specific growth or decay information. In this case, we know $q(t) = 15{,}000$ when $t = 10$. Thus, $q(10) = 5000e^{c(10)} = 15{,}000 \Rightarrow (e^c)^{10} = 3 \Rightarrow e^c = \pm 3^{1/10} = 3^{1/10}$. Note on the last step we merely took the 10th root of both sides of the equation. In future problems, we will omit the step using the "$\pm$" symbol since exponential expressions are positive. We could now solve for c, but we will stop at solving for e^c since this expression can be substituted into the expression for $q(t)$, and this will save us some algebraic manipulations. Thus, $q(t) = 5000e^{ct} = 5000(3^{1/10})^t = 5000(3)^{t/10}$.

The number of bacteria at the end of 20 hours corresponds to $t = 20$ and $q(20) = 5000(3)^2 = 45{,}000$.

The number of bacteria will be 50,000 when $q(t) = 50{,}000$.
$q(t) = 50{,}000 \Rightarrow 5000(3)^{t/10} = 50{,}000 \Rightarrow 3^{t/10} = 10 \Rightarrow \ln(3^{t/10}) = \ln 10 \Rightarrow$

$$(t/10)\ln 3 = \ln 10 \Rightarrow t = \frac{10\ln 10}{\ln 3} \approx 20.96 \text{ hr.}$$

5 Let $P(t)$ denote the population at time t and $t = 0$ correspond to Jan. 1, 1980. Since the population is increasing at a rate of 2% per year, $c = 0.02$ in (7.33). Thus, $dP/dt = cP \Rightarrow dP/dt = 0.02P \Rightarrow P(t) = ke^{0.02t}$. Since $P(0) = 4.5 \times 10^9$, the population (in billions) is given by $P(t) = 4.5e^{0.02t}$. $P(t) = 40 \Rightarrow 4.5e^{0.02t} = 40$

$\Rightarrow e^{0.02t} = \frac{40}{4.5} \Rightarrow 0.02t = \ln\left(\frac{40}{4.5}\right) \Rightarrow t = \dfrac{\ln(40/4.5)}{0.02} \approx 109.24$ yr (March 29, 2089).

Note: We will use the following step in the solutions.

$$\boxed{y = e^x \Leftrightarrow \ln y = x}$$

Sometimes this relatively simple step is the one that seems to be the most confusing. For example, we used this step above when going from

$$e^{0.02t} = \tfrac{40}{4.5} \quad \text{to} \quad 0.02t = \ln\left(\tfrac{40}{4.5}\right).$$

7 *Hint:* See Example 3.

11 $\dfrac{dP}{dz} = \dfrac{-0.0342\,P}{T}$ { given }

       $= \dfrac{-0.0342\,P}{288 - 0.01z}$ { definition of T }

$\dfrac{dP}{P} = (-0.0342)\dfrac{dz}{288 - 0.01z}$ { separating the variables }

$\displaystyle\int \tfrac{1}{P}\,dP = (-0.0342)\int \dfrac{1}{288 - 0.01z}\,dz$ { integrate both sides }

$\ln P = \dfrac{-0.0342}{-0.01}\ln(288 - 0.01z) + C$ { find antiderivatives, $z < 28{,}800$ }

      $= 3.42\ln(288 - 0.01z) + C$ { simplify }

      $= \ln(288 - 0.01z)^{3.42} + C$ { $r\ln x = \ln x^r$ }

$P = e^{\ln(288 - 0.01z)^{3.42} + C}$ { $\ln x = y \Leftrightarrow x = e^y$ }

      $= e^{\ln(288 - 0.01z)^{3.42}} e^C$ { $e^{a+b} = e^a e^b$ }

$P(z) = e^C(2.88 - 0.01z)^{3.42}$ { $e^{\ln x} = x$ }

$P(0) = 1 \Rightarrow e^C(2.88 - 0)^{3.42} = 1 \Rightarrow e^C = 288^{-3.42} \Rightarrow$

$P(z) = 288^{-3.42}(2.88 - 0.01z)^{3.42} = \dfrac{(2.88 - 0.01z)^{3.42}}{288^{3.42}} = \left[\dfrac{288 - 0.01z}{288}\right]^{3.42}.$

13 This problem is somewhat different than some of the others because we are given a
relative amount, $2.5S$, for the initial amount of ^{90}Sr, rather than a static amount.

Let $q(t)$ be the quantity of ^{90}Sr at time t. Then, by (7.33), $q(t) = ke^{ct}$ and
$q(0) = 2.5S \Rightarrow k = 2.5S$. A half-life of 29 years $\Rightarrow q(29) = (2.5S)e^{c(29)} = \frac{1}{2}(2.5S)$
$\Rightarrow e^{c(29)} = \frac{1}{2} \Rightarrow e^c = \left(\frac{1}{2}\right)^{1/29}$. Thus, $q(t) = (2.5S)\left(\frac{1}{2}\right)^{t/29}$.

The safe level occurs when $q(t) = S$, i.e., $S = (2.5S)\left(\frac{1}{2}\right)^{t/29} \Rightarrow \left(\frac{1}{2}\right)^{t/29} = \frac{1}{2.5} \Rightarrow$

$$\ln\left(\tfrac{1}{2}\right)^{t/29} = \ln\left(\tfrac{1}{2.5}\right) \Rightarrow (t/29)\ln\left(\tfrac{1}{2}\right) = \ln\left(\tfrac{1}{5/2}\right) \Rightarrow t = \frac{29\ln(2/5)}{\ln(1/2)} \approx 38.34 \text{ yr.}$$

15 Let $q(t)$ be the amount of sodium pentobarbital at time t.

Then, by (7.33), $q(t) = ke^{ct}$. A half-life of 4 hours $\Rightarrow e^{c(4)} = \frac{1}{2} \Rightarrow e^c = \left(\frac{1}{2}\right)^{1/4}$.

Thus, $q(t) = k\left(\frac{1}{2}\right)^{t/4}$. We want enough initial sodium pentobarbital so that the dog
will still be anesthetized after 45 minutes. We must have

$$20 \text{ kilograms} \times \frac{30 \text{ mg of sodium pentobarbital}}{1 \text{ kg of body weight}} = 600 \text{ mg}$$

after 45 minutes, hence we desire $q\left(\frac{45}{60}\right) = 600 \Rightarrow k\left(\frac{1}{2}\right)^{(45/60)/4} = 600 \Rightarrow$

$$k = \frac{600}{\left(\frac{1}{2}\right)^{3/16}} \approx 683.27 \text{ mg.}$$

17 $v\dfrac{dv}{dy} = -\dfrac{k}{y^2} \Rightarrow v\,dv = -k\dfrac{dy}{y^2} \Rightarrow \displaystyle\int v\,dv = -k\int y^{-2}\,dy \Rightarrow \frac{1}{2}v^2 = \frac{k}{y} + C_1 \Rightarrow$

$v = \sqrt{\dfrac{2k}{y} + C_2}\ \{2C_1 = C_2\}$. $v(y_0) = v_0 = \sqrt{\dfrac{2k}{y_0} + C_2} \Rightarrow v_0^2 = \dfrac{2k}{y_0} + C_2 \Rightarrow$

$$C_2 = v_0^2 - \frac{2k}{y_0}. \text{ Thus, } v(y) = \sqrt{\frac{2k}{y} + \left(v_0^2 - \frac{2k}{y_0}\right)} = \sqrt{2k\left(\frac{1}{y} - \frac{1}{y_0}\right) + v_0^2}.$$

19 Let $q(t)$ be the quantity of ^{14}C at time t. Then, by (7.33), $q(t) = q_0 e^{ct}$.

A half-life of 5700 yr $\Rightarrow q(5700) = q_0 e^{c(5700)} = \frac{1}{2}q_0 \Rightarrow e^{c(5700)} = \frac{1}{2} \Rightarrow$

$e^c = \left(\frac{1}{2}\right)^{1/5700}$. Thus, $q(t) = q_0\left(\frac{1}{2}\right)^{t/5700}$. $q(t) = 20\% \, q_0 \Rightarrow 0.2q_0 = q_0\left(\frac{1}{2}\right)^{t/5700} \Rightarrow$

$0.2 = \left(\frac{1}{2}\right)^{t/5700} \Rightarrow \ln(0.2) = \ln\left(\frac{1}{2}\right)^{t/5700} \Rightarrow \ln(0.2) = (t/5700)\ln\left(\frac{1}{2}\right) \Rightarrow$

$$t = \frac{5700\ln(0.2)}{\ln(1/2)} \approx 13{,}235 \text{ yr.}$$

21 $I = I_0 e^{-f(x)} \Rightarrow dI/dx = I_0 e^{-f(x)} \cdot D_x[-f(x)] = I_0 e^{-f(x)} \cdot D_x\left[-k\int_0^x \rho(h)\,dh\right] =$

$$I_0 e^{-f(x)} \cdot \left[-k\rho(x)\right]\ \{\text{by } (5.35)\} = -k\rho(x) I_0 e^{-f(x)} = -k\rho(x) I.$$

$\boxed{23}$ $V = \frac{4}{3}\pi r^3 \Rightarrow r = (\frac{3}{4\pi} V)^{1/3}$. Thus, $S = 4\pi r^2 = 4\pi(\frac{3}{4\pi} V)^{2/3} = \left[4\pi(\frac{3}{4\pi})^{2/3}\right] V^{2/3}$

$= k_1 V^{2/3}$, for a constant k_1. $\frac{dV}{dt}$ is proportional to the surface area of the cell $\Rightarrow$

$\frac{dV}{dt} = k_2 S = k_2 k_1 V^{2/3} = kV^{2/3}$, where k_2 and $k = k_1 k_2$ are constants. $k > 0$ since

the rate of growth, $\frac{dV}{dt}$, increases as the surface area S increases and $V^{2/3} > 0$.

Solving for V: $\frac{dV}{dt} = kV^{2/3} \Rightarrow V^{-2/3} dV = k\, dt \Rightarrow \int V^{-2/3} dV = \int k\, dt \Rightarrow$

$3V^{1/3} = kt + C \Rightarrow V = \left[\frac{1}{3}(kt + C)\right]^3 \Rightarrow V(t) = \frac{1}{27}(kt + C)^3.$

7.7 Review Exercises

$\boxed{1}$ $y = 10 - 15x \Rightarrow 15x = 10 - y \Rightarrow$

$$x = \frac{10 - y}{15} \Rightarrow f^{-1}(y) = \frac{10 - y}{15} \Rightarrow f^{-1}(x) = \frac{10 - x}{15}$$

$\boxed{3}$ $f'(x) = 6x^2 - 8 < 0$ for $-1 \le x \le 1 \Rightarrow f$ is $\downarrow \Rightarrow f^{-1}$ exists by (7.6). Since

$$f(0) = 5, f^{-1}(5) = 0. \text{ By (7.7), } D_x f^{-1}(5) = \frac{1}{f'(f^{-1}(5))} = \frac{1}{f'(0)} = \frac{1}{-8} = -\frac{1}{8}.$$

$\boxed{5}$ $f(x) = \ln|4 - 5x^3|^5 = 5\ln|4 - 5x^3| \Rightarrow f'(x) = 5 \cdot \frac{1}{4 - 5x^3} \cdot (-15x^2) = \frac{75x^2}{5x^3 - 4}$

$\boxed{9}$ $f(x) = \ln\dfrac{(3x + 2)^4 \sqrt{6x - 5}}{8x - 7} = \ln(3x + 2)^4 + \ln\sqrt{6x - 5} - \ln(8x - 7)$

$= 4\ln(3x + 2) + \frac{1}{2}\ln(6x - 5) - \ln(8x - 7) \Rightarrow$

$f'(x) = 4 \cdot \dfrac{1}{3x + 2} \cdot 3 + \dfrac{1}{2} \cdot \dfrac{1}{6x - 5} \cdot 6 - \dfrac{1}{8x - 7} \cdot 8 = \dfrac{12}{3x + 2} + \dfrac{3}{6x - 5} - \dfrac{8}{8x - 7}$

$\boxed{11}$ Using the reciprocal rule, $f(x) = \dfrac{1}{\ln(2x^2 + 3)} \Rightarrow$

$$f'(x) = -\dfrac{D_x\left[\ln(2x^2 + 3)\right]}{\left[\ln(2x^2 + 3)\right]^2} = -\dfrac{\left[1/(2x^2 + 3)\right] \cdot 4x}{\left[\ln(2x^2 + 3)\right]^2} = \dfrac{-4x}{(2x^2 + 3)\left[\ln(2x^2 + 3)\right]^2}$$

$\boxed{15}$ $f(x) = e^{\ln(x^2+1)} = x^2 + 1 \Rightarrow f'(x) = 2x$

$\boxed{19}$ $f(x) = 10^x \log x \Rightarrow$

$$f'(x) = 10^x D_x(\log x) + D_x(10^x)\log x = \dfrac{10^x}{x\ln 10} + 10^x(\ln 10)\log x$$

$\boxed{21}$ $f(x) = \sqrt{\ln\sqrt{x}} = (\ln\sqrt{x})^{1/2} \Rightarrow f'(x) = \frac{1}{2}(\ln\sqrt{x})^{-1/2} D_x(\ln\sqrt{x}) =$

$$\frac{1}{2}(\ln\sqrt{x})^{-1/2}\frac{1}{\sqrt{x}} D_x(\sqrt{x}) = \frac{1}{2}(\ln\sqrt{x})^{-1/2}\left(\frac{1}{\sqrt{x}} \cdot \frac{1}{2\sqrt{x}}\right) = \frac{1}{4x\sqrt{\ln\sqrt{x}}}$$

$\boxed{25}$ $f(x) = \sqrt{e^{3x} + e^{-3x}} \Rightarrow$

$$f'(x) = \frac{1}{2}(e^{3x} + e^{-3x})^{-1/2}(3e^{3x} - 3e^{-3x}) = \dfrac{3(e^{3x} - e^{-3x})}{2\sqrt{e^{3x} + e^{-3x}}}$$

29 $y = x^{\ln x} \Rightarrow \ln y = \ln\left(x^{\ln x}\right) = \ln x\,(\ln x) = (\ln x)^2 \Rightarrow$

$$\frac{y'}{y} = 2(\ln x)^1 \cdot \frac{1}{x} \Rightarrow y' = \frac{2\ln x}{x} \cdot y = \frac{2\ln x\,\left(x^{\ln x}\right)}{x}$$

31 $f(x) = \ln|\tan x - \sec x| \Rightarrow f'(x) = \dfrac{D_x\,(\tan x - \sec x)}{\tan x - \sec x} = \dfrac{\sec^2 x - \sec x \tan x}{\tan x - \sec x} =$

$$\frac{\sec x(\sec x - \tan x)}{\tan x - \sec x} = \frac{-\sec x(\tan x - \sec x)}{\tan x - \sec x} = -\sec x$$

33 $f(x) = \csc e^{-2x}\cot e^{-2x} \Rightarrow$

$$f'(x) = \csc e^{-2x}\,D_x\left(\cot e^{-2x}\right) + \cot e^{-2x}\,D_x\left(\csc e^{-2x}\right)$$

$$= \csc e^{-2x}\left[(-\csc^2 e^{-2x})(-2e^{-2x})\right] + \cot e^{-2x}\left[(-\csc e^{-2x}\cot e^{-2x})(-2e^{-2x})\right]$$

$$= 2e^{-2x}\csc e^{-2x}(\csc^2 e^{-2x} + \cot^2 e^{-2x})$$

37 $f(x) = (\sin x)^{\cos x} = e^{\cos x\,\ln\sin x} \Rightarrow f'(x) = e^{\cos x\,\ln\sin x}\,D_x\,(\cos x\,\ln\sin x) =$

$$e^{\cos x\,\ln\sin x}\left[\cos x\cdot\frac{\cos x}{\sin x} + (-\sin x)\ln\sin x\right] = (\sin x)^{\cos x}\,(\cos x\cot x - \sin x\ln\sin x)$$

39 $1 + xy = e^{xy} \Rightarrow D_x\,(1 + xy) = D_x\left(e^{xy}\right) \Rightarrow 0 + (xy' + 1\cdot y) = e^{xy}\,D_x\,(xy) \Rightarrow$

$$xy' + y = e^{xy}(xy' + y) \Rightarrow y'(x - xe^{xy}) = ye^{xy} - y \Rightarrow y' = \frac{y(e^{xy} - 1)}{x(1 - e^{xy})} = -\frac{y}{x}.$$

41 $y = (x + 2)^{4/3}(x - 3)^{3/2} \Rightarrow \ln y = \ln(x + 2)^{4/3} + \ln(x - 3)^{3/2} =$

$$\tfrac{4}{3}\ln(x + 2) + \tfrac{3}{2}\ln(x - 3) \Rightarrow \frac{y'}{y} = \frac{4}{3(x + 2)} + \frac{3}{2(x - 3)} \Rightarrow$$

$$y' = \left[\frac{4}{3(x + 2)} + \frac{3}{2(x - 3)}\right]\cdot y = \left[\frac{4}{3(x + 2)} + \frac{3}{2(x - 3)}\right](x + 2)^{4/3}(x - 3)^{3/2}$$

43 (a) $u = \sqrt{x},\ 2\,du = \frac{1}{\sqrt{x}}\,dx \Rightarrow$

$$\int \frac{1}{\sqrt{x}\,e^{\sqrt{x}}}\,dx = \int e^{-\sqrt{x}}\frac{1}{\sqrt{x}}\,dx = 2\int e^{-u}\,du = -2e^{-u} + C = -2e^{-\sqrt{x}} + C$$

(b) $\displaystyle\int_1^4 \frac{1}{\sqrt{x}\,e^{\sqrt{x}}}\,dx = -2\left[e^{-\sqrt{x}}\right]_1^4 = -2(e^{-2} - e^{-1}) \approx 0.465$

45 (a) $u = -x^2,\ -\frac{1}{2}\,du = x\,dx \Rightarrow$

$$\int x\,4^{-x^2}\,dx = -\frac{1}{2}\int 4^u\,du = -\frac{4^u}{2\ln 4} + C = -\frac{4^{-x^2}}{2\ln 4} + C$$

(b) $\displaystyle\int_0^1 x\,4^{-x^2}\,dx = -\frac{1}{2\ln 4}\left[4^{-x^2}\right]_0^1 = -\frac{1}{2\ln 4}\left(\frac{1}{4} - 1\right) = \frac{3}{8\ln 4} \approx 0.271$

47 $u = x^2,\ \frac{1}{2}\,du = x\,dx \Rightarrow$

$$\int x\tan x^2\,dx = \frac{1}{2}\int \tan u\,du = -\frac{1}{2}\ln|\cos u| + C = -\frac{1}{2}\ln\left|\cos x^2\right| + C$$

$\boxed{51}$ $u = 1 - \ln x$, $-du = \frac{dx}{x} \Rightarrow$

$$\int \frac{1}{x - x \ln x} \, dx = \int \frac{1}{x(1 - \ln x)} \, dx = -\int \frac{1}{u} \, du = -\ln|u| + C = -\ln|1 - \ln x| + C$$

$\boxed{55}$ Using long division,

$$\int \frac{x^2}{x + 2} \, dx = \int \left(x - 2 + \frac{4}{x + 2} \right) dx = \frac{1}{2}x^2 - 2x + 4\ln|x + 2| + C$$

$\boxed{59}$ $u = x^2 + 1$, $\frac{1}{2} du = x \, dx \Rightarrow$

$$\int \frac{x}{x^4 + 2x^2 + 1} \, dx = \int \frac{x}{(x^2 + 1)^2} \, dx = \frac{1}{2} \int \frac{1}{u^2} \, du = -\frac{1}{2u} + C = -\frac{1}{2(x^2 + 1)} + C$$

$\boxed{63}$ $\int 5^x e^x \, dx = \int (5e)^x \, dx \{ \text{treat } 5e \text{ as a constant} \} = \dfrac{(5e)^x}{\ln(5e)} + C$

$\boxed{65}$ $u = \log x$, $du = \dfrac{1}{x \ln 10} \, dx \Rightarrow \displaystyle\int \frac{1}{x\sqrt{\log x}} \, dx = \ln 10 \int u^{-1/2} \, du =$

$$2 \ln 10 \sqrt{u} + C = 2 \ln 10 \sqrt{\log x} + C$$

$\boxed{71}$ $u = 1 - 2\sin 2x$, $-\frac{1}{4} du = \cos 2x \, dx \Rightarrow$

$$\int \frac{\cos 2x}{1 - 2\sin 2x} \, dx = -\frac{1}{4} \int \frac{1}{u} \, du = -\frac{1}{4} \ln|u| + C = -\frac{1}{4} \ln|1 - 2\sin 2x| + C$$

$\boxed{75}$ $u = 3x$, $\frac{1}{3} du = dx \Rightarrow \int (\csc 3x + 1)^2 \, dx =$

$$\tfrac{1}{3} \int (\csc u + 1)^2 \, du = \tfrac{1}{3} \int (\csc^2 u + 2\csc u + 1) \, du$$

$$= \tfrac{1}{3}(-\cot u + 2\ln|\csc u - \cot u| + u) + C$$

$$= -\tfrac{1}{3}\cot 3x + \tfrac{2}{3}\ln|\csc 3x - \cot 3x| + x + C$$

$\boxed{79}$ $y'' = -e^{-3x} \Rightarrow \int y'' \, dy = \int -e^{-3x} \, dx \Rightarrow y' = \frac{1}{3}e^{-3x} + C$. $y' = 2$ if $x = 0 \Rightarrow$

$2 = \frac{1}{3} + C \Rightarrow C = \frac{5}{3}$. $y' = \frac{1}{3}e^{-3x} + \frac{5}{3} \Rightarrow \int y' \, dy = \int (\frac{1}{3}e^{-3x} + \frac{5}{3}) \, dx \Rightarrow$

$y = -\frac{1}{9}e^{-3x} + \frac{5}{3}x + C$. $y = -1$ if $x = 0 \Rightarrow -\frac{1}{9} + C = -1 \Rightarrow C = -\frac{8}{9}$.

$$\text{Thus, } y = -\tfrac{1}{9}e^{-3x} + \tfrac{5}{3}x - \tfrac{8}{9}.$$

$\boxed{81}$ $a(t) = v'(t) = e^{t/2} \Rightarrow \int v'(t) \, dt = \int e^{t/2} \, dt \Rightarrow v(t) = 2e^{t/2} + C$.

$v(0) = 6 \Rightarrow C = 4$ and $v(t) = 2e^{t/2} + 4$. Since $v(t) > 0$, the distance traveled

from $t = 0$ to $t = 4$ is $s(4) - s(0) = \displaystyle\int_0^4 v(t) \, dt = \int_0^4 (2e^{t/2} + 4) \, dt = \left[4e^{t/2} + 4t \right]_0^4$

$$= (4e^2 + 16) - (4 + 0) = 4e^2 + 12 \approx 41.56 \text{ cm.}$$

$\boxed{83}$ $y = xe^{1/x^3} + \ln|2 - x^2| \Rightarrow y' = xe^{1/x^3}(-3/x^4) + e^{1/x^3} - \dfrac{2x}{2 - x^2}$.

At $x = 1$, $y' = -3e + e - 2 = -2(1 + e)$. Tangent line at $P(1, e)$:

$$y - e = -2(1 + e)(x - 1), \text{ or approximately, } y = -7.44x + 10.15.$$

85 Using disks,

$$V = \pi \int_{-3}^{-2} (e^{4x})^2 \, dx = \pi \int_{-3}^{-2} e^{8x} \, dx = \pi \left[\tfrac{1}{8} e^{8x} \right]_{-3}^{-2} = \tfrac{\pi}{8}(e^{-16} - e^{-24}) \approx 4.42 \times 10^{-8}.$$

87 Let $q(t)$ be the quantity of radioactive substance at time t. Then, $q(t) = Ae^{ct}$.

A half-life of 5 days $\Rightarrow q(5) = Ae^{c(5)} = \tfrac{1}{2}A \Rightarrow e^{c(5)} = \tfrac{1}{2} \Rightarrow e^c = \left(\tfrac{1}{2}\right)^{1/5}$.

Thus, $q(t) = A\left(\tfrac{1}{2}\right)^{t/5}$. $q(t) = 1\% A \Rightarrow 0.01A = A\left(\tfrac{1}{2}\right)^{t/5} \Rightarrow \left(\tfrac{1}{2}\right)^{t/5} = \tfrac{1}{100} \Rightarrow$

$$\ln\left(\tfrac{1}{2}\right)^{t/5} = \ln\left(\tfrac{1}{100}\right) \Rightarrow (t/5)\ln\left(\tfrac{1}{2}\right) = \ln\left(\tfrac{1}{100}\right) \Rightarrow t = \frac{5\ln(1/100)}{\ln(1/2)} \approx 33.2 \text{ days.}$$

89 The rate at which sugar *does not dissolve* is also directly proportional to the amount
that does not dissolve. Thus, $q(t) = q_0 e^{ct}$, where q represents the amount of sugar
that remains _undissolved_. $q(0) = q_0 e^0 = 10 \Rightarrow q_0 = 10$. Since after 3 hours one-
half of the sugar remains undissolved, $q(3) = q_0 e^{c(3)} = \tfrac{1}{2}q_0 \Rightarrow e^{c(3)} = \tfrac{1}{2} \Rightarrow$

$$e^c = \left(\tfrac{1}{2}\right)^{1/3} \Rightarrow q(t) = q_0 e^{ct} = 10\left(\tfrac{1}{2}\right)^{t/3}.$$

(a) If 2 more pounds dissolve, 3 pounds will remain undissolved.

$$q(t) = 3 \Rightarrow \left(\tfrac{1}{2}\right)^{t/3} = \tfrac{3}{10} \Rightarrow t = \frac{3\ln(3/10)}{\ln(1/2)} \approx 5.2 \text{ or } 2.2 \text{ additional hr.}$$

(b) $t = 7$ for 8:00 P.M. Since the total amount of sugar is 10 lb,

the amount of dissolved sugar when $t = 7$ is $10 - q(7) =$

$$10 - 10\left(\tfrac{1}{2}\right)^{7/3} = 10\left[1 - \left(\tfrac{1}{2}\right)^{7/3}\right] \approx 8.016 \text{ lb.}$$

Chapter 8: Inverse Trigonometric and Hyperbolic Functions

Exercises 8.1

1 (a) $\sin^{-1}\left(-\frac{\sqrt{2}}{2}\right) = -\frac{\pi}{4}$ since $\sin\left(-\frac{\pi}{4}\right) = -\frac{\sqrt{2}}{2}$ and $-\frac{\pi}{2} \leq -\frac{\pi}{4} \leq \frac{\pi}{2}$.

A common mistake is to think that the answer is $\frac{7\pi}{4}$, however,

this is outside the range of the arcsine function, which is $[-\frac{\pi}{2}, \frac{\pi}{2}]$.

(b) $\cos^{-1}\left(-\frac{1}{2}\right) = \frac{2\pi}{3}$ since $\cos\frac{2\pi}{3} = -\frac{1}{2}$ and $0 \leq \frac{2\pi}{3} \leq \pi$

(c) $\tan^{-1}\left(-\sqrt{3}\right) = -\frac{\pi}{3}$ since $\tan\left(-\frac{\pi}{3}\right) = -\sqrt{3}$ and $-\frac{\pi}{2} < -\frac{\pi}{3} < \frac{\pi}{2}$

5 (a) A common mistake is to confuse this problem with the fact that $\sin\frac{\pi}{3} = \frac{\sqrt{3}}{2}$,

however, $\sin^{-1}\frac{\pi}{3}$ is not defined since $\frac{\pi}{3} > 1$, i.e., $\frac{\pi}{3} \notin [-1, 1]$.

(b) $\cos^{-1}\frac{\pi}{2}$ is not defined since $\frac{\pi}{2} > 1$, i.e., $\frac{\pi}{2} \notin [-1, 1]$

(c) $\tan^{-1} 1 = \frac{\pi}{4}$ since $\tan\frac{\pi}{4} = 1$ and $-\frac{\pi}{2} < \frac{\pi}{4} < \frac{\pi}{2}$

7 (a) $\sin\left[\arcsin\left(-\frac{3}{10}\right)\right] = -\frac{3}{10}$ since $-1 \leq -\frac{3}{10} \leq 1$

(b) $\cos\left(\arccos\frac{1}{2}\right) = \frac{1}{2}$ since $-1 \leq \frac{1}{2} \leq 1$

(c) $\tan\left(\arctan 14\right) = 14$ since 14 is a real number

9 (a) $\sin^{-1}\left(\sin\frac{\pi}{3}\right) = \frac{\pi}{3}$ since $-\frac{\pi}{2} \leq \frac{\pi}{3} \leq \frac{\pi}{2}$

(b) $\cos^{-1}\left[\cos\left(\frac{5\pi}{6}\right)\right] = \frac{5\pi}{6}$ since $0 \leq \frac{5\pi}{6} \leq \pi$

(c) $\tan^{-1}\left[\tan\left(-\frac{\pi}{6}\right)\right] = -\frac{\pi}{6}$ since $-\frac{\pi}{2} < -\frac{\pi}{6} < \frac{\pi}{2}$

13 (a) $\sin\left[\cos^{-1}\left(-\frac{1}{2}\right)\right] = \sin\frac{2\pi}{3} = \frac{\sqrt{3}}{2}$

(b) $\cos\left(\tan^{-1} 1\right) = \cos\frac{\pi}{4} = \frac{\sqrt{2}}{2}$

(c) $\tan\left[\sin^{-1}\left(-1\right)\right] = \tan\left(-\frac{\pi}{2}\right)$, which is not defined.

15 (a) Let $\theta = \sin^{-1}\frac{2}{3}$. From *Figure 15a*, $\cot\left(\sin^{-1}\frac{2}{3}\right) = \cot\theta = \frac{x}{y} = \frac{\sqrt{5}}{2}$.

(b) Let $\theta = \tan^{-1}\left(-\frac{3}{5}\right)$. From *Figure 15b*, $\sec\left[\tan^{-1}\left(-\frac{3}{5}\right)\right] = \sec\theta = \frac{r}{x} = \frac{\sqrt{34}}{5}$.

(c) Let $\theta = \cos^{-1}\left(-\frac{1}{4}\right)$. From *Figure 15c*, $\csc\left[\cos^{-1}\left(-\frac{1}{4}\right)\right] = \csc\theta = \frac{r}{y} = \frac{4}{\sqrt{15}}$.

Note: Triangles could be used for the figures, similar to those in Exercises 19 & 21.

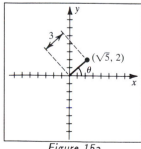

Figure 15a

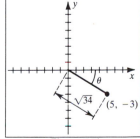

Figure 15b

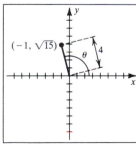

Figure 15c

$\boxed{17}$ (a) $\sin\left(\arcsin\frac{1}{2} + \arccos 0\right) = \sin\left(\frac{\pi}{6} + \frac{\pi}{2}\right) = \sin\frac{2\pi}{3} = \frac{\sqrt{3}}{2}$.

(b) Remember, $\arctan\left(-\frac{3}{4}\right)$ and $\arcsin\frac{4}{5}$ are angles. To abbreviate this solution,

we let $\alpha = \arctan\left(-\frac{3}{4}\right)$ and $\beta = \arcsin\frac{4}{5}$. Using the difference identity for the

cosine and a figure as in Exercise 15, we have $\cos\left[\arctan\left(-\frac{3}{4}\right) - \arcsin\frac{4}{5}\right]$

$$= \cos(\alpha - \beta) = \cos\alpha\cos\beta + \sin\alpha\sin\beta = \frac{4}{5}\cdot\frac{3}{5} + \left(-\frac{3}{5}\right)\cdot\frac{4}{5} = 0.$$

(c) Let $\alpha = \arctan\frac{4}{3}$ and $\beta = \arccos\frac{8}{17}$. $\tan\left(\arctan\frac{4}{3} + \arccos\frac{8}{17}\right) =$

$$\tan(\alpha + \beta) = \frac{\tan\alpha + \tan\beta}{1 - \tan\alpha\tan\beta} = \frac{\frac{4}{3} + \frac{15}{8}}{1 - \frac{4}{3}\cdot\frac{15}{8}}\cdot\frac{24}{24} = \frac{32 + 45}{24 - 60} = -\frac{77}{36}.$$

$\boxed{19}$ Let $\alpha = \tan^{-1}x$. From *Figure 19*, $\sin\left(\tan^{-1}x\right) = \sin\alpha = \dfrac{x}{\sqrt{x^2 + 1}}$.

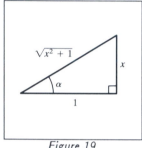

Figure 19

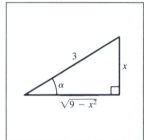

Figure 21

$\boxed{21}$ Let $\alpha = \sin^{-1}\frac{x}{3}$. From *Figure 21*, $\sec\left(\sin^{-1}\frac{x}{3}\right) = \sec\alpha = \dfrac{3}{\sqrt{9 - x^2}}$.

$\boxed{23}$ $y = \sin^{-1}2x$ • Horizontally compress $y = \sin^{-1}x$ by a factor of 2.

Note that the domain changes from $[-1, 1]$ to $\left[-\frac{1}{2}, \frac{1}{2}\right]$.

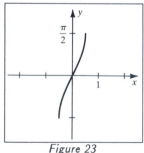

Figure 23

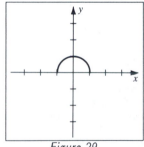

Figure 29

$\boxed{29}$ Sketching a triangle as in Exercises 19–22, we see that $\sin\left(\arccos x\right) = \sqrt{1 - x^2}$.

Thus, we have the graph of the semicircle $y = \sqrt{1 - x^2}$ on the interval $[-1, 1]$.

$\boxed{33}$ (a) Using the quadratic formula, $2\tan^2 t + 9\tan t + 3 = 0 \Rightarrow$

$$\tan t = \frac{-9 \pm \sqrt{81 - 24}}{4} \Rightarrow t = \arctan\frac{1}{4}\left(-9 \pm \sqrt{57}\right).$$

(b) $\arctan\frac{1}{4}\left(-9 - \sqrt{57}\right) \approx -1.3337$, $\arctan\frac{1}{4}\left(-9 + \sqrt{57}\right) \approx -0.3478$

$\boxed{37}$ (a) Let β denote the angle by the sailboat with opposite side d and hypotenuse k.

Now $\sin \beta = \frac{d}{k} \Rightarrow \beta = \sin^{-1} \frac{d}{k}$. Using alternate interior angles,

we see that $\alpha + \beta = \theta$. Thus, $\alpha = \theta - \beta = \theta - \sin^{-1} \frac{d}{k}$.

(b) From part (a), $\alpha = \theta - \sin^{-1} \frac{d}{k} = 53.4° - \sin^{-1} \frac{50}{210} \approx 39.63°$ or $40°$.

$\boxed{\text{Exercises 8.2}}$

$\boxed{1}$ Using (8.8)(i) with $u = \sqrt{x}$, $f(x) = \sin^{-1} \sqrt{x} \Rightarrow$

$$f'(x) = \frac{1}{\sqrt{1 - (\sqrt{x})^2}} \cdot D_x (\sqrt{x}) = \frac{1}{\sqrt{1 - x}} \cdot \frac{1}{2\sqrt{x}} = \frac{1}{2\sqrt{x}\sqrt{1 - x}}.$$

$\boxed{3}$ Using (8.8)(iii) with $u = 3x - 5$, $f(x) = \tan^{-1}(3x - 5) \Rightarrow$

$$f'(x) = \frac{1}{1 + (3x - 5)^2} \cdot D_x (3x - 5) = \frac{3}{9x^2 - 30x + 26}.$$

$\boxed{5}$ Using the product rule, $f(x) = e^{-x} \operatorname{arcsec} e^{-x} \Rightarrow f'(x)$

$= e^{-x} D_x (\operatorname{arcsec} e^{-x}) + D_x (e^{-x}) \operatorname{arcsec} e^{-x}$

$= e^{-x} \left[\dfrac{1}{e^{-x} \sqrt{(e^{-x})^2 - 1}} \cdot (-e^{-x}) \right] \{ \text{use (8.8)(iv)}, u = e^{-x} \} + (-e^{-x}) \operatorname{arcsec} e^{-x}$

$= \dfrac{-e^{-x}}{\sqrt{e^{-2x} - 1}} - e^{-x} \operatorname{arcsec} e^{-x}$

$\boxed{9}$ Using (8.8)(iv) with $u = (x^2 - 1)^{1/2}$, $f(x) = \sec^{-1} \sqrt{x^2 - 1} \Rightarrow$

$f'(x) = \dfrac{1}{\sqrt{x^2 - 1} \sqrt{(\sqrt{x^2 - 1})^2 - 1}} \left[\frac{1}{2}(x^2 - 1)^{-1/2} (2x) \right]$

$= \dfrac{1}{\sqrt{x^2 - 1} \sqrt{(x^2 - 1) - 1}} \cdot \dfrac{x}{\sqrt{x^2 - 1}} = \dfrac{x}{(x^2 - 1) \sqrt{x^2 - 2}}.$

$\boxed{13}$ $f(x) = (1 + \cos^{-1} 3x)^3 \Rightarrow$

$f'(x) = 3(1 + \cos^{-1} 3x)^2 D_x (1 + \cos^{-1} 3x)$

$= 3(1 + \cos^{-1} 3x)^2 \left(-\dfrac{1}{\sqrt{1 - (3x)^2}} \cdot 3 \right) \{ \text{use (8.8)(ii) with } u = 3x \}$

$= -\dfrac{9(1 + \cos^{-1} 3x)^2}{\sqrt{1 - 9x^2}}$

$\boxed{17}$ $f(x) = \cos(x^{-1}) + (\cos x)^{-1} + \cos^{-1} x = \cos(\frac{1}{x}) + \frac{1}{\cos x} + \cos^{-1} x \Rightarrow$

$f'(x) = -\sin(\frac{1}{x}) D_x (\frac{1}{x}) + D_x (\sec x) + D_x (\cos^{-1} x)$

$= -\sin(\frac{1}{x})(-\frac{1}{x^2}) + \sec x \tan x + \left(-\dfrac{1}{\sqrt{1 - x^2}} \right)$

$= (1/x^2) \sin(1/x) + \sec x \tan x - \dfrac{1}{\sqrt{1 - x^2}}$

19 We recognize f as being of the form a^u, with $u = \arcsin(x^3)$.

Thus, $f(x) = 3^{\arcsin(x^3)} \Rightarrow f'(x) = \left(3^{\arcsin(x^3)} \ln 3\right) D_x\left[\arcsin(x^3)\right] =$

$$\left(3^{\arcsin(x^3)} \ln 3\right) \cdot \frac{1}{\sqrt{1-(x^3)^2}} \cdot (3x^2) = 3^{\arcsin(x^3)} \frac{(3\ln 3)\,x^2}{\sqrt{1-x^6}}.$$

23 Using the product rule, $f(x) = \sqrt{x}\,\sec^{-1}\sqrt{x} \Rightarrow$

$f'(x) = \sqrt{x}\,D_x\left(\sec^{-1}\sqrt{x}\right) + D_x\left(\sqrt{x}\right)\sec^{-1}\sqrt{x}$

$$= \sqrt{x} \cdot \frac{1}{\sqrt{x}\,\sqrt{(\sqrt{x})^2 - 1}}\left(\frac{1}{2\sqrt{x}}\right) + \frac{1}{2\sqrt{x}}\cdot\sec^{-1}\sqrt{x} = \frac{1}{2\sqrt{x}}\left(\frac{1}{\sqrt{x-1}} + \sec^{-1}\sqrt{x}\right).$$

25 Since f has a variable base and a variable exponent, we will rewrite it as an

exponential expression with base e, and then differentiate.

$$f(x) = (\tan x)^{\arctan x} = e^{\arctan x\,\ln(\tan x)} \Rightarrow$$

$$f'(x) = e^{\arctan x\,\ln(\tan x)}\,D_x\left[\arctan x\,\ln(\tan x)\right]$$

$$= e^{\arctan x\,\ln(\tan x)}\left(\arctan x\cdot\frac{1}{\tan x}\cdot\sec^2 x + \frac{1}{1+x^2}\cdot\ln\tan x\right)$$

$$= (\tan x)^{\arctan x}\left(\arctan x\,\cot x\,\sec^2 x + \frac{\ln\tan x}{1+x^2}\right)$$

27 $x^2 + x\sin^{-1}y = ye^x \Rightarrow 2x + \left(x\cdot\dfrac{1}{\sqrt{1-y^2}}\cdot y' + 1\cdot\sin^{-1}y\right) = ye^x + y'e^x \Rightarrow$

$$y'\left(\frac{x}{\sqrt{1-y^2}} - e^x\right) = ye^x - 2x - \sin^{-1}y \Rightarrow y' = \frac{ye^x - 2x - \sin^{-1}y}{\dfrac{x}{\sqrt{1-y^2}} - e^x}$$

29 (a) Using (8.9)(ii), $\displaystyle\int\frac{1}{x^2+16}\,dx = \int\frac{1}{4^2+x^2}\,dx = \frac{1}{4}\tan^{-1}\frac{x}{4} + C$

(b) $\displaystyle\int_0^4\frac{1}{x^2+16}\,dx = \left[\frac{1}{4}\tan^{-1}\frac{x}{4}\right]_0^4 = \frac{1}{4}(\tan^{-1}1 - \tan^{-1}0) = \frac{1}{4}\left(\frac{\pi}{4} - 0\right) = \frac{\pi}{16}$

31 (a) Using (8.9)(i) with $u = x^2$, $\frac{1}{2}du = x\,dx \Rightarrow$

$$\int\frac{x}{\sqrt{1-(x^2)^2}}\,dx = \frac{1}{2}\int\frac{1}{\sqrt{1-u^2}}\,du = \frac{1}{2}\sin^{-1}u + C = \frac{1}{2}\sin^{-1}(x^2) + C.$$

(b) $\displaystyle\int_0^{\sqrt{2}/2}\frac{x}{\sqrt{1-x^4}}\,dx = \left[\frac{1}{2}\sin^{-1}(x^2)\right]_0^{\sqrt{2}/2} = \frac{1}{2}(\sin^{-1}\frac{1}{2} - \sin^{-1}0) = \frac{1}{2}\left(\frac{\pi}{6} - 0\right) = \frac{\pi}{12}$

33 $u = \cos x$, $-du = \sin x\,dx \Rightarrow$

$$\int\frac{\sin x}{\cos^2 x + 1}\,dx = -\int\frac{1}{u^2+1}\,du = -\tan^{-1}u + C = -\tan^{-1}(\cos x) + C$$

37 $u = e^x$, $du = e^x\,dx \Rightarrow$

$$\int\frac{e^x}{\sqrt{16-e^{2x}}}\,dx = \int\frac{1}{\sqrt{4^2-u^2}}\,du = \sin^{-1}\frac{u}{4} + C = \sin^{-1}\frac{e^x}{4} + C$$

$\boxed{39}$ $u = x^3$, $\frac{1}{3} du = x^2 dx \Rightarrow \displaystyle\int \frac{1}{x \sqrt{x^6 - 4}} dx = \frac{1}{3} \int \frac{x^2}{x \cdot x^2 \sqrt{(x^3)^2 - 2^2}} dx =$

$$\frac{1}{3} \int \frac{1}{u \sqrt{u^2 - 2^2}} du = \frac{1}{3} \cdot \frac{1}{2} \sec^{-1} \frac{u}{2} + C \{ (8.9)(\text{iii}) \} = \frac{1}{6} \sec^{-1} \frac{x^3}{2} + C$$

$\boxed{43}$ $u = e^x$, $du = e^x dx$ and $dx = \frac{1}{e^x} du = \frac{1}{u} du$.

$$\int \frac{1}{\sqrt{e^{2x} - 25}} dx = \int \frac{1}{u \sqrt{u^2 - 5^2}} du = \frac{1}{5} \sec^{-1} \frac{u}{5} + C$$

$\boxed{45}$ The 7 foot measurement is exact. The 10 foot measurement is not, so we let x denote the length of the side opposite angle θ. Our "Find when knowing" statement is

Find $\underline{d\theta}$ when $\underline{x = 10'}$ knowing $\underline{dx = \pm 0.5''}$.

To keep all measurements in feet, we will use $dx = \pm 0.5 \left(\frac{1}{12}\right)$ ft. Next, we need a relationship involving the angle θ, and the sides 7 and x.

$\tan \theta = \frac{x}{7} \Rightarrow \theta = \tan^{-1} \frac{x}{7}$ and $d\theta = D_x \left(\tan^{-1} \frac{x}{7}\right) dx =$

$$\left[\frac{1}{1 + (x/7)^2} \cdot \frac{1}{7} \right] dx = \frac{1}{7 \left[1 + \left(\frac{10}{7}\right)^2 \right]} (\pm 0.5 \cdot \frac{1}{12}) = \pm \frac{49}{7 \cdot 149 \cdot 24} = \pm \frac{7}{3576} \text{ rad.}$$

$\boxed{49}$ See *Figure 49*. Since θ is to be maximized, we will first write θ in terms of x.

$\theta = \alpha - \beta = \tan^{-1}(80/x) - \tan^{-1}(60/x) \Rightarrow$

$$\frac{d\theta}{dx} = \frac{1}{1 + (80/x)^2} D_x (80/x) - \frac{1}{1 + (60/x)^2} D_x (60/x)$$

$$= \frac{-80/x^2}{1 + (80/x)^2} - \frac{-60/x^2}{1 + (60/x)^2} = \frac{60}{x^2 + 3600} - \frac{80}{x^2 + 6400}.$$

$$\frac{d\theta}{dx} = 0 \Rightarrow \frac{60}{x^2 + 3600} = \frac{80}{x^2 + 6400} \Rightarrow$$

$60(x^2 + 6400) - 80(x^2 + 3600) = 0 \Rightarrow 20x^2 = 96{,}000 \Rightarrow x = \sqrt{4800} \approx 69.3 \text{ ft.}$

This is a maximum since $\frac{d\theta}{dx} > 0$ on $(0, \sqrt{4800})$ and $\frac{d\theta}{dx} < 0$ on $(\sqrt{4800}, \infty)$.

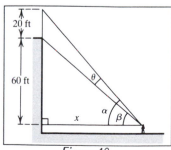

Figure 49

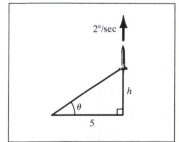

Figure 51

51   Find $\frac{dh}{dt}$ when $\theta = 30°$ knowing $\frac{d\theta}{dt} = 2°/\text{sec}$.

From *Figure 51*, $\tan\theta = \frac{h}{5}$ and $\theta = \tan^{-1}\frac{h}{5}$. Using the chain rule to differentiate

$h = 5\tan\theta$ with respect to t, we have $\frac{dh}{dt} = \frac{dh}{d\theta}\cdot\frac{d\theta}{dt}$. We know $\frac{d\theta}{dt}$ is $(2\cdot\frac{\pi}{180}$ rad/sec$)$,

but need to find $\frac{dh}{d\theta}$. If the problem didn't say "use inverse trigonometric functions",

we could use the relationship $h = 5\tan\theta$ to easily find $\frac{dh}{d\theta}$. However, we will use

$\theta = \tan^{-1}\frac{h}{5}$ to find $\frac{d\theta}{dh}$ and then use (7.8) to realize that $\frac{dh}{d\theta} = \frac{1}{d\theta/dh}$. Thus,

$$\frac{d\theta}{dh} = \frac{1}{1 + (h/5)^2}\cdot\frac{1}{5} = \frac{1}{1 + (5\tan\theta/5)^2}\cdot\frac{1}{5}$$

$$= \frac{1}{1 + \tan^2\theta}\cdot\frac{1}{5} = \frac{1}{5(1 + \frac{1}{3})}\;\{\theta = 30°\} = \frac{3}{20}.$$

Hence, $\frac{dh}{dt} = \frac{dh}{d\theta}\cdot\frac{d\theta}{dt} = \frac{20}{3}(2\cdot\frac{\pi}{180}) = \frac{2\pi}{27} \approx 0.233$ mi/sec.

Exercises 8.3

1   (a) Using (8.10) with $x = 4$, $\sinh 4 = \frac{e^4 - e^{-4}}{2} \approx 27.2899$.

(b) Using (8.10) with $x = \ln 4$,

$$\cosh\ln 4 = \frac{e^{\ln 4} + e^{-\ln 4}}{2} = \frac{4 + (e^{\ln 4})^{-1}}{2} = \frac{4 + 4^{-1}}{2} = \frac{17}{8} = 2.1250.$$

Note: Cosh and sech are even functions, whereas sinh, tanh, coth, and csch are odd.

{See Figures 8.13, 8.14, and 8.18.}

(c) Since tanh is an odd function, $\tanh(-3) = -\tanh 3$.

$$\text{Using (8.12)(i) with } x = 3,\ -\tanh 3 = -\frac{e^3 - e^{-3}}{e^3 + e^{-3}} \approx -0.9951.$$

(d) Using (8.12)(ii) with $x = 10$, $\coth 10 = \frac{e^{10} + e^{-10}}{e^{10} - e^{-10}} \approx 1.0000$.

(e) Using (8.12)(iii) with $x = 2$, $\text{sech}\, 2 = \frac{2}{e^2 + e^{-2}} \approx 0.2658$.

(f) Since csch is an odd function, $\text{csch}(-1) = -\text{csch}\, 1$.

$$\text{Using (8.12)(iv) with } x = 1,\ -\text{csch}\, 1 = -\frac{2}{e^1 - e^{-1}} \approx -0.8509.$$

3   Using (8.14)(i) with $u = 5x$, $f(x) = \sinh 5x \Rightarrow$

$$f'(x) = \cosh 5x \cdot D_x(5x) = \cosh 5x \cdot 5 = 5\cosh 5x.$$

7   Using the product rule, $f(x) = \sqrt{x}\,\tanh\sqrt{x} \Rightarrow$

$f'(x) = \sqrt{x}\,D_x(\tanh\sqrt{x}) + D_x(\sqrt{x})\tanh\sqrt{x}$

$$= \sqrt{x}\,\text{sech}^2\sqrt{x}\left(\frac{1}{2\sqrt{x}}\right) + \left(\frac{1}{2\sqrt{x}}\right)\tanh\sqrt{x}$$

$$= \frac{1}{2\sqrt{x}}(\sqrt{x}\,\text{sech}^2\sqrt{x} + \tanh\sqrt{x}).$$

$\boxed{9}$ $f(x) = \coth(1/x) \Rightarrow f'(x) = -\operatorname{csch}^2(1/x) \cdot (-1/x^2) = (1/x^2)\operatorname{csch}^2(1/x)$

$\boxed{11}$ Using the quotient rule,

$$f(x) = \frac{\operatorname{sech}(x^2)}{x^2 + 1} \Rightarrow f'(x) = \frac{(x^2 + 1)\left[-\operatorname{sech}(x^2)\tanh(x^2) \cdot 2x\right] - \operatorname{sech}(x^2) \cdot 2x}{(x^2 + 1)^2}$$

$$= \frac{-2x\operatorname{sech}(x^2)\left[(x^2 + 1)\tanh(x^2) + 1\right]}{(x^2 + 1)^2}.$$

$\boxed{13}$ $f(x) = \operatorname{csch}^2 6x = (\operatorname{csch} 6x)^2 \Rightarrow f'(x) = 2(\operatorname{csch} 6x)^1 D_x(\operatorname{csch} 6x) =$

$$2\operatorname{csch} 6x \cdot (-\operatorname{csch} 6x \coth 6x) \cdot 6 = -12\operatorname{csch}^2 6x \coth 6x$$

$\boxed{15}$ $f(x) = \ln \sinh 2x \Rightarrow$

$$f'(x) = \frac{1}{\sinh 2x} D_x(\sinh 2x) = \frac{1}{\sinh 2x} \cdot 2\cosh 2x = 2 \cdot \frac{\cosh x}{\sinh x} = 2\coth 2x$$

$\boxed{17}$ $f(x) = \cosh\sqrt{4x^2 + 3} \Rightarrow f'(x) = \sinh\sqrt{4x^2 + 3}\ D_x(\sqrt{4x^2 + 3}) =$

$$\sinh\sqrt{4x^2 + 3} \cdot \tfrac{1}{2}(4x^2 + 3)^{-1/2}(8x) = \frac{4x\sinh\sqrt{4x^2 + 3}}{\sqrt{4x^2 + 3}}$$

$\boxed{21}$ $f(x) = \coth(\ln x) \Rightarrow f'(x) = -\operatorname{csch}^2(\ln x)\ D_x(\ln x) = -\operatorname{csch}^2 \ln x \cdot \tfrac{1}{x} =$

$$-\frac{1}{x}\left(\frac{2}{e^{\ln x} - e^{-\ln x}}\right)^2 = -\frac{1}{x}\left[\frac{2}{x - (1/x)}\right]^2 = -\frac{1}{x}\left[\frac{2}{x - (1/x)} \cdot \frac{x}{x}\right]^2 =$$

$$-\frac{1}{x}\left[\frac{2x}{x^2 - 1}\right]^2 = -\frac{1}{x} \cdot \frac{4x^2}{(x^2 - 1)^2} = -\frac{4x}{(x^2 - 1)^2}$$

Note: $f(x) = \coth(\ln x) = \dfrac{e^{\ln x} + e^{-\ln x}}{e^{\ln x} - e^{-\ln x}} = \dfrac{x + (e^{\ln x})^{-1}}{x - (e^{\ln x})^{-1}} = \dfrac{x + x^{-1}}{x - x^{-1}}$

$= \dfrac{x + (1/x)}{x - (1/x)} \cdot \dfrac{x}{x} = \dfrac{x^2 + 1}{x^2 - 1}$. We could now differentiate this form of $f(x)$ to obtain

the same answer, but this method is not preferred in this section since the emphasis is

to learn how to differentiate the hyperbolic functions.

$\boxed{25}$ $f(x) = \tan^{-1}(\operatorname{csch} x) \Rightarrow f'(x) = \dfrac{1}{1 + (\operatorname{csch} x)^2} D_x(\operatorname{csch} x)$

$$= \frac{1}{1 + (\operatorname{csch} x)^2}(-\operatorname{csch} x \coth x) = -\frac{\operatorname{csch} x \coth x}{1 + (\coth^2 x - 1)}$$

$$= \frac{-\operatorname{csch} x}{\coth x} = -\frac{1}{\sinh x} \cdot \frac{\sinh x}{\cosh x} = -\operatorname{sech} x$$

$\boxed{27}$ $u = x^3$, $\tfrac{1}{3} du = x^2\, dx \Rightarrow$

$$\int x^2 \cosh(x^3)\, dx = \tfrac{1}{3}\int \cosh u\, du = \tfrac{1}{3}\sinh u + C = \tfrac{1}{3}\sinh(x^3) + C$$

$\boxed{29}$ $u = \sqrt{x}$, $2\, du = \dfrac{dx}{\sqrt{x}} \Rightarrow \displaystyle\int \frac{\sinh\sqrt{x}}{\sqrt{x}}\, dx = 2\int \sinh u\, du = 2\cosh u + C = 2\cosh\sqrt{x} + C$

$\boxed{33}$ $u = \tfrac{1}{2}x$, $2\, du = dx \Rightarrow \displaystyle\int \operatorname{csch}^2(\tfrac{1}{2}x)\, dx = 2\int \operatorname{csch}^2 u\, du = -2\int -\operatorname{csch}^2 u\, du =$

$$-2\coth u + C = -2\coth(\tfrac{1}{2}x) + C$$

$\boxed{35}$ $u = 3x,\ \frac{1}{3}\,du = dx \Rightarrow$

$$\int \tanh 3x \operatorname{sech} 3x\,dx = -\frac{1}{3}\int -\operatorname{sech} u \tanh u\,du = -\frac{1}{3}\operatorname{sech} u + C = -\frac{1}{3}\operatorname{sech} 3x + C$$

$\boxed{37}$ $\displaystyle \int \cosh x \operatorname{csch}^2 x\,dx = \int \cosh x \cdot \frac{1}{\sinh^2 x}\,dx = \int \frac{\cosh x}{\sinh x}\cdot\frac{1}{\sinh x}\,dx = \int \coth x \operatorname{csch} x\,dx =$

$$-\int -\coth x \operatorname{csch} x\,dx = -\operatorname{csch} x + C$$

$\boxed{39}$ $\displaystyle \int \coth x\,dx = \int \frac{\cosh x}{\sinh x}\,dx.\ \ u = \sinh x,\ du = \cosh x\,dx \Rightarrow$

$$I = \int \frac{1}{u}\,du = \ln|u| + C = \ln|\sinh x| + C$$

$\boxed{41}$ $u = \sinh x,\ du = \cosh x\,dx \Rightarrow \int \sinh x \cosh x\,dx = \int u\,du = \frac{1}{2}u^2 + C = \frac{1}{2}\sinh^2 x + C$

Alternate solution:

$$u = \cosh x,\ du = \sinh x\,dx \Rightarrow \int \sinh x \cosh x\,dx = \int u\,du = \frac{1}{2}u^2 + C = \frac{1}{2}\cosh^2 x + C.$$

Note that since $\cosh^2 x - \sinh^2 x = 1$, the two answers differ by a constant.

$\boxed{43}$ The graphs of $y = \sinh 3x$ and $y = 0$ intersect at $x = 0$. On $[0,\,1]$, $\sinh 3x \geq 0$.

$$A = \int_0^1 \sinh 3x\,dx = \left[\tfrac{1}{3}\cosh 3x\right]_0^1 = \tfrac{1}{3}(\cosh 3 - 1) \approx 3.023.$$

$\boxed{45}$ $y' = \cosh x = \dfrac{e^x + e^{-x}}{2} = 2 \Rightarrow e^x + e^{-x} = 4.$

Since $e^{-x} = \frac{1}{e^x}$, we multiply the equation by e^x to eliminate all denominators.

$e^x(e^x + e^{-x}) = e^x(4) \Rightarrow e^{2x} + 1 = 4e^x \Rightarrow (e^x)^2 - 4e^x + 1 = 0.$

We recognize this equation as a quadratic equation in e^x. Let $y = e^x$.

Then $y^2 - 4y + 1 = 0$ and we can solve for y using the quadratic formula.

$y = \dfrac{4 \pm \sqrt{16 - 4}}{2} = 2 \pm \sqrt{3} \Rightarrow e^x = 2 \pm \sqrt{3} \Rightarrow x = \ln(2 \pm \sqrt{3}).$

At $x = \ln(2 + \sqrt{3})$, $y = \sinh\left[\ln(2 + \sqrt{3})\right] = \dfrac{e^{\ln(2+\sqrt{3})} - e^{-\ln(2+\sqrt{3})}}{2} =$

$$\dfrac{(2 + \sqrt{3}) - \left[e^{\ln(2+\sqrt{3})}\right]^{-1}}{2} = \dfrac{(2 + \sqrt{3}) - (2 + \sqrt{3})^{-1}}{2} =$$

$$\dfrac{(2 + \sqrt{3}) - \dfrac{1}{2 + \sqrt{3}}\cdot\dfrac{2 - \sqrt{3}}{2 - \sqrt{3}}}{2} =$$

$$\dfrac{(2 + \sqrt{3}) - \dfrac{2 - \sqrt{3}}{1}}{2} = \dfrac{2\sqrt{3}}{2} = \sqrt{3}.\ \ \text{Similarly, at } x = \ln(2 - \sqrt{3}),\ y = -\sqrt{3}.$$

Thus, the points are $(\ln(2 \pm \sqrt{3}),\ \pm\sqrt{3})$.

[47] From *Figure 47*, $\triangle ODQ$ is a right triangle with base $\cosh t$ and height $\sinh t$. Its area is $A_1 = \frac{1}{2}(\cosh t)(\sinh t)$. The area of the nonshaded part of the triangle is the area bounded by the curve $y = \sqrt{x^2 - 1}$, the x-axis, and the vertical line $x = \cosh t$.

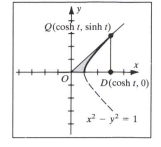

Figure 47

This area is $A_2 = \int_1^{\cosh t} \sqrt{x^2 - 1}\, dx$. The area of the shaded region is $A = A_1 - A_2$. It follows that

$$\frac{dA}{dt} = D_t\left(\tfrac{1}{2}\cosh t \sinh t\right) - D_t\left(\int_1^{\cosh t} \sqrt{x^2 - 1}\, dx\right)$$

$$= \tfrac{1}{2}(\cosh t \cdot \cosh t + \sinh t \cdot \sinh t) - \sqrt{\cosh^2 t - 1}\, D_t\,(\cosh t) \;\{\text{Exercise 51, §5.6}\}$$

$$= \tfrac{1}{2}(\cosh^2 t + \sinh^2 t) - \sqrt{\cosh^2 t - 1}\,(\sinh t)$$

$$= \tfrac{1}{2}(\cosh^2 t + \sinh^2 t) - \sinh^2 t = \tfrac{1}{2}(\cosh^2 t - \sinh^2 t) = \tfrac{1}{2}(1) = \tfrac{1}{2}.$$

Therefore, since $\frac{dA}{dt} = \frac{1}{2}$, $A = \frac{1}{2}t + C$ and $C = 0$ since $A = 0$ if $t = 0$.

Thus, $A = \tfrac{1}{2}t$, or, equivalently, $t = 2A$.

[49] (a) By symmetry, $A = 2\displaystyle\int_0^{315}\left(-127.7\cosh\tfrac{x}{127.7} + 757.7\right)dx =$

$$2\left[-(127.7)^2\sinh\tfrac{x}{127.7} + (757.7)x\right]_0^{315} \approx 286{,}574 \text{ ft}^2.$$

(b) Let $y = f(x)$ and use (6.14). $y' = -\sinh\tfrac{x}{127.7} \Rightarrow$

$$1 + (y')^2 = 1 + \sinh^2\tfrac{x}{127.7} = \cosh^2\tfrac{x}{127.7} \Rightarrow L = 2\int_0^{315}\sqrt{\cosh^2\tfrac{x}{127.7}}\, dx =$$

$$2\int_0^{315}\cosh\tfrac{x}{127.7}\, dx = 2\left[127.7\sinh\tfrac{x}{127.7}\right]_0^{315} \approx 1494 \text{ ft.}$$

[51] (a) $\displaystyle\lim_{h\to\infty} v^2 = \lim_{h\to\infty}\frac{gL}{2\pi}\tanh\frac{2\pi h}{L} = \frac{gL}{2\pi}\lim_{h\to\infty}\tanh\frac{2\pi h}{L} = \frac{gL}{2\pi}(1) = \frac{gL}{2\pi}$

Note: $\displaystyle\lim_{x\to\infty}\tanh x = \lim_{x\to\infty}\frac{e^x - e^{-x}}{e^x + e^{-x}} = \lim_{x\to\infty}\left(\frac{e^x - e^{-x}}{e^x + e^{-x}}\cdot\frac{e^{-x}}{e^{-x}}\right) =$

$$\lim_{x\to\infty}\frac{1 - e^{-2x}}{1 + e^{-2x}} = \frac{1 - 0}{1 + 0} = 1. \text{ See Figure 8.18.}$$

(b) Let L be arbitrary and define $f(h) = [v(h)]^2$. Then,

$$f(0) = 0,\ f'(h) = \frac{gL}{2\pi}\,\text{sech}^2\frac{2\pi h}{L}\cdot\frac{2\pi}{L} = g\,\text{sech}^2\frac{2\pi h}{L} \text{ and } f'(0) = g\cdot(1)^2 = g.$$

Now for small enough h, and hence, small $\frac{h}{L}$ we have, $f(h) - f(0) \approx f'(0)(h) \Rightarrow$

$v^2 - 0 \approx (g)(h) \Rightarrow v \approx \sqrt{gh}$. Thus, the rate of change in the position of the

wave with respect to the depth is independent of its length. (cont.)

Alternate solution: Let $f(x) = \tanh x$. Then $f(x) - f(0) \approx f'(0)\, x \Rightarrow$

$\tanh x - 0 \approx \operatorname{sech}^2(0)\, x \Rightarrow \tanh x \approx x$ when $x \approx 0$. Let $x = \frac{2\pi h}{L}$.

Then, if $x \approx 0$ and hence $\frac{h}{L} \approx 0$, $v^2 = \frac{gL}{2\pi}\tanh\left(\frac{2\pi h}{L}\right) \approx \left(\frac{gL}{2\pi}\right)\left(\frac{2\pi h}{L}\right) = gh \Rightarrow$

$$v \approx \sqrt{gh} \text{ for } x \approx 0.$$

$\boxed{59}$ Start with the RHS since it would be difficult to recognize

how to start with the LHS and make any progress toward obtaining the RHS.

RHS $= \sinh x \cosh y + \cosh x \sinh y$

$$= \frac{e^x - e^{-x}}{2} \cdot \frac{e^y + e^{-y}}{2} + \frac{e^x + e^{-x}}{2} \cdot \frac{e^y - e^{-y}}{2}$$

$$= \frac{(e^x - e^{-x})(e^y + e^{-y})}{4} + \frac{(e^x + e^{-x})(e^y - e^{-y})}{4}$$

$$= \frac{(e^{x+y} + e^{x-y} - e^{-x+y} - e^{-x-y}) + (e^{x+y} - e^{x-y} + e^{-x+y} - e^{-x-y})}{4}$$

$$= \frac{2e^{x+y} - 2e^{-x-y}}{4} = \frac{e^{x+y} - e^{-(x+y)}}{2} = \sinh(x+y) = \text{LHS}$$

$\boxed{61}$ We could start with the RHS and proceed as in Exercise 59,

but we choose to start with the LHS and make use of previous exercise results.

$\quad$ LHS $= \sinh(x - y) = \sinh(x + (-y))$

$$= \sinh x \cosh(-y) + \cosh x \sinh(-y) \text{ (Exercise 59)}$$

$$= \sinh x \cosh y - \cosh x \sinh y \text{ (Exercises 57 and 58)} = \text{RHS}$$

$\boxed{63}$ LHS $= \tanh(x + y) = \dfrac{\sinh(x+y)}{\cosh(x+y)} = \dfrac{\sinh x \cosh y + \cosh x \sinh y}{\cosh x \cosh y + \sinh x \sinh y}$ (Exer. 59 & 60).

At this point, we appear to be stuck — so we will look at the RHS to see what terms

we want to get. To "force" the term $\cosh x \cosh y$ in the denominator to equal 1, we

will divide each term by $\cosh x \cosh y$. Thus,

$$\frac{\sinh x \cosh y + \cosh x \sinh y}{\cosh x \cosh y + \sinh x \sinh y} \cdot \frac{1/(\cosh x \cosh y)}{1/(\cosh x \cosh y)}$$

$$= \frac{\dfrac{\sinh x \cosh y}{\cosh x \cosh y} + \dfrac{\cosh x \sinh y}{\cosh x \cosh y}}{\dfrac{\cosh x \cosh y}{\cosh x \cosh y} + \dfrac{\sinh x \sinh y}{\cosh x \cosh y}}$$

$$= \frac{\tanh x + \tanh y}{1 + \tanh x \tanh y} = \text{RHS}.$$

$\boxed{65}$ $\quad$ LHS $= \sinh(2x) = \sinh(x + x)$

$$= \sinh x \cosh x + \cosh x \sinh x \text{ (Exercise 59)}$$

$$= 2\sinh x \cosh x = \text{RHS}$$

67 From Exercise 66,

$$\cosh 2y = \cosh^2 y + \sinh^2 y$$
$$= (1 + \sinh^2 y) + \sinh^2 y$$
$$= 1 + 2\sinh^2 y,$$

and hence

$$\sinh^2 y = \frac{\cosh 2y - 1}{2}.$$

Let $y = \frac{x}{2}$ to obtain the identity.

71 LHS $= \cosh nx + \sinh nx$ $\qquad$ { given }

$\qquad = \dfrac{e^{nx} + e^{-nx}}{2} + \dfrac{e^{nx} - e^{-nx}}{2}$ $\qquad$ { definition of cosh and sinh }

$\qquad = e^{nx}$ $\qquad$ { simplify }

$\qquad = (e^x)^n$ $\qquad$ { change form }

$\qquad = (\cosh x + \sinh x)^n$ $\qquad$ { apply Exercise 55 }

Exercises 8.4

1 (a) Using (8.16)(i), $\sinh^{-1} 1 = \ln\left[(1) + \sqrt{(1)^2 + 1}\right] = \ln(1 + \sqrt{2}) \approx 0.8814$.

(b) Using (8.16)(ii), $\cosh^{-1} 2 = \ln\left[(2) + \sqrt{(2)^2 - 1}\right] = \ln(2 + \sqrt{3}) \approx 1.3170$.

(c) Using (8.16)(iii),

$$\tanh^{-1}\left(-\frac{1}{2}\right) = \frac{1}{2}\ln\left[\frac{1 + (-1/2)}{1 - (-1/2)}\right] = \frac{1}{2}\ln\frac{1/2}{3/2} = \frac{1}{2}\ln\frac{1}{3} \approx -0.5493.$$

(d) Using (8.16)(iv),

$$\operatorname{sech}^{-1}\frac{1}{2} = \ln\frac{1 + \sqrt{1 - (1/2)^2}}{1/2} = \ln\frac{1 + (\sqrt{3}/2)}{1/2} = \ln(2 + \sqrt{3}) \approx 1.3170.$$

Some calculators have a $\cosh^{-1}$ function, but not a sech^{-1} function.

Use the relationship $\operatorname{sech}^{-1}(x) = \cosh^{-1}(1/x)$ to compute values of sech^{-1}.

Note that parts (b) and (d) are equal.

3 Using (8.17)(i) with $u = 5x$, $f(x) = \sinh^{-1} 5x \Rightarrow$

$$f'(x) = \frac{1}{\sqrt{(5x)^2 + 1}} \cdot 5 = \frac{5}{\sqrt{25x^2 + 1}}.$$

5 Using (8.17)(ii) with $u = \sqrt{x}$, $f(x) = \cosh^{-1}\sqrt{x} \Rightarrow$

$$f'(x) = \frac{1}{\sqrt{(\sqrt{x})^2 - 1}} \cdot \frac{1}{2}x^{-1/2} = \frac{1}{2\sqrt{x}\sqrt{x - 1}}.$$

9 Using (8.17)(iv) with $u = x^2$, $f(x) = \operatorname{sech}^{-1} x^2 \Rightarrow$

$$f'(x) = \frac{-1}{x^2\sqrt{1 - (x^2)^2}} \cdot 2x = -\frac{2}{x\sqrt{1 - x^4}}.$$

$\boxed{13}$ $f(x) = \ln \cosh^{-1} 4x \Rightarrow f'(x) = \dfrac{1}{\cosh^{-1} 4x} \cdot D_x \left(\cosh^{-1} 4x \right) =$

$$\dfrac{1}{\cosh^{-1} 4x} \cdot \dfrac{1}{\sqrt{(4x)^2 - 1}} \cdot 4 = \dfrac{4}{\sqrt{16x^2 - 1}\ \cosh^{-1} 4x}$$

$\boxed{15}$ Using (8.17)(iii) with $u = x + 1$, $f(x) = \tanh^{-1}(x + 1) \Rightarrow$

$$f'(x) = \dfrac{1}{1 - (x + 1)^2} \cdot 1 = \dfrac{1}{-x^2 - 2x} = -\dfrac{1}{x^2 + 2x}.$$

$\boxed{19}$ Using (8.18)(i) with $u = 4x$, $\frac{1}{4} du = dx \Rightarrow$

$$\int \dfrac{1}{\sqrt{81 + 16x^2}}\, dx = \dfrac{1}{4} \int \dfrac{1}{\sqrt{9^2 + u^2}}\, du = \dfrac{1}{4} \sinh^{-1} \dfrac{u}{9} + C = \dfrac{1}{4} \sinh^{-1} \dfrac{4x}{9} + C.$$

$\boxed{21}$ Using (8.18)(iii) with $u = 2x$, $\frac{1}{2} du = dx \Rightarrow$

$$\int \dfrac{1}{49 - 4x^2}\, dx = \dfrac{1}{2} \int \dfrac{1}{7^2 - u^2}\, du = \dfrac{1}{2} \cdot \dfrac{1}{7} \tanh^{-1} \dfrac{u}{7} + C = \dfrac{1}{14} \tanh^{-1} \dfrac{2x}{7} + C.$$

$\boxed{23}$ Using (8.18)(ii) with $u = e^x$, $du = e^x\, dx \Rightarrow$

$$\int \dfrac{e^x}{\sqrt{e^{2x} - 16}}\, dx = \int \dfrac{1}{\sqrt{u^2 - 4^2}}\, du = \cosh^{-1} \dfrac{u}{4} + C = \cosh^{-1} \dfrac{e^x}{4} + C.$$

$\boxed{25}$ Using (8.18)(iv) with $u = x^2$, $\frac{1}{2} du = x\, dx \Rightarrow \int \dfrac{1}{x \sqrt{9 - x^4}}\, dx = \int \dfrac{x}{x \cdot x \sqrt{9 - (x^2)^2}}\, dx =$

$$\dfrac{1}{2} \int \dfrac{1}{u \sqrt{3^2 - u^2}}\, du = -\dfrac{1}{2} \cdot \dfrac{1}{3} \operatorname{sech}^{-1} \dfrac{|u|}{3} + C = -\dfrac{1}{6} \operatorname{sech}^{-1} \dfrac{x^2}{3} + C.$$

$\boxed{27}$ Let the particle have coordinates $P(1, y)$. Its distance from the origin is $\sqrt{1 + y^2}$ and

its velocity is given by $\dfrac{dy}{dt} = k \sqrt{1 + y^2}$, where k is a constant of proportionality.

$\dfrac{dy}{dt} = 3$ at $y = 0 \Rightarrow 3 = k \sqrt{1 + 0^2} \Rightarrow k = 3$.

Then, $\dfrac{dy}{dt} = 3 \sqrt{1 + y^2} \Rightarrow \dfrac{dy}{\sqrt{1 + y^2}} = 3\, dt \Rightarrow \int \dfrac{1}{\sqrt{1 + y^2}}\, dy = \int 3\, dt \Rightarrow$

$\sinh^{-1} y = 3t + C \Rightarrow \sinh(\sinh^{-1} y) = \sinh(3t + C) \Rightarrow y = \sinh(3t + C)$.

At $t = 0$, $y = 0$ { since initially $(x, y) = (1, 0)$ }, and $y = \sinh\big[3(0) + C\big] =$

$\qquad$ $\sinh C$. Therefore, $0 = \sinh C$, or, equivalently, $C = 0$. Thus, $y = \sinh 3t$.

$\boxed{29}$ Refer to Figure 8.14 in the text. We can obtain the
graph of $y = \sinh^{-1} x$ by reflecting the graph of
$y = f(x) = \sinh x$ through the line $y = x$ (see Figure
7.6) since f is one-to-one. From this reflection, we see
that $y = \sinh^{-1} x = \ln\left(x + \sqrt{x^2 + 1}\right)$ is an odd
function with domain $D = \text{range } R = \mathbb{R}$. Since
$\displaystyle \lim_{x \to \infty} \sqrt{x^2 + 1} = x$, $y = \sinh^{-1} x$ is asymptotic to
$y = \ln(x + x) = \ln(2x) = \ln x + \ln 2$ for positive x.

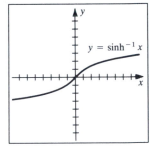

Figure 29

$\boxed{33}$ $D_x \cosh^{-1} u = D_x \ln\left(u + \sqrt{u^2 - 1}\right)$ { definition of $\cosh^{-1} u$ }

$$= \frac{1}{u + \sqrt{u^2 - 1}} D_x\left(u + \sqrt{u^2 - 1}\right) \quad \{\text{apply (7.11)(i)}\}$$

$$= \frac{1}{u + \sqrt{u^2 - 1}} \cdot \left[1 + \tfrac{1}{2}(u^2 - 1)^{-1/2}(2u)\right] D_x u$$
                                         { differentiate $u + \sqrt{u^2 - 1}$ }

$$= \frac{1 + u/\sqrt{u^2 - 1}}{u + \sqrt{u^2 - 1}} D_x u \quad \{\text{simplify}\}$$

$$= \frac{1 + u/\sqrt{u^2 - 1}}{u + \sqrt{u^2 - 1}} \cdot \frac{\sqrt{u^2 - 1}}{\sqrt{u^2 - 1}} D_x u \quad \{\text{multiply by lcd}\}$$

$$= \frac{\sqrt{u^2 - 1} + u}{\sqrt{u^2 - 1}\left(u + \sqrt{u^2 - 1}\right)} D_x u \quad \{\text{simplify}\}$$

$$= \frac{1}{\sqrt{u^2 - 1}} D_x u, \ u > 1 \quad \{\text{cancel } \sqrt{u^2 - 1} + u\}$$

$\boxed{37}$ $D_u\left(\frac{1}{a}\tanh^{-1}\frac{u}{a}\right) = \frac{1}{a}\left[\frac{1}{1 - (u/a)^2} \cdot \frac{1}{a}\right] = \frac{1}{a}\left(\frac{a^2}{a^2 - u^2} \cdot \frac{1}{a}\right) = \frac{1}{a^2 - u^2} \Rightarrow$

$$\int \frac{1}{a^2 - u^2}\, du = \frac{1}{a}\tanh^{-1}\frac{u}{a} + C$$

$\boxed{39}$ $y = \cosh^{-1}x \Rightarrow \cosh y = \cosh\left(\cosh^{-1}x\right) \Rightarrow x = \cosh y = \frac{1}{2}(e^y + e^{-y}) \ (x \geq 1) \Rightarrow$

$2x = e^y + e^{-y} \Rightarrow e^y(2x) = e^y(e^y + e^{-y}) \Rightarrow e^{2y} - 2xe^y + 1 = 0.$

We recognize this equation as a quadratic equation in e^y. Let $z = e^y$.

Then $z^2 - 2xz + 1 = 0$ and we can solve for z using the quadratic formula.

$$z = \frac{2x \pm \sqrt{(2x)^2 - 4(1)(1)}}{2(1)} = \tfrac{1}{2}(2x \pm \sqrt{4x^2 - 4}) = \tfrac{1}{2}\left(2x \pm \sqrt{4(x^2 - 1)}\right) =$$

$x \pm \sqrt{x^2 - 1}$. The $+$ sign must be chosen since $y \geq 0 \Rightarrow e^y \geq 1$, and

$x - \sqrt{x^2 - 1} < 1$ if $x > 1$. Hence, $z = e^y = x + \sqrt{x^2 - 1} \Rightarrow y = \ln\left(x + \sqrt{x^2 - 1}\right),$

$$x \geq 1.$$

8.5 Review Exercises

$\boxed{1}$ $f(x) = \arctan\sqrt{x - 1} \Rightarrow$

$$f'(x) = \frac{1}{1 + (\sqrt{x - 1})^2} \cdot \tfrac{1}{2}(x - 1)^{-1/2} = \frac{1}{1 + (x - 1)} \cdot \frac{1}{2\sqrt{x - 1}} = \frac{1}{2x\sqrt{x - 1}}$$

$\boxed{3}$ $f(x) = x^2 \operatorname{arcsec}(x^2) \Rightarrow$

$$f'(x) = x^2 \cdot \frac{1}{x^2\sqrt{(x^2)^2 - 1}} \cdot 2x + (2x) \cdot \operatorname{arcsec}(x^2) = \frac{2x}{\sqrt{x^4 - 1}} + 2x\operatorname{arcsec}(x^2)$$

$\boxed{5}$ $f(x) = 2^{\arctan 2x} \Rightarrow f'(x) = (2^{\arctan 2x}\ln 2) \cdot \frac{1}{1 + (2x)^2} \cdot 2 = 2^{\arctan 2x}\left(\frac{2\ln 2}{1 + 4x^2}\right)$

$\boxed{9}$ $f(x) = \sin^{-1}\sqrt{1 - x^2} \Rightarrow$

$$f'(x) = \frac{1}{\sqrt{1 - (\sqrt{1 - x^2})^2}} \cdot \tfrac{1}{2}(1 - x^2)^{-1/2}(-2x) = \frac{1}{\sqrt{x^2}} \cdot \frac{-x}{\sqrt{1 - x^2}} = \frac{-x}{\sqrt{x^2(1 - x^2)}}$$

$\boxed{13}$ $f(x) = \tan^{-1}(\tan^{-1} x) \Rightarrow f'(x) = \frac{1}{1 + (\tan^{-1} x)^2} \cdot \frac{1}{1 + x^2} = \frac{1}{(1 + x^2)\left[1 + (\tan^{-1} x)^2\right]}$

$\boxed{17}$ $f(x) = e^{-x}\sinh e^{-x} \Rightarrow$

$$f'(x) = e^{-x}\cosh e^{-x} \cdot (-e^{-x}) + (-e^{-x})\sinh e^{-x} = -e^{-x}(e^{-x}\cosh e^{-x} + \sinh e^{-x})$$

$\boxed{19}$ $f(x) = \frac{\sinh x}{\cosh x - \sinh x} \Rightarrow f'(x) = \frac{(\cosh x - \sinh x)\cosh x - \sinh x(\sinh x - \cosh x)}{(\cosh x - \sinh x)^2}$

$$= \frac{\cosh^2 x - \sinh^2 x}{(\cosh x - \sinh x)^2} = \frac{1}{(\cosh x - \sinh x)^2} = \frac{1}{(e^{-x})^2} \; \{\text{see } 8.3.56\} = \frac{1}{e^{-2x}} = e^{2x}$$

$\boxed{23}$ $f(x) = \tanh^{-1}(\tanh \sqrt[3]{x}) = \sqrt[3]{x} \Rightarrow f'(x) = \tfrac{1}{3}x^{-2/3} = \frac{1}{3x^{2/3}}$

$\boxed{25}$ $\int \frac{1}{4 + 9x^2}\,dx = \int \frac{1}{9(\frac{4}{9} + x^2)}\,dx = \tfrac{1}{9}\int \frac{1}{(\frac{2}{3})^2 + x^2}\,dx = \tfrac{1}{9} \cdot \frac{1}{\frac{2}{3}}\tan^{-1}\frac{x}{\frac{2}{3}} + C =$

$$\tfrac{1}{6}\tan^{-1}\frac{3x}{2} + C$$

$\boxed{29}$ $u = x^2, \; \tfrac{1}{2}\,du = x\,dx \Rightarrow \int \frac{x}{\text{sech}(x^2)}\,dx = \tfrac{1}{2}\int \frac{1}{\text{sech } u}\,du = \tfrac{1}{2}\int \cosh u\,du =$

$$\tfrac{1}{2}\sinh u + C = \tfrac{1}{2}\sinh(x^2) + C$$

$\boxed{31}$ Since $\frac{1}{\sqrt{1 - x^2}}$ is an even function, $\int_{-1/2}^{1/2}\frac{1}{\sqrt{1 - x^2}}\,dx = 2\int_0^{1/2}\frac{1}{\sqrt{1 - x^2}}\,dx =$

$$2\left[\sin^{-1} x\right]_0^{1/2} = 2(\sin^{-1}\tfrac{1}{2} - \sin^{-1} 0) = 2(\tfrac{\pi}{6} - 0) = \tfrac{\pi}{3}.$$

$\boxed{37}$ $u = 2x, \; \tfrac{1}{2}\,du = dx \Rightarrow$

$$\int \frac{1}{x\sqrt{9 - 4x^2}}\,dx = \tfrac{1}{2}\int \frac{1}{(\frac{1}{2}u)\sqrt{3^2 - u^2}}\,du = -\tfrac{1}{3}\text{sech}^{-1}\frac{|u|}{3} + C = -\tfrac{1}{3}\text{sech}^{-1}\left(\tfrac{2}{3}|x|\right) + C$$

$\boxed{39}$ $u = 25x^2 + 36, \; \tfrac{1}{50}\,du = x\,dx \Rightarrow \int \frac{x}{\sqrt{25x^2 + 36}}\,dx = \tfrac{1}{50}\int u^{-1/2}\,du =$

$$\tfrac{1}{50}(2u^{1/2}) + C = \tfrac{1}{25}\sqrt{u} + C = \tfrac{1}{25}\sqrt{25x^2 + 36} + C$$

$\boxed{41}$ $m_{AB} = \frac{7 - (-3)}{4 - 2} = 5. \;\; y = \sin^{-1} 3x \Rightarrow y' = \frac{1}{\sqrt{1 - (3x)^2}} \cdot 3 = \frac{3}{\sqrt{1 - 9x^2}}.$

$y' = 5 \Rightarrow \tfrac{3}{5} = \sqrt{1 - 9x^2} \Rightarrow \tfrac{9}{25} = 1 - 9x^2 \Rightarrow 9x^2 = \tfrac{16}{25} \Rightarrow x^2 = \tfrac{16}{9 \cdot 25} \Rightarrow x = \pm\tfrac{4}{15}.$

The points are $(\pm\tfrac{4}{15}, \sin^{-1}(\pm\tfrac{4}{5}))$.

$\boxed{43}$ $f(x) = 8\sec x + \csc x \Rightarrow f'(x) = 8\sec x \tan x - \csc x \cot x = \dfrac{8\sin x}{\cos^2 x} - \dfrac{\cos x}{\sin^2 x} =$

$\dfrac{8\sin^3 x - \cos^3 x}{\cos^2 x \sin^2 x}$. $f'(x) = 0 \Rightarrow 8\sin^3 x - \cos^3 x = 0 \Rightarrow 8\sin^3 x = \cos^3 x \Rightarrow$

$2\sin x = \cos x \Rightarrow \tan x = \frac{1}{2} \Rightarrow x = c = \tan^{-1}\frac{1}{2}$. $f'(x) < 0$ on $(0, c)$ and $f'(x) > 0$

on $(c, \frac{\pi}{2})$. Hence, $f \downarrow$ on $(0, c]$ and $\uparrow$ on $[c, \frac{\pi}{2})$. f has a *LMIN* of $f(c) =$

$$f(\tan^{-1}\tfrac{1}{2}) = 8\sec(\tan^{-1}\tfrac{1}{2}) + \csc(\tan^{-1}\tfrac{1}{2}) = 8(\tfrac{\sqrt{5}}{2}) + \tfrac{\sqrt{5}}{1} = 5\sqrt{5}.$$

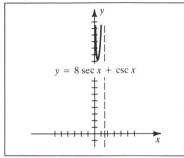

Figure 43

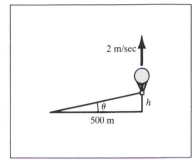

Figure 47

$\boxed{47}$ From *Figure 47*, $\tan\theta = \dfrac{h}{500} \Rightarrow \theta = \tan^{-1}\dfrac{h}{500}$. $\dfrac{d\theta}{dt} = \dfrac{d\theta}{dh} \cdot \dfrac{dh}{dt} = \dfrac{500}{500^2 + h^2} \cdot \dfrac{dh}{dt}$.

When $h = 100$ and $\dfrac{dh}{dt} = 2$, $\dfrac{d\theta}{dt} = \dfrac{500}{500^2 + 100^2}(2) =$

$$\tfrac{1}{260}\,\text{rad/sec} = (\tfrac{1}{260} \cdot \tfrac{180}{\pi})^\circ/\text{sec} \approx 0.22^\circ/\text{sec}.$$

$\boxed{49}$ Let s denote the altitude of the stunt man.

Then $s(t) = -\frac{1}{2}gt^2 + v_0 t + s_0$ { Exercise 5.1.63 } $= -16t^2 + 100$.

Note that $g = 32$ ft/sec^2, the stunt man's initial velocity

is $v_0 = 0$ ft/sec, and his initial altitude is $s_0 = 100$ ft.

Now, $\tan\theta = \dfrac{s}{200} = \dfrac{100 - 16t^2}{200} = \dfrac{25 - 4t^2}{50} \Rightarrow \theta = \tan^{-1}\left(\dfrac{25 - 4t^2}{50}\right)$. Thus,

$\dfrac{d\theta}{dt} = \dfrac{1}{1 + \left[\frac{1}{50}(25 - 4t^2)\right]^2} \cdot (-\frac{8}{50}t)$. At $t = 2$, $\dfrac{d\theta}{dt} = -\dfrac{800}{2581} \approx -0.31$ rad/sec.

Chapter 9: Techniques of Integration

Note: The symbol $\triangleq$ indicates that integration by parts has taken place using substitution A. The substitutions for u, du, dv, and v are listed at the end of the problem.

1 When choosing u and dv, it may take a few tries to get the right combination. As a general rule, try to pick the most complicated part of the integrand that you know how to integrate, and let it equal dv. Since we know how to integrate e^{-x}, let $dv = e^{-x}\,dx$ and $u = x$. Then by (9.1),

$$\int xe^{-x}\,dx = \int (x)(e^{-x}\,dx) = \int u\,dv = uv - \int v\,du \stackrel{A}{=} -xe^{-x} - \int -e^{-x}\,dx$$

$$= -xe^{-x} - e^{-x} + C$$

$$= -(x+1)e^{-x} + C$$

A. $u = x$, $du = dx$, $dv = e^{-x}\,dx$, $v = -e^{-x}$

3 As in Exercise 1, we let $dv = e^{3x}\,dx$ and $u = x^2$.

$$\int x^2 e^{3x}\,dx = \int (x^2)(e^{3x}\,dx) = \int u\,dv = uv - \int v\,du \stackrel{A}{=} \tfrac{1}{3}x^2 e^{3x} - \tfrac{2}{3}\int xe^{3x}\,dx$$

We now have an integral that is similar in form to the one we started with. We will use the same substitution scheme, with u containing the algebraic portion and dv containing the exponential portion. Continuing,

$$I = \tfrac{1}{3}x^2 e^{3x} - \tfrac{2}{3}\left[uv - \int v\,du\right]$$

$$\stackrel{B}{=} \tfrac{1}{3}x^2 e^{3x} - \tfrac{2}{3}\left[\tfrac{1}{3}xe^{3x} - \tfrac{1}{3}\int e^{3x}\,dx\right]$$

$$= \tfrac{1}{3}x^2 e^{3x} - \tfrac{2}{9}xe^{3x} + \tfrac{2}{9}\int e^{3x}\,dx$$

$$= \tfrac{1}{3}x^2 e^{3x} - \tfrac{2}{9}xe^{3x} + \tfrac{2}{27}e^{3x} + C$$

$$= \tfrac{1}{27}e^{3x}(9x^2 - 6x + 2) + C$$

A. $u = x^2$, $du = 2x\,dx$, $dv = e^{3x}\,dx$, $v = \tfrac{1}{3}e^{3x}$

B. $u = x$, $du = dx$, $dv = e^{3x}\,dx$, $v = \tfrac{1}{3}e^{3x}$

7 Since we know the antiderivative of $\sec x \tan x$, let $dv = \sec x \tan x\,dx$.

$$\int x \sec x \tan x\,dx = uv - \int v\,du \stackrel{A}{=} x \sec x - \int \sec x\,dx$$

$$= x \sec x - \ln|\sec x + \tan x| + C$$

A. $u = x$, $du = dx$, $dv = \sec x \tan x\,dx$, $v = \sec x$

11 Since we do not know the antiderivative of $\tan^{-1}x$, our only choice for dv is dx.

$$\int \tan^{-1}x\,dx = uv - \int v\,du \overset{A}{=} x\tan^{-1}x - \int \frac{x}{1+x^2}\,dx$$

$$\left\{ \text{Use substitution to find } \int \frac{x}{1+x^2}\,dx \text{ with } u = 1+x^2. \right\}$$

$$= x\tan^{-1}x - \tfrac{1}{2}\ln(1+x^2) + C$$

A. $u = \tan^{-1}x,\ du = \dfrac{1}{1+x^2}\,dx,\ dv = dx,\ v = x$

13 Since we do not know the antiderivative of $\ln x$, but we do know the antiderivative of

$\sqrt{x}$, let $dv = \sqrt{x}\,dx$. $\int \sqrt{x}\,\ln x\,dx = uv - \int v\,du \overset{A}{=} \tfrac{2}{3}x^{3/2}\ln x - \tfrac{2}{3}\int x^{1/2}\,dx$

$$= \tfrac{2}{3}x^{3/2}\ln x - \tfrac{2}{3}(\tfrac{2}{3}x^{3/2}) + C$$

$$= \tfrac{2}{9}x^{3/2}(3\ln x - 2) + C$$

A. $u = \ln x,\ du = \dfrac{1}{x}\,dx,\ dv = x^{1/2}\,dx,\ v = \tfrac{2}{3}x^{3/2}$

17 Exercise 17 is similar to Example 5. We will need to use integration by parts twice
and then solve for I. Our choice for dv could be either $e^{-x}\,dx$ or $\sin x\,dx$, since we
know their antiderivatives. However, once you choose dv, do not change your pattern
for the second application of integration by parts. We have picked $dv = e^{-x}\,dx$ for
both substitutions.

$$\int e^{-x}\sin x\,dx \overset{A}{=} -e^{-x}\sin x + \int e^{-x}\cos x\,dx$$

$$\overset{B}{=} -e^{-x}\sin x + \left[-e^{-x}\cos x - \int e^{-x}\sin x\,dx \right]$$

$\Rightarrow$ $\text{I} = -e^{-x}\sin x - e^{-x}\cos x - \text{I}$

$\Rightarrow$ $2\text{I} = -e^{-x}\sin x - e^{-x}\cos x$

$\Rightarrow$ $\text{I} = -\tfrac{1}{2}e^{-x}(\sin x + \cos x) + C$

A. $u = \sin x,\ du = \cos x\,dx,\ dv = e^{-x}\,dx,\ v = -e^{-x}$

B. $u = \cos x,\ du = -\sin x\,dx,\ dv = e^{-x}\,dx,\ v = -e^{-x}$

19 If we let $y = \cos x$, then $dy = -\sin x\,dx$, and

$\int \sin x \ln \cos x\,dx = -\int \ln \cos x\,(-\sin x)\,dx = -\int \ln y\,dy$. By Example 3,

$$\text{I} = -\int \ln y\,dy = -y\ln y + y + C = y(1 - \ln y) + C = \cos x(1 - \ln \cos x) + C.$$

[21] We do not know the antiderivative of $\csc^3 x$, however, we do know the antiderivative of $\csc^2 x$ and $\csc x$. Since the antiderivative of $\csc^2 x$ is the simplest of the two, let $dv = \csc^2 x \, dx$.

$$\int \csc^3 x \, dx \overset{A}{=} -\csc x \cot x - \int \csc x \cot^2 x \, dx$$

$$= -\csc x \cot x - \int \csc x (\csc^2 x - 1) \, dx$$

$$= -\csc x \cot x - \int \csc^3 x \, dx + \int \csc x \, dx$$

$$\Rightarrow \quad I = -\csc x \cot x - I + \int \csc x \, dx$$

$$\Rightarrow \quad 2I = -\csc x \cot x + \ln|\csc x - \cot x|$$

$$\Rightarrow \quad I = -\tfrac{1}{2}\csc x \cot x + \tfrac{1}{2}\ln|\csc x - \cot x| + C$$

A. $u = \csc x$, $du = -\csc x \cot x \, dx$, $dv = \csc^2 x \, dx$, $v = -\cot x$

[23] If we rewrite the integrand as $x^2 \cdot \dfrac{x}{\sqrt{x^2 + 1}}$, then we can integrate the second factor

using the method of substitution. Let $dv = \dfrac{x}{\sqrt{x^2 + 1}} \, dx$.

$$\int_0^1 \frac{x^3}{\sqrt{x^2 + 1}} \, dx \overset{A}{=} \left[x^2 \sqrt{x^2 + 1} \right]_0^1 - 2\int_0^1 x\sqrt{x^2 + 1} \, dx$$

$$\{ \text{use the method of substitution to integrate } x\sqrt{x^2 + 1} \}$$

$$= \left[x^2 \sqrt{x^2 + 1} \right]_0^1 - \left[\tfrac{2}{3}(x^2 + 1)^{3/2} \right]_0^1$$

$$= (\sqrt{2} - 0) - \tfrac{2}{3}(2\sqrt{2} - 1) = \tfrac{1}{3}(2 - \sqrt{2}) \approx 0.20$$

A. $u = x^2$, $du = 2x \, dx$, $dv = \dfrac{x}{\sqrt{x^2 + 1}} \, dx$, $v = \sqrt{x^2 + 1}$

[29] Exercise 29 is similar to Example 5 and Exercise 17—the arithmetic is slighty more complicated.

$$\int e^{4x} \sin 5x \, dx \overset{A}{=} -\tfrac{1}{5}e^{4x} \cos 5x + \tfrac{4}{5}\int e^{4x} \cos 5x \, dx$$

$$\overset{B}{=} -\tfrac{1}{5}e^{4x} \cos 5x + \tfrac{4}{5}\left[\tfrac{1}{5}e^{4x} \sin 5x - \tfrac{4}{5}\int e^{4x} \sin 5x \, dx \right]$$

$$= -\tfrac{1}{5}e^{4x} \cos 5x + \tfrac{4}{25}e^{4x} \sin 5x - \tfrac{16}{25}\int e^{4x} \sin 5x \, dx$$

$$\Rightarrow \quad I = -\tfrac{1}{5}e^{4x} \cos 5x + \tfrac{4}{25}e^{4x} \sin 5x - \tfrac{16}{25}I$$

$$\Rightarrow \quad \tfrac{41}{25}I = -\tfrac{1}{5}e^{4x} \cos 5x + \tfrac{4}{25}e^{4x} \sin 5x$$

$$\Rightarrow \quad I = \tfrac{1}{41}e^{4x}(4\sin 5x - 5\cos 5x) + C$$

A. $u = e^{4x}$, $du = 4e^{4x} \, dx$, $dv = \sin 5x \, dx$, $v = -\tfrac{1}{5}\cos 5x$

B. $u = e^{4x}$, $du = 4e^{4x} \, dx$, $dv = \cos 5x \, dx$, $v = \tfrac{1}{5}\sin 5x$

$\boxed{33}$ In this exercise, integration by parts is performed three times.

Each time we let $dv = \sinh x \, dx$, and the power of x is reduced by one.

$\int x^3 \sinh x \, dx \overset{A}{=} x^3 \cosh x - 3 \int x^2 \cosh x \, dx$

$\qquad \overset{B}{=} x^3 \cosh x - 3 \left[x^2 \sinh x - 2 \int x \sinh x \, dx \right]$

$\qquad \overset{C}{=} x^3 \cosh x - 3x^2 \sinh x + 6(x \cosh x - \int \cosh x \, dx)$

$\qquad = x^3 \cosh x - 3x^2 \sinh x + 6x \cosh x - 6 \sinh x + C$

A. $u = x^3$, $du = 3x^2 \, dx$, $dv = \sinh x \, dx$, $v = \cosh x$

B. $u = x^2$, $du = 2x \, dx$, $dv = \cosh x \, dx$, $v = \sinh x$

C. $u = x$, $du = dx$, $dv = \sinh x \, dx$, $v = \cosh x$

$\boxed{35}$ In this exercise, we must rewrite the integrand in a special way: $\sqrt{x} \cdot \dfrac{\cos \sqrt{x}}{\sqrt{x}}$.

Then we can integrate $\dfrac{\cos \sqrt{x}}{\sqrt{x}}$ by using the method of substitution.

$\int \cos \sqrt{x} \, dx = \int \sqrt{x} \cdot \dfrac{\cos \sqrt{x}}{\sqrt{x}} \, dx$

$\qquad \overset{A}{=} 2\sqrt{x} \sin \sqrt{x} - \int \dfrac{\sin \sqrt{x}}{\sqrt{x}} \, dx$

$\qquad = 2\sqrt{x} \sin \sqrt{x} + 2 \cos \sqrt{x} + C$

A. $u = \sqrt{x}$, $du = \dfrac{1}{2\sqrt{x}} \, dx$, $dv = \dfrac{\cos \sqrt{x}}{\sqrt{x}} \, dx$, $v = 2 \sin \sqrt{x}$

$\boxed{39}$ $\int x^m e^x \, dx = uv - \int v \, du \overset{A}{=} x^m e^x - m \int x^{m-1} e^x \, dx$

A. $u = x^m$, $du = mx^{m-1} \, dx$, $dv = e^x \, dx$, $v = e^x$

$\boxed{43}$ We will need to use the reduction formula five times.

Each time the power of x will be reduced by one.

$I = x^5 e^x - 5 \int x^4 e^x \, dx$

$\quad = x^5 e^x - 5 \left[x^4 e^x - 4 \int x^3 e^x \, dx \right]$

$\quad = x^5 e^x - 5x^4 e^x + 20 \left[x^3 e^x - 3 \int x^2 e^x \, dx \right]$

$\quad = x^5 e^x - 5x^4 e^x + 20x^3 e^x - 60 \left[x^2 e^x - 2 \int x e^x \, dx \right]$

$\quad = x^5 e^x - 5x^4 e^x + 20x^3 e^x - 60x^2 e^x + 120 \left[x e^x - \int e^x \, dx \right]$

$\quad = e^x (x^5 - 5x^4 + 20x^3 - 60x^2 + 120x - 120) + C$

45 The technique of integration by parts in this exercise is done in a manner similar to that in Exercise 35.

$$A = \int_0^{\pi^2} \sin \sqrt{x} \, dx = \int_0^{\pi^2} \sqrt{x} \cdot \frac{\sin \sqrt{x}}{\sqrt{x}} \, dx$$

$$\overset{A}{=} \left[-2\sqrt{x} \cos \sqrt{x} \right]_0^{\pi^2} + \int_0^{\pi^2} \frac{\cos \sqrt{x}}{\sqrt{x}} \, dx$$

$$= \left[-2\sqrt{x} \cos \sqrt{x} + 2 \sin \sqrt{x} \right]_0^{\pi^2}$$

$$= (2\pi + 0) - (0 - 0) = 2\pi$$

A. $u = \sqrt{x}$, $du = \frac{1}{2\sqrt{x}} \, dx$, $dv = \frac{\sin \sqrt{x}}{\sqrt{x}} \, dx$, $v = -2 \cos \sqrt{x}$

49 Let $f(x) = e^x$, $g(x) = 0$, and $\rho = 1$ in (6.25).

$$m = \int_0^{\ln 3} e^x \, dx = \left[e^x \right]_0^{\ln 3} = 3 - 1 = 2.$$

$$M_x = \frac{1}{2} \int_0^{\ln 3} (e^x)^2 \, dx = \frac{1}{2} \int_0^{\ln 3} e^{2x} \, dx = \frac{1}{2} \left[\frac{1}{2} e^{2x} \right]_0^{\ln 3} = \frac{1}{4}(9 - 1) = 2.$$

$$M_y = \int_0^{\ln 3} x e^x \, dx \overset{A}{=} x e^x - \int e^x \, dx = \left[x e^x - e^x \right]_0^{\ln 3} =$$

$$\left[(\ln 3)(3) - 3 \right] - \left[0 - 1 \right] = 3 \ln 3 - 2.$$

$$\bar{x} = \frac{M_y}{m} = \frac{3 \ln 3 - 2}{2} \approx 0.65 \text{ and } \bar{y} = \frac{M_x}{m} = \frac{2}{2} = 1.$$

A. $u = x$, $du = dx$, $dv = e^x \, dx$, $v = e^x$

51 Substituting $(v + C)$ for v in (9.1) yields $\int u \, dv = u(v + C) - \int (v + C) \, du =$
$$uv + uC - \int v \, du - Cu = uv - \int v \, du, \text{ which is (9.1).}$$

53 $\left| \frac{1}{x} \, dx = 1 + \right| \frac{1}{x} \, dx \Rightarrow \ln |x| + C_1 = 1 + \ln |x| + C_2 \Rightarrow C_1 - C_2 = 1.$

Note that indefinite integrals represent a class of functions which differ by a constant.

Thus, $\int \frac{1}{x} \, dx - \int \frac{1}{x} \, dx \neq 0$, and the logic is incorrect. Also, $u = \frac{1}{x}$ does not have a continuous first derivative and hence, does not satisfy the condition stated in (9.1).

Exercises 9.2

1 We will follow (2) of Guidelines (9.2) since the exponent on $\cos x$ is odd.

$$\int \cos^3 x \, dx = \int (\cos^2 x) \cos x \, dx = \int (1 - \sin^2 x) \cos x \, dx; \; u = \sin x, \; du = \cos x \, dx \Rightarrow$$

$$I = \int (1 - u^2) \, du = u - \frac{1}{3} u^3 + C = \sin x - \frac{1}{3} \sin^3 x + C$$

3 We will follow (3) of Guidelines (9.2) since the exponents on $\sin x$ and $\cos x$ are even. Use the half-angle formulas on the back end papers to simplify $\sin^2 x$ and $\cos^2 x$.

$$\int \sin^2 x \cos^2 x \, dx = \int \frac{1 - \cos 2x}{2} \cdot \frac{1 + \cos 2x}{2} \, dx \qquad \{\text{half-angle formulas}\}$$

$$= \int \tfrac{1}{4}(1 - \cos^2 2x) \, dx \qquad \{\text{simplify}\}$$

$$= \int \tfrac{1}{4}\left(1 - \frac{1 + \cos 4x}{2}\right) dx \qquad \{\text{half-angle formula}\}$$

$$= \int \left[\tfrac{1}{4} - \tfrac{1}{8}(1 + \cos 4x)\right] dx \qquad \{\text{simplify}\}$$

$$= \int (\tfrac{1}{8} - \tfrac{1}{8}\cos 4x) \, dx \qquad \{\text{simplify}\}$$

$$= \tfrac{1}{8}x - \tfrac{1}{32}\sin 4x + C \qquad \{\text{find antiderivative}\}$$

7 $$\int \sin^6 x \, dx = \int (\sin^2 x)^3 \, dx \qquad \{\text{change integrand form}\}$$

$$= \int \left(\frac{1 - \cos 2x}{2}\right)^3 dx \qquad \{\text{half-angle formula for } \sin^2 x\}$$

$$= \tfrac{1}{8}\int (1 - 3\cos 2x + 3\cos^2 2x - \cos^3 2x) \, dx \qquad \{\text{simplify}\}$$

$$= \tfrac{1}{8}\int \left[1 - 3\cos 2x + 3\left(\frac{1 + \cos 4x}{2}\right) - (\cos^2 2x)\cos 2x\right] dx$$

$$\{\text{half-angle formula for } \cos^2 2x\}$$

$$= \tfrac{1}{8}\int \left[1 - 3\cos 2x + \tfrac{3}{2}(1 + \cos 4x) - (1 - \sin^2 2x)\cos 2x\right] dx \quad \{\text{simplify}\}$$

$$= \tfrac{1}{8}\int (\tfrac{5}{2} - 4\cos 2x + \tfrac{3}{2}\cos 4x + \sin^2 2x \cos 2x) \, dx \qquad \{\text{combine like terms}\}$$

$$= \tfrac{1}{8}(\tfrac{5}{2}x - 2\sin 2x + \tfrac{3}{8}\sin 4x + \tfrac{1}{6}\sin^3 2x) + C$$

$$\{\text{find antiderivative; for last term, use } u = \sin 2x \text{ and } \tfrac{1}{2}\, du = \cos 2x \, dx\}$$

9 We will use (2) of Guidelines (9.3) — note that (1) could also be used.

$\int \tan^3 x \sec^4 x \, dx = \int \tan^3 x (\sec^2 x) \sec^2 x \, dx = \int \tan^3 x (1 + \tan^2 x) \sec^2 x \, dx =$

$\int (\tan^3 x + \tan^5 x)\sec^2 x \, dx; \; u = \tan x, \; du = \sec^2 x \, dx \Rightarrow$

$$I = \int (u^3 + u^5) \, du = \tfrac{1}{4}u^4 + \tfrac{1}{6}u^6 + C = \tfrac{1}{4}\tan^4 x + \tfrac{1}{6}\tan^6 x + C$$

11 We will use (1) of Guidelines (9.3) since the exponent on $\tan x$ is odd.

$\int \tan^3 x \sec^3 x \, dx = \int (\tan^2 x)(\sec^2 x)\sec x \tan x \, dx = \int (\sec^2 x - 1)\sec^2 x \sec x \tan x \, dx =$

$\int (\sec^4 x - \sec^2 x)\sec x \tan x \, dx; \; u = \sec x, \; du = \sec x \tan x \, dx \Rightarrow$

$$I = \int (u^4 - u^2) \, du = \tfrac{1}{5}u^5 - \tfrac{1}{3}u^3 + C = \tfrac{1}{5}\sec^5 x - \tfrac{1}{3}\sec^3 x + C$$

[13] $\int \tan^6 x \, dx = \int \tan^4 x (\tan^2 x) \, dx$

$= \int \tan^4 x (\sec^2 x - 1) \, dx$

$= \int \left[\tan^4 x \sec^2 x - \tan^2 x (\tan^2 x) \right] dx$

$= \int \left[\tan^4 x \sec^2 x - \tan^2 x (\sec^2 x - 1) \right] dx$

$= \int \left(\tan^4 x \sec^2 x - \tan^2 x \sec^2 x + \sec^2 x - 1 \right) dx$

$= \frac{1}{5} \tan^5 x - \frac{1}{3} \tan^3 x + \tan x - x + C$

[17] $\int (\tan x + \cot x)^2 \, dx = \int (\tan^2 x + 2 \tan x \cot x + \cot^2 x) \, dx$

$= \int (\tan^2 x + 2 + \cot^2 x) \, dx$

$= \int \left[(\tan^2 x + 1) + (1 + \cot^2 x) \right] dx$

$= \int (\sec^2 x + \csc^2 x) \, dx$

$= \tan x - \cot x + C$

[23] The product-to-sum formulas can be found on the back end papers of the text.

We will use the identity $\sin u \cos v = \frac{1}{2} \left[\sin(u + v) + \sin(u - v) \right]$ with $u = 3x$ and

$v = 2x$. Thus,

$\int_0^{\pi/2} \sin 3x \cos 2x \, dx = \int_0^{\pi/2} \frac{1}{2} \left[\sin(3x + 2x) + \sin(3x - 2x) \right] dx$

$= \frac{1}{2} \int_0^{\pi/2} (\sin 5x + \sin x) \, dx$

$= \frac{1}{2} \left[-\frac{1}{5} \cos 5x - \cos x \right]_0^{\pi/2} = \frac{1}{2} \left[0 - \left(-\frac{6}{5} \right) \right] = \frac{3}{5}.$

[27] The method of substitution can be used in this exercise. Let $u = 2 - \sin x$ and

$du = -\cos x \, dx$. Thus, $\int \frac{\cos x}{2 - \sin x} \, dx = -\int \frac{1}{2 - \sin x} (-\cos x) \, dx = -\int \frac{1}{u} \, du =$

$-\ln |u| + C = -\ln(2 - \sin x) + C.$ Note that $2 - \sin x$ is always positive.

[31] Refer to (6.5) for the disk method. Using symmetry and disks,

$$V = \int_0^{2\pi} \pi (\cos^2 x)^2 \, dx$$

$$= 4\pi \int_0^{\pi/2} (\cos^2 x)^2 \, dx$$

$$= 4\pi \int_0^{\pi/2} \left[\tfrac{1}{2}(1 + \cos 2x)\right]^2 \, dx$$

$$= \pi \int_0^{\pi/2} (1 + 2\cos 2x + \cos^2 2x) \, dx$$

$$= \pi \int_0^{\pi/2} \left[1 + 2\cos 2x + \tfrac{1}{2}(1 + \cos 4x)\right] dx$$

$$= \pi \int_0^{\pi/2} (\tfrac{3}{2} + 2\cos 2x + \tfrac{1}{2}\cos 4x) \, dx$$

$$= \pi \left[\tfrac{3}{2}x + \sin 2x + \tfrac{1}{8}\sin 4x\right]_0^{\pi/2} = \tfrac{3\pi^2}{4} \approx 7.40.$$

[33] Since $v(t) = \cos^2 \pi t \geq 0$, we know that the point does not move to the left on the coordinate line. $v(t)$ has a period equal to 1 second. Thus, the distance traveled in *any* 5-second interval is given by

$$s(x + 5) - s(x) = \int_x^{x+5} v(t) \, dt = \int_x^{x+5} \cos^2 \pi t \, dt = \int_x^{x+5} \tfrac{1}{2}(1 + \cos 2\pi t) \, dt$$

$$= \left[\tfrac{1}{2}t + \tfrac{1}{4\pi}\sin 2\pi t\right]_x^{x+5}$$

$$= \left[\tfrac{1}{2}(x + 5) + \tfrac{1}{4\pi}\sin\left[2\pi(x + 5)\right]\right] - \left[\tfrac{1}{2}x + \tfrac{1}{4\pi}\sin(2\pi x)\right]$$

$$= \tfrac{5}{2} + \tfrac{1}{4\pi}\left[\sin(2\pi x + 10\pi) - \sin(2\pi x)\right]$$

$$= \tfrac{5}{2}, \text{ since } \sin(2\pi x + 10\pi) = \sin(2\pi x).$$

[35] Using the trigonometric product-to-sum formulas from the back end papers of the text, we have the following:

(a) $I = \int \sin mx \sin nx \, dx = \tfrac{1}{2}\int \left[\cos(m - n)x - \cos(m + n)x\right] dx.$

If $m \neq n$, then $I = \dfrac{\sin(m - n)x}{2(m - n)} - \dfrac{\sin(m + n)x}{2(m + n)} + C.$

If $m = n$, then $I = \tfrac{1}{2}\int (1 - \cos 2mx) \, dx = \tfrac{x}{2} - \dfrac{\sin 2mx}{4m} + C.$

(b) $I_1 = \int \sin mx \cos nx\, dx = \frac{1}{2}\int \big[\sin(m+n)x + \sin(m-n)x \big] dx.$

If $m \neq n$, then $I_1 = -\dfrac{\cos(m+n)x}{2(m+n)} - \dfrac{\cos(m-n)x}{2(m-n)} + C.$

If $m = n$, then $I_1 = \dfrac{1}{2}\displaystyle\int \sin 2mx\, dx = -\dfrac{\cos 2mx}{4m} + C.$

$I_2 = \int \cos mx \cos nx\, dx = \frac{1}{2}\int \big[\cos(m+n)x + \cos(m-n)x \big] dx.$

If $m \neq n$, then $I_2 = \dfrac{\sin(m+n)x}{2(m+n)} + \dfrac{\sin(m-n)x}{2(m-n)} + C.$

If $m = n$, then $I_2 = \dfrac{1}{2}\displaystyle\int (\cos 2mx + 1)\, dx = \dfrac{x}{2} + \dfrac{\sin 2mx}{4m} + C.$

Exercises 9.3

$\boxed{1}$ See *Figure 1*. Let $x = 2\sin\theta$ and $dx = 2\cos\theta\, d\theta$ for $-\frac{\pi}{2} < \theta < \frac{\pi}{2}$.

Then $\sqrt{4 - x^2} = \sqrt{4 - 4\sin^2\theta} = \sqrt{4(1 - \sin^2\theta)} = \sqrt{4\cos^2\theta} = 2|\cos\theta| = 2\cos\theta$

$\{\cos\theta > 0 \text{ for } -\frac{\pi}{2} < \theta < \frac{\pi}{2}\}. \displaystyle\int \frac{1}{x\sqrt{4 - x^2}}\, dx =$

$\displaystyle\int \frac{1}{(2\sin\theta)(2\cos\theta)}\, 2\cos\theta\, d\theta = \frac{1}{2}\int \csc\theta\, d\theta = \frac{1}{2}\ln|\csc\theta - \cot\theta| + C.$

Since $\csc\theta = \dfrac{2}{x}$ and $\cot\theta = \dfrac{\sqrt{4 - x^2}}{x}$, it follows that $I = \dfrac{1}{2}\ln\left| \dfrac{2}{x} - \dfrac{\sqrt{4 - x^2}}{x} \right| + C.$

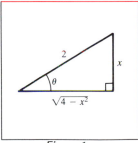

Figure 1

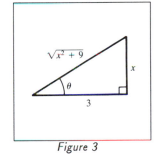

Figure 3

$\boxed{3}$ See *Figure 3*. Let $x = 3\tan\theta$ and $dx = 3\sec^2\theta\, d\theta$, where $-\frac{\pi}{2} < \theta < \frac{\pi}{2}$.

Then $\sqrt{9 + x^2} = \sqrt{9 + 9\tan^2\theta} = \sqrt{9(1 + \tan^2\theta)} = \sqrt{9\sec^2\theta} = 3|\sec\theta| = 3\sec\theta$

$\{\sec\theta > 0 \text{ for } -\frac{\pi}{2} < \theta < \frac{\pi}{2}\}. \displaystyle\int \frac{1}{x\sqrt{9 + x^2}}\, dx = \int \frac{1}{(3\tan\theta)(3\sec\theta)}\, 3\sec^2\theta\, d\theta =$

$\dfrac{1}{3}\displaystyle\int \cot\theta \sec\theta\, d\theta = \dfrac{1}{3}\int \csc\theta\, d\theta = \dfrac{1}{3}\ln|\csc\theta - \cot\theta| + C = \dfrac{1}{3}\ln\left| \dfrac{\sqrt{x^2 + 9}}{x} - \dfrac{3}{x} \right| + C$

$\boxed{5}$ See *Figure 5*. Let $x = 5 \sec \theta$ and $dx = 5 \sec \theta \tan \theta \, d\theta$, where $\theta \in (0, \frac{\pi}{2}) \cup (\pi, \frac{3\pi}{2})$.

Then $\sqrt{x^2 - 25} = \sqrt{25 \sec^2 \theta - 25} = \sqrt{25(\sec^2 \theta - 1)} = \sqrt{25 \tan^2 \theta} = 5|\tan \theta| =$

$5 \tan \theta \; \{ \tan \theta > 0 \text{ for } \theta \in (0, \frac{\pi}{2}) \cup (\pi, \frac{3\pi}{2}) \}. \; \int \dfrac{1}{x^2 \sqrt{x^2 - 25}} \, dx =$

$\int \dfrac{1}{(25 \sec^2 \theta)(5 \tan \theta)} 5 \sec \theta \tan \theta \, d\theta = \dfrac{1}{25} \int \cos \theta \, d\theta = \dfrac{1}{25} \sin \theta + C = \dfrac{\sqrt{x^2 - 25}}{25x} + C$

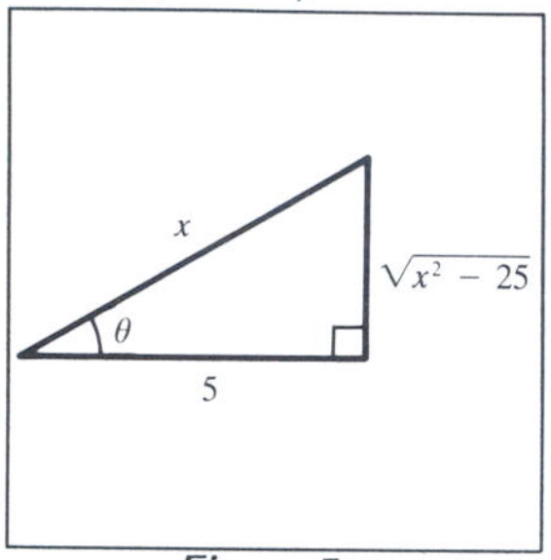

Figure 5

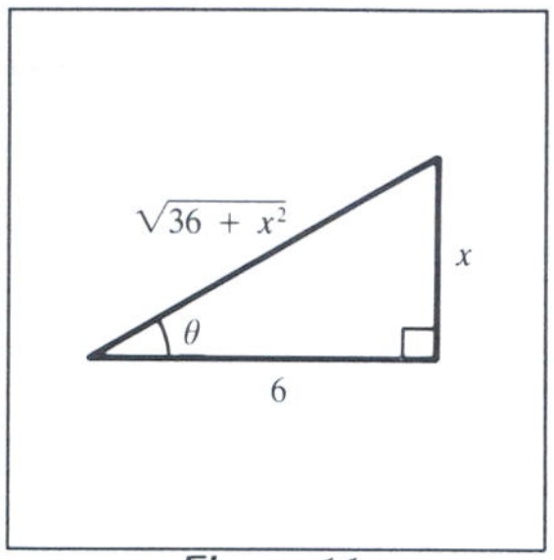

Figure 11

Note: In (9.4), the resulting expressions are $a \cos \theta$, $a \sec \theta$, and $a \tan \theta$, respectively. We have gone through the derivation of each of the three expressions in Exercises 1, 3, and 5, and will now use these results without mention.

$\boxed{7}$ The most straightforward way to evaluate this integral is to use the method of substitution. $u = 4 - x^2$, $-\frac{1}{2} du = x \, dx \Rightarrow \int \dfrac{x}{\sqrt{4 - x^2}} \, dx = -\frac{1}{2} \int u^{-1/2} \, du =$

$-\sqrt{u} + C = -\sqrt{4 - x^2} + C$. This integral could also be evaluated by using the same trigonometric substitution that we used in Exercise 1.

$\boxed{11}$ See *Figure 11*. $x = 6 \tan \theta$, $dx = 6 \sec^2 \theta \, d\theta \Rightarrow$

$\int \dfrac{1}{(36 + x^2)^2} \, dx = \int \dfrac{6 \sec^2 \theta}{(36 \sec^2 \theta)^2} \, d\theta = \dfrac{1}{216} \int \cos^2 \theta \, d\theta = \dfrac{1}{216} \int \dfrac{1 + \cos 2\theta}{2} \, d\theta$

$\qquad = \dfrac{1}{432} (\theta + \frac{1}{2} \sin 2\theta) + C$

$\qquad = \dfrac{1}{432} \left[\theta + \frac{1}{2}(2 \sin \theta \cos \theta) \right] + C$

$\qquad = \dfrac{1}{432} (\theta + \sin \theta \cos \theta) + C$

$\qquad = \dfrac{1}{432} \left[\tan^{-1} \left(\frac{x}{6} \right) + \dfrac{x}{\sqrt{x^2 + 36}} \cdot \dfrac{6}{\sqrt{x^2 + 36}} \right] + C$

$\qquad\qquad\qquad \left\{ x = 6 \tan \theta \Rightarrow \tan \theta = \frac{x}{6} \Rightarrow \theta = \tan^{-1} \left(\frac{x}{6} \right) \right\}$

$\qquad = \dfrac{1}{432} \left[\tan^{-1} \left(\frac{x}{6} \right) + \dfrac{6x}{x^2 + 36} \right] + C$

$\boxed{13}$ $\int \dfrac{1}{\sqrt{9 - x^2}} \, dx = \int \dfrac{1}{\sqrt{3^2 - x^2}} \, dx = \sin^{-1} \left(\frac{x}{3} \right) + C$, by (8.9)(i). This integral could also be evaluated by using the trigonometric substitution $x = 3 \sin \theta$, $dx = 3 \cos \theta \, d\theta$.

Note: In the first printing, the radicand was $9 - 4x^2$.

The answer is then $\frac{1}{2} \sin^{-1} \left(\frac{2}{3} x \right) + C$.

$\boxed{17}$ See *Figure 17.* $3x = 7\tan\theta$ or $x = \frac{7}{3}\tan\theta$, $dx = \frac{7}{3}\sec^2\theta\,d\theta \Rightarrow$

$$\int \frac{x^3}{\sqrt{9x^2 + 49}}\,dx = \int \frac{(\frac{7}{3}\tan\theta)^3(\frac{7}{3}\sec^2\theta)}{7\sec\theta}\,d\theta$$

$$= \frac{7^3}{3^4}\int \tan^2\theta\,\sec\theta\,\tan\theta\,d\theta$$

$$= \frac{7^3}{3^4}\int (\sec^2\theta - 1)\sec\theta\,\tan\theta\,d\theta$$

$$= \frac{7^3}{3^4}\left[\frac{1}{3}\sec^3\theta - \sec\theta\right] + C$$

$$= \frac{7^3}{3^4}\left[\frac{(9x^2 + 49)^{3/2}}{3(7)^3} - \frac{\sqrt{9x^2 + 49}}{7}\right] + C$$

$$= \frac{1}{243}(9x^2 + 49)^{3/2} - \frac{49}{81}\sqrt{9x^2 + 49} + C$$

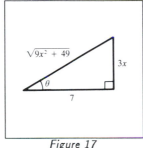

Figure 17

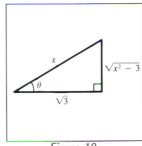

Figure 19

$\boxed{19}$ See *Figure 19.* $x = \sqrt{3}\sec\theta$, $dx = \sqrt{3}\sec\theta\,\tan\theta\,d\theta \Rightarrow$

$$\int \frac{1}{x^4\sqrt{x^2 - 3}}\,dx = \int \frac{\sqrt{3}\sec\theta\,\tan\theta}{(9\sec^4\theta)(\sqrt{3}\tan\theta)}\,d\theta = \frac{1}{9}\int \cos^3\theta\,d\theta = \frac{1}{9}\int (\cos^2\theta)\cos\theta\,d\theta =$$

$$\frac{1}{9}\int (1 - \sin^2\theta)\cos\theta\,d\theta = \frac{1}{9}(\sin\theta - \frac{1}{3}\sin^3\theta) + C$$

$$= \frac{1}{9}\left[\frac{\sqrt{x^2 - 3}}{x} - \frac{(x^2 - 3)^{3/2}}{3x^3}\right] + C$$

$$= \frac{1}{9}\left[\frac{3x^2(x^2 - 3)^{1/2} - (x^2 - 3)^{3/2}}{3x^3}\right] + C$$

$$= \frac{\left[3x^2 - (x^2 - 3)\right](x^2 - 3)^{1/2}}{27x^3} + C\,\{\text{factor }(x^2 - 3)^{1/2}\}$$

$$= \frac{(3 + 2x^2)\sqrt{x^2 - 3}}{27x^3} + C$$

[23] Let $f(x) = \dfrac{x}{\sqrt{x^2 + 25}}$. Then $f(0) = 0$, $f(x) \geq 0$ on $[0, 5]$, and (6.11) applies.

Using shells, $V = \displaystyle\int_0^5 2\pi x \dfrac{x}{\sqrt{x^2 + 25}}\, dx = 2\pi \int_0^5 \dfrac{x^2}{\sqrt{x^2 + 25}}\, dx$. $x = 5\tan\theta \Rightarrow$

$dx = 5\sec^2\theta\, d\theta$. $x = 0, 5 \Rightarrow \theta = 0, \frac{\pi}{4}$. Thus, $V = 2\pi \displaystyle\int_0^{\pi/4} \dfrac{(25\tan^2\theta)(5\sec^2\theta)}{5\sec\theta}\, d\theta$

$= 50\pi \displaystyle\int_0^{\pi/4} \tan^2\theta \sec\theta\, d\theta = 50\pi \int_0^{\pi/4} (\sec^2\theta - 1)\sec\theta\, d\theta$

$= 50\pi \displaystyle\int_0^{\pi/4} (\sec^3\theta - \sec\theta)\, d\theta \; \{\text{Example 6, §9.1}\,\}$

$= 50\pi \left[(\tfrac{1}{2}\sec\theta \tan\theta + \tfrac{1}{2}\ln|\sec\theta + \tan\theta|) - \ln|\sec\theta + \tan\theta| \right]_0^{\pi/4}$

$= 25\pi \left[\sec\theta \tan\theta - \ln|\sec\theta + \tan\theta| \right]_0^{\pi/4} = 25\pi \left[\sqrt{2} - \ln(\sqrt{2} + 1) \right] \approx 41.85$

[25] $x\, dy - \sqrt{x^2 - 16}\, dx \Rightarrow \dfrac{dy}{dx} = f'(x) = \dfrac{\sqrt{x^2 - 16}}{x} \Rightarrow f(x) = \displaystyle\int \dfrac{\sqrt{x^2 - 16}}{x}\, dx.$

See *Figure 25*. $x = 4\sec\theta$, $dx = 4\sec\theta\tan\theta\, d\theta \Rightarrow$

$f(x) = \displaystyle\int \dfrac{(4\tan\theta)(4\sec\theta\tan\theta)}{4\sec\theta}\, d\theta = 4\int \tan^2\theta\, d\theta = 4\int (\sec^2\theta - 1)\, d\theta =$

$4(\tan\theta - \theta) + C = 4\left[\tfrac{1}{4}\sqrt{x^2 - 16} - \sec^{-1}\tfrac{x}{4} \right] + C = \sqrt{x^2 - 16} - 4\sec^{-1}\tfrac{x}{4} + C.$

$f(4) = \sqrt{4^2 - 16} - 4\sec^{-1}(1) + C = 0 - 4(0) + C = C.$ $f(4) = 0 \Rightarrow C = 0,$

and hence, $f(x) = y = \sqrt{x^2 - 16} - 4\sec^{-1}\tfrac{x}{4}.$

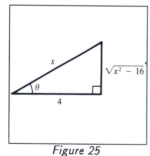

Figure 25

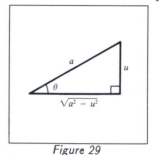

Figure 27   *Figure 29*

[27] See *Figure 27*. $u = a\tan\theta$, $du = a\sec^2\theta\, d\theta \Rightarrow$

$\displaystyle\int \sqrt{a^2 + u^2}\, du = \int (a\sec\theta)(a\sec^2\theta)\, d\theta = a^2\int \sec^3\theta\, d\theta$

$= \tfrac{1}{2}a^2(\sec\theta \tan\theta + \ln|\sec\theta + \tan\theta|) + C \; \{\text{see Example 6, §9.1}\}$

$= \tfrac{1}{2}a^2\left(\dfrac{\sqrt{a^2 + u^2}}{a} \cdot \dfrac{u}{a} + \ln\left| \dfrac{\sqrt{a^2 + u^2}}{a} + \dfrac{u}{a} \right| \right) + C_1$

$= \dfrac{u^2}{2}\sqrt{a^2 + u^2} + \tfrac{1}{2}a^2\left[\ln\left| \sqrt{a^2 + u^2} + u \right| - \ln a \right] + C_1 \; \{\ln(p/q) = \ln p - \ln q\}$

$= \dfrac{u}{2}\sqrt{a^2 + u^2} + \dfrac{a^2}{2}\ln\left| u + \sqrt{a^2 + u^2} \right| + C$, where $C = C_1 - \tfrac{1}{2}a^2\ln a$

[29] See *Figure 29.* $u = a\sin\theta,\ du = a\cos\theta\,d\theta \Rightarrow$

$$\int u^2 \sqrt{a^2 - u^2}\,du = \int (a\sin\theta)^2 (a\cos\theta)(a\cos\theta)\,d\theta = a^4 \int \sin^2\theta\,\cos^2\theta\,d\theta$$

$$= a^4(\tfrac{1}{8}\theta - \tfrac{1}{32}\sin 4\theta) + C \text{ \{see Exercise 3, §9.2\}}$$

$$= a^4 \Big[\tfrac{1}{8}\theta - \tfrac{1}{32}(2\sin 2\theta\,\cos 2\theta)\Big] + C$$

$$= a^4 \Big[\tfrac{1}{8}\theta - \tfrac{1}{16}(2\sin\theta\,\cos\theta)(1 - 2\sin^2\theta)\Big] + C$$

$$= \tfrac{a^4}{8}\sin^{-1}\tfrac{u}{a} - \tfrac{a^4}{8}\Big(\tfrac{u}{a}\Big)\Big(\tfrac{\sqrt{a^2 - u^2}}{a}\Big)\Big(1 - \tfrac{2u^2}{a^2}\Big) + C$$

$$= \tfrac{a^4}{8}\sin^{-1}\tfrac{u}{a} - \tfrac{a^2}{8}(u)\sqrt{a^2 - u^2}\Big(\tfrac{a^2 - 2u^2}{a^2}\Big) + C$$

$$= \tfrac{a^4}{8}\sin^{-1}\tfrac{u}{a} + \tfrac{u}{8}(2u^2 - a^2)\sqrt{a^2 - u^2} + C$$

Exercises 9.4

Note: In this section, K denotes the constant of integration.

[1] By **Rule a** of (9.5) and as in Example 1, let $\dfrac{5x - 12}{x(x - 4)} = \dfrac{A}{x} + \dfrac{B}{x - 4}$. Then, if we multiply this equation by $x(x - 4)$, we obtain $5x - 12 = A(x - 4) + Bx$. Since this equation must hold for every value of x, we choose values of x that will make it easy to determine A and B, namely, the values that make the denominators zero.

$x = 0$: $5(0) - 12 = A(0 - 4) + B(0) \Rightarrow -12 = -4A \Rightarrow A = 3$

$x = 4$: $5(4) - 12 = A(4 - 4) + B(4) \Rightarrow 8 = 4B \Rightarrow B = 2$

$$\text{Thus, I} = \int \Big[\tfrac{3}{x} + \tfrac{2}{x - 4}\Big]\,dx = 3\ln|x| + 2\ln|x - 4| + K.$$

[3] $\dfrac{37 - 11x}{(x + 1)(x - 2)(x - 3)} = \dfrac{A}{x + 1} + \dfrac{B}{x - 2} + \dfrac{C}{x - 3}.$

Multiply this equation by $(x + 1)(x - 2)(x - 3)$ to obtain

$$37 - 11x = A(x - 2)(x - 3) + B(x + 1)(x - 3) + C(x + 1)(x - 2).$$

$x = -1$: $37 - 11(-1) = A(-3)(-4) + B(0) + C(0) \Rightarrow 48 = 12A \Rightarrow A = 4$

$x = 2$: $37 - 11(2) = A(0) + B(3)(-1) + C(0) \Rightarrow 15 = -3B \Rightarrow B = -5$

$x = 3$: $37 - 11(3) = A(0) + B(0) + C(4)(1) \Rightarrow 4 = 4C \Rightarrow C = 1.$ Thus,

$$I = \int \Big[\tfrac{4}{x + 1} + \tfrac{-5}{x - 2} + \tfrac{1}{x - 3}\Big]\,dx = 4\ln|x + 1| - 5\ln|x - 2| + \ln|x - 3| + K.$$

5 By Rule **a** of (9.5) and as in Example 2, let $\dfrac{6x - 11}{(x - 1)^2} = \dfrac{A}{x - 1} + \dfrac{B}{(x - 1)^2}$.

Multiply this equation by $(x - 1)^2$ to obtain $6x - 11 = A(x - 1) + B$.

$x = 1$: $6(1) - 11 = A(0) + B \Rightarrow B = -5$.

Equating x terms of the equation $6x - 11 = Ax + (B - A)$ implies that $6 = A$.

$$\text{Thus, I} = \int \left[\frac{6}{x - 1} + \frac{-5}{(x - 1)^2} \right] dx = 6\ln|x - 1| + \frac{5}{x - 1} + K.$$

9 $\dfrac{5x^2 - 10x - 8}{x^3 - 4x} = \dfrac{5x^2 - 10x - 8}{x(x - 2)(x + 2)} = \dfrac{A}{x} + \dfrac{B}{x - 2} + \dfrac{C}{x + 2} \Rightarrow$

$5x^2 - 10x - 8 = A(x - 2)(x + 2) + Bx(x + 2) + Cx(x - 2).$

$x = 0 \Rightarrow A = 2$, $x = 2 \Rightarrow B = -1$, and $x = -2 \Rightarrow C = 4$.

$$\text{Thus, I} = \int \left[\frac{2}{x} + \frac{-1}{x - 2} + \frac{4}{x + 2} \right] dx = 2\ln|x| - \ln|x - 2| + 4\ln|x + 2| + K.$$

13 By Rule **a** of (9.5) and as in Example 2,

let $\dfrac{9x^4 + 17x^3 + 3x^2 - 8x + 3}{x^4(x + 3)} = \dfrac{A}{x} + \dfrac{B}{x^2} + \dfrac{C}{x^3} + \dfrac{D}{x^4} + \dfrac{E}{x + 3}$.

Multiply this equation by $x^4(x + 3)$ to obtain

$9x^4 + 17x^3 + 3x^2 - 8x + 3 =$

$$Ax^3(x + 3) + Bx^2(x + 3) + Cx(x + 3) + D(x + 3) + Ex^4.$$

$x = 0$: $3 = A(0) + B(0) + C(0) + D(3) + E(0) \Rightarrow 3 = 3D \Rightarrow D = 1$

$x = -3$: $9(-3)^4 + 17(-3)^3 + 3(-3)^2 - 8(-3) + 3 =$

$$A(0) + B(0) + C(0) + D(0) + E(-3)^4 \Rightarrow 324 = 81E \Rightarrow E = 4$$

By equating coefficients of like powers we have:

$9x^4 + 17x^3 + 3x^2 - 8x + 3 =$

$$(A + E)x^4 + (3A + B)x^3 + (3B + C)x^2 + (3C + D)x + 3D,$$

which indicates that x^4: $9 = A + E \Rightarrow A = 5$

x^3: $17 = 3A + B \Rightarrow B = 2$

x^2: $3 = 3B + C \Rightarrow C = -3$.

$$\text{Thus, I} = \int \left[\frac{5}{x} + \frac{2}{x^2} + \frac{-3}{x^3} + \frac{1}{x^4} + \frac{4}{x + 3} \right] dx =$$

$$5\ln|x| - \frac{2}{x} + \frac{3}{2x^2} - \frac{1}{3x^3} + 4\ln|x + 3| + K.$$

15 Since the degree of the numerator is greater than or equal to the degree of the denominator, we must first use long division to change the form of the integrand.

$$\frac{x^3 + 6x^2 + 3x + 16}{x^3 + 4x} = 1 + \frac{6x^2 - x + 16}{x^3 + 4x}.$$

By Rule b of (9.5) and as in Example 3, let $\dfrac{6x^2 - x + 16}{x(x^2 + 4)} = \dfrac{A}{x} + \dfrac{Bx + C}{x^2 + 4} \Rightarrow$

$6x^2 - x + 16 = A(x^2 + 4) + (Bx + C)x.$ $x = 0 \Rightarrow 16 = 4A \Rightarrow A = 4.$

Expanding gives $6x^2 - x + 16 = (A + B)x^2 + Cx + 4A.$

Equating x terms: $-1 = C$, and x^2 terms: $6 = A + B \Rightarrow B = 2.$

Thus, $I = \displaystyle\int \left[1 + \frac{4}{x} + \frac{2x - 1}{x^2 + 4} \right] dx = \int \left[1 + \frac{4}{x} + \frac{2x}{x^2 + 4} - \frac{1}{x^2 + 4} \right] dx =$

$$x + 4\ln|x| + \ln(x^2 + 4) - \frac{1}{2}\tan^{-1}\left(\frac{x}{2}\right) + K.$$

19 By Rule b of (9.5) and as in Example 3, let $\dfrac{x^2 + 3x + 1}{(x^2 + 4)(x^2 + 1)} = \dfrac{Ax + B}{x^2 + 4} + \dfrac{Cx + D}{x^2 + 1}.$

$x^2 + 3x + 1 = (Ax + B)(x^2 + 1) + (Cx + D)(x^2 + 4).$

Expanding gives $x^2 + 3x + 1 =$

$$(A + C)x^3 + (B + D)x^2 + (A + 4C)x + (B + 4D).$$

Equating coefficients $\{ x^0$ is used for constants $\}$ we have:

$x^3 \colon 0 = A + C$ and $x \colon 3 = A + 4C \Rightarrow C = 1$ and $A = -1;$

$x^2 \colon 1 = B + D$ and $x^0 \colon 1 = B + 4D \Rightarrow D = 0$ and $B = 1.$

Thus, $I = \displaystyle\int \left[\frac{-x + 1}{x^2 + 4} + \frac{x}{x^2 + 1} \right] dx = \int \left[-\frac{x}{x^2 + 4} + \frac{1}{x^2 + 4} + \frac{x}{x^2 + 1} \right] dx =$

$$-\frac{1}{2}\ln(x^2 + 4) + \frac{1}{2}\tan^{-1}\left(\frac{x}{2}\right) + \frac{1}{2}\ln(x^2 + 1) + K.$$

21 By Rule b of (9.5) and as in Example 4, let $\dfrac{2x^3 + 10x}{(x^2 + 1)^2} = \dfrac{Ax + B}{x^2 + 1} + \dfrac{Cx + D}{(x^2 + 1)^2}.$

Multiply this equation by $(x^2 + 1)^2$ to obtain

$$2x^3 + 10x = (Ax + B)(x^2 + 1) + Cx + D.$$

Expanding gives $2x^3 + 0x^2 + 10x + 0 = Ax^3 + Bx^2 + (A + C)x + (B + D).$

Equating coefficients we have:

$x^3 \colon 2 = A; \ x^2 \colon 0 = B; \ x^0 \colon 0 = B + D \Rightarrow D = 0; \ x \colon 10 = A + C \Rightarrow C = 8.$

Thus, $I = \displaystyle\int \left[\frac{2x}{x^2 + 1} + \frac{8x}{(x^2 + 1)^2} \right] dx = \ln(x^2 + 1) - \frac{4}{x^2 + 1} + K.$

25 Since the degree of the numerator is greater than the degree of the denominator, we must first use long division to change the form of the integrand.

$$\frac{x^6 - x^3 + 1}{x^4 + 9x^2} = x^2 - 9 + \frac{-x^3 + 81x^2 + 1}{x^4 + 9x^2}.$$

Using both **Rule a** and **Rule b** of (9.5), let $\dfrac{-x^3 + 81x^2 + 1}{x^2(x^2 + 9)} = \dfrac{A}{x} + \dfrac{B}{x^2} + \dfrac{Cx + D}{x^2 + 9}$.

Multiply this equation by $x^2(x^2 + 9)$ to obtain

$$-x^3 + 81x^2 + 1 = Ax(x^2 + 9) + B(x^2 + 9) + (Cx + D)x^2.$$

Expanding gives $-1x^3 + 81x^2 + 0x + 1 = (A + C)x^3 + (B + D)x^2 + 9Ax + 9B$.

Equating coefficients we have: x^0: $1 = 9B \Rightarrow B = \frac{1}{9}$; x: $0 = 9A \Rightarrow A = 0$;

$$x^2: 81 = B + D \Rightarrow D = \tfrac{728}{9}; \ x^3: -1 = A + C \Rightarrow C = -1.$$

Thus, $I = \displaystyle\int \left[x^2 - 9 + \frac{\frac{1}{9}}{x^2} + \frac{-x + \frac{728}{9}}{x^2 + 9} \right] dx$

$$= \int \left[x^2 - 9 + \frac{\frac{1}{9}}{x^2} - \frac{x}{x^2 + 9} + \frac{\frac{728}{9}}{x^2 + 9} \right] dx$$

$$= \tfrac{1}{3}x^3 - 9x - \frac{1}{9x} - \tfrac{1}{2}\ln(x^2 + 9) + \tfrac{728}{27}\tan^{-1}(x/3) + K.$$

29 As in Example 2,

let $\dfrac{4x^3 + 2x^2 - 5x - 18}{(x - 4)(x + 1)^3} = \dfrac{A}{x - 4} + \dfrac{B}{x + 1} + \dfrac{C}{(x + 1)^2} + \dfrac{D}{(x + 1)^3}$.

Multiply this equation by $(x - 4)(x + 1)^3$ to obtain

$4x^3 + 2x^2 - 5x - 18 =$

$$A(x + 1)^3 + B(x - 4)(x + 1)^2 + C(x - 4)(x + 1) + D(x - 4).$$

$x = -1 \Rightarrow -15 = -5D \Rightarrow D = 3$ and $x = 4 \Rightarrow 250 = 125A \Rightarrow A = 2$.

Equating coefficients we have: x^3: $4 = A + B = 2 + B \Rightarrow B = 2$;

x^0: $-18 = A - 4B - 4C - 4D = 2 - 4(2) - 4C - 4(3) \Rightarrow C = 0$. Thus, $I =$

$$\int \left[\frac{2}{x - 4} + \frac{2}{x + 1} + \frac{3}{(x + 1)^3} \right] dx = 2\ln|x - 4| + 2\ln|x + 1| - \frac{3}{2(x + 1)^2} + K.$$

31 $\dfrac{x^3 + 3x^2 + 3x + 63}{(x - 3)^2(x + 3)^2} = \dfrac{A}{x - 3} + \dfrac{B}{(x - 3)^2} + \dfrac{C}{x + 3} + \dfrac{D}{(x + 3)^2}$.

Multiply this equation by $(x - 3)^2(x + 3)^2$ to obtain

$x^3 + 3x^2 + 3x + 63 =$

$$A(x - 3)(x + 3)^2 + B(x + 3)^2 + C(x - 3)^2(x + 3) + D(x - 3)^2.$$

$x = 3 \Rightarrow 126 = 36B \Rightarrow B = \frac{7}{2}$ and $x = -3 \Rightarrow 54 = 36D \Rightarrow D = \frac{3}{2}$.

Equating x^3 terms: $1 = A + C$, and x^0 terms: $63 = -27A + 9B + 27C + 9D \Rightarrow$

$7 = -3A + B + 3C + D = -3A + \frac{7}{2} + 3C + \frac{3}{2} \Rightarrow 3A - 3C = -2$.

Solving $A + C = 1$ and $3A - 3C = -2$ yields $A = \frac{1}{6}$ and $C = \frac{5}{6}$.   (cont.)

Thus, $I = \int \left[\dfrac{\frac{1}{6}}{x-3} + \dfrac{\frac{7}{2}}{(x-3)^2} + \dfrac{\frac{5}{6}}{x+3} + \dfrac{\frac{3}{2}}{(x+3)^2} \right] dx =$

$$\frac{1}{6} \ln |x-3| - \frac{7}{2(x-3)} + \frac{5}{6} \ln |x+3| - \frac{3}{2(x+3)} + K.$$

[33] $\dfrac{1}{a^2 - u^2} = \dfrac{1}{(a+u)(a-u)} = \dfrac{A}{a+u} + \dfrac{B}{a-u} \Rightarrow 1 = A(a-u) + B(a+u).$

$u = -a \Rightarrow 1 = A(2a) + B(0) \Rightarrow A = \dfrac{1}{2a}$ and

$u = a \Rightarrow 1 = A(0) + B(2a) \Rightarrow B = \dfrac{1}{2a}.$ Thus, $I =$

$$\frac{1}{2a} \int \left[\frac{1}{a+u} + \frac{1}{a-u} \right] du = \frac{1}{2a} (\ln |a+u| - \ln |a-u|) + K = \frac{1}{2a} \ln \left| \frac{a+u}{a-u} \right| + K.$$

[37] $f(x) = \dfrac{x}{x^2 - 2x - 3} = \dfrac{x}{(x-3)(x+1)} = \dfrac{A}{x-3} + \dfrac{B}{x+1} \Rightarrow$

$x = A(x+1) + B(x-3).$ $x = 3 \Rightarrow A = \frac{3}{4}$ and $x = -1 \Rightarrow B = \frac{1}{4}.$

Since $f(x) \le 0$ on $[0, 2]$, $A = -\displaystyle\int_0^2 f(x)\, dx = -\int_0^2 \left[\dfrac{\frac{3}{4}}{x-3} + \dfrac{\frac{1}{4}}{x+1} \right] dx =$

$$-\left[\tfrac{3}{4} \ln |x-3| + \tfrac{1}{4} \ln |x+1| \right]_0^2 = -\left[(\tfrac{1}{4} \ln 3) - (\tfrac{3}{4} \ln 3) \right] = \tfrac{1}{2} \ln 3 \approx 0.55.$$

[41] $I = \displaystyle\int \left(\dfrac{1}{ax^2 + bx} \cdot \dfrac{1/x^2}{1/x^2} \right) dx = \int \dfrac{1/x^2}{a + b/x} dx;$ $u = a + \dfrac{b}{x},$ $-\dfrac{1}{b} du = \dfrac{1}{x^2} dx \Rightarrow$

$$I = -\frac{1}{b} \int \frac{1}{u} du = -\frac{1}{b} \ln |u| + K = -\frac{1}{b} \ln \left| a + \frac{b}{x} \right| + K.$$

[43] By (9.5), $\dfrac{f(x)}{g(x)} = \dfrac{f(x)}{(x - c_1)(x - c_2) \cdots (x - c_n)} =$

$\dfrac{A_1}{x - c_1} + \dfrac{A_2}{x - c_2} + \cdots + \dfrac{A_k}{x - c_k} + \cdots + \dfrac{A_n}{x - c_n} \Rightarrow$

$(x - c_k) \dfrac{f(x)}{g(x)} = \dfrac{A_1(x - c_k)}{x - c_1} + \dfrac{A_2(x - c_k)}{x - c_2} + \cdots + A_k + \cdots + \dfrac{A_n(x - c_k)}{x - c_n}$ for

$k = 1, 2, \ldots, n.$ As $x \to c_k$, all terms on the right-hand side approach zero except A_k.

Thus, $A_k = \displaystyle\lim_{x \to c_k} \left[\dfrac{(x - c_k) f(x)}{g(x)} \right] = \lim_{x \to c_k} \left[\dfrac{f(x)}{\dfrac{g(x) - g(c_k)}{x - c_k}} \right]$

(since $g(c_k) = 0$) $= \dfrac{f(x)}{g'(c_k)},$ for $k = 1, 2, \ldots, n.$

Exercises 9.5

Note: $\{PF\}$ indicates that the partial fractions method was used.

[1] $u = x + 1,$ $du = dx \Rightarrow$

$$\int \frac{1}{(x+1)^2 + 4} dx = \int \frac{1}{u^2 + 2^2} du = \tfrac{1}{2} \tan^{-1} \frac{u}{2} + C = \tfrac{1}{2} \tan^{-1} \frac{x+1}{2} + C.$$

$\boxed{3}$ $x^2 - 4x + 8 = \left[(x^2 - 4x + 4) - 4\right] + 8 = (x - 2)^2 + 4.$ $u = x - 2,\ du = dx \Rightarrow$

$$\int \frac{1}{x^2 - 4x + 8}\, dx = \int \frac{1}{(x - 2)^2 + 4}\, dx = \int \frac{1}{u^2 + 2^2}\, du =$$

$$\tfrac{1}{2} \tan^{-1} \tfrac{u}{2} + C = \tfrac{1}{2} \tan^{-1} \tfrac{x - 2}{2} + C.$$

$\boxed{5}$ $4x - x^2 = -(x^2 - 4x) = -(x^2 - 4x + 4) + 4 = 4 - (x^2 - 4x + 4) =$

$4 - (x - 2)^2.$ $u = x - 2,\ du = dx \Rightarrow \displaystyle\int \frac{1}{\sqrt{4x - x^2}}\, dx = \int \frac{1}{\sqrt{4 - (x - 2)^2}}\, dx =$

$$\int \frac{1}{\sqrt{2^2 - u^2}}\, du = \sin^{-1} \tfrac{u}{2} + C = \sin^{-1} \tfrac{x - 2}{2} + C.$$

$\boxed{9}$ $x^2 + 4x + 5 = (x^2 + 4x + 4) + 1 = (x + 2)^2 + 1.$

$u = x + 2,\ du = dx \Rightarrow$

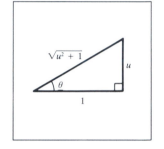

$$\int \frac{1}{(x^2 + 4x + 5)^2}\, dx = \int \frac{1}{(u^2 + 1)^2}\, du. \text{ Let } u = \tan \theta,$$

$u^2 + 1 = \tan^2 \theta + 1 = \sec^2 \theta,$ and $du = \sec^2 \theta\, d\theta.$

Thus, $I = \displaystyle\int \frac{\sec^2 \theta}{(\sec^2 \theta)^2}\, d\theta = \int \frac{1}{\sec^2 \theta}\, d\theta = \int \cos^2 \theta\, d\theta =$

$\tfrac{1}{2}\displaystyle\int (1 + \cos 2\theta)\, d\theta = \tfrac{1}{2}\theta + \tfrac{1}{4}\sin 2\theta + C = $

Figure 9

$\tfrac{1}{2}\theta + \tfrac{1}{4}(2 \sin \theta \cos \theta) + C = \tfrac{1}{2}\theta + \tfrac{1}{2}\sin \theta \cos \theta + C$

$$= \tfrac{1}{2}\left(\tan^{-1} u + \frac{u}{\sqrt{u^2 + 1}} \cdot \frac{1}{\sqrt{u^2 + 1}}\right) + C$$

$$= \tfrac{1}{2}\left(\tan^{-1} u + \frac{u}{u^2 + 1}\right) + C$$

$$= \tfrac{1}{2}\left[\tan^{-1}(x + 2) + \frac{x + 2}{(x + 2)^2 + 1}\right] + C$$

$$= \tfrac{1}{2}\left[\tan^{-1}(x + 2) + \frac{x + 2}{x^2 + 4x + 5}\right] + C. \text{ See } \textit{Figure 9.}$$

$\boxed{13}$ *Hint*: Factor out the 2 in the denominator.

$\boxed{15}$ $u = e^x,\ du = e^x\, dx \Rightarrow \displaystyle\int \frac{e^x}{e^{2x} + 3e^x + 2}\, dx = \int \frac{e^x}{(e^x)^2 + 3e^x + 2}\, dx =$

$$\int \frac{e^x}{(e^x + 2)(e^x + 1)}\, dx = \int \frac{1}{(u + 2)(u + 1)}\, du = \int \left[\frac{1}{u + 1} - \frac{1}{u + 2}\right] du \ \{\text{PF}\} =$$

$$\ln \left|\frac{u + 1}{u + 2}\right| + C = \ln \left(\frac{e^x + 1}{e^x + 2}\right) + C.$$

$\boxed{17}$ $\dfrac{x^2 - 4x + 6}{x^2 - 4x + 5} = \dfrac{(x^2 - 4x + 5) + 1}{x^2 - 4x + 5} = 1 + \dfrac{1}{x^2 - 4x + 5}.$

$x^2 - 4x + 5 = (x^2 - 4x + 4) + 1 = (x - 2)^2 + 1.$

$u = x - 2,\ du = dx;\ x = 2, 3 \Rightarrow u = 0, 1.\ \displaystyle\int_2^3 \frac{x^2 - 4x + 6}{x^2 - 4x + 5}\,dx =$

$\displaystyle\int_2^3 \left[1 + \frac{1}{(x-2)^2 + 1}\right] dx = \int_0^1 \left[1 + \frac{1}{u^2 + 1}\right] du = \Big[u + \tan^{-1} u\Big]_0^1 = 1 + \frac{\pi}{4} \approx 1.79.$

$\boxed{19}$ $x^2 + 4x + 29 = (x^2 + 4x + 4) + 25 = (x + 2)^2 + 25.$ $u = x + 2,\ du = dx;$

$x = -2, 3 \Rightarrow u = 0, 5.$ Since $y > 0$ on $[-2, 3]$, $A = \displaystyle\int_{-2}^3 \frac{1}{x^2 + 4x + 29}\,dx =$

$\displaystyle\int_{-2}^3 \frac{1}{(x+2)^2 + 25}\,dx = \int_0^5 \frac{1}{u^2 + 5^2}\,du = \Big[\tfrac{1}{5}\tan^{-1} \tfrac{u}{5}\Big]_0^5 = \tfrac{1}{5}\big(\tfrac{\pi}{4} - 0\big) = \tfrac{\pi}{20} \approx 0.16.$

Exercises 9.6

Note: $\{\text{IP}\}$ indicates that integration by parts was used.

$\boxed{1}$ Let $u = (x + 9)^{1/3}.$ $u = (x + 9)^{1/3} \Rightarrow u^3 = x + 9 \Rightarrow x = u^3 - 9.$

Now $dx = 3u^2\,du$ and $\displaystyle\int x\,\sqrt[3]{x + 9}\,dx = \int (u^3 - 9)(u)(3u^2)\,du = \int (3u^6 - 27u^3)\,du =$

$\tfrac{3}{7}u^7 - \tfrac{27}{4}u^4 + C = \tfrac{3}{7}(x + 9)^{7/3} - \tfrac{27}{4}(x + 9)^{4/3} + C.$

$\boxed{5}$ $u = \sqrt{x} + 4,\ x = (u - 4)^2,$ and $dx = 2(u - 4)\,du;\ x = 4, 9 \Rightarrow u = 6, 7 \Rightarrow$

$\displaystyle\int_4^9 \frac{1}{\sqrt{x} + 4}\,dx = 2\int_6^7 \frac{u - 4}{u}\,du = 2\int_6^7 \left(1 - \frac{4}{u}\right) du = 2\Big[u - 4\ln|u|\Big]_6^7 =$

$2\Big[(7 - 4\ln 7) - (6 - 4\ln 6)\Big] = 2(1 + 4\ln\tfrac{6}{7}) \approx 0.767.$

$\boxed{7}$ As in Example 2, $u = x^{1/6},\ x = u^6,\ \sqrt{x} = u^3,\ \sqrt[3]{x} = u^2,$ and $dx = 6u^5\,du \Rightarrow$

$\displaystyle\int \frac{\sqrt{x}}{1 + \sqrt[3]{x}}\,dx = \int \frac{(u^3)(6u^5)}{1 + u^2}\,du = 6\int \frac{u^8}{1 + u^2}\,du$

$= 6\int \left(u^6 - u^4 + u^2 - 1 + \frac{1}{1 + u^2}\right) du \ \{\text{long division}\}$

$= \tfrac{6}{7}u^7 - \tfrac{6}{5}u^5 + 2u^3 - 6u + 6\tan^{-1} u + C$

$= \tfrac{6}{7}x^{7/6} - \tfrac{6}{5}x^{5/6} + 2x^{1/2} - 6x^{1/6} + 6\tan^{-1}\big(x^{1/6}\big) + C.$

$\boxed{9}$ $u = \sqrt{x - 2},\ x = u^2 + 2,$ and $dx = 2u\,du \Rightarrow$

$\displaystyle\int \frac{1}{(x + 1)\sqrt{x - 2}}\,dx = \int \frac{2u}{(u^2 + 3)(u)}\,du = 2\int \frac{1}{u^2 + 3}\,du = 2\int \frac{1}{u^2 + (\sqrt{3})^2}\,du =$

$\tfrac{2}{\sqrt{3}}\tan^{-1}\frac{u}{\sqrt{3}} + C = \tfrac{2}{\sqrt{3}}\tan^{-1}\sqrt{\frac{x - 2}{3}} + C.$

$\boxed{13}$ $u = 1 + e^x$, $e^x = u - 1$, and $e^x\, dx = du \Rightarrow$

$$\int e^{3x}\sqrt{1 + e^x}\, dx = \int e^{2x}\sqrt{1 + e^x}\,(e^x)\, dx$$

$$= \int (u - 1)^2 (u^{1/2})\, du = \int (u^2 - 2u + 1)(u^{1/2})\, du$$

$$= \int (u^{5/2} - 2u^{3/2} + u^{1/2})\, du = \tfrac{2}{7}u^{7/2} - \tfrac{4}{5}u^{5/2} + \tfrac{2}{3}u^{3/2} + C$$

$$= \tfrac{2}{7}(1 + e^x)^{7/2} - \tfrac{4}{5}(1 + e^x)^{5/2} + \tfrac{2}{3}(1 + e^x)^{3/2} + C.$$

$\boxed{17}$ $u = \sqrt{x + 4}$, $x = u^2 - 4$, and $dx = 2u\, du \Rightarrow \displaystyle\int \sin\sqrt{x + 4}\, dx = \int \sin u\,(2u)\, du =$

$$2\int u\sin u\, du = 2\sin u - 2u\cos u + C\,\{\text{IP with } u = u \text{ and } dv = \sin u\, du\} =$$

$$2\sin\sqrt{x + 4} - 2\sqrt{x + 4}\cos\sqrt{x + 4} + C.$$

$\boxed{21}$ $u = \cos x$, $-du = \sin x\, dx \Rightarrow \displaystyle\int \frac{\sin x}{\cos x(\cos x - 1)}\, dx = -\int \frac{1}{u(u - 1)}\, du =$

$$-\int\left[\frac{-1}{u} + \frac{1}{u - 1}\right] du\,\{\text{PF}\} = \int\left(\frac{1}{u} + \frac{1}{1 - u}\right) du = \ln|u| - \ln|1 - u| + C =$$

$$\ln|\cos x| - \ln(1 - \cos x) + C \text{ since } (1 - \cos x) \text{ is always nonnegative.}$$

$\boxed{23}$ $u = e^x$, $du = e^x\, dx \Rightarrow$

$$\int \frac{e^x}{e^{2x} - 1}\, dx = \int \frac{1}{u^2 - 1}\, du = \int \frac{1}{(u - 1)(u + 1)}\, du = \int\left[\frac{\frac{1}{2}}{u - 1} + \frac{-\frac{1}{2}}{u + 1}\right] du\,\{\text{PF}\}$$

$$= \tfrac{1}{2}\ln|u - 1| - \tfrac{1}{2}\ln|u + 1| + C = \tfrac{1}{2}\ln|e^x - 1| - \tfrac{1}{2}\ln(e^x + 1) + C.$$

$\boxed{25}$ $u = \sin x$, $du = \cos x \Rightarrow \displaystyle\int \frac{\sin 2x}{\sin^2 x - 2\sin x - 8}\, dx = \int \frac{2\sin x\cos x}{\sin^2 x - 2\sin x - 8}\, dx =$

$$\int \frac{2u}{u^2 - 2u - 8}\, du = 2\int \frac{u}{(u - 4)(u + 2)}\, du = 2\int\left[\frac{\frac{2}{3}}{u - 4} + \frac{\frac{1}{3}}{u + 2}\right] du\,\{\text{PF}\}$$

$$= \tfrac{4}{3}\ln|u - 4| + \tfrac{2}{3}\ln|u + 2| + C$$

$$= \tfrac{4}{3}\ln|\sin x - 4| + \tfrac{2}{3}\ln|\sin x + 2| + C$$

$$= \tfrac{4}{3}\ln|4 - \sin x| + \tfrac{2}{3}\ln|\sin x + 2| + C\,\{|a - b| = |b - a|\}$$

$$= \tfrac{4}{3}\ln(4 - \sin x) + \tfrac{2}{3}\ln(\sin x + 2) + C,$$

$$\text{since } (4 - \sin x) \text{ and } (\sin x + 2) \text{ are always positive.}$$

27 Refer to (9.6) for the substitutions for $\sin x$, $\cos x$, dx, and u.

$$\int \frac{1}{2 + \sin x}\, dx = \int \frac{1}{2 + \frac{2u}{1 + u^2}} \cdot \frac{2}{1 + u^2}\, du = \int \frac{2}{2(1 + u^2) + 2u}\, du =$$

$$\int \frac{1}{u^2 + u + 1}\, du.$$ We recognize the denominator as being quadratic in u and

complete the square. Thus, $I = \int \frac{1}{(u + \frac{1}{2})^2 + \frac{3}{4}}\, du.$ Let $v = u + \frac{1}{2}$, $dv = du$.

Then, $I = \int \frac{1}{v^2 + (\sqrt{3}/2)^2}\, dv = \frac{2}{\sqrt{3}} \tan^{-1} \frac{2v}{\sqrt{3}} + C = \frac{2}{\sqrt{3}} \tan^{-1} \frac{2(u + \frac{1}{2})}{\sqrt{3}} + C =$

$$\frac{2}{\sqrt{3}} \tan^{-1} \frac{2 \tan (x/2) + 1}{\sqrt{3}} + C.$$

31 $\int \frac{\sec x}{4 - 3 \tan x}\, dx = \int \frac{1}{\cos x \left(4 - 3 \frac{\sin x}{\cos x}\right)}\, dx = \int \frac{1}{4 \cos x - 3 \sin x}\, dx =$

$$\int \frac{1}{4 \cdot \frac{1 - u^2}{1 + u^2} - 3 \cdot \frac{2u}{1 + u^2}} \cdot \frac{2}{1 + u^2}\, du = \int \frac{2}{4(1 - u^2) - 6u}\, du =$$

$$\int \frac{2}{-2(2u^2 + 3u - 2)}\, du = -\int \frac{1}{2u^2 + 3u - 2}\, du = -\int \frac{1}{(2u - 1)(u + 2)}\, du =$$

$$\int \left[\frac{-\frac{2}{5}}{2u - 1} + \frac{\frac{1}{5}}{u + 2}\right] du \ \{PF\} = -\tfrac{1}{5} \ln|2u - 1| + \tfrac{1}{5} \ln|u + 2| + C =$$

$$-\tfrac{1}{5} \ln|2 \tan (x/2) - 1| + \tfrac{1}{5} \ln|\tan (x/2) + 2| + C.$$

33 $\int \sec x\, dx = \int \frac{1}{\cos x}\, dx = \int \frac{1 + u^2}{1 - u^2} \cdot \frac{2}{1 + u^2}\, du = \int \frac{2}{1 - u^2}\, du =$

$$\int \frac{2}{(1 + u)(1 - u)}\, du = \int \left[\frac{1}{1 + u} + \frac{1}{1 - u}\right] du \ \{PF\} =$$

$$\ln|1 + u| - \ln|1 - u| + C = \ln \left|\frac{1 + u}{1 - u}\right| + C = \ln \left|\frac{1 + \tan\frac{1}{2}x}{1 - \tan\frac{1}{2}x}\right| + C.$$

Exercises 9.7

1 The integrand contains the form $\sqrt{a^2 + u^2}$. Let $u = 3x$, $x = \frac{1}{3}u$, and $dx = \frac{1}{3}\, du$.

Thus, $\int \frac{\sqrt{4 + 9x^2}}{x}\, dx = \int \frac{\sqrt{4 + u^2}}{u}\, du.$ Using Formula 23 with $a = 2$,

$$I = \sqrt{4 + u^2} - 2 \ln \left|\frac{2 + \sqrt{4 + u^2}}{u}\right| + C = \sqrt{4 + 9x^2} - 2 \ln \left|\frac{2 + \sqrt{4 + 9x^2}}{3x}\right| + C.$$

5 The integrand contains the form $a + bu$. Using Formula 54 with $a = 2$, $b = -3$,

and $u = x$, $\displaystyle\int x\sqrt{2 - 3x}\,dx = \frac{2}{15(-3)^2}\big[3(-3)x - 2(2)\big]\big[2 + (-3)x\big]^{3/2} + C =$

$$\tfrac{2}{135}(-9x - 4)(2 - 3x)^{3/2} + C = -\tfrac{2}{135}(9x + 4)(2 - 3x)^{3/2} + C.$$

7 The integrand is of the form $\sin^n u$. Let $u = 3x$, $\tfrac{1}{3}\,du = dx \Rightarrow \displaystyle\int \sin^6 3x\,dx =$

$\tfrac{1}{3}\displaystyle\int \sin^6 u\,du$. Using Formula 73 three times and starting with $n = 6$ gives

$$I = \tfrac{1}{3}\left[-\tfrac{1}{6}\sin^5 u \cos u + \tfrac{5}{6}\int \sin^4 u\,du\right]$$

$$= -\tfrac{1}{18}\sin^5 u \cos u + \tfrac{5}{18}\left[-\tfrac{1}{4}\sin^3 u \cos u + \tfrac{3}{4}\int \sin^2 u\,du\right]$$

$$= -\tfrac{1}{18}\sin^5 u \cos u - \tfrac{5}{72}\sin^3 u \cos u + \tfrac{5}{24}\left[-\tfrac{1}{2}\sin u \cos u + \tfrac{1}{2}u\right] + C$$

$$= -\tfrac{1}{18}\sin^5 u \cos u - \tfrac{5}{72}\sin^3 u \cos u - \tfrac{5}{48}\sin u \cos u + \tfrac{5}{48}u + C$$

$$= -\tfrac{1}{18}\sin^5 3x \cos 3x - \tfrac{5}{72}\sin^3 3x \cos 3x - \tfrac{5}{48}\sin 3x \cos 3x + \tfrac{5}{16}x + C.$$

11 The integrand contains an inverse trigonometric function.

Using Formula 90 with $u = x$, $\displaystyle\int x \sin^{-1} x\,dx = \frac{2x^2 - 1}{4}\sin^{-1} x + \frac{x\sqrt{1 - x^2}}{4} + C.$

15 The integrand contains the form $\sqrt{2au - u^2}$. To begin, since u^2 must equal $9x^2$,

let $u = 3x$, $x = \tfrac{1}{3}u$, and $dx = \tfrac{1}{3}\,du$. Then, $\displaystyle\int \frac{\sqrt{5x - 9x^2}}{x}\,dx = \int \frac{\sqrt{5(\tfrac{1}{3}u) - u^2}}{\tfrac{1}{3}u}(\tfrac{1}{3}\,du) =$

$\displaystyle\int \frac{\sqrt{\tfrac{5}{3}u - u^2}}{u}\,du = \int \frac{\sqrt{2(\tfrac{5}{6})u - u^2}}{u}\,du.$ Using Formula 115 with $a = \tfrac{5}{6}$,

$$I = \sqrt{\tfrac{5}{3}u - u^2} + \tfrac{5}{6}\cos^{-1}\!\left(\frac{\tfrac{5}{6} - u}{\tfrac{5}{6}}\right) + C = \sqrt{5x - 9x^2} + \tfrac{5}{6}\cos^{-1}\frac{5 - 18x}{5} + C.$$

17 The integrand contains the form $\dfrac{1}{u^2 - a^2}$. To begin, since $5x^4$ must equal u^2,

let $u = \sqrt{5}x^2$, $\dfrac{1}{2\sqrt{5}}\,du = x\,dx$. Then, $\displaystyle\int \frac{x}{5x^4 - 3}\,dx = \frac{1}{2\sqrt{5}}\int \frac{1}{u^2 - 3}\,du.$

Negating both sides of Formula 19 gives us

$$-\int \frac{1}{a^2 - u^2}\,du = -\frac{1}{2a}\ln\left|\frac{u + a}{u - a}\right| + C \Rightarrow \int \frac{1}{u^2 - a^2}\,du = \frac{1}{2a}\ln\left|\frac{u - a}{u + a}\right| + C.$$

Using this formula with $a = \sqrt{3}$,

$$I = \frac{1}{2\sqrt{5}}\left[\frac{1}{2\sqrt{3}}\ln\left|\frac{u - \sqrt{3}}{u + \sqrt{3}}\right|\right] + C = \frac{1}{4\sqrt{15}}\ln\left|\frac{\sqrt{5}x^2 - \sqrt{3}}{\sqrt{5}x^2 + \sqrt{3}}\right| + C.$$

19 The integrand contains an inverse trigonometric function. Let $u = e^x$ and

$$du = e^x \, dx. \text{ Then, } \int e^{2x} \cos^{-1} e^x \, dx = \int e^x \cos^{-1} e^x \, (e^x) \, dx = \int u \cos^{-1} u \, du.$$

Using Formula 91, $I = \frac{1}{4}(2u^2 - 1)\cos^{-1} u - \frac{1}{4}(u\sqrt{1 - u^2}) + C =$

$$\frac{1}{4}(2e^{2x} - 1)\cos^{-1} e^x - \frac{1}{4}e^x \sqrt{1 - e^{2x}} + C.$$

21 The integrand contains the form $a + bu$, where $a = 2$, $b = 1$, and $u = x$.

Using Formula 60 three times and starting with $n = 3$ gives $\int x^3 \sqrt{2 + x} \, dx$

$$= \frac{2}{(1)\big[2(3) + 3\big]}\left[x^3(2 + 1 \cdot x)^{3/2} - 3(2)\int x^2 \sqrt{2 + 1 \cdot x} \, dx\right]$$

$$= \frac{2}{9}\left[x^3(2 + x)^{3/2} - 6\int x^2 \sqrt{2 + x} \, dx\right]$$

$$= \frac{2}{9}x^3(2 + x)^{3/2} - \frac{4}{3} \cdot \left\{\frac{2}{7}\left[x^2(2 + x)^{3/2} - 4\int x\sqrt{2 + x} \, dx\right]\right\} \{n = 2\}$$

$$= \frac{2}{9}x^3(2 + x)^{3/2} - \frac{8}{21}x^2(2 + x)^{3/2} + \frac{32}{21} \cdot \left\{\frac{2}{5}\left[x(2 + x)^{3/2} - 2\int \sqrt{2 + x} \, dx\right]\right\}$$
$$\{n = 1\}$$

$$= \frac{2}{9}x^3(2 + x)^{3/2} - \frac{8}{21}x^2(2 + x)^{3/2} + \frac{64}{105}x(2 + x)^{3/2} - \frac{256}{315}(2 + x)^{3/2} + C$$

$$= (\frac{2}{9}x^3 - \frac{8}{21}x^2 + \frac{64}{105}x - \frac{256}{315})(2 + x)^{3/2} + C$$

$$= \frac{2}{315}(35x^3 - 60x^2 + 96x - 128)(2 + x)^{3/2} + C \, \{\text{factor out } \frac{2}{315}\}.$$

27 The integrand contains the form $a + bu$. To see this, let $u = x^{1/3}$, $x = u^3$, and

$$dx = 3u^2 \, du. \text{ Then, } \int \frac{1}{x(4 + \sqrt[3]{x})} \, dx = 3\int \frac{1}{u(4 + u)} \, du. \text{ Using Formula 49 with } a =$$

$$4 \text{ and } b = 1, I = 3\left[\frac{1}{4}\ln\left|\frac{u}{4 + 1 \cdot u}\right|\right] + C = \frac{3}{4}\ln\left|\frac{u}{4 + u}\right| + C = \frac{3}{4}\ln\left|\frac{\sqrt[3]{x}}{4 + \sqrt[3]{x}}\right| + C.$$

29 The integrand contains the form $\sqrt{a^2 - u^2}$. Then $u = \sec x$,

$$du = \sec x \tan x \, dx \left\{\frac{du}{u} = \tan x \, dx\right\} \Rightarrow \int \sqrt{16 - \sec^2 x} \, \tan x \, dx = \int \frac{\sqrt{16 - u^2}}{u} \, du.$$

Using Formula 32 with $a = 4$, $I = \sqrt{16 - u^2} - 4\ln\left|\frac{4 + \sqrt{16 - u^2}}{u}\right| + C =$

$$\sqrt{16 - \sec^2 x} - 4\ln\left|\frac{4 + \sqrt{16 - \sec^2 x}}{\sec x}\right| + C.$$

| 9.8 Review Exercises |

1 $\int x \sin^{-1} x \, dx = uv - \int v \, du \overset{A}{=} \frac{1}{2} x^2 \sin^{-1} x - \frac{1}{2} \int \dfrac{x^2}{\sqrt{1 - x^2}} \, dx.$ In the last integral, I_1,

use the trigonometric substitution $x = \sin\theta \ \{-\frac{\pi}{2} \le \theta \le \frac{\pi}{2}\}$, $dx = \cos\theta \, d\theta$, and

$$\sqrt{1 - x^2} = \sqrt{1 - \sin^2\theta} = \sqrt{\cos^2\theta} = \cos\theta. \quad I_1 = \int \dfrac{\sin^2\theta}{\cos\theta} \cos\theta \, d\theta = \int \sin^2\theta \, d\theta =$$

$$\frac{1}{2}\int (1 - \cos 2\theta) \, d\theta = \frac{1}{2}(\theta - \frac{1}{2}\sin 2\theta) + C = \frac{1}{2}\theta - \frac{1}{4}(2\sin\theta \, \cos\theta) + C =$$

$$\frac{1}{2}(\theta - \sin\theta \, \cos\theta) + C = \frac{1}{2}(\sin^{-1} x - x\sqrt{1 - x^2}) + C.$$

$$\text{Thus, } I = \frac{1}{2}x^2 \sin^{-1} x - \frac{1}{2}I_1 = \frac{1}{2}x^2 \sin^{-1} x - \frac{1}{4}\sin^{-1} x + \frac{1}{4}x\sqrt{1 - x^2} + C.$$

A. $u = \sin^{-1} x, \ du = \dfrac{1}{\sqrt{1 - x^2}} \, dx, \ dv = x \, dx, \ v = \frac{1}{2}x^2$

Note: We could use the table of integrals for many of the exercises in this section, but we will concentrate on the techniques in §9.1–§9.6.

3 $\displaystyle\int_0^1 \ln(1 + x) \, dx = \Big[uv\Big]_0^1 - \int_0^1 v \, du \overset{A}{=} \Big[x \ln(1 + x)\Big]_0^1 - \int_0^1 \dfrac{x}{1 + x} \, dx$

$$= \Big[x \ln(1 + x)\Big]_0^1 - \int_0^1 \Big[1 - \dfrac{1}{1 + x}\Big] \, dx \ \{\text{long division}\}$$

$$= \ln 2 - \Big[x - \ln(1 + x)\Big]_0^1$$

$$= \ln 2 - (1 - \ln 2) = 2\ln 2 - 1 \approx 0.39.$$

A. $u = \ln(1 + x), \ du = \dfrac{1}{1 + x} \, dx, \ dv = dx, \ v = x$

5 $u = \sin 2x, \ \frac{1}{2} du = \cos 2x \, dx \Rightarrow \displaystyle\int \cos^3 2x \sin^2 2x \, dx = \int (\cos^2 2x)\sin^2 2x \cos 2x \, dx =$

$$\int (1 - \sin^2 2x)\sin^2 2x \cos 2x \, dx = \frac{1}{2}\int (1 - u^2) u^2 \, du = \frac{1}{6}u^3 - \frac{1}{10}u^5 + C =$$

$$\frac{1}{6}\sin^3 2x - \frac{1}{10}\sin^5 2x + C.$$

7 $u = \sec x, \ du = \sec x \tan x \, dx \Rightarrow$

$$\int \tan x \sec^5 x \, dx = \int \sec^4 x (\sec x \tan x) \, dx = \int u^4 \, du = \frac{1}{5}u^5 + C = \frac{1}{5}\sec^5 x + C.$$

9   See *Figure 9.* $x = 5\tan\theta, \ dx = 5\sec^2\theta \, d\theta$, and

$$\sqrt{x^2 + 25} = \sqrt{25\tan^2\theta + 25} = 5\sec\theta \Rightarrow$$

$$\int \dfrac{1}{(x^2 + 25)^{3/2}} \, dx = \int \dfrac{5\sec^2\theta}{(5\sec\theta)^3} \, d\theta = \frac{1}{25}\int \cos\theta \, d\theta =$$

$$\frac{1}{25}\sin\theta + C = \dfrac{x}{25\sqrt{x^2 + 25}} + C.$$

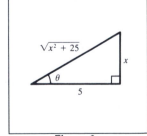

Figure 9

$\boxed{13}$ $\dfrac{x^3+1}{x(x-1)^3} = \dfrac{A}{x} + \dfrac{B}{x-1} + \dfrac{C}{(x-1)^2} + \dfrac{D}{(x-1)^3}$ $\Rightarrow$

$x^3 + 1 = A(x-1)^3 + Bx(x-1)^2 + Cx(x-1) + Dx.$

$x = 1 \Rightarrow D = 2$ and $x = 0 \Rightarrow A = -1.$

Expanding gives $x^3 + 0x^2 + 0x + 1 =$

$$(A+B)x^3 + (-3A - 2B + C)x^2 + (3A + B - C + D)x + (-A).$$

Equating x^3 terms: $1 = A + B \Rightarrow B = 2,$

and x^2 terms: $0 = -3A - 2B + C \Rightarrow C = 1.$ Thus,

$$\text{I} = \int \left[\frac{-1}{x} + \frac{2}{x-1} + \frac{1}{(x-1)^2} + \frac{2}{(x-1)^3} \right] dx$$

$$= -\ln|x| + 2\ln|x-1| - \frac{1}{x-1} - \frac{1}{(x-1)^2} + K$$

$$= 2\ln|x-1| - \ln|x| - \frac{x}{(x-1)^2} + K.$$

$\boxed{17}$ $4 + 4x - x^2 = 4 - (x^2 - 4x + 4) + 4 = 8 - (x-2)^2;$ $u = x - 2,$ $du = dx \Rightarrow$

$$\int \frac{x}{\sqrt{4 + 4x - x^2}}\, dx = \int \frac{x}{\sqrt{8 - (x-2)^2}}\, dx = \int \frac{u+2}{\sqrt{8 - u^2}}\, du =$$

$$\int \frac{u}{\sqrt{8-u^2}}\, du + \int \frac{2}{\sqrt{8-u^2}}\, du = -(8 - u^2)^{1/2} + 2\sin^{-1}\frac{u}{\sqrt{8}} + C =$$

$$-\sqrt{4 + 4x - x^2} + 2\sin^{-1}\frac{x-2}{\sqrt{8}} + C.$$

$\boxed{19}$ $u = \sqrt[3]{x+8} \Rightarrow u^3 = x + 8;$ $3u^2\, du = dx \Rightarrow$

$$\int \frac{\sqrt[3]{x+8}}{x}\, dx = \int \frac{u(3u^2)}{u^3 - 8}\, du = 3\int \left[1 + \frac{8}{u^3 - 8} \right] du.\ \ u^3 - 8 =$$

$(u-2)(u^2 + 2u + 4) = (u-2)\left[(u+1)^2 + 3\right].$ $z = u + 1,$ $dz = du \Rightarrow$

$$\text{I} = 3\int \left[1 + \frac{8}{(z-3)(z^2+3)} \right] dz = 3\int dz + 24\int \left[\frac{\frac{1}{12}}{z-3} - \frac{\frac{1}{12}z + \frac{1}{4}}{z^2 + 3} \right] dz\ \{\text{PF}\}$$

$$= 3\int dz + 2\int \frac{1}{z-3}\, dz - 2\int \frac{z}{z^2+3}\, dz - 6\int \frac{1}{z^2 + (\sqrt{3})^2}\, dz$$

$$= 3z + 2\ln|z-3| - \ln|z^2 + 3| - \frac{6}{\sqrt{3}}\tan^{-1}\frac{z}{\sqrt{3}} + C$$

$$= 3(u+1) + 2\ln|u-2| - \ln|u^2 + 2u + 4| - \frac{6}{\sqrt{3}}\tan^{-1}\frac{u+1}{\sqrt{3}} + C$$

$$= 3(x+8)^{1/3} + \ln\left[(x+8)^{1/3} - 2\right]^2 - \ln\left|(x+8)^{2/3} + 2(x+8)^{1/3} + 4\right| -$$

$$\frac{6}{\sqrt{3}}\tan^{-1}\frac{(x+8)^{1/3} + 1}{\sqrt{3}} + C.$$

$\boxed{21}$ $\int e^{2x} \sin 3x \, dx = uv - \int v \, du \overset{A}{=} \frac{1}{2} e^{2x} \sin 3x - \frac{3}{2} \int e^{2x} \cos 3x \, dx$

$$\overset{B}{=} \frac{1}{2} e^{2x} \sin 3x - \frac{3}{2}\left[\frac{1}{2} e^{2x} \cos 3x + \frac{3}{2} \int e^{2x} \sin 3x \, dx \right]$$

$$= \frac{1}{2} e^{2x} \sin 3x - \frac{3}{2}\left[\frac{1}{2} e^{2x} \cos 3x + \frac{3}{2} I \right] \Rightarrow$$

$$I = \frac{1}{2} e^{2x} \sin 3x - \frac{3}{4} e^{2x} \cos 3x - \frac{9}{4} I$$

$$\frac{13}{4} I = \frac{1}{2} e^{2x} \sin 3x - \frac{3}{4} e^{2x} \cos 3x$$

$$I = \frac{1}{13} e^{2x} (2 \sin 3x - 3 \cos 3x) + C.$$

A. $u = \sin 3x$, $du = 3 \cos 3x \, dx$, $dv = e^{2x} \, dx$, $v = \frac{1}{2} e^{2x}$

B. $u = \cos 3x$, $du = -3 \sin 3x \, dx$, $dv = e^{2x} \, dx$, $v = \frac{1}{2} e^{2x}$

$\boxed{25}$ $u = 4 - x^2$, $-\frac{1}{2} du = x \, dx \Rightarrow$

$$\int \frac{x}{\sqrt{4 - x^2}} \, dx = -\frac{1}{2} \int u^{-1/2} \, du = -\sqrt{u} + C = -\sqrt{4 - x^2} + C.$$

$\boxed{31}$ $u = e^x$, $du = e^x \, dx \Rightarrow$

$$\int e^x \sec e^x \, dx = \int \sec u \, du = \ln|\sec u + \tan u| + C = \ln|\sec e^x + \tan e^x| + C.$$

$\boxed{33}$ $\int x^2 \sin 5x \, dx = uv - \int v \, du \overset{A}{=} -\frac{1}{5} x^2 \cos 5x + \frac{2}{5} \int x \cos 5x \, dx$

$$\overset{B}{=} -\frac{1}{5} x^2 \cos 5x + \frac{2}{5}\left[\frac{1}{5} x \sin 5x - \frac{1}{5} \int \sin 5x \, dx \right]$$

$$= -\frac{1}{5} x^2 \cos 5x + \frac{2}{25} x \sin 5x - \frac{2}{25}\left(-\frac{1}{5} \cos 5x \right) + C$$

$$= -\frac{1}{5} x^2 \cos 5x + \frac{2}{25} x \sin 5x + \frac{2}{125} \cos 5x + C$$

$$= \frac{1}{125}\left[10x \sin 5x - (25x^2 - 2) \cos 5x \right] + C.$$

A. $u = x^2$, $du = 2x \, dx$, $dv = \sin 5x \, dx$, $v = -\frac{1}{5} \cos 5x$

B. $u = x$, $du = dx$, $dv = \cos 5x \, dx$, $v = \frac{1}{5} \sin 5x$

$\boxed{39}$ See *Figure 39*.

$2x = 5 \tan \theta$ or $x = \frac{5}{2} \tan \theta$, $dx = \frac{5}{2} \sec^2 \theta \, d\theta$,

and $\sqrt{4x^2 + 25} = \sqrt{25 \tan^2 \theta + 25} = 5 \sec \theta \Rightarrow$

$$\int \frac{x^2}{\sqrt{4x^2 + 25}} \, dx = \int \frac{\left(\frac{5}{2} \tan \theta \right)^2 \left(\frac{5}{2} \sec^2 \theta \right)}{5 \sec \theta} \, d\theta =$$

$$\frac{25}{8} \int \tan^2 \theta \sec \theta \, d\theta = \frac{25}{8} \int (\sec^2 \theta - 1) \sec \theta \, d\theta$$

(cont.)

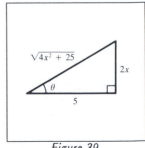

Figure 39

$$= \tfrac{25}{8}\left[\tfrac{1}{2}\sec\theta\,\tan\theta + \tfrac{1}{2}\ln|\sec\theta + \tan\theta| - \ln|\sec\theta + \tan\theta|\,\right] + C_1 \ \{\text{Exam. 6, §9.1}\}$$

$$= \tfrac{25}{8}\left[\tfrac{1}{2}\sec\theta\,\tan\theta - \tfrac{1}{2}\ln|\sec\theta + \tan\theta|\,\right] + C_1$$

$$= \tfrac{25}{16}\left[\sec\theta\,\tan\theta - \ln|\sec\theta + \tan\theta|\,\right] + C_1$$

$$= \frac{25}{16}\left[\frac{\sqrt{4x^2+25}}{5}\cdot\frac{2x}{5} - \ln\left|\frac{\sqrt{4x^2+25}}{5} + \frac{2x}{5}\right|\,\right] + C_1$$

$$= \frac{25}{16}\left[\frac{2x\sqrt{4x^2+25}}{25} - \left(\ln\left|\sqrt{4x^2+25} + 2x\right| - \ln 5\right)\right] + C_1$$

$$= \tfrac{1}{16}\left[2x\sqrt{4x^2+25} - 25\ln\left(\sqrt{4x^2+25} + 2x\right)\right] + \tfrac{25}{16}\ln 5 + C_1$$

$$= \tfrac{1}{16}\left[2x\sqrt{4x^2+25} - 25\ln\left(\sqrt{4x^2+25} + 2x\right)\right] + C, \text{ where } C = \tfrac{25}{16}\ln 5 + C_1.$$

$\boxed{43}$ $\displaystyle\int x\cot x\,\csc x\,dx = uv - \int v\,du \overset{A}{=} -x\csc x + \int \csc x\,dx$

$$= -x\csc x + \ln|\csc x - \cot x| + C.$$

A. $u = x$, $du = dx$, $dv = \cot x\,\csc x\,dx$, $v = -\csc x$

$\boxed{47}$ $z = \sqrt{x}$, $z^2 = x$, and $2z\,dz = dx \Rightarrow \displaystyle\int \sqrt{x}\,\sin\sqrt{x}\,dx = \int z\sin z\,(2z)\,dz = 2\int z^2\sin z\,dz.$

$$2\left[uv - \int v\,du\right] \overset{A}{=} 2\left[-z^2\cos z + 2\int z\cos z\,dz\right]$$

$$= -2z^2\cos z + 4\int z\cos z\,dz$$

$$\overset{B}{=} -2z^2\cos z + 4\left[z\sin z - \int \sin z\,dz\right]$$

$$= -2z^2\cos z + 4z\sin z - 4\int \sin z\,dz$$

$$= -2z^2\cos z + 4z\sin z + 4\cos z + C$$

$$= -2x\cos\sqrt{x} + 4\sqrt{x}\,\sin\sqrt{x} + 4\cos\sqrt{x} + C.$$

A. $u = z^2$, $du = 2z\,dz$, $dv = \sin z\,dz$, $v = -\cos z$

B. $u = z$, $du = dz$, $dv = \cos z\,dz$, $v = \sin z$

$\boxed{53}$ $u = \sqrt{16-x^2}$, $u^2 = 16 - x^2$, and $-u\,du = x\,dx \Rightarrow$

$$\int \frac{x^3}{\sqrt{16-x^2}}\,dx = \int \frac{x^2}{\sqrt{16-x^2}}\,x\,dx = \int \frac{16-u^2}{u}(-u)\,du = \int (u^2 - 16)\,du =$$

$$\tfrac{1}{3}u^3 - 16u + C = \tfrac{1}{3}(16-x^2)^{3/2} - 16(16-x^2)^{1/2} + C.$$

$\boxed{59}$ $u = \tan x$, $du = \sec^2 x\,dx \Rightarrow$

$$\int \frac{e^{\tan x}}{\cos^2 x}\,dx = \int e^{\tan x}\,\sec^2 x\,dx = \int e^u\,du = e^u + C = e^{\tan x} + C.$$

$\boxed{61}$ See *Figure 61*. $\sqrt{5}x = \sqrt{7}\tan\theta$ or $x = \sqrt{\frac{7}{5}}\tan\theta$,

$dx = \sqrt{\frac{7}{5}}\sec^2\theta\, d\theta$, and $\sqrt{7 + 5x^2} = \sqrt{7 + 7\tan^2\theta} =$

$\sqrt{7}\sec\theta \Rightarrow \displaystyle\int \frac{1}{\sqrt{7 + 5x^2}}\, dx = \int \frac{\sqrt{\frac{7}{5}}\sec^2\theta}{\sqrt{7}\sec\theta}\, d\theta =$

$\dfrac{1}{\sqrt{5}}\displaystyle\int \sec\theta\, d\theta = \dfrac{1}{\sqrt{5}}\ln|\sec\theta + \tan\theta| + C_1$

$= \dfrac{1}{\sqrt{5}}\ln\left|\dfrac{\sqrt{7 + 5x^2}}{\sqrt{7}} + \dfrac{\sqrt{5}x}{\sqrt{7}}\right| + C_1$

Figure 61

$= \dfrac{1}{\sqrt{5}}\left[\ln\left|\sqrt{7 + 5x^2} + \sqrt{5}x\right| - \ln\sqrt{7}\right] + C_1$

$= \dfrac{1}{\sqrt{5}}\ln\left|\sqrt{7 + 5x^2} + \sqrt{5}x\right| + \left(C_1 - \dfrac{1}{\sqrt{5}}\ln\sqrt{7}\right)$

$= \dfrac{1}{\sqrt{5}}\ln\left|\sqrt{7 + 5x^2} + \sqrt{5}x\right| + C$, where $C = C_1 - \dfrac{1}{\sqrt{5}}\ln\sqrt{7}$.

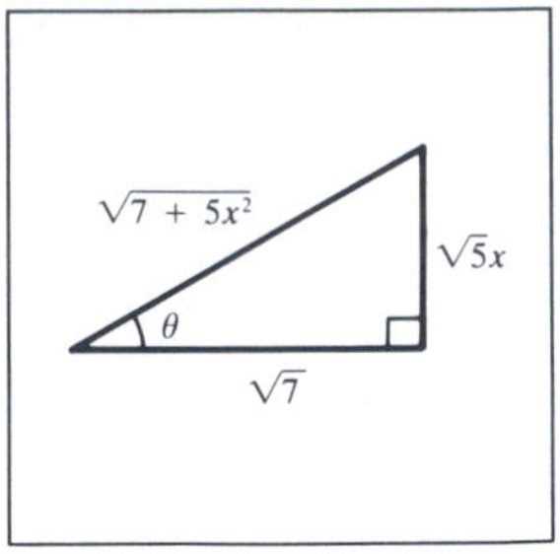

$\boxed{63}$ $\displaystyle\int \cot^6 x\, dx = \int \cot^4 x(\cot^2 x)\, dx = \int \cot^4 x(\csc^2 x - 1)\, dx$

$= \displaystyle\int \left[\cot^4 x\,\csc^2 x - \cot^2 x(\cot^2 x)\right] dx$

$= \displaystyle\int \left[\cot^4 x\,\csc^2 x - \cot^2 x(\csc^2 x - 1)\right] dx$

$= \displaystyle\int \left[\cot^4 x\,\csc^2 x - \cot^2 x\,\csc^2 x + (\cot^2 x)\right] dx$

$= \displaystyle\int \left[\cot^4 x\,\csc^2 x - \cot^2 x\,\csc^2 x + \csc^2 x - 1\right] dx$

$= -\frac{1}{5}\cot^5 x + \frac{1}{3}\cot^3 x - \cot x - x + C.$

$\boxed{67}$ $\displaystyle\int (x^2 - \mathrm{sech}^2 4x)\, dx = \int x^2\, dx - \int \mathrm{sech}^2 4x\, dx$

$= \frac{1}{3}x^3 - \frac{1}{4}\displaystyle\int \mathrm{sech}^2 u\, du\ \{\, u = 4x,\ \frac{1}{4}\, du = dx\,\}$

$= \frac{1}{3}x^3 - \frac{1}{4}\tanh u + C = \frac{1}{3}x^3 - \frac{1}{4}\tanh 4x + C.$

$\boxed{71}$ $11 - 10x - x^2 = 11 - (x^2 + 10x + 25) + 25 = 36 - (x + 5)^2$. $u = x + 5$,

$du = dx \Rightarrow \displaystyle\int \frac{3}{\sqrt{11 - 10x - x^2}}\, dx = \int \frac{3}{\sqrt{36 - (x + 5)^2}}\, dx = \int \frac{3}{\sqrt{6^2 - u^2}}\, du =$

$3\sin^{-1}\frac{u}{6} + C = 3\sin^{-1}\frac{x + 5}{6} + C.$

$\boxed{75}$ $\dfrac{4x^2 - 12x - 10}{(x-2)(x-1)(x-3)} = \dfrac{A}{x-1} + \dfrac{B}{x-2} + \dfrac{C}{x-3} \Rightarrow$

$$4x^2 - 12x - 10 = A(x-2)(x-3) + B(x-1)(x-3) + C(x-1)(x-2).$$

$$x = 1:\ -18 = 2A \Rightarrow A = -9;$$

$$x = 2:\ -18 = -B \Rightarrow B = 18;$$

$$x = 3:\ -10 = 2C \Rightarrow C = -5.$$

$$\text{I} = \int \frac{4x^2 - 12x - 10}{(x-2)(x-1)(x-3)}\, dx$$

$$= \int \left[\frac{-9}{x-1} + \frac{18}{x-2} + \frac{-5}{x-3} \right] dx$$

$$= -9\ln|x-1| + 18\ln|x-2| - 5\ln|x-3| + C.$$

$\boxed{77}$ $\displaystyle\int (x^3 + 1)\cos x\, dx = \int x^3 \cos x\, dx + \int \cos x\, dx = \text{I}_1 + \text{I}_2.$

$$\text{I}_1 = uv - \int v\, du \overset{\text{A}}{=} x^3 \sin x - 3\int x^2 \sin x\, dx$$

$$\overset{\text{B}}{=} x^3 \sin x - 3\left[-x^2 \cos x + 2\int x \cos x\, dx \right]$$

$$= x^3 \sin x + 3x^2 \cos x - 6\int x \cos x\, dx$$

$$\overset{\text{C}}{=} x^3 \sin x + 3x^2 \cos x - 6\left[x \sin x - \int \sin x\, dx \right]$$

$$= x^3 \sin x + 3x^2 \cos x - 6x \sin x - 6\cos x + C_1.$$

$$\text{I}_2 = \int \cos x\, dx = \sin x + C_2.$$

$$\text{Thus, I} = \text{I}_1 + \text{I}_2 = x^3 \sin x + 3x^2 \cos x - 6x \sin x - 6\cos x + \sin x + C.$$

A. $u = x^3,\ du = 3x^2\, dx,\ dv = \cos x\, dx,\ v = \sin x$

B. $u = x^2,\ du = 2x\, dx,\ dv = \sin x\, dx,\ v = -\cos x$

C. $u = x,\ du = dx,\ dv = \cos x\, dx,\ v = \sin x$

$\boxed{79}$ See *Figure 79*. $2x = 3\sin\theta$ or $x = \frac{3}{2}\sin\theta$, $dx =$

$\frac{3}{2}\cos\theta\, d\theta$, and $\sqrt{9 - 4x^2} = \sqrt{9 - 9\sin^2\theta} = 3\cos\theta \Rightarrow$

$$\int \frac{\sqrt{9 - 4x^2}}{x^2}\, dx = \int \frac{3\cos\theta}{(\frac{3}{2}\sin\theta)^2}\left(\tfrac{3}{2}\cos\theta\right) d\theta =$$

$$2\int \cot^2\theta\, d\theta = 2\int (\csc^2\theta - 1)\, d\theta =$$

$$-2\cot\theta - 2\theta + C = -\frac{\sqrt{9 - 4x^2}}{x} - 2\sin^{-1}\frac{2x}{3} + C.$$

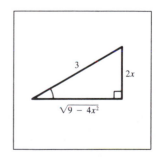

Figure 79

$\boxed{83}$ $u = \sqrt[4]{x}$, $u^2 = \sqrt{x}$, $u^4 = x$, and $4u^3\,du = dx \Rightarrow \int \dfrac{1}{x(\sqrt{x} + \sqrt[4]{x})}\,dx =$

$$\int \frac{4u^3}{u^4(u^2 + u)}\,du = 4\int \frac{1}{u^2(u + 1)}\,du = 4\int\left[-\frac{1}{u} + \frac{1}{u^2} + \frac{1}{u + 1}\right]du\ \{PF\} =$$

$$-4\ln|u| - \frac{4}{u} + 4\ln|u + 1| + C = -\ln x - \frac{4}{\sqrt[4]{x}} + 4\ln(\sqrt[4]{x} + 1) + C.$$

$\boxed{87}$ $\displaystyle\int \frac{x^2}{(25 + x^2)^2}\,dx = uv - \int v\,du \overset{A}{=} -\frac{x}{2(25 + x^2)} + \frac{1}{2}\int \frac{1}{25 + x^2}\,dx$

$$= -\frac{x}{2(25 + x^2)} + \frac{1}{10}\tan^{-1}\frac{x}{5} + C.$$

A. $u = x$, $du = dx$, $dv = \dfrac{x}{(25 + x^2)^2}\,dx$, $v = -\dfrac{1}{2(25 + x^2)}$

$\boxed{93}$ $\displaystyle\int \frac{(x^2 - 2)^2}{x}\,dx = \int \frac{x^4 - 4x^2 + 4}{x}\,dx = \int\left(\frac{x^4}{x} - \frac{4x^2}{x} + \frac{4}{x}\right)dx =$

$$\int (x^3 - 4x + 4x^{-1})\,dx = \tfrac{1}{4}x^4 - 2x^2 + 4\ln|x| + C.$$

$\boxed{95}$ $\displaystyle\int x^{3/2}\ln x\,dx = uv - \int v\,du \overset{A}{=} \tfrac{2}{5}x^{5/2}\ln x - \tfrac{2}{5}\int x^{3/2}\,dx =$

$$\tfrac{2}{5}x^{5/2}\ln x - \tfrac{2}{5}(\tfrac{2}{5}x^{5/2}) + C = \tfrac{2}{5}x^{5/2}\ln x - \tfrac{4}{25}x^{5/2} + C.$$

A. $u = \ln x$, $du = \dfrac{1}{x}\,dx$, $dv = x^{3/2}\,dx$, $v = \tfrac{2}{5}x^{5/2}$

$\boxed{97}$ $u = (2x + 3)^{1/3}$, $x = \tfrac{1}{2}(u^3 - 3)$, and $dx = \tfrac{3}{2}u^2\,du \Rightarrow \displaystyle\int \frac{x^2}{\sqrt[3]{2x + 3}}\,dx =$

$$\int \frac{\tfrac{1}{4}(u^3 - 3)^2}{u}(\tfrac{3}{2}u^2)\,du = \frac{3}{8}\int(u^6 - 6u^3 + 9)u\,du = \frac{3}{8}\int(u^7 - 6u^4 + 9u)\,du =$$

$$\tfrac{3}{64}u^8 - \tfrac{9}{20}u^5 + \tfrac{27}{16}u^2 + C = \tfrac{3}{64}(2x + 3)^{8/3} - \tfrac{9}{20}(2x + 3)^{5/3} + \tfrac{27}{16}(2x + 3)^{2/3} + C.$$

$\boxed{99}$ $z = x^2$, $\tfrac{1}{2}\,dz = x\,dx \Rightarrow$

$$\int x^3\,e^{(x^2)}\,dx = \int x^2\,e^{(x^2)}\,x\,dx = \tfrac{1}{2}\int ze^z\,dz$$

$$\overset{A}{=} \tfrac{1}{2}ze^z - \tfrac{1}{2}\int e^z\,dz + C$$

$$= \tfrac{1}{2}ze^z - \tfrac{1}{2}e^z + C$$

$$= \tfrac{1}{2}x^2\,e^{(x^2)} - \tfrac{1}{2}e^{(x^2)} + C = \tfrac{1}{2}e^{(x^2)}(x^2 - 1) + C.$$

A. $u = z$, $du = dz$, $dv = e^z\,dz$, $v = e^z$

Chapter 10: Indeterminate Forms and Improper Integrals

Note: Let L denote the indicated limit, and DNE denote *does not exist.* The notation $\{\frac{0}{0}\}$ or $\{\frac{\infty}{\infty}\}$ indicates the form of the limit, and that L'Hôpital's rule was applied to obtain the next limit, as opposed to a simplification of the limit.

1 Since $\lim\limits_{x \to 0} \sin x = 0$ and $\lim\limits_{x \to 0} 2x = 0$, $\frac{\sin x}{2x}$ has the indeterminate form $\frac{0}{0}$ at $x = 0$.

Thus, L'Hôpital's rule applies. (Always be sure to check if L'Hôpital's rule applies.)

$$\lim_{x \to 0} \frac{\sin x}{2x} \; \{\tfrac{0}{0}\} = \lim_{x \to 0} \frac{D_x(\sin x)}{D_x(2x)} = \lim_{x \to 0} \frac{\cos x}{2} = \frac{1}{2}$$

3 Since $\lim\limits_{x \to 5} (\sqrt{x-1} - 2) = 0$ and $\lim\limits_{x \to 5} (x^2 - 25) = 0$,

$\dfrac{\sqrt{x-1} - 2}{x^2 - 25}$ has the indeterminate form $\frac{0}{0}$ at $x = 5$. Thus, L'Hôpital's rule applies.

$$\lim_{x \to 5} \frac{\sqrt{x-1} - 2}{x^2 - 25} \; \{\tfrac{0}{0}\} = \lim_{x \to 5} \frac{D_x(\sqrt{x-1} - 2)}{D_x(x^2 - 25)} = \lim_{x \to 5} \frac{\frac{1}{2}(x-1)^{-1/2}}{2x} = \frac{1/4}{10} = \frac{1}{40}$$

5 $\lim\limits_{x \to 2} \dfrac{2x^2 - 5x + 2}{5x^2 - 7x - 6} \; \{\tfrac{0}{0}\} = \lim\limits_{x \to 2} \dfrac{D_x(2x^2 - 5x + 2)}{D_x(5x^2 - 7x - 6)} = \lim\limits_{x \to 2} \dfrac{4x - 5}{10x - 7} = \dfrac{3}{13}$

7 Since $\lim\limits_{x \to 1} (x^3 - 3x + 2) = 0$ and $\lim\limits_{x \to 1} (x^2 - 2x - 1) = -2$,

$\dfrac{x^3 - 3x + 2}{x^2 - 2x - 1}$ does *not* have an indeterminate form at $x = 1$.

Thus, (2.8)(iii) applies, and $\lim\limits_{x \to 1} \dfrac{x^3 - 3x + 2}{x^2 - 2x - 1} = \dfrac{1 - 3 + 2}{1 - 2 - 1} = \dfrac{0}{-2} = 0.$

11 In this exercise, we will apply L'Hôpital's rule twice since the second ratio also has an

indeterminate form. $\lim\limits_{x \to 0} \dfrac{x + 1 - e^x}{x^2} \; \{\tfrac{0}{0}\} = \lim\limits_{x \to 0} \dfrac{D_x(x + 1 - e^x)}{D_x(x^2)} =$

$$\lim_{x \to 0} \frac{1 - e^x}{2x} \; \{\tfrac{0}{0}\} = \lim_{x \to 0} \frac{D_x(1 - e^x)}{D_x(2x)} = \lim_{x \to 0} \frac{-e^x}{2} = \frac{-1}{2} = -\frac{1}{2}$$

15 Since $(\cos^2 x) \to 0^+$ and $(1 + \sin x) \to 2$ as $x \to \frac{\pi}{2}$, $\dfrac{1 + \sin x}{\cos^2 x}$ does *not* have an

indeterminate form at $x = \frac{\pi}{2}$, and L'Hôpital's rule does not apply.

$$\lim_{x \to (\pi/2)} \frac{1 + \sin x}{\cos^2 x} = \infty.$$

17 Since $\lim\limits_{x \to (\pi/2)^-} (2 + \sec x) = +\infty$ and $\lim\limits_{x \to (\pi/2)^-} (3 \tan x) = +\infty$,

$\dfrac{2 + \sec x}{3 \tan x}$ has an indeterminate form of $\frac{\infty}{\infty}$ at $x = \frac{\pi}{2}$ and L'Hôpital's rule applies.

$$\lim_{x \to (\pi/2)^-} \frac{2 + \sec x}{3 \tan x} \; \{\tfrac{\infty}{\infty}\} = \lim_{x \to (\pi/2)^-} \frac{D_x(2 + \sec x)}{D_x(3 \tan x)} = \lim_{x \to (\pi/2)^-} \frac{\sec x \tan x}{3 \sec^2 x} =$$

$$\frac{1}{3} \lim_{x \to (\pi/2)^-} \sin x = \frac{1}{3}(1) = \frac{1}{3}$$

$\boxed{21}$ Since $\lim\limits_{x \to 0^+} \ln \sin x = -\infty$ and $\lim\limits_{x \to 0^+} \ln \sin 2x = -\infty$,

$\dfrac{\ln \sin x}{\ln \sin 2x}$ has an indeterminate form of $\frac{\infty}{\infty}$ and L'Hôpital's rule applies.

Note: No distinction is made between ∞ and $-\infty$ for use in L'Hôpital's rule.

$$\lim_{x \to 0^+} \frac{\ln \sin x}{\ln \sin 2x} \left\{ \tfrac{\infty}{\infty} \right\} = \lim_{x \to 0^+} \frac{D_x (\ln \sin x)}{D_x (\ln \sin 2x)} =$$

$$\lim_{x \to 0^+} \frac{\cos x / \sin x}{2 \cos 2x / \sin 2x} = \lim_{x \to 0^+} \frac{\cos x \sin 2x}{2 \cos 2x \sin x} =$$

$$\lim_{x \to 0^+} \frac{(\cos x)(2 \sin x \cos x)}{2 \cos 2x \sin x} = \lim_{x \to 0^+} \frac{\cos^2 x}{\cos 2x} = \frac{1}{1} = 1$$

$\boxed{25}$ Since $(x \cos x + e^{-x}) \to 1$ and $(x^2) \to 0^+$ as $x \to 0$, $\dfrac{x \cos x + e^{-x}}{x^2}$

does not have an indeterminate form at $x = 0$. $\lim\limits_{x \to 0} \dfrac{x \cos x + e^{-x}}{x^2} = \infty$.

$\boxed{27}$ $\dfrac{2x^2 + 3x + 1}{5x^2 + x + 4}$ has an indeterminate form of $\frac{\infty}{\infty}$ as $x \to \infty$. In this exercise,

we must apply L'Hôpital's rule twice since the second ratio also has an indeterminate

form of $\frac{\infty}{\infty}$. $\lim\limits_{x \to \infty} \dfrac{2x^2 + 3x + 1}{5x^2 + x + 4} \left\{ \tfrac{\infty}{\infty} \right\} = \lim\limits_{x \to \infty} \dfrac{D_x (2x^2 + 3x + 1)}{D_x (5x^2 + x + 4)} =$

$\lim\limits_{x \to \infty} \dfrac{4x + 3}{10x + 1} \left\{ \tfrac{\infty}{\infty} \right\} = \lim\limits_{x \to \infty} \dfrac{D_x (4x + 3)}{D_x (10x + 1)} = \lim\limits_{x \to \infty} \dfrac{4}{10} = \dfrac{2}{5}$. Note that this is the same

type of limit that you studied in earlier chapters when finding horizontal asymptotes.

$\boxed{31}$ (*Note:* The nth derivative of x^n is $n!$. For example, if $f(x) = x^4$, then $f'(x) = 4x^3$,

$f''(x) = 12x^2$, $f'''(x) = 24x$, and $f^{(4)}(x) = 24 = 4!$.) Let n be an integer > 0.

$\dfrac{x^n}{e^x}$ has the indeterminate form $\frac{\infty}{\infty}$ as $x \to \infty$. $\lim\limits_{x \to \infty} \dfrac{x^n}{e^x} \left\{ \tfrac{\infty}{\infty} \right\} = \lim\limits_{x \to \infty} \dfrac{D_x (x^n)}{D_x (e^x)} =$

$\lim\limits_{x \to \infty} \dfrac{n x^{n-1}}{e^x} \left\{ \tfrac{\infty}{\infty} \right\} = \lim\limits_{x \to \infty} \dfrac{n(n - 1) x^{n-2}}{e^x} \left\{ \tfrac{\infty}{\infty} \right\}$. This scheme can be repeated n times,

and after n applications of L'Hôpital's rule, $L = \lim\limits_{x \to \infty} \dfrac{n!}{e^x} = 0$.

$\boxed{35}$ $\lim\limits_{x \to 0} \dfrac{\sin^{-1} 2x}{\sin^{-1} x} \left\{ \tfrac{0}{0} \right\} = \lim\limits_{x \to 0} \dfrac{D_x (\sin^{-1} 2x)}{D_x (\sin^{-1} x)} = \lim\limits_{x \to 0} \dfrac{2/\sqrt{1 - 4x^2}}{1/\sqrt{1 - x^2}} = \dfrac{2}{1} = 2$

$\boxed{41}$ $\dfrac{x^4 - x^3 - 3x^2 + 5x - 2}{x^4 - 5x^3 + 9x^2 - 7x + 2}$ has the indeterminate form of $\frac{0}{0}$ at $x = 1$. In this

exercise, L'Hôpital's rule must be applied three times. Each time the ratio has the

indeterminate form of $\frac{0}{0}$. Make sure that you verify this at each step before applying

L'Hôpital's rule. $\lim\limits_{x \to 1} \dfrac{x^4 - x^3 - 3x^2 + 5x - 2}{x^4 - 5x^3 + 9x^2 - 7x + 2} \left\{ \tfrac{0}{0} \right\} =$

$\lim\limits_{x \to 1} \dfrac{4x^3 - 3x^2 - 6x + 5}{4x^3 - 15x^2 + 18x - 7} \left\{ \tfrac{0}{0} \right\} = \lim\limits_{x \to 1} \dfrac{12x^2 - 6x - 6}{12x^2 - 30x + 18} \left\{ \tfrac{0}{0} \right\} =$

$\lim\limits_{x \to 1} \dfrac{24x - 6}{24x - 30} = \dfrac{18}{-6} = -3$

$\boxed{43}$ $\lim\limits_{x \to 0} \dfrac{x - \tan^{-1}x}{x \sin x}$ $\{\frac{0}{0}\}$ $= \lim\limits_{x \to 0} \dfrac{D_x \left(x - \tan^{-1}x\right)}{D_x \left(x \sin x\right)}$

$$= \lim\limits_{x \to 0} \dfrac{1 - \dfrac{1}{1 + x^2}}{x \cos x + \sin x} = \lim\limits_{x \to 0} \dfrac{1 - \dfrac{1}{1 + x^2}}{x \cos x + \sin x} \cdot \dfrac{1 + x^2}{1 + x^2}$$

$$= \lim\limits_{x \to 0} \dfrac{(1 + x^2) - 1}{(1 + x^2)(x \cos x + \sin x)} = \lim\limits_{x \to 0} \dfrac{x^2}{(1 + x^2)(x \cos x + \sin x)} \ \{\tfrac{0}{0}\}$$

$$= \lim\limits_{x \to 0} \dfrac{D_x\left(x^2\right)}{(1 + x^2)\, D_x \left(x \cos x + \sin x\right) + (x \cos x + \sin x)\, D_x\left(1 + x^2\right)}$$

$$= \lim\limits_{x \to 0} \dfrac{2x}{(1 + x^2)(-x \sin x + \cos x + \cos x) + (x \cos x + \sin x)(2x)} = \dfrac{0}{2} = 0$$

$\boxed{45}$ $\lim\limits_{x \to \infty} \dfrac{x^{3/2} + 5x - 4}{x \ln x}$ $\{\frac{\infty}{\infty}\}$ $= \lim\limits_{x \to \infty} \dfrac{D_x \left(x^{3/2} + 5x - 4\right)}{D_x \left(x \ln x\right)} =$

$\lim\limits_{x \to \infty} \dfrac{\frac{3}{2}x^{1/2} + 5}{1 + \ln x}$ $\{\frac{\infty}{\infty}\}$ $= \lim\limits_{x \to \infty} \dfrac{D_x\left(\frac{3}{2}x^{1/2} + 5\right)}{D_x \left(1 + \ln x\right)} =$

$$\lim\limits_{x \to \infty} \dfrac{3/(4\sqrt{x})}{1/x} = \tfrac{3}{4} \lim\limits_{x \to \infty} \sqrt{x} = \infty$$

$\boxed{49}$ $\lim\limits_{x \to \infty} \dfrac{2e^{3x} + \ln x}{e^{3x} + x^2}$ $\{\frac{\infty}{\infty}\}$ $= \lim\limits_{x \to \infty} \dfrac{D_x \left(2e^{3x} + \ln x\right)}{D_x \left(e^{3x} + x^2\right)} =$

$$= \lim\limits_{x \to \infty} \dfrac{6e^{3x} + 1/x}{3e^{3x} + 2x} \ \{\tfrac{\infty}{\infty}\} = \lim\limits_{x \to \infty} \dfrac{D_x \left(6e^{3x} + 1/x\right)}{D_x \left(3e^{3x} + 2x\right)} =$$

$$= \lim\limits_{x \to \infty} \dfrac{18e^{3x} - 1/x^2}{9e^{3x} + 2} \ \{\tfrac{\infty}{\infty}\} = \lim\limits_{x \to \infty} \dfrac{D_x \left(18e^{3x} - 1/x^2\right)}{D_x \left(9e^{3x} + 2\right)} =$$

$$\lim\limits_{x \to \infty} \dfrac{54e^{3x} + 2/x^3}{27e^{3x}} = \lim\limits_{x \to \infty} \left(2 + \dfrac{2}{27x^3 e^{3x}}\right) = 2 + 0 = 2$$

$\boxed{51}$ $\lim\limits_{x \to \infty} \left(\dfrac{x - \cos x}{x}\right) = \lim\limits_{x \to \infty} \left(\dfrac{x}{x} - \dfrac{\cos x}{x}\right) = \lim\limits_{x \to \infty} \left(1 - \dfrac{\cos x}{x}\right) = 1 - 0 = 1.$

Note: The ratio $\dfrac{x - \cos x}{x}$ has the indeterminate form $\dfrac{\infty}{\infty}$ as $x \to \infty$.

If we apply L'Hôpital's rule we find that

$$L = \lim\limits_{x \to \infty} \dfrac{D_x \left(x - \cos x\right)}{D_x \left(x\right)} = \lim\limits_{x \to \infty} \dfrac{1 + \sin x}{1} = \lim\limits_{x \to \infty} \left(1 + \sin x\right) \text{ and DNE.}$$

For L'Hôpital's rule to apply, this limit $\left\{ \lim\limits_{x \to \infty} \dfrac{f'(x)}{g'(x)} = \lim\limits_{x \to \infty} \dfrac{D_x \left(x - \cos x\right)}{D_x \left(x\right)} \right\}$ must

exist or equal $\pm\infty$. Since L does not exist, L'Hôpital's rule does not apply.

$\boxed{53}$ Let $f(x) = \dfrac{\ln\left(\tan x + \cos x\right)}{\sqrt{\ln\left(x^2 + 1\right)}}$. $f(10^{-1}) \approx 0.9129$, $f(10^{-2}) \approx 0.9901$,

$f(10^{-3}) \approx 0.9990$, $f(10^{-4}) \approx 0.9999$. We predict that $\lim\limits_{x \to 0^+} f(x) \approx 1.$

$\boxed{55}$ $\frac{mg}{k}(1 - e^{-(k/m)t}) = \frac{mg(1 - e^{-(k/m)t})}{k}$ has the indeterminate form $\frac{0}{0}$ at $k = 0$

since $\lim\limits_{k \to 0^+} (1 - e^{-(k/m)t}) = 1 - e^0 = 1 - 1 = 0.$

$$\lim\limits_{k \to 0^+} v(t) = \lim\limits_{k \to 0^+} \frac{mg(1 - e^{-kt/m})}{k} \{\tfrac{0}{0}\} = mg \lim\limits_{k \to 0^+} \frac{D_k(1 - e^{-(t/m)k})}{D_k(k)} =$$

$$mg \lim\limits_{k \to 0^+} \frac{0 - (-\tfrac{t}{m})e^{-(t/m)k}}{1} = mg \lim\limits_{k \to 0^+} (\tfrac{t}{m}e^{-(t/m)k}) = mg(\tfrac{t}{m})e^0 = gt.$$

$\boxed{59}$ *Note:* By (5.35), $D_x\big[Si(x)\big] = D_x \int_0^x \frac{\sin u}{u}\, du = \frac{\sin x}{x}.$

Also note that $\lim\limits_{x \to 0} Si(x) = \lim\limits_{x \to 0} \int_0^x \frac{\sin u}{u}\, du = \int_0^0 \frac{\sin u}{u}\, du = 0.$

(a) $\frac{Si(x)}{x}$ has the indeterminate form $\frac{0}{0}$ at $x = 0.$

$$\lim\limits_{x \to 0} \frac{Si(x)}{x} \{\tfrac{0}{0}\} = \lim\limits_{x \to 0} \frac{D_x\big[Si(x)\big]}{D_x(x)} = \lim\limits_{x \to 0} \frac{(\sin x)/x}{1} = \lim\limits_{x \to 0} \frac{\sin x}{x} = 1$$

(b) $\frac{Si(x) - x}{x^3}$ has the indeterminate form $\frac{0}{0}$ at $x = 0.$

In this part of the exercise, we must apply L'Hôpital's rule three times.

$$\lim\limits_{x \to 0} \frac{Si(x) - x}{x^3} \{\tfrac{0}{0}\} = \lim\limits_{x \to 0} \frac{D_x\big[Si(x) - x\big]}{D_x(x^3)}$$

$$= \lim\limits_{x \to 0} \frac{\frac{\sin x}{x} - 1}{3x^2} \{\tfrac{0}{0}\} = \lim\limits_{x \to 0} \frac{D_x\left(\frac{\sin x}{x} - 1\right)}{D_x(3x^2)} = \lim\limits_{x \to 0} \frac{(x \cos x - \sin x)/x^2}{6x}$$

$$= \tfrac{1}{6} \lim\limits_{x \to 0} \frac{x \cos x - \sin x}{x^3} \{\tfrac{0}{0}\} = \tfrac{1}{6} \lim\limits_{x \to 0} \frac{D_x(x \cos x - \sin x)}{D_x(x^3)} =$$

$$\tfrac{1}{6} \lim\limits_{x \to 0} \frac{-x \sin x}{3x^2} = -\tfrac{1}{18} \lim\limits_{x \to 0} \frac{\sin x}{x} = -\tfrac{1}{18}(1) = -\tfrac{1}{18}$$

$\boxed{63}$ $\lim\limits_{n \to -1} \int_1^x t^n\, dt = \lim\limits_{n \to -1} \left[\frac{t^{n+1}}{n+1}\right]_1^x = \lim\limits_{n \to -1} \frac{x^{n+1} - 1}{n+1}.$ The ratio $\frac{x^{n+1} - 1}{n+1}$

has the indeterminate form $\frac{0}{0}$ at $n = -1$ since $x^0 = 1.$ To apply L'Hôpital's rule,

we must differentiate with respect to n. In this exercise, x is a constant and n is the

variable. Thus, by (7.28)(i), $D_n(x^{n+1} - 1) = x^{n+1}(\ln x).$ $\lim\limits_{n \to -1} \frac{x^{n+1} - 1}{n+1} \{\tfrac{0}{0}\} =$

$$\lim\limits_{n \to -1} \frac{D_n(x^{n+1} - 1)}{D_n(n+1)} = \lim\limits_{n \to -1} \frac{x^{n+1} \ln x}{1} = x^0 \ln x = \ln x = \int_1^x t^{-1}\, dt$$

Note: As in §10.1, the notation $\{\frac{0}{0}\}$ or $\{\frac{\infty}{\infty}\}$ indicates the form of the limit and that L'Hôpital's rule was applied to obtain the next limit. We have also included the notations $\{0 \cdot \infty\}$, $\{0^0\}$, $\{\infty^0\}$, $\{1^\infty\}$, and $\{\infty - \infty\}$ to indicate the form of the limit.

$\boxed{1}$ Since $\lim\limits_{x \to 0^+} x = 0$ and $\lim\limits_{x \to 0^+} \ln x = -\infty$, $(x \ln x)$ has the indeterminate form $0 \cdot \infty$ at $x = 0$. Thus, we will apply (10.3) to rewrite $(x \ln x)$ as $\dfrac{\ln x}{1/x}$.

This ratio has an indeterminate form of $\frac{\infty}{\infty}$ at $x = 0$ and L'Hôpital's rule applies.

$$\lim_{x \to 0^+} (x \ln x) \ \{0 \cdot \infty\} = \lim_{x \to 0^+} \frac{\ln x}{1/x} \ \{\tfrac{\infty}{\infty}\} = \lim_{x \to 0^+} \frac{D_x (\ln x)}{D_x (1/x)} = \lim_{x \to 0^+} \frac{1/x}{-1/x^2}$$

$$= \lim_{x \to 0^+} (-x) = 0$$

$\boxed{5}$ Since $\lim\limits_{x \to 0} e^{-x} = 1$ and $\lim\limits_{x \to 0} \sin x = 0$, $e^{-x} \sin x$ does not have an indeterminate form

at $x = 0$. Thus, $\lim\limits_{x \to 0} e^{-x} \sin x = 1 \cdot 0 = 0$.

$\boxed{7}$ Since $\lim\limits_{x \to 0^+} \sin x = 0$ and $\lim\limits_{x \to 0^+} \ln \sin x = -\infty$, $\sin x \ln \sin x$ has the indeterminate

form $0 \cdot \infty$ at $x = 0$. Use (10.3) by writing $\sin x$ as $\dfrac{1}{\csc x}$.

Then, $\lim\limits_{x \to 0^+} \sin x \ln \sin x \ \{0 \cdot \infty\} = \lim\limits_{x \to 0^+} \dfrac{\ln \sin x}{\csc x} \ \{\tfrac{\infty}{\infty}\} = \lim\limits_{x \to 0^+} \dfrac{D_x (\ln \sin x)}{D_x (\csc x)} =$

$$\lim_{x \to 0^+} \frac{\cos x / \sin x}{-\csc x \cot x} = \lim_{x \to 0^+} (-\sin x) = 0.$$

$\boxed{9}$ $\lim\limits_{x \to \infty} x \sin \dfrac{1}{x} \ \{\infty \cdot 0\} = \lim\limits_{x \to \infty} \dfrac{\sin (1/x)}{1/x} \ \{\tfrac{0}{0}\} = \lim\limits_{x \to \infty} \dfrac{D_x \left[\sin (1/x)\right]}{D_x (1/x)} =$

$$\lim_{x \to \infty} \frac{(-1/x^2) \cos (1/x)}{-1/x^2} = \lim_{x \to \infty} \cos (1/x) = \cos 0 = 1$$

$\boxed{13}$ $\lim\limits_{x \to \infty} (1 + \tfrac{1}{x})^{5x}$ is of the form 1^∞. We will follow the guidelines given in (10.4).

Let $y = (1 + \tfrac{1}{x})^{5x}$ and hence $\ln y = \ln (1 + \tfrac{1}{x})^{5x} = 5x \ln (1 + \tfrac{1}{x})$.

$$\lim_{x \to \infty} \ln y \ \{\infty \cdot 0\} = \lim_{x \to \infty} \frac{\ln (1 + \tfrac{1}{x})}{1/(5x)} \ \{\tfrac{0}{0}\} = \lim_{x \to \infty} \frac{D_x \left[\ln (1 + \tfrac{1}{x})\right]}{D_x \left[1/(5x)\right]} =$$

$$\lim_{x \to \infty} \frac{(-1/x^2)/(1 + \tfrac{1}{x})}{-1/(5x^2)} = \lim_{x \to \infty} \frac{5}{1 + \tfrac{1}{x}} = \frac{5}{1 + 0} = 5.$$

Thus, $\lim\limits_{x \to \infty} \ln y = \ln \lim\limits_{x \to \infty} y = 5 \Rightarrow \lim\limits_{x \to \infty} y = e^5$ by (10.4)(3a).

$\boxed{15}$ $\lim\limits_{x \to 0^+} (e^x - 1)^x$ is of the form 0^0 and we will follow (10.4). $y = (e^x - 1)^x \Rightarrow$

$\ln y = \ln(e^x - 1)^x = x \ln(e^x - 1)$. $\lim\limits_{x \to 0^+} \ln y \ \{0 \cdot \infty\} = \lim\limits_{x \to 0^+} \dfrac{\ln(e^x - 1)}{1/x} \ \{\frac{\infty}{\infty}\} =$

$\lim\limits_{x \to 0^+} \dfrac{D_x[\ln(e^x - 1)]}{D_x(1/x)} = \lim\limits_{x \to 0^+} \dfrac{e^x/(e^x - 1)}{-1/x^2} = \lim\limits_{x \to 0^+} \dfrac{-x^2 e^x}{e^x - 1} \ \{\frac{0}{0}\} =$

$\lim\limits_{x \to 0^+} \dfrac{D_x(-x^2 e^x)}{D_x(e^x - 1)} = \lim\limits_{x \to 0^+} \dfrac{-x^2 e^x - 2x e^x}{e^x} = \lim\limits_{x \to 0^+} (-x^2 - 2x) = 0.$

Thus, by (10.4)(3a), L $= e^0 = 1$.

$\boxed{17}$ $\lim\limits_{x \to \infty} x^{1/x}$ is of the form ∞^0 and we will follow (10.4).

$y = x^{1/x} \Rightarrow \ln y = \ln x^{1/x} = \frac{1}{x} \ln x$. $\lim\limits_{x \to \infty} \ln y \ \{0 \cdot \infty\} = \lim\limits_{x \to \infty} \dfrac{\ln x}{x} \ \{\frac{\infty}{\infty}\} =$

$\lim\limits_{x \to \infty} \dfrac{D_x(\ln x)}{D_x(x)} = \lim\limits_{x \to \infty} \dfrac{1/x}{1} = 0.$ Thus, L $= e^0 = 1$.

$\boxed{19}$ $\lim\limits_{x \to (\pi/2)^-} (\tan x)^x = \infty$, since it is of the form $\infty^{\pi/2}$,

which is not an indeterminate form.

$\boxed{21}$ $\lim\limits_{x \to 0^+} (2x + 1)^{\cot x}$ is of the form 1^∞ and we will follow (10.4).

$y = (2x + 1)^{\cot x} \Rightarrow \ln y = \ln(2x + 1)^{\cot x} = \cot x \ln(2x + 1).$

$\lim\limits_{x \to 0^+} \ln y \ \{\infty \cdot 0\} = \lim\limits_{x \to 0^+} \dfrac{\ln(2x + 1)}{1/\cot x} = \lim\limits_{x \to 0^+} \dfrac{\ln(2x + 1)}{\tan x} \ \{\frac{0}{0}\} =$

$\lim\limits_{x \to 0^+} \dfrac{D_x[\ln(2x + 1)]}{D_x(\tan x)} = \lim\limits_{x \to 0^+} \dfrac{2/(2x + 1)}{\sec^2 x} = \dfrac{2}{1} = 2.$ Thus, L $= e^2$.

$\boxed{23}$ Since $\lim\limits_{x \to \infty} \dfrac{x^2}{x - 1} = \infty$ and $\lim\limits_{x \to \infty} \dfrac{x^2}{x + 1} = \infty$, the expression has the indeterminate

form $\infty - \infty$. $\lim\limits_{x \to \infty} \left(\dfrac{x^2}{x - 1} - \dfrac{x^2}{x + 1} \right) \{\infty - \infty\} = \lim\limits_{x \to \infty} \left[\dfrac{x^2(x + 1) - x^2(x - 1)}{(x - 1)(x + 1)} \right]$

$= \lim\limits_{x \to \infty} \dfrac{2x^2}{x^2 - 1} \ \{\frac{\infty}{\infty}\} = \lim\limits_{x \to \infty} \dfrac{D_x(2x^2)}{D_x(x^2 - 1)} = \lim\limits_{x \to \infty} \dfrac{4x}{2x} = \lim\limits_{x \to \infty} 2 = 2$

$\boxed{27}$ $\lim\limits_{x \to 1^-} (1 - x)^{\ln x}$ is of the form 0^0 and we will follow (10.4). $y = (1 - x)^{\ln x} \Rightarrow$

$\ln y = \ln x \ln(1 - x)$. $\lim\limits_{x \to 1^-} \ln y \ \{0 \cdot \infty\} = \lim\limits_{x \to 1^-} \dfrac{\ln(1 - x)}{1/\ln x} \ \{\frac{\infty}{\infty}\} =$

$\lim\limits_{x \to 1^-} \dfrac{D_x[\ln(1 - x)]}{D_x(1/\ln x)} = \lim\limits_{x \to 1^-} \dfrac{-1/(1 - x)}{-1/[x(\ln x)^2]} = \lim\limits_{x \to 1^-} \dfrac{x(\ln x)^2}{1 - x} \ \{\frac{0}{0}\} =$

$\lim\limits_{x \to 1^-} \dfrac{D_x[x(\ln x)^2]}{D_x(1 - x)} = \lim\limits_{x \to 1^-} \dfrac{2 \ln x + (\ln x)^2}{-1} = \dfrac{2(0) + (0)^2}{-1} = 0.$ Thus, L $= e^0 = 1$.

31 $\lim\limits_{x \to 0^+} \cot 2x \, \tan^{-1} x \, \{\infty \cdot 0\} = \lim\limits_{x \to 0^+} \dfrac{\tan^{-1} x}{1/\cot 2x} = \lim\limits_{x \to 0^+} \dfrac{\tan^{-1} x}{\tan 2x} \, \{\frac{0}{0}\} =$

$$\lim\limits_{x \to 0^+} \dfrac{D_x(\tan^{-1} x)}{D_x(\tan 2x)} = \lim\limits_{x \to 0^+} \dfrac{1/(1+x^2)}{2\sec^2 2x} = \dfrac{1}{2}$$

33 $\lim\limits_{x \to 0} \cot^2 x = \infty$ and $\lim\limits_{x \to 0} e^{-x} = 1$. Thus,

$$\cot^2 x - e^{-x} \text{ does } not \text{ have an indeterminate form and } \lim\limits_{x \to 0} (\cot^2 x - e^{-x}) = \infty.$$

35 $\lim\limits_{x \to (\pi/2)^-} (1 + \cos x)^{\tan x}$ has the form 1^∞ and we will follow (10.4).

$$y = (1 + \cos x)^{\tan x} \Rightarrow \ln y = \tan x \, \ln(1 + \cos x). \qquad \lim\limits_{x \to (\pi/2)^-} \ln y \, \{\infty \cdot 0\} =$$

$$\lim\limits_{x \to (\pi/2)^-} \dfrac{\ln(1 + \cos x)}{\cot x} \, \{\frac{0}{0}\} = \lim\limits_{x \to (\pi/2)^-} \dfrac{D_x\big[\ln(1 + \cos x)\big]}{D_x(\cot x)} =$$

$$\lim\limits_{x \to (\pi/2)^-} \dfrac{-\sin x/(1 + \cos x)}{-\csc^2 x} = \dfrac{-1/1}{-1} = 1. \text{ Thus, } L = e^1.$$

41 Since $\lim\limits_{x \to \infty} \sinh x = \infty$ and $\lim\limits_{x \to \infty} x = \infty$, $\sinh x - x$ has the indeterminate form of

$\infty - \infty$. $\lim\limits_{x \to \infty} (\sinh x - x) \, \{\infty - \infty\}$

$$= \lim\limits_{x \to \infty} \sinh x \Big(1 - \dfrac{x}{\sinh x}\Big) \, \{\text{factor out the fastest growing function}\}$$

$$= \lim\limits_{x \to \infty} \sinh x \cdot \lim\limits_{x \to \infty} \Big(1 - \dfrac{x}{\sinh x}\Big) \, \{\text{limit of a product} = \text{the product of limits}\}$$

$$= \Big(\lim\limits_{x \to \infty} \sinh x\Big)(1 - 0) \, \Big\{\text{since } \lim\limits_{x \to \infty} \dfrac{x}{\sinh x} \, \{\frac{\infty}{\infty}\} = \lim\limits_{x \to \infty} \dfrac{1}{\cosh x} = 0\Big\}$$

$$= \lim\limits_{x \to \infty} \sinh x = \infty$$

45 (a) $f(x) = x^{1/x} = e^{(1/x)\ln x} \Rightarrow f'(x) =$

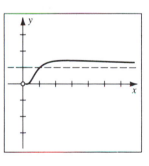

$$e^{(1/x)\ln x}\Big(\dfrac{1}{x^2} - \dfrac{1}{x^2}\ln x\Big) = \dfrac{1}{x^2} e^{(1/x)\ln x}(1 - \ln x).$$

$f'(x) = 0 \Rightarrow 1 - \ln x = 0 \Rightarrow x = e.$

$f'(x) > 0$ on $(0, e)$ and $f'(x) < 0$ on $(e, \infty) \Rightarrow$

$f(e) = e^{1/e} \approx 1.44$ is a *LMAX*.

$y = x^{1/x} \Rightarrow \ln y = \dfrac{1}{x}\ln x$ and $\lim\limits_{x \to 0^+} \dfrac{\ln x}{x} = -\infty.$

Hence, $\lim\limits_{x \to 0^+} x^{1/x} = 0$ by (10.4)(3c).

Figure 45

(b) From Exercise 17, $\lim\limits_{x \to \infty} x^{1/x} = 1$. Thus, $y = 1$ is a horizontal asymptote.

(c) *Note*: There are *PI* at $x \approx 0.58$ and $x \approx 4.37$. (You are not expected to calculate

this fact, but you may wish to use it as an aid when sketching your graph.)

$\boxed{47}$ $\left(\dfrac{a^{1/x} + b^{1/x}}{2}\right)^x$ has the indeterminate form 1^{∞} as $x \to \infty$. We will follow (10.4).

$$y = \left(\dfrac{a^{1/x} + b^{1/x}}{2}\right)^x \Rightarrow \ln y = x \ln\left[\tfrac{1}{2}(a^{1/x} + b^{1/x})\right]. \quad \lim_{x \to \infty} \ln y \; \{\infty \cdot 0\}$$

$$= \lim_{x \to \infty} \dfrac{\ln\left[\tfrac{1}{2}(a^{1/x} + b^{1/x})\right]}{1/x} \; \{\tfrac{0}{0}\} = \lim_{x \to \infty} \dfrac{D_x\left\{\ln\left[\tfrac{1}{2}(a^{1/x} + b^{1/x})\right]\right\}}{D_x(1/x)}$$

$$= \lim_{x \to \infty} \dfrac{1/\left[\tfrac{1}{2}(a^{1/x} + b^{1/x})\right] \cdot \tfrac{1}{2}\left[a^{1/x} \ln a\left(-\tfrac{1}{x^2}\right) + b^{1/x} \ln b\left(-\tfrac{1}{x^2}\right)\right]}{-1/x^2}$$

$$= \lim_{x \to \infty} \left[\dfrac{-x^2}{1} \cdot \dfrac{2}{a^{1/x} + b^{1/x}} \cdot \dfrac{a^{1/x} \ln a + b^{1/x} \ln b}{-2x^2}\right]$$

$$= \lim_{x \to \infty} \dfrac{a^{1/x} \ln a + b^{1/x} \ln b}{a^{1/x} + b^{1/x}} = \dfrac{(1)\ln a + (1)\ln b}{1 + 1} =$$

$$\tfrac{1}{2}(\ln a + \ln b) = \tfrac{1}{2}\ln(ab) = \ln(ab)^{1/2} = \ln\sqrt{ab}. \text{ Thus, } L = e^{\ln\sqrt{ab}} = \sqrt{ab}.$$

Exercises 10.3

Note: C denotes that the integral converges, D denotes that it diverges.

$\boxed{1}$ Apply (10.5)(i) with $f(x) = x^{-4/3}$ and $a = 1$. $\displaystyle\int_1^{\infty} \dfrac{1}{x^{4/3}}\,dx = \lim_{t \to \infty} \int_1^t x^{-4/3}\,dx =$

$$\lim_{t \to \infty} \left[-3x^{-1/3}\right]_1^t = (-3)\lim_{t \to \infty}\left[\dfrac{1}{\sqrt[3]{t}} - 1\right] = -3(0 - 1) = 3; \text{ C}$$

$\boxed{3}$ Apply (10.5)(i) with $f(x) = x^{-3/4}$ and $a = 1$.

$$\int_1^{\infty} \dfrac{1}{x^{3/4}}\,dx = \lim_{t \to \infty} \int_1^t x^{-3/4}\,dx = \lim_{t \to \infty}\left[4x^{1/4}\right]_1^t = 4\lim_{t \to \infty}(t^{1/4} - 1) = \infty; \text{ D}$$

$\boxed{7}$ $\displaystyle\int_0^{\infty} e^{-2x}\,dx = \lim_{t \to \infty} \int_0^t e^{-2x}\,dx = \lim_{t \to \infty}\left[-\tfrac{1}{2}e^{-2x}\right]_0^t =$

$$-\tfrac{1}{2}\lim_{t \to \infty}(e^{-2t} - 1) = -\tfrac{1}{2}(0 - 1) = \tfrac{1}{2}; \text{ C}$$

$\boxed{11}$ Apply (10.5)(ii) with $f(x) = (x - 8)^{-2/3}$ and $a = 0$.

$$\int_{-\infty}^0 \dfrac{1}{(x - 8)^{2/3}}\,dx = \lim_{t \to -\infty} \int_t^0 (x - 8)^{-2/3}\,dx = \lim_{t \to -\infty}\left[3(x - 8)^{1/3}\right]_t^0 =$$

$$3\lim_{t \to -\infty}\left[-2 - (t - 8)^{1/3}\right] = 3\left[-2 - (-\infty)\right] = \infty; \text{ D}$$

[15] Apply (10.6) with $f(x) = xe^{-x^2}$ and $a = 0$ {you could pick any number for a}.

$$\int_{-\infty}^{\infty} xe^{-x^2} \, dx = \int_{-\infty}^{0} xe^{-x^2} \, dx + \int_{0}^{\infty} xe^{-x^2} \, dx$$

$$= \lim_{t \to -\infty} \int_{t}^{0} xe^{-x^2} \, dx + \lim_{s \to \infty} \int_{0}^{s} xe^{-x^2} \, dx$$

$$= \lim_{t \to -\infty} \left[-\tfrac{1}{2} e^{-x^2} \right]_{t}^{0} + \lim_{s \to \infty} \left[-\tfrac{1}{2} e^{-x^2} \right]_{0}^{s}$$

$$= -\tfrac{1}{2} \lim_{t \to -\infty} (1 - e^{-t^2}) - \tfrac{1}{2} \lim_{s \to \infty} (e^{-s^2} - 1)$$

$$= -\tfrac{1}{2}(1 - 0) - \tfrac{1}{2}(0 - 1) = 0; \ \text{C}$$

[19] Apply (10.5)(i). $\displaystyle \int_{0}^{\infty} \cos x \, dx = \lim_{t \to \infty} \int_{0}^{t} \cos x \, dx = \lim_{t \to \infty} \left[\sin x \right]_{0}^{t} = \lim_{t \to \infty} \sin t.$

This limit DNE since the sine function oscillates between $+1$ and -1. D

[21] Apply (10.6).

$$\int_{-\infty}^{\infty} \operatorname{sech} x \, dx = \lim_{t \to -\infty} \int_{t}^{0} \operatorname{sech} x \, dx + \lim_{s \to \infty} \int_{0}^{s} \operatorname{sech} x \, dx$$

$$= 2 \lim_{t \to \infty} \int_{0}^{t} \operatorname{sech} x \, dx \ \{\text{even integrand}\}$$

$$= 2 \lim_{t \to \infty} \left[\tan^{-1} (\sinh x) \right]_{0}^{t} \ \{\text{Formula 107}\}$$

$$= 2 \lim_{t \to \infty} \left[\tan^{-1} (\sinh t) - 0 \right]$$

$$= 2(\tfrac{\pi}{2} - 0) \ \{ \lim_{t \to \infty} \sinh t = \infty \text{ and } \lim_{t \to \infty} \tan^{-1} t = \tfrac{\pi}{2} \}$$

$$= \pi; \ \text{C}$$

[23] $\displaystyle \int_{-\infty}^{0} \frac{1}{x^2 - 3x + 2} \, dx = \lim_{t \to -\infty} \int_{t}^{0} \frac{1}{(x - 2)(x - 1)} \, dx =$

$$\lim_{t \to -\infty} \int_{t}^{0} \left[\frac{1}{x - 2} - \frac{1}{x - 1} \right] dx \ \{\text{PF}\} = \lim_{t \to -\infty} \left[\ln|x - 2| - \ln|x - 1| \right]_{t}^{0} =$$

$$\lim_{t \to -\infty} \left[\ln \left| \frac{x - 2}{x - 1} \right| \right]_{t}^{0} = \lim_{t \to -\infty} \left[\ln|2| - \ln \left| \frac{t - 2}{t - 1} \right| \right] = \ln 2 - \ln 1 = \ln 2; \ \text{C}$$

[25] It may not be obvious whether the first integral converges or diverges. Since $1/x^4$ can be integrated and $0 \le \dfrac{1}{1 + x^4} \le \dfrac{1}{x^4}$ on $[1, \infty)$, we will let $g(x) = 1/x^4$ and $a = 1$. If $\displaystyle \int_{1}^{\infty} \frac{1}{x^4} \, dx$ converges, then $\displaystyle \int_{1}^{\infty} \frac{1}{1 + x^4} \, dx$ converges also. If $\displaystyle \int_{1}^{\infty} \frac{1}{x^4} \, dx$ were to diverge, then we could say *nothing* about the convergence or divergence of $\displaystyle \int_{1}^{\infty} \frac{1}{1 + x^4} \, dx$. $\displaystyle \int_{a}^{\infty} g(x) \, dx = \int_{1}^{\infty} \frac{1}{x^4} \, dx = \lim_{t \to \infty} \left[-\frac{1}{3x^3} \right]_{1}^{t} = -\tfrac{1}{3}(0 - 1) = \tfrac{1}{3} \Rightarrow$

$\displaystyle \int_{1}^{\infty} \frac{1}{1 + x^4} \, dx$ *converges* by (i). *Note:* Other functions like $g(x) = 1/x^2$ or $1/x^3$ could have also been used. The selection of g is not unique.

$\boxed{27}$ Let $f(x) = \frac{1}{x}$ and $a = 2$. Then since $0 \le \frac{1}{x} \le \frac{1}{\ln x}$ on $[2, \infty)$,

if $\displaystyle\int_2^\infty \frac{1}{x}\, dx$ diverges, so does $I = \displaystyle\int_2^\infty \frac{1}{\ln x}\, dx$. $\displaystyle\int_2^\infty \frac{1}{x}\, dx = \lim_{t \to \infty}\Big[\ln |x|\Big]_2^t = \infty \Rightarrow$

I *diverges* by (ii).

$\boxed{29}$ R is the unbounded region between the x-axis and the graph $y = 1/x$ for $x \ge 1$.

See the figure for Exercise 35 in the text.

(a) $A = \displaystyle\int_1^\infty \frac{1}{x}\, dx = \lim_{t \to \infty}\Big[\ln |x|\Big]_1^t = \lim_{t \to \infty} \ln |t| - \ln 1 = \infty.$ Not possible.

(b) Using disks (6.5), $V = \pi \displaystyle\int_1^\infty \left(\frac{1}{x}\right)^2 dx = \pi \lim_{t \to \infty}\left[-\frac{1}{x}\right]_1^t =$

$$-\pi\left[\left(\lim_{t \to \infty} \frac{1}{t}\right) - (1)\right] = -\pi(0 - 1) = \pi.$$

$\boxed{33}$ Using shells (6.11) with $f(x) = e^{-x^2}$, $V = \displaystyle\int_0^\infty 2\pi\, x\, e^{-x^2}\, dx = \lim_{t \to \infty}\int_0^t 2\pi\, x\, e^{-x^2}\, dx =$

$$\lim_{t \to \infty}\left[-\pi e^{-x^2}\right]_0^t = -\pi\left[\left(\lim_{t \to \infty} e^{-t^2}\right) - (1)\right] = -\pi(0 - 1) = \pi.$$

$\boxed{35}$ (a) See Exercise 29(b).

(b) Using shells, $V = \displaystyle\int_1^\infty 2\pi\, x\frac{1}{x}\, dx = \lim_{t \to \infty}\int_1^t 2\pi\, dx = \lim_{t \to \infty}\Big[2\pi x\Big]_1^t = \infty.$ No.

(c) By (6.19) with $f(x) = \frac{1}{x}$ and $f'(x) = -\frac{1}{x^2}$, $S = \displaystyle\int_1^\infty 2\pi f(x)\sqrt{1 + \big[f'(x)\big]^2}\, dx =$

$\displaystyle\int_1^\infty 2\pi \frac{1}{x}\sqrt{1 + (-\frac{1}{x^2})^2}\, dx = \int_1^\infty 2\pi \frac{1}{x}\sqrt{1 + \frac{1}{x^4}}\, dx.$ In order to show that this

integral diverges, we need to pick a function f such that $0 \le f(x) \le \frac{2\pi}{x}\sqrt{1 + \frac{1}{x^4}}$,

$\displaystyle\int_1^\infty f(x)\, dx$ diverges, and we know how to integrate $f(x)$. Let $f(x) = \frac{2\pi}{x}$.

Then, $0 \le \frac{2\pi}{x} \le \frac{2\pi}{x}\sqrt{1 + \frac{1}{x^4}}$ and $\displaystyle\int_1^\infty \frac{2\pi}{x}\, dx = \lim_{t \to \infty}\Big[2\pi \ln |x|\Big]_1^t = \infty \Rightarrow S$ DNE.

$\boxed{39}$ (a) $R(t) = e^{-kt} \Rightarrow R'(t) = -ke^{-kt}$.

Average time $= \displaystyle\int_0^\infty (-t)\, R'(t)\, dt = \int_0^\infty kte^{-kt}\, dt =$

$\displaystyle\lim_{s \to \infty}\left[-te^{-kt} - \frac{1}{k}e^{-kt}\right]_0^s \{\text{IP; see }\#1, \S 9.1\} = \lim_{s \to \infty}\left[\frac{-s}{e^{ks}} - \frac{1}{ke^{ks}}\right] + \frac{1}{k} = \frac{1}{k}.$

(b) If it is possible, then $R(t) = \frac{1}{t + 1} \Rightarrow R'(t) = -\frac{1}{(t + 1)^2}$. Average time $=$

$\displaystyle\int_0^\infty \frac{t}{(t + 1)^2}\, dt = \int_0^\infty \left[\frac{1}{t + 1} - \frac{1}{(t + 1)^2}\right] dt \{\text{PF}\} = \lim_{s \to \infty}\left[\ln |t + 1| + \frac{1}{t + 1}\right]_0^s$

$= \infty.$ This is a contradiction since it indicates that the average repair time is

unbounded. Thus, it is not possible.

45 $L\left[\cos x\right] = \displaystyle\int_0^\infty e^{-sx}\left(\cos x\right)dx$ { IP; see Exam. 5, §9.1 or use Formula 99 on page A25 }

$$= \lim_{t\to\infty}\left[\frac{e^{-sx}(-s\cos x + \sin x)}{s^2 + 1}\right]_0^t$$

$$= \frac{1}{s^2 + 1}\cdot\left\{\left[\lim_{t\to\infty}e^{-st}(-s\cos t + \sin t)\right] - \left[e^0(-s + 0)\right]\right\}$$

$$= \frac{1}{s^2 + 1}\left[0 - (-s)\right] = \frac{s}{s^2 + 1},\ s > 0$$

49 (a) $\Gamma(1) = \displaystyle\int_0^\infty e^{-x}\,dx = \lim_{t\to\infty}\left[-e^{-x}\right]_0^t = 0 - (-1) = 1.$

$\Gamma(2) = \displaystyle\int_0^\infty xe^{-x}\,dx$ { IP } $= \lim_{t\to\infty}\left[-xe^{-x} - e^{-x}\right]_0^t = (0 - 0) - (0 - 1) = 1.$

$\Gamma(3) = \displaystyle\int_0^\infty x^2\,e^{-x}\,dx$ { IP } $= \lim_{t\to\infty}\left[-x^2\,e^{-x} - 2xe^{-x} - 2e^{-x}\right]_0^t =$

$$(0 - 0 - 0) - (0 - 0 - 2) = 2.$$

(b) $\Gamma(n + 1) = \displaystyle\int_0^\infty x^n\,e^{-x}\,dx \overset{A}{=} \lim_{t\to\infty}\left[-x^n\,e^{-x}\right]_0^t + \int_0^\infty nx^{n-1}\,e^{-x}\,dx$

$$= \left[\left(-\lim_{t\to\infty}\frac{t^n}{e^t}\right) - 0\right] + n\int_0^\infty x^{n-1}\,e^{-x}\,dx$$

$$= 0\ \{\text{see \#31, §10.1}\} + n\int_0^\infty x^{n-1}\,e^{-x}\,dx$$

$$= 0 + n\,\Gamma(n) = n\,\Gamma(n).$$

A. $u = x^n,\ du = nx^{n-1}\,dx,\ dv = e^{-x}\,dx,\ v = -e^{-x}$

(c) $\Gamma(1) = 1 = 1!$ Assume $\Gamma(n + 1) = n!$ for some $n \geq 1.$

Then $\Gamma(n + 2) = (n + 1)\,\Gamma(n + 1)$ { by part (b) }

$$= (n + 1)\,n!\ \{\text{by the induction hypothesis}\} = (n + 1)!.$$

51 $u = \frac{1}{x} \Rightarrow x = \frac{1}{u}$ and $dx = -\frac{1}{u^2}\,du;\ x = 2 \Rightarrow u = \frac{1}{2}$ and $x\to\infty \Rightarrow u\to 0^+.$

$$I = \int_2^\infty \frac{1}{\sqrt{x^4 + x}}\,dx = \int_{1/2}^0 \frac{1}{\sqrt{(1/u^4) + (1/u)}}\left(-\frac{1}{u^2}\right)du = \int_0^{1/2}\frac{1}{\sqrt{1 + u^3}}\,du.$$

Let $f(u) = \frac{1}{\sqrt{1 + u^3}}.$ $I \approx S = \frac{0.5 - 0}{3(4)}\left[f(0) + 4f(\tfrac{1}{8}) + 2f(\tfrac{1}{4}) + 4f(\tfrac{3}{8}) + f(\tfrac{1}{2})\right] \approx$

$\frac{1}{24}\left[1 + 4(0.9990) + 2(0.9923) + 4(0.9746) + 0.9428\right] = \frac{1}{24}(11.8218) \approx 0.4926.$

Exercises 10.4

1 Since $\frac{1}{\sqrt[3]{x}}$ is discontinuous at the left end point, $x = 0$, apply (10.7)(ii).

$$\int_0^8 \frac{1}{\sqrt[3]{x}}\,dx = \lim_{t\to 0^+}\int_t^8 x^{-1/3}\,dx = \lim_{t\to 0^+}\left[\frac{3}{2}x^{2/3}\right]_t^8 =$$

$$\frac{3}{2}\left[8^{2/3} - \left(\lim_{t\to 0^+}t^{2/3}\right)\right] = \frac{3}{2}(4 - 0) = 6;\ \text{C}$$

$\boxed{3}$ Since $\frac{1}{x^2}$ is discontinuous at $x = 0$, which is in the interval $[-3, -1]$, apply (10.8).

$$\int_{-3}^{1} \frac{1}{x^2}\,dx = \int_{-3}^{0} \frac{1}{x^2}\,dx + \int_{0}^{1} \frac{1}{x^2}\,dx = I_1 + I_2.$$

For I to converge, *both* I_1 and I_2 must converge. By (10.7)(i),

$$I_1 = \lim_{t \to 0^-} \int_{-3}^{t} x^{-2}\,dx = \lim_{t \to 0^-}\left[-\frac{1}{x}\right]_{-3}^{t} = \lim_{t \to 0^-}\left(-\frac{1}{t}\right) - \left(\frac{1}{3}\right) = \infty;\ D$$

Note: It is only necessary that one integral diverges for I to diverge.

In future solutions, we will compute one integral, and if it converges,

continue until we can draw a conclusion.

$\boxed{5}$ Since $\sec^2 x$ is discontinuous at the right end point, $x = \frac{\pi}{2}$, apply (10.7)(i).

$$\int_{0}^{\pi/2} \sec^2 x\,dx = \lim_{t \to (\pi/2)^-} \int_{0}^{t} \sec^2 x\,dx = \lim_{t \to (\pi/2)^-}\left[\tan x\right]_{0}^{t} =$$

$$\lim_{t \to (\pi/2)^-}(\tan t) - 0 = \infty;\ D$$

$\boxed{11}$ Since $\dfrac{1}{(x+1)^3}$ is discontinuous at $x = -1$, which is in the interval $[-2, 2]$,

apply (10.8). $\displaystyle\int_{-2}^{2} \frac{1}{(x+1)^3}\,dx = \int_{-2}^{-1} \frac{1}{(x+1)^3}\,dx + \int_{-1}^{2} \frac{1}{(x+1)^3}\,dx = I_1 + I_2.$

$$I_1 = \lim_{t \to -1^-} \int_{-2}^{t} (x+1)^{-3}\,dx = \lim_{t \to -1^-}\left[\frac{-2}{(x+1)^2}\right]_{-2}^{t} = \left(\lim_{t \to -1^-} \frac{-2}{(t+1)^2}\right) - (-2)$$

$$= -\infty;\ D$$

$\boxed{13}$ $\dfrac{1}{\sqrt{4 - x^2}}$ is discontinuous at $x = -2$. $\displaystyle\int_{-2}^{0} \frac{1}{\sqrt{4 - x^2}}\,dx = \lim_{t \to -2^+} \int_{t}^{0} \frac{1}{\sqrt{4 - x^2}}\,dx =$

$$\lim_{t \to -2^+}\left[\sin^{-1}\frac{x}{2}\right]_{t}^{0} = \sin^{-1} 0 - \lim_{t \to -2^+} \sin^{-1}\frac{t}{2} = 0 - \sin^{-1}(-1) = -\left(-\frac{\pi}{2}\right) = \frac{\pi}{2};\ C$$

$\boxed{17}$ $x \ln x$ is discontinuous at $x = 0$.

$$\int_{0}^{1} x \ln x\,dx = \lim_{t \to 0^+} \int_{t}^{1} x \ln x\,dx$$

$$= \lim_{t \to 0^+}\left[\tfrac{1}{2}x^2 \ln x - \tfrac{1}{4}x^2\right]_{t}^{1} \quad \{\text{IP or Formula 101, page A25}\,\}$$

$$= \tfrac{1}{4}\lim_{t \to 0^+}\left[x^2(2 \ln x - 1)\right]_{t}^{1}$$

$$= -\tfrac{1}{4} - \tfrac{1}{4}\lim_{t \to 0^+}\left[t^2(2 \ln t - 1)\right] \quad \{0 \cdot \infty\}$$

$$= -\tfrac{1}{4} - \tfrac{1}{4}\lim_{t \to 0^+}\frac{2 \ln t - 1}{1/t^2} \quad \{\tfrac{\infty}{\infty}\}$$

$$= -\tfrac{1}{4} - \tfrac{1}{4}\lim_{t \to 0^+}\frac{2/t}{-2/t^3}$$

$$= -\tfrac{1}{4} - \tfrac{1}{4}\lim_{t \to 0^+}(-t^2) = -\tfrac{1}{4} - 0 = -\tfrac{1}{4};\ C$$

21 Since $x^2 - 5x + 4 = (x - 1)(x - 4) = 0$ at $x = 4$, the integrand is discontinuous

at $x = 4$. $\displaystyle\int_2^4 \frac{x - 2}{x^2 - 5x + 4}\, dx = \lim_{t \to 4^-} \int_2^t \frac{x - 2}{(x - 1)(x - 4)}\, dx$

$$= \lim_{t \to 4^-} \int_2^t \left[\frac{\frac{1}{3}}{x - 1} + \frac{\frac{2}{3}}{x - 4} \right] dx \ \{\text{PF}\}$$

$$= \lim_{t \to 4^-} \left[\tfrac{1}{3}\ln|x - 1| + \tfrac{2}{3}\ln|x - 4| \right]_2^t$$

$$= \lim_{t \to 4^-} \left(\tfrac{1}{3}\ln|t - 1| + \tfrac{2}{3}\ln|t - 4| \right) - \left(\tfrac{1}{3}\ln 1 + \tfrac{2}{3}\ln 2 \right)$$

$$= \tfrac{1}{3}\ln 3 + \tfrac{2}{3}\lim_{t \to 4^-} \ln|t - 4| - \tfrac{2}{3}\ln 2 = -\infty; \ \text{D}$$

23 $\dfrac{1}{x^2}\cos\dfrac{1}{x}$ is discontinuous at $x = 0$.

$$I = \int_{-1}^2 \frac{1}{x^2}\cos\frac{1}{x}\, dx = \int_{-1}^0 \frac{1}{x^2}\cos\frac{1}{x}\, dx + \int_0^2 \frac{1}{x^2}\cos\frac{1}{x}\, dx = I_1 + I_2.$$

$$I_1 = \lim_{t \to 0^-} \int_{-1}^t \frac{1}{x^2}\cos\frac{1}{x}\, dx = \lim_{t \to 0^-} \left[-\sin\frac{1}{x} \right]_{-1}^t =$$

$$\left[\lim_{t \to 0^-} \left(-\sin\frac{1}{t} \right) \right] - \left[-\sin(-1) \right]. \ \text{The limit DNE. \ D}$$

25 The integrand is discontinuous at $x = \frac{\pi}{2}$.

$$I = \int_0^\pi \frac{\cos x}{\sqrt{1 - \sin x}}\, dx = \int_0^{\pi/2} \frac{\cos x}{\sqrt{1 - \sin x}}\, dx + \int_{\pi/2}^\pi \frac{\cos x}{\sqrt{1 - \sin x}}\, dx = I_1 + I_2.$$

$\{\,\text{To integrate, let } u = 1 - \sin x \text{ and } du = -\cos x\, dx.\,\}$

$$I_1 = \lim_{t \to (\pi/2)^-} \int_0^t \frac{\cos x}{\sqrt{1 - \sin x}}\, dx = \lim_{t \to (\pi/2)^-} \left[-2\sqrt{1 - \sin x} \right]_0^t = 0 - (-2) = 2.$$

$$I_2 = \lim_{t \to (\pi/2)^+} \int_t^\pi \frac{\cos x}{\sqrt{1 - \sin x}}\, dx = \lim_{t \to (\pi/2)^+} \left[-2\sqrt{1 - \sin x} \right]_t^\pi = -2 - 0 = -2.$$

$$I = I_1 + I_2 = 2 + (-2) = 0. \ \text{C}$$

29 The integrand is discontinuous at $x = 4$ and also has an infinite limit of integration.

It will be necessary to break the given integral I into three integrals. The choice of

breaking the integral at $x = 5$ is arbitrary. Any value of $x > 4$ would work.

$$I = \int_0^\infty \frac{1}{(x - 4)^2}\, dx = \int_0^4 \frac{1}{(x - 4)^2}\, dx + \int_4^5 \frac{1}{(x - 4)^2}\, dx + \int_5^\infty \frac{1}{(x - 4)^2}\, dx =$$

$$I_1 + I_2 + I_3.$$

$$I_1 = \int_0^4 \frac{1}{(x - 4)^2}\, dx = \lim_{t \to 4^-} \int_0^t \frac{1}{(x - 4)^2}\, dx = \lim_{t \to 4^-} \left[-\frac{1}{x - 4} \right]_0^t =$$

$$\lim_{t \to 4^-} \left(-\frac{1}{t - 4} \right) - \frac{1}{4} = \infty; \ \text{D}$$

31 Since $|\sin x| \le 1$, it follows that $0 \le \dfrac{\sin x}{\sqrt{x}} \le \dfrac{1}{\sqrt{x}}$ on $(0, \pi]$ and

$$\int_0^\pi \frac{1}{\sqrt{x}}\, dx = \lim_{t \to 0^+} \int_t^\pi x^{-1/2}\, dx = \lim_{t \to 0^+} \left[2\sqrt{x} \right]_t^\pi = 2\sqrt{\pi} \ \Rightarrow \ \int_0^\pi \frac{\sin x}{\sqrt{x}}\, dx \ \text{converges by (i)},$$

where $f(x) = \dfrac{\sin x}{\sqrt{x}}$, $g(x) = \dfrac{1}{\sqrt{x}}$, and $[a, b] = [0, \pi]$.

33 Since $\cosh x \geq 1$, it follows that $0 \leq \dfrac{1}{(x-2)^2} \leq \dfrac{\cosh x}{(x-2)^2}$ on $[0, 2)$ and

$$\int_0^2 \frac{1}{(x-2)^2}\, dx = \lim_{t \to 2^-} \int_0^t \frac{1}{(x-2)^2}\, dx = \lim_{t \to 2^-}\left[-\frac{1}{x-2}\right]_0^t = \infty \Rightarrow$$

$$\int_0^2 \frac{\cosh x}{(x-2)^2}\, dx \text{ diverges by (ii), where } f(x) = \frac{1}{(x-2)^2},\ g(x) = \frac{\cosh x}{(x-2)^2},$$

and $[a,\, b] = [0,\, 2]$.

35 Let n be a real number.

    (i)   If $n \geq 0$, $I = \displaystyle\int_0^1 x^n\, dx = \left[\frac{x^{n+1}}{n+1}\right]_0^1 = \frac{1}{n+1}$; C

    (ii) If $-1 < n < 0$, then $n + 1 > 0$ and $I = \displaystyle\lim_{t \to 0^+}\int_t^1 x^n\, dx = \lim_{t \to 0^+}\left[\frac{x^{n+1}}{n+1}\right]_t^1 =$

$$\frac{1}{n+1}\left(1 - \lim_{t \to 0^+} t^{n+1}\right) = \frac{1}{n+1}(1 - 0) = \frac{1}{n+1};\ \text{C}.$$

        Note: Since $n + 1 > 0$, $\displaystyle\lim_{t \to 0^+} t^{n+1} = 0$.

    (iii) If $n = -1$, $I = \displaystyle\lim_{t \to 0^+}\int_t^1 \frac{1}{x}\, dx = \lim_{t \to 0^+}\Big[\ln |x|\Big]_t^1 = \ln 1 - \lim_{t \to 0^+} \ln |t| = \infty$; D

    (iv) If $n < -1$, then $n + 1 < 0$ and $I = \displaystyle\lim_{t \to 0^+}\left[\frac{x^{n+1}}{n+1}\right]_t^1 =$

$$\frac{1}{n+1}\left(1 - \lim_{t \to 0^+} t^{n+1}\right) = \frac{1}{n+1}(1 - \infty) = \infty;\ \text{D}.$$

        Note: Since $n + 1 < 0$, $\displaystyle\lim_{t \to 0^+} t^{n+1} = \infty$ and $\dfrac{1}{n+1} < 0$.

Thus, I converges iff $n > -1$.

37 R is the region between the graph of $y = \dfrac{1}{\sqrt{x}}$, the x-axis,

        and the lines $x = 0$ and $x = 1$. $\dfrac{1}{\sqrt{x}}$ is discontinuous at $x = 0$.

    (a) $A = \displaystyle\int_0^1 \frac{1}{\sqrt{x}}\, dx = \lim_{t \to 0^+}\int_t^1 \frac{1}{\sqrt{x}}\, dx = \lim_{t \to 0^+}\Big[2\sqrt{x}\Big]_t^1 = 2(1 - 0) = 2.$

    (b) Using disks, $V = \pi\displaystyle\int_0^1 \left(\frac{1}{\sqrt{x}}\right)^2 dx = \pi \lim_{t \to 0^+}\int_t^1 \frac{1}{x}\, dx = \pi \lim_{t \to 0^+}\Big[\ln |x|\Big]_t^1 = \infty.$

Not possible.

41 $u = \sqrt{x}$, $u^2 = x$, and $2u\, du = dx$; $x = 0, 1 \Rightarrow u = 0, 1$.

    Thus, $\displaystyle\int_0^1 \frac{\cos x}{\sqrt{x}}\, dx = \int_0^1 \frac{\cos(u^2)}{u} 2u\, du = 2\int_0^1 \cos(u^2)\, du$. Let $f(x) = \cos(x^2)$.

    $T = (2)\dfrac{1-0}{2(4)}\left\{f(0) + 2\left[f(\tfrac{1}{4}) + f(\tfrac{1}{2}) + f(\tfrac{3}{4})\right] + f(1)\right\}$

        $\approx \tfrac{1}{4}\Big[1 + 2(0.9980) + 0.9689 + 0.8459) + 0.5403\Big] = \tfrac{1}{4}(7.1659) \approx 1.7915.$

[45] (a) From the formula for T, $0 \leq y \leq y_0$. Note that t can be written as a function of y. If t is undefined for some value of y when $0 \leq y \leq y_0$, then t is discontinuous and the integral is improper. Now, $y = 0 \Rightarrow y_0 e^{-kt} = 0 \Rightarrow e^{-kt} = 0$ $\{y_0 \neq 0\}$ and t is undefined at $y = 0$ since $e^{-kt} \neq 0$ for any finite value of t.

(b) First we must express t as a function of y. Solving for t gives the following.

$$y = y_0 e^{-kt} \Rightarrow \frac{y}{y_0} = e^{-kt} \Rightarrow t = -\frac{1}{k} \ln \frac{y}{y_0} \Rightarrow T = \frac{1}{y_0} \int_0^{y_0} -\frac{1}{k} \ln \frac{y}{y_0} \, dy.$$

$$z = \frac{y}{y_0} \text{ and } dz = \frac{1}{y_0} dy \Rightarrow$$

$$T = -\frac{1}{k} \int_0^1 \ln z \, dz = -\frac{1}{k} \lim_{t \to 0^+} \left[z \ln z - z \right]_t^1 \{ \text{Example 3, §9.1} \}$$

$$= \frac{1}{k} + \frac{1}{k} \lim_{t \to 0^+} t \ln t \; \{ 0 \cdot \infty \}$$

$$= \frac{1}{k} + \frac{1}{k} \lim_{t \to 0^+} \frac{\ln t}{1/t} \; \{ \frac{\infty}{\infty} \}$$

$$= \frac{1}{k} + \frac{1}{k} \lim_{t \to 0^+} \frac{1/t}{-1/t^2}$$

$$= \frac{1}{k} + \frac{1}{k} \lim_{t \to 0^+} (-t) = \frac{1}{k} + 0 = \frac{1}{k}.$$

Time τ occurs when $\frac{y}{y_0} = \frac{1}{2}$ or $t \{ = \tau \} = -\frac{1}{k} \ln \frac{1}{2} = \frac{\ln 2}{k} \Rightarrow k = \frac{\ln 2}{\tau}$.

$$\text{Thus, } T = \frac{1}{k} = \frac{1}{\ln 2/\tau} = \frac{\tau}{\ln 2}.$$

10.5 Review Exercises

[3] Since $\lim_{x \to \infty} x^2 + 2x + 3 = \infty$ and $\lim_{x \to \infty} \ln(x + 1) = \infty$,

$\dfrac{x^2 + 2x + 3}{\ln(x + 1)}$ has the indeterminate form $\frac{\infty}{\infty}$ and L'Hôpital's rule applies.

$$\lim_{x \to \infty} \frac{x^2 + 2x + 3}{\ln(x + 1)} \{ \tfrac{\infty}{\infty} \} = \lim_{x \to \infty} \frac{D_x(x^2 + 2x + 3)}{D_x[\ln(x + 1)]} =$$

$$\lim_{x \to \infty} \frac{2x + 2}{1/(x + 1)} = \lim_{x \to \infty} (2x + 2)(x + 1) = \infty$$

[5] Since $\lim_{x \to 0} (e^{2x} - e^{-2x} - 4x) = 1 - 1 - 0 = 0$ and $\lim_{x \to 0} x^3 = 0$,

$\dfrac{e^{2x} - e^{-2x} - 4x}{x^3}$ has the indeterminate form $\frac{0}{0}$ at $x = 0$.

Thus, L'Hôpital's rule applies and we will need to use it three times.

$$\lim_{x \to 0} \frac{e^{2x} - e^{-2x} - 4x}{x^3} \{ \tfrac{0}{0} \} = \lim_{x \to 0} \frac{2e^{2x} + 2e^{-2x} - 4}{3x^2} \{ \tfrac{0}{0} \} =$$

$$\lim_{x \to 0} \frac{4e^{2x} - 4e^{-2x}}{6x} \{ \tfrac{0}{0} \} = \lim_{x \to 0} \frac{8e^{2x} + 8e^{-2x}}{6} = \frac{16}{6} = \frac{8}{3}$$

boxed{11} Since $\lim_{x \to 0} (1 + 8x^2) = 1$ and $\lim_{x \to 0} \frac{1}{x^2} = \infty$, $\lim_{x \to 0} (1 + 8x^2)^{1/x^2}$ has the form 1^∞.

$y = (1 + 8x^2)^{1/x^2} \Rightarrow \ln y = \frac{1}{x^2} \ln(1 + 8x^2)$. $\lim_{x \to 0} \ln y \{\infty \cdot 0\} =$

$\lim_{x \to 0} \frac{\ln(1 + 8x^2)}{x^2} \{\frac{0}{0}\} = \lim_{x \to 0} \frac{16x/(1 + 8x^2)}{2x} = \lim_{x \to 0} \frac{8}{1 + 8x^2} = 8$. Thus, $L = e^8$.

boxed{13} Since $\lim_{x \to \infty} (e^x + 1) = \infty$ and $\lim_{x \to \infty} \frac{1}{x} = 0$, $\lim_{x \to \infty} (e^x + 1)^{1/x}$ has the form ∞^0.

$y = (e^x + 1)^{1/x} \Rightarrow \ln y = \frac{1}{x} \ln(e^x + 1)$. $\lim_{x \to \infty} \ln y \{0 \cdot \infty\} =$

$\lim_{x \to \infty} \frac{\ln(e^x + 1)}{x} \{\frac{\infty}{\infty}\} = \lim_{x \to \infty} \frac{e^x}{e^x + 1} \{\frac{\infty}{\infty}\} = \lim_{x \to \infty} \frac{e^x}{e^x} = 1$. Thus, $L = e^1$.

boxed{15} This limit can be calculated without L'Hôpital's rule. To do this, multiply both

numerator and denominator by $1/x$. $\left(\text{Since } x > 0, \frac{1}{x} = \sqrt{\frac{1}{x^2}}.\right)$ $\lim_{x \to \infty} \frac{\sqrt{x^2 + 1}}{x} =$

$\lim_{x \to \infty} \frac{(1/x)\sqrt{x^2 + 1}}{(1/x) \cdot x} = \lim_{x \to \infty} \frac{\sqrt{(1/x^2)(x^2 + 1)}}{1} = \lim_{x \to \infty} \sqrt{1 + 1/x^2} = \sqrt{1 + 0} = 1$

boxed{17} $\int_4^\infty \frac{1}{\sqrt{x}} \, dx = \lim_{t \to \infty} \int_4^t x^{-1/2} \, dx = \lim_{t \to \infty} \left[2\sqrt{x}\right]_4^t = \left(\lim_{t \to \infty} 2\sqrt{t}\right) - 4 = \infty$; D

boxed{19} The integrand is discontinuous at $x = -2$ and the integral has an infinite limit.

$\int_{-\infty}^0 \frac{1}{x + 2} \, dx = \int_{-\infty}^{-3} \frac{1}{x + 2} \, dx + \int_{-3}^{-2} \frac{1}{x + 2} \, dx + \int_{-2}^0 \frac{1}{x + 2} \, dx = I_1 + I_2 + I_3$.

$I_3 = \lim_{t \to -2+} \int_t^0 \frac{1}{x + 2} \, dx = \lim_{t \to -2+} \left[\ln|x + 2|\right]_t^0 = \ln 2 - \lim_{t \to -2+} \ln|t + 2| = \infty$; D

Note: If any of the integrals diverge, then I diverges. In this exercise,

           all three integrals diverge. You may evaluate any one of them.

boxed{21} The integrand $\frac{1}{\sqrt[3]{x}}$ is discontinuous at $x = 0$.

$I = \int_{-8}^1 \frac{1}{\sqrt[3]{x}} \, dx = \int_{-8}^0 x^{-1/3} \, dx + \int_0^1 x^{-1/3} \, dx = I_1 + I_2$.

$I_1 = \lim_{t \to 0-} \int_{-8}^t x^{-1/3} \, dx = \lim_{t \to 0-} \left[\frac{3}{2}x^{2/3}\right]_{-8}^t = \frac{3}{2}(0 - 4) = -6$.

$I_2 = \lim_{t \to 0+} \int_t^1 x^{-1/3} \, dx = \lim_{t \to 0+} \left[\frac{3}{2}x^{2/3}\right]_t^1 = \frac{3}{2}(1 - 0) = \frac{3}{2}$.

$I = I_1 + I_2 = -6 + \frac{3}{2} = -\frac{9}{2}$. C

[25] Note that the integrand is an even function, and hence is symmetric with respect to

the y-axis. $\displaystyle\int_{-\infty}^{\infty} \frac{1}{e^x + e^{-x}}\, dx = \int_{-\infty}^{0} \frac{1}{e^x + e^{-x}}\, dx + \int_{0}^{\infty} \frac{1}{e^x + e^{-x}}\, dx =$

$\displaystyle 2\int_{0}^{\infty} \frac{1}{e^x + e^{-x}}\, dx = 2\int_{0}^{\infty} \frac{1}{e^x + e^{-x}} \cdot \frac{e^x}{e^x}\, dx = 2\lim_{t \to \infty} \int_{0}^{t} \frac{e^x}{e^{2x} + 1}\, dx =$

$\{$ To integrate, let $u = e^x$, $u^2 = e^{2x}$, and $du = e^x\, dx.\}$

$$2\lim_{t \to \infty}\left[\tan^{-1} e^x\right]_0^t = 2\left[\left(\lim_{t \to \infty} \tan^{-1} e^t\right) - (\tan^{-1} 1)\right] = 2(\tfrac{\pi}{2} - \tfrac{\pi}{4}) = \tfrac{\pi}{2};\ C$$

[29] $u = \frac{1}{x}$, $x = \frac{1}{u}$, and $dx = -\frac{1}{u^2}\, du$. $x = 1 \Rightarrow u = 1$ and $x \to \infty \Rightarrow u = 0^+$.

$$I = \int_{1}^{\infty} e^{-x^2}\, dx = \lim_{t \to 0^+} \int_{1}^{t} e^{-1/u^2}\left(-\frac{1}{u^2}\right) du = \lim_{t \to 0^+} \int_{t}^{1} \frac{e^{-1/u^2}}{u^2}\, du.$$

Let $f(u) = \dfrac{e^{-1/u^2}}{u^2}$. Note that $f(0)$ is undefined. However, f has a removable

discontinuity at $u = 0$. To remove it, we must find $\lim\limits_{u \to 0} f(u)$ and define $f(0)$ to

equal this limit. Since $\displaystyle\lim_{u \to 0^+} \frac{e^{-1/u^2}}{u^2} = \lim_{v \to \infty} v^2 e^{-v^2} = \lim_{v \to \infty} \frac{v^2}{e^{v^2}} = 0$

(where $v = \frac{1}{u}$ and L'Hôpital's rule was used), define $f(0) = 0$.

$S = \frac{1-0}{3(4)}\left[f(0) + 4f(\tfrac{1}{4}) + 2f(\tfrac{1}{2}) + 4f(\tfrac{3}{4}) + f(1)\right]$

$\approx \frac{1}{12}\left[0 + 4(0) + 2(0.0733) + 4(0.3005) + 0.3679\right] = \frac{1}{12}(1.7165) \approx 0.1430.$

[31] Since $(\sin t)^{2/3} \geq 0$ and generates an infinite area between its graph and the x-axis as

$x \to \infty$, $\displaystyle\lim_{x \to \infty} f(x) = \int_{1}^{x} (\sin t)^{2/3}\, dt = \infty$ and $\displaystyle\lim_{x \to \infty} g(x) = \infty.$

Thus, $\dfrac{f(x)}{g(x)}$ has an indeterminate form of $\frac{\infty}{\infty}$ as $x \to \infty$ and L'Hôpital's rule applies.

$\displaystyle\lim_{x \to \infty} \frac{f(x)}{g(x)}\ \{\tfrac{\infty}{\infty}\} = \lim_{x \to \infty} \frac{f'(x)}{g'(x)}\ \{$ by (5.35), $f'(x) = (\sin x)^{2/3}\ \} =$

$$\lim_{x \to \infty} \frac{(\sin x)^{2/3}}{2x} = 0.\ \text{Note that } \left|(\sin x)^{2/3}\right| \leq 1 \text{ and } 2x \to \infty.$$

Chapter 11: Infinite Series

Note: In Exercises 1–16, the first four terms are found by

substituting 1, 2, 3, and 4 for n in the nth term.

[1] $a_n = \dfrac{n}{3n+2} \Rightarrow a_1 = \dfrac{1}{3(1)+2} = \dfrac{1}{5}, \; a_2 = \dfrac{2}{3(2)+2} = \dfrac{2}{8} = \dfrac{1}{4}$,

$$a_3 = \dfrac{3}{3(3)+2} = \dfrac{3}{11}, \; a_4 = \dfrac{4}{3(4)+2} = \dfrac{4}{14} = \dfrac{2}{7}.$$

$$\lim_{n \to \infty} a_n = \lim_{n \to \infty} \dfrac{n}{3n+2} = \lim_{n \to \infty} \left(\dfrac{n}{3n+2} \cdot \dfrac{1/n}{1/n} \right) = \lim_{n \to \infty} \dfrac{1}{3+2/n} = \dfrac{1}{3+0} = \dfrac{1}{3}$$

[5] Since $a_n = -5$ for every n, it follows that $a_1 = -5$, $a_2 = -5$, $a_3 = -5$,

and $a_4 = -5$. $\displaystyle \lim_{n \to \infty} a_n = \lim_{n \to \infty} (-5) = -5$

[11] $a_n = (-1)^{n+1} \dfrac{3n}{n^2 + 4n + 5}$. First, consider the factor $(-1)^{n+1}$. $(-1)^{n+1}$ is equal to

$+1$ when $n+1$ is an even integer and is equal to -1 when $n+1$ is an odd integer.
$(-1)^{n+1}$ has the effect of causing the terms of the sequence to alternate in sign.

Thus, $a_1 = (-1)^{1+1} \dfrac{3(1)}{1^2 + 4(1) + 5} = \dfrac{3}{10}, \; a_2 = (-1)^{2+1} \dfrac{3(2)}{2^2 + 4(2) + 5} = -\dfrac{6}{17}$,

$$a_3 = (-1)^{3+1} \dfrac{3(3)}{3^2 + 4(3) + 5} = \dfrac{9}{26}, \; a_4 = (-1)^{4+1} \dfrac{3(4)}{4^2 + 4(4) + 5} = -\dfrac{12}{37}.$$

Since the terms alternate in sign, we will examine $\displaystyle \lim_{n \to \infty} |a_n|$.

$$\lim_{n \to \infty} |a_n| = \lim_{n \to \infty} \left| (-1)^{n+1} \dfrac{3n}{n^2 + 4n + 5} \right| = \lim_{n \to \infty} \dfrac{3n}{n^2 + 4n + 5} = 0,$$

since the highest power of n is greater in the denominator than in the numerator.

Thus, by (11.8), $\displaystyle \lim_{n \to \infty} a_n = \lim_{n \to \infty} (-1)^{n+1} \dfrac{3n}{n^2 + 4n + 5} = 0.$

[15] $a_n = 1 + (-1)^{n+1} \Rightarrow a_1 = 1 + (-1)^{1+1} = 1 + 1 = 2,$

$a_2 = 1 + (-1)^{2+1} = 1 + (-1) = 0,$

$a_3 = 1 + (-1)^{3+1} = 1 + 1 = 2,$

$a_4 = 1 + (-1)^{4+1} = 1 + (-1) = 0.$

The sequence follows the pattern 2, 0, 2, 0, 2, 0, ..., and a_n will never approach one

value. Thus, $\displaystyle \lim_{n \to \infty} a_n = \lim_{n \to \infty} \left[1 + (-1)^{n+1} \right]$ DNE since it does not converge to a

real number L.

Note: Let C denote *converges* and D denote *diverges.*

Also, let L denote the limit of the sequence, if it exists.

$\boxed{17}$ $\lim\limits_{n\to\infty} a_n = \lim\limits_{n\to\infty} 6(-\frac{5}{6})^n = 6 \lim\limits_{n\to\infty} (-\frac{5}{6})^n = 6 \cdot 0 = 0$ by (11.6)(i) with $r = -\frac{5}{6}$; C

$\boxed{19}$ In order to determine the limit, refer to the graph of $y = \arctan x$ (Figure 8.6, page

429). From the graph, we can see that $\lim\limits_{x\to\infty} \arctan x = \frac{\pi}{2}$. Thus, by (11.5)(i),

$\lim\limits_{n\to\infty} a_n = \lim\limits_{n\to\infty} \arctan n = \frac{\pi}{2}$; C.

$\boxed{23}$ Using (11.5)(i) and L'Hôpital's rule (10.2), $\lim\limits_{x\to\infty} \frac{\ln x}{x} \{\frac{\infty}{\infty}\} = \lim\limits_{x\to\infty} \frac{1/x}{1} = 0$,

and hence $\lim\limits_{n\to\infty} |a_n| = \lim\limits_{n\to\infty} \left|(-1)^n \frac{\ln n}{n}\right| = \lim\limits_{n\to\infty} \frac{\ln n}{n} = 0$.

Thus, by (11.8), $\lim\limits_{n\to\infty} a_n = \lim\limits_{n\to\infty} (-1)^n \frac{\ln n}{n} = 0$; C.

$\boxed{25}$ Using (11.5)(ii), $\lim\limits_{x\to\infty} \frac{4x^4 + 1}{2x^2 - 1} \{\frac{\infty}{\infty}\} = \lim\limits_{x\to\infty} \frac{16x^3}{4x} = \lim\limits_{x\to\infty} 4x^2 = \infty$.

Thus, $\lim\limits_{n\to\infty} a_n = \lim\limits_{n\to\infty} \frac{4n^4 + 1}{2n^2 - 1} = \infty$; D.

$\boxed{29}$ By (7.32)(ii), $\lim\limits_{n\to\infty} a_n = \lim\limits_{n\to\infty} (1 + \frac{1}{n})^n = e$; C.

$\boxed{31}$ This exercise is similar to Example 8.

Since $\lim\limits_{n\to\infty} \sin n$ DNE, we will apply (11.7) to determine $\lim\limits_{n\to\infty} a_n = \lim\limits_{n\to\infty} \frac{\sin n}{2^n}$.

Now, $-1 \le \sin n \le 1$ for all $n \Rightarrow -\frac{1}{2^n} \le \frac{\sin n}{2^n} \le \frac{1}{2^n}$.

Since $\lim\limits_{n\to\infty} -\frac{1}{2^n} = 0$ and $\lim\limits_{n\to\infty} \frac{1}{2^n} = 0$, it follows that $\lim\limits_{n\to\infty} \frac{\sin n}{2^n} = 0$ by (11.7); C.

$\boxed{33}$ $\lim\limits_{n\to\infty} a_n = \lim\limits_{n\to\infty} \left(\frac{n^2}{2n - 1} - \frac{n^2}{2n + 1}\right) = \lim\limits_{n\to\infty} \frac{n^2(2n + 1) - n^2(2n - 1)}{(2n - 1)(2n + 1)} =$

$\lim\limits_{n\to\infty} \frac{2n^2}{4n^2 - 1} = \lim\limits_{n\to\infty} \frac{2n^2/n^2}{4n^2/n^2 - 1/n^2} = \frac{1}{2}$; C.

$\boxed{35}$ $a_n = \cos \pi n \Rightarrow a_1 = \cos \pi = -1$, $a_2 = \cos 2\pi = 1$, $a_3 = \cos 3\pi = -1$, $a_4 =$

$\cos 4\pi = 1$. The sequence follows the pattern $-1, 1, -1, 1, -1, 1, \ldots$, and does not

approach one value for large values of n. Thus, $\lim\limits_{n\to\infty} a_n = \lim\limits_{n\to\infty} \cos \pi n$ DNE; D.

$\boxed{37}$ By Exercise 17, §10.2, $\lim\limits_{x\to\infty} x^{1/x} = 1$.

Thus, by (11.5)(i), $\lim\limits_{n\to\infty} a_n = \lim\limits_{n\to\infty} n^{1/n} = 1$; C.

$\boxed{41}$ $\lim\limits_{n\to\infty} a_n = \lim\limits_{n\to\infty} (\sqrt{n + 1} - \sqrt{n}) = \lim\limits_{n\to\infty} \left(\frac{\sqrt{n + 1} - \sqrt{n}}{1} \cdot \frac{\sqrt{n + 1} + \sqrt{n}}{\sqrt{n + 1} + \sqrt{n}}\right) =$

$\lim\limits_{n\to\infty} \frac{(n + 1) - (n)}{\sqrt{n + 1} + \sqrt{n}} = \lim\limits_{n\to\infty} \frac{1}{\sqrt{n + 1} + \sqrt{n}} = 0$; C.

43 (a) Next year's bird population on island A is determined by the number of birds that stay (90% of its present population) and the number of birds that migrate (5% of island C's present population). Thus, $A_{n+1} = 0.90A_n + 0.05C_n$. In a similar manner, $B_{n+1} = 0.80B_n + 0.10A_n$ and $C_{n+1} = 0.95C_n + 0.20B_n$.

(b) Let $\lim\limits_{n \to \infty} A_n = a$, $\lim\limits_{n \to \infty} B_n = b$, and $\lim\limits_{n \to \infty} C_n = c$. Then, as $n \to \infty$,

$$\lim_{n \to \infty} A_{n+1} = \lim_{n \to \infty} 0.90A_n + \lim_{n \to \infty} 0.05C_n \Rightarrow$$

$$a = 0.90a + 0.05c \Rightarrow 0.1a = 0.05c \Rightarrow c = 2a,$$

$$\lim_{n \to \infty} B_{n+1} = \lim_{n \to \infty} 0.80B_n + \lim_{n \to \infty} 0.10A_n \Rightarrow$$

$$b = 0.80b + 0.10a \Rightarrow 0.2b = 0.1a \Rightarrow a = 2b,$$

and $\quad \lim\limits_{n \to \infty} C_{n+1} = \lim\limits_{n \to \infty} 0.95C_n + \lim\limits_{n \to \infty} 0.20B_n \Rightarrow$

$$c = 0.95c + 0.20b \Rightarrow 0.05c = 0.2b \Rightarrow c = 4b.$$

Now, $35{,}000 = a + b + c = 2b + b + 4b = 7b \Rightarrow b = 5000$.

Thus, there will be 5000 birds on B, 10,000 birds on A, and 20,000 birds on C.

47 (a) $a_1 = 1$, $a_2 = \cos a_1 = \cos 1 \approx 0.540302$, $a_3 = \cos a_2 \approx 0.857553$,

$a_4 = \cos a_3 \approx 0.654290$, $a_5 = \cos a_4 \approx 0.793480$. As this process is repeated *many* times, the sequence appears to converge to approximately 0.7390851.

(b) $a_{k+1} = \cos a_k \Rightarrow \lim\limits_{k \to \infty} a_{k+1} = \lim\limits_{k \to \infty} \cos a_k \Rightarrow \lim\limits_{k \to \infty} a_{k+1} = \cos\left(\lim\limits_{k \to \infty} a_k\right) \Rightarrow$

$$L = \cos L.$$

51 (a) $f(x) = \frac{1}{4}\sin x \cos x + 1 = \frac{1}{8}(2\sin x \cos x) + 1 = \frac{1}{8}\sin 2x + 1 \Rightarrow$
$\left|f'(x)\right| = \frac{1}{4}|\cos 2x| \le \frac{1}{4} < 1$, since $|\cos 2x| \le 1$ for any x. Let $B = \frac{1}{4}$.

Thus, the sequence converges for any a_1.

(b) Letting $a_{k+1} = f(a_k) = \frac{1}{4}\sin a_k \cos a_k + 1 = \frac{1}{8}\sin 2a_k + 1$ with
$a_1 = 1$ and then $a_1 = -100$ gives the following results.

$a_1 = 1$	$a_1 = -100$
$a_2 = 1.113662$	$a_2 = 1.109162$
$a_3 = 1.099015$	$a_3 = 1.099697$
$a_4 = 1.101207$	$a_4 = 1.101107$
$a_5 = 1.100884$	$a_5 = 1.100899$

It appears that $\lim\limits_{n \to \infty} a_n \approx 1.10$.

| Exercises 11.2 |

Note: It is important that you understand the difference between a sequence and a series. Carefully compare (11.1) and (11.11). In a series, the terms are added together, whereas in a sequence, they are not.

$\boxed{1}$ (a) $a_n = \dfrac{-2}{(2n+5)(2n+3)} \Rightarrow S_1 = a_1 = \dfrac{-2}{(2\cdot 1 + 5)(2\cdot 1 + 3)} = \dfrac{-2}{7\cdot 5} = -\dfrac{2}{35};$

$S_2 = (a_1) + a_2 = S_1 + a_2 = -\dfrac{2}{35} + \dfrac{-2}{63} = -\dfrac{4}{45};$

$S_3 = (a_1 + a_2) + a_3 = S_2 + a_3 = -\dfrac{4}{45} + \dfrac{-2}{99} = -\dfrac{6}{55}.$

(b) Using partial fractions, $a_n = \dfrac{-2}{(2n+5)(2n+3)} = \dfrac{1}{2n+5} - \dfrac{1}{2n+3}.$

$S_n = a_1 + a_2 + a_3 + \cdots + a_n$

$\quad = \left(\dfrac{1}{7} - \dfrac{1}{5}\right) + \left(\dfrac{1}{9} - \dfrac{1}{7}\right) + \left(\dfrac{1}{11} - \dfrac{1}{9}\right) + \cdots + \left(\dfrac{1}{2n+5} - \dfrac{1}{2n+3}\right)$

$\quad = -\dfrac{1}{5} + \left(\dfrac{1}{7} - \dfrac{1}{7}\right) + \left(\dfrac{1}{9} - \dfrac{1}{9}\right) + \cdots + \left(\dfrac{1}{2n+3} - \dfrac{1}{2n+3}\right) + \dfrac{1}{2n+5}$

$\quad = \dfrac{1}{2n+5} - \dfrac{1}{5} = -\dfrac{2n}{5(2n+5)}.$

Note: It may be easier to compute the values in part (a)

after finding the general formula in part (b).

(c) $S = \lim_{n\to\infty} S_n$

$\quad = \lim_{n\to\infty}\left[-\dfrac{2n}{5(2n+5)}\right]$ {consider only the terms with the highest powers of n}

$\quad = -\dfrac{2}{5\cdot 2} = -\dfrac{1}{5}.$

$\boxed{5}$ (a) $S_1 = a_1 = \ln\dfrac{1}{2} = \ln 1 - \ln 2 = 0 - \ln 2 = -\ln 2;$

$S_2 = S_1 + a_2 = -\ln 2 + \ln\dfrac{2}{3} = -\ln 2 + (\ln 2 - \ln 3) = -\ln 3;$

$S_3 = S_2 + a_3 = -\ln 3 + \ln\dfrac{3}{4} = -\ln 3 + (\ln 3 - \ln 4) = -\ln 4.$

(b) $a_n = \ln\dfrac{n}{n+1} = \ln n - \ln(n+1).$

$S_n = (\ln 1 - \ln 2) + (\ln 2 - \ln 3) + (\ln 3 - \ln 4) + \cdots + \left[\ln n - \ln(n+1)\right]$

$\quad = \ln 1 + (\ln 2 - \ln 2) + (\ln 3 - \ln 3) + \cdots + (\ln n - \ln n) - \ln(n+1)$

$\quad = \ln 1 - \ln(n+1) = -\ln(n+1).$

(c) $S = \lim_{n\to\infty} S_n = \lim_{n\to\infty}\left[-\ln(n+1)\right] = -\lim_{n\to\infty}\ln(n+1) = -\infty.$

Since $\{S_n\}$ diverges, the series $\sum a_n$ diverges and has no sum.

Note: From (11.15), the series converges if $|r| < 1$ and diverges is $|r| \geq 1$.

$\boxed{7}$ $3 + \frac{3}{4} + \cdots + \frac{3}{4^{n-1}} + \cdots = 3 + 3(\frac{1}{4}) + 3(\frac{1}{4})^2 + \cdots + 3(\frac{1}{4})^{n-1} + \cdots$.

From this series, we see that $a = 3$ and $r = \frac{1}{4}$. Thus, $S = \frac{a}{1 - r} = \frac{3}{1 - \frac{1}{4}} = 4$; C.

$\boxed{11}$ $0.37 + 0.0037 + \cdots + \frac{37}{(100)^n} + \cdots =$

$$0.37 + 0.37(\frac{1}{100}) + 0.37(\frac{1}{100})^2 + \cdots + 0.37(\frac{1}{100})^{n-1} + \cdots.$$

From this series, we see that $a = 0.37$ and $r = \frac{1}{100}$.

$$\text{Thus, } S = \frac{a}{1 - r} = \frac{0.37}{1 - \frac{1}{100}} = \frac{37}{99}; \text{ C.}$$

$\boxed{13}$ $\displaystyle\sum_{n=1}^{\infty} 2^{-n}3^{n-1} = \sum_{n=1}^{\infty} \frac{3^{n-1}}{2^n} = \sum_{n=1}^{\infty} \frac{1}{2}(\frac{3}{2})^{n-1} =$

$$\frac{1}{2} + \frac{1}{2}(\frac{3}{2}) + \frac{1}{2}(\frac{3}{2})^2 + \cdots + \frac{1}{2}(\frac{3}{2})^{n-1} + \cdots \Rightarrow a = \frac{1}{2} \text{ and } r = \frac{3}{2} > 1 \Rightarrow D$$

$\boxed{17}$ $1 - x + x^2 - x^3 + \cdots + (-1)^n x^n + \cdots =$

$$1 + 1(-x) + 1(-x)^2 + 1(-x)^3 + \cdots.$$

This is a geometric series with $a = 1$ and $r = -x$. S converges if $|r| < 1$.

Thus, S converges if $|-x| < 1$, or, equivalently, $-1 < x < 1$.

$$a = 1, r = -x \Rightarrow S = \frac{a}{1 - r} = \frac{1}{1 - (-x)} = \frac{1}{1 + x}.$$

$\boxed{19}$ $\frac{1}{2} + \frac{(x - 3)}{4} + \frac{(x - 3)^2}{8} + \cdots + \frac{(x - 3)^n}{2^{n+1}} + \cdots =$

$$\frac{1}{2} + \frac{1}{2}\left(\frac{x - 3}{2}\right) + \frac{1}{2}\left(\frac{x - 3}{2}\right)^2 + \cdots + \frac{1}{2}\left(\frac{x - 3}{2}\right)^n + \cdots.$$

This is a geometric series with $a = \frac{1}{2}$ and $r = \frac{x - 3}{2}$.

$$|r| < 1 \Rightarrow \left|\frac{x - 3}{2}\right| < 1 \Rightarrow |x - 3| < 2 \Rightarrow -2 < x - 3 < 2 \Rightarrow 1 < x < 5.$$

$$S = \frac{a}{1 - r} = \frac{1/2}{1 - \left(\frac{x - 3}{2}\right)} = \frac{1}{5 - x}.$$

$\boxed{21}$ $0.\overline{23} = 0.23 + 0.0023 + 0.000023 + \cdots$

$$= \frac{23}{100} + \frac{23}{(100)^2} + \frac{23}{(100)^3} + \cdots$$

$$= \frac{23}{100} + \frac{23}{100}(\frac{1}{100}) + \frac{23}{100}(\frac{1}{100})^2 + \cdots$$

$$= \frac{\frac{23}{100}}{1 - \frac{1}{100}} = \frac{23}{99} \text{ since } a = \frac{23}{100} \text{ and } r = \frac{1}{100}$$

$\boxed{23}$ $3.2\overline{394} = 3.2 + 0.0394 + 0.0000394 + 0.0000000394 + \cdots$

$$= 3.2 + \frac{394}{10^4} + \frac{394}{10^7} + \frac{394}{10^{10}} + \cdots$$

$$= 3.2 + \frac{394}{10{,}000} + \frac{394}{10{,}000}\left(\frac{1}{1000}\right) + \frac{394}{10{,}000}\left(\frac{1}{1000}\right)^2 + \cdots$$

$$= 3.2 + \frac{\frac{394}{10{,}000}}{1 - \frac{1}{1000}} = 3.2 + \frac{394}{9990} = \frac{16{,}181}{4995}$$

$\boxed{25}$ $\dfrac{1}{4\cdot 5} + \dfrac{1}{5\cdot 6} + \cdots + \dfrac{1}{(n+3)(n+4)} + \cdots = \displaystyle\sum_{n=1}^{\infty} \dfrac{1}{(n+3)(n+4)} = \sum_{n=4}^{\infty} \dfrac{1}{n(n+1)}.$

By Example 1, $\displaystyle\sum_{n=1}^{\infty} \dfrac{1}{n(n+1)} = 1.$

By (11.19) with $k = 3$, $\displaystyle\sum_{n=4}^{\infty} \dfrac{1}{n(n+1)}$ also converges.

$\boxed{29}$ $\dfrac{1}{4} + \dfrac{1}{5} + \cdots + \dfrac{1}{n+3} + \cdots = \displaystyle\sum_{n=1}^{\infty} \dfrac{1}{n+3} = \sum_{n=4}^{\infty} \dfrac{1}{n}.$

Since $\displaystyle\sum_{n=1}^{\infty} \dfrac{1}{n}$ diverges by (11.14), so does $\displaystyle\sum_{n=4}^{\infty} \dfrac{1}{n}$ by (11.19).

$\boxed{33}$ Since $\displaystyle\lim_{n\to\infty} a_n = \lim_{n\to\infty} \dfrac{3n}{5n-1} = \dfrac{3}{5} \neq 0$, the series diverges.

$\boxed{35}$ Since $\displaystyle\lim_{n\to\infty} a_n = \lim_{n\to\infty} \dfrac{1}{n^2+3} = 0$, further investigation is necessary.

Note: The nth-term test *never* indicates that a series is convergent.

$\boxed{39}$ Since $\displaystyle\lim_{n\to\infty} a_n = \lim_{n\to\infty} \dfrac{n}{\ln(n+1)} \left\{\dfrac{\infty}{\infty}\right\} = \lim_{n\to\infty} \dfrac{1}{1/(n+1)} = \infty \neq 0,$

the series diverges.

$\boxed{41}$ Since $\displaystyle\sum_{n=3}^{\infty} (\tfrac{1}{4})^n = \sum_{n=1}^{\infty} (\tfrac{1}{4})^3(\tfrac{1}{4})^{n-1}$ and $\displaystyle\sum_{n=3}^{\infty} (\tfrac{3}{4})^n = \sum_{n=1}^{\infty} (\tfrac{3}{4})^3(\tfrac{3}{4})^{n-1}$ are both convergent

geometric series, $\displaystyle\sum_{n=3}^{\infty} (\tfrac{1}{4})^n + \sum_{n=3}^{\infty} (\tfrac{3}{4})^n = \sum_{n=3}^{\infty} \left[(\tfrac{1}{4})^n + (\tfrac{3}{4})^n\right]$ is also convergent.

$$\sum_{n=3}^{\infty} \left[(\tfrac{1}{4})^n + (\tfrac{3}{4})^n\right] = \sum_{n=3}^{\infty} (\tfrac{1}{4})^n \left\{ a = (\tfrac{1}{4})^3,\, r = \tfrac{1}{4} \right\} + \sum_{n=3}^{\infty} (\tfrac{3}{4})^n \left\{ a = (\tfrac{3}{4})^3,\, r = \tfrac{3}{4} \right\}$$

$$= \frac{1/4^3}{1 - \frac{1}{4}} + \frac{3^3/4^3}{1 - \frac{3}{4}} = \frac{1}{48} + \frac{27}{16} = \frac{41}{24};\ \text{C}$$

$\boxed{45}$ Since $\displaystyle\sum_{n=1}^{\infty} \dfrac{1}{8^n} = \sum_{n=1}^{\infty} (\tfrac{1}{8})(\tfrac{1}{8})^{n-1} \left\{ \text{of the form } \sum_{n=1}^{\infty} ar^{n-1} \text{ with } a = \tfrac{1}{8} \text{ and } r = \tfrac{1}{8} \right\}$

$$= \frac{\frac{1}{8}}{1 - \frac{1}{8}} = \frac{1}{7} \text{ and by Example 1, } \sum_{n=1}^{\infty} \frac{1}{n(n+1)} = 1,$$

the series $\displaystyle\sum_{n=1}^{\infty} \left[\frac{1}{8^n} + \frac{1}{n(n+1)}\right]$ converges to $\frac{1}{7} + 1 = \frac{8}{7}.$

[47] $\displaystyle\sum_{n=1}^{\infty}\left(\frac{5}{n+2}-\frac{5}{n+3}\right) = \sum_{n=3}^{\infty}\left(\frac{5}{n}-\frac{5}{n+1}\right)$ { replace $n+2$ with n, or, equivalently,

n with $n-2$ }

$$= 5\sum_{n=3}^{\infty}\left(\frac{1}{n}-\frac{1}{n+1}\right) = 5\sum_{n=3}^{\infty}\frac{1}{n(n+1)}$$

$$= 5\left[\sum_{n=1}^{\infty}\frac{1}{n(n+1)} - \frac{1}{1\cdot 2} - \frac{1}{2\cdot 3}\right]$$ { subtract the first two

terms from the sum of the series in Example 1 }

$$= 5\left[1 - \frac{1}{2} - \frac{1}{6}\right] = 5\left(\frac{1}{3}\right) = \frac{5}{3}; \text{ C}$$

[51] We must find an m such that $S_m = 1 + \frac{1}{2} + \frac{1}{3} + \cdots + \frac{1}{m} \geq 3$.

$S_m = S_{2^k} > (k+1)\left(\frac{1}{2}\right)$ { from Example 3 } $= 3 \Rightarrow k+1 = 6 \Rightarrow k = 5$ and

$$m = 2^k = 2^5 = 32.\ S_{32} = 1 + \frac{1}{2} + \frac{1}{3} + \cdots + \frac{1}{32} \approx 4.058495.$$

[55] The ball initially falls 10 m, after which, in each up and down cycle it travels:

5 m up, 5 m down; $\frac{5}{2}$ m up, $\frac{5}{2}$ m down; $\frac{5}{4}$ m up, $\frac{5}{4}$ m down; etc. This total distance

traveled is $d = 10 + 2(5) + 2\left(\frac{5}{2}\right) + 2\left(\frac{5}{4}\right) + \cdots$

$$= 10 + 10 + 10\left(\frac{1}{2}\right) + 10\left(\frac{1}{4}\right) + \cdots$$

$$= 10 + \sum_{n=0}^{\infty} 10\left(\frac{1}{2}\right)^n = 10 + \frac{10}{1-\frac{1}{2}} = 10 + 20 = 30 \text{ m}.$$

[57] (a) Immediately after the first dose, there are Q units of the drug in the bloodstream.

$A(1) = Q$. After the second dose, there is a new Q units plus $A(1)e^{-cT} = Qe^{-cT}$ units from the first dose. Thus, $A(2) = Q + Qe^{-cT}$. Similarly,

$A(3) = Q + A(2)e^{-cT} = Q + (Q + Qe^{-cT})e^{-cT} = Q + Qe^{-cT} + Qe^{-2cT}$,

$A(4) = Q + A(3)e^{-cT} = Q + (Q + Qe^{-cT} + Qe^{-2cT})e^{-cT} =$

$$Q + Qe^{-cT} + Qe^{-2cT} + Qe^{-3cT}, \ldots \Rightarrow A(k) = \sum_{n=0}^{k-1} Qe^{-ncT}.$$

(b) Since all terms are positive, $A(k)$ is an increasing sequence with the form

$a + ar + ar^2 + \cdots + ar^{k-1}$, where $a = Q$ and $r = e^{-cT}$.

Since $c > 0$, $|r| = \left|e^{-cT}\right| < 1$. Thus, by (11.15)(ii),

$$\text{the upper bound is } \lim_{k\to\infty} A(k) = \sum_{n=0}^{\infty} Q(e^{-cT})^n = \frac{Q}{1 - e^{-cT}}.$$

(c) We must find T such that the upper bound found in (b) is less than M. Thus,

$$\frac{Q}{1 - e^{-cT}} < M \Rightarrow Q < M(1 - e^{-cT}) \Rightarrow Q - M < -Me^{-cT} \Rightarrow$$

$$Me^{-cT} < M - Q \Rightarrow e^{-cT} < \frac{M-Q}{M} \Rightarrow -cT < \ln\frac{M-Q}{M} \Rightarrow$$

$$T > -\frac{1}{c}\ln\frac{M-Q}{M} \ \{-c < 0\}.$$

61 (a) From the second figure we see that $(\frac{1}{4}a_k)^2 + (\frac{3}{4}a_k)^2 = (a_{k+1})^2 \Rightarrow$

$$\tfrac{10}{16}a_k^2 = a_{k+1}^2 \Rightarrow a_{k+1} = \tfrac{1}{4}\sqrt{10}\,a_k.$$

(b) From part (a), $a_n = (\frac{1}{4}\sqrt{10})a_{n-1}$

$$= \tfrac{1}{4}\sqrt{10}(\tfrac{1}{4}\sqrt{10})a_{n-2} = (\tfrac{1}{4}\sqrt{10})^2 a_{n-2}$$

$$= (\tfrac{1}{4}\sqrt{10})^2(\tfrac{1}{4}\sqrt{10})a_{n-3} = (\tfrac{1}{4}\sqrt{10})^3 a_{n-3} = \cdots = (\tfrac{1}{4}\sqrt{10})^{n-1} a_1.$$

$A_{k+1} = a_{k+1}^2 = \frac{10}{16}a_k^2 = \frac{5}{8}A_k$ since $A_k = a_k^2$. In a manner similar to finding an

expression for a_n, $A_n = \frac{5}{8}A_{n-1} = (\frac{5}{8})^2 A_{n-2} = (\frac{5}{8})^3 A_{n-3} = \cdots$,

and hence $A_n = (\frac{5}{8})^{n-1}A_1$.

$P_{k+1} = 4a_{k+1} = 4 \cdot \frac{1}{4}\sqrt{10}\,a_k = \sqrt{10}\,a_k = \sqrt{10}\,(\frac{1}{4}P_k)$.

As with a_n, it follows that $P_n = (\frac{1}{4}\sqrt{10})^{n-1}P_1$.

(c) $\sum_{n=1}^{\infty} P_n = \sum_{n=1}^{\infty} P_1(\frac{1}{4}\sqrt{10})^{n-1}$ is an infinite geometric series with first term $a =$

P_1 and $r = \frac{1}{4}\sqrt{10} < 1$. $\sum_{n=1}^{\infty} P_n = \dfrac{a}{1-r} = \dfrac{P_1}{1 - \frac{1}{4}\sqrt{10}} = \dfrac{4P_1}{4 - \sqrt{10}} = \dfrac{16a_1}{4 - \sqrt{10}}.$

$\sum_{n=1}^{\infty} A_n = \sum_{n=1}^{\infty} (\tfrac{5}{8})^{n-1} A_1 = \sum_{n=1}^{\infty} (\tfrac{5}{8})^{n-1} a_1^2 = \dfrac{a_1^2}{1 - \frac{5}{8}} = \tfrac{8}{3}a_1^2.$

Exercises 11.3

1 (a) In (11.23), let $f(n) = a_n = \dfrac{1}{(3 + 2n)^2}$ and $f(x) = \dfrac{1}{(3 + 2x)^2}.$

(i) Since $(3 + 2x)^2 > 0$ if $x \geq 1$, f is a positive-valued function.

(ii) f is continuous on $\mathbb{R} - \{-\frac{3}{2}\}$ and hence is continuous on $[1, \infty)$.

(iii) By the reciprocal rule, $f'(x) = -\dfrac{2(3 + 2x) \cdot 2}{\left[(3 + 2x)^2\right]^2} = -\dfrac{4}{(2x + 3)^3}.$ If $x \geq 1$,

then $(2x + 3)^3 > 0$ and f' is negative. Thus, f is decreasing on $[1, \infty)$.

It now follows that f satisfies the hypotheses of (11.23).

(b) $\displaystyle\int_1^{\infty} f(x)\,dx = \int_1^{\infty} (3 + 2x)^{-2}\,dx = \lim_{t \to \infty} \int_1^t (3 + 2x)^{-2}\,dx$

$$= \lim_{t \to \infty} \left[\frac{1}{2} \cdot \frac{(3 + 2x)^{-1}}{-1}\right]_1^t = \lim_{t \to \infty} \left[-\frac{1}{2(3 + 2x)}\right]_1^t$$

$$= \left[\lim_{t \to \infty} -\frac{1}{2(3 + 2t)}\right] - \left[-\frac{1}{2(5)}\right] = 0 - \left(-\frac{1}{10}\right); \text{ C}$$

It is important to realize that $S = \sum_{n=1}^{\infty} \dfrac{1}{(3 + 2n)^2} \neq \frac{1}{10}$, but $\displaystyle\int_1^{\infty} f(x)\,dx = \frac{1}{10}$

and, therefore, S converges. We did not find out what S actually converges to.

3 (a) In (11.23), let $f(n) = \dfrac{1}{4n + 7}$ and $f(x) = \dfrac{1}{4x + 7}$.

(i) Since $4x + 7 > 0$ if $x \geq 1$, f is a positive-valued function.

(ii) f is continuous on $\mathbb{R} - \{-\frac{7}{4}\}$ and hence is continuous on $[1, \infty)$.

(iii) $f(x) = \dfrac{1}{4x + 7} \Rightarrow f'(x) = -\dfrac{4}{(4x + 7)^2} < 0$ if $x \geq 1$.

Thus, f is decreasing on $[1, \infty)$.

It now follows that f satisfies the hypotheses of (11.23).

(b) $\displaystyle\int_1^\infty f(x)\,dx = \int_1^\infty (4x + 7)^{-1}\,dx = \lim_{t \to \infty} \int_1^t (4x + 7)^{-1}\,dx =$

$\displaystyle\lim_{t \to \infty} \left[\frac{1}{4}\ln|4x + 7|\right]_1^t = \left[\lim_{t \to \infty} \frac{1}{4}\ln|4t + 7|\right] - \left[\frac{1}{4}\ln|11|\right] = \infty$; D

11 (a) In (11.23), let $f(n) = \dfrac{\arctan n}{1 + n^2}$ and $f(x) = \dfrac{\arctan x}{1 + x^2}$.

(i) (Refer to Figure 8.6 for the graph of $y = \arctan x$.)

Since $\arctan x > 0$ and $1 + x^2 > 0$ for $x \geq 1$, f is a positive-valued function.

(ii) Since $1 + x^2 \neq 0$ and both $\arctan x$ and $1 + x^2$ are continuous,

f is continuous on $[1, \infty)$.

(iii) $f(x) = \dfrac{\arctan x}{1 + x^2} \Rightarrow$

$f'(x) = \dfrac{(1 + x^2) \cdot 1/(1 + x^2) - (\arctan x)(2x)}{(1 + x^2)^2} = \dfrac{1 - 2x \arctan x}{(1 + x^2)^2} < 0$ if

$x \geq 1$ since $(\arctan x) \geq \frac{\pi}{4}$ on $[1, \infty)$. Thus, f is decreasing on $[1, \infty)$.

(b) $\displaystyle\int_1^\infty f(x)\,dx = \lim_{t \to \infty} \int_1^t \arctan x \left(\dfrac{1}{1 + x^2}\right) dx = \lim_{t \to \infty} \left[\frac{1}{2}(\arctan x)^2\right]_1^t =$

$\displaystyle\frac{1}{2}\lim_{t \to \infty} (\arctan t)^2 - \frac{1}{2}(\arctan 1)^2 = \frac{1}{2}(\frac{\pi}{2})^2 - \frac{1}{2}(\frac{\pi}{4})^2 = \frac{3\pi^2}{32}$; C

Note: Exer. 13–20: S (the given series) converges by (11.26)(i) or diverges by (11.26)(ii).

13 Generally, if the highest power of n in the denominator is greater than the highest power of n in the numerator by *more* than one, the series will converge. It is often a good idea to choose a p-series (11.25) to use in the basic comparison test.

In this problem, we believe the given series converges because the degree of the numerator is zero and the degree of the denominator is four, so we will pick a p-series that converges and is greater (term-by-term) than the given series. In

(11.26)(i), let $a_n = \dfrac{1}{n^4 + n^2 + 1}$ and $b_n = \dfrac{1}{n^4}$. Then, S converges since

$\dfrac{1}{n^4 + n^2 + 1} < \dfrac{1}{n^4}$ for $n \geq 1$, and $\displaystyle\sum_{n=1}^\infty \dfrac{1}{n^4}$ converges by (11.25) with $p = 4$.

$\boxed{17}$ For $n \geq 1$, $\arctan n$ increases from a value of $\frac{\pi}{4}$. We can see that for large n, the terms of $\frac{\arctan n}{n}$ become larger than those of the divergent harmonic series (11.14). Therefore, we suspect that the given series diverges. Let $a_n = \frac{\arctan n}{n}$ and $b_n = \frac{\pi/4}{n}$ in (11.26)(ii). Then, S diverges since $\frac{\arctan n}{n} \geq \frac{\pi/4}{n}$ and $\frac{\pi}{4} \sum_{n=1}^{\infty} \frac{1}{n}$ diverges.

$\boxed{19}$ Since n^n increases much faster than n^2, $\frac{1}{n^n}$ decreases much faster than $\frac{1}{n^2}$, and we suspect that the given series converges. Let $a_n = \frac{1}{n^n}$ and $b_n = \frac{1}{n^2}$ in (11.26)(i). Then, S converges since $\frac{1}{n^n} \leq \frac{1}{n^2}$ for $n \geq 1$ and $\sum_{n=1}^{\infty} \frac{1}{n^2}$ converges by (11.25). *Note:* We could have picked n^3 or n^4 instead of n^2—the choice was really arbitrary—we just wanted a known convergent series.

$\boxed{21}$ Let $a_n = \frac{\sqrt{n}}{n + 4}$. By deleting all terms in the numerator and denominator of a_n except those that have the greatest effect on the magnitude, we obtain $b_n = \frac{\sqrt{n}}{n} = \frac{1}{\sqrt{n}}$, which is a divergent p-series with $p = \frac{1}{2}$. Using the limit comparison test (11.27), we have $\lim_{n \to \infty} \frac{a_n}{b_n} = \lim_{n \to \infty} \frac{\sqrt{n}}{n + 4} \cdot \frac{\sqrt{n}}{1} = \lim_{n \to \infty} \frac{n}{n + 4} = 1 > 0$. Thus, S diverges by (11.27).

$\boxed{23}$ Since the numerator is a constant and the term having the greatest effect in the denominator is $\sqrt{n^3}$ (ignore the coefficient 4), let $a_n = \frac{1}{\sqrt{4n^3 - 5n}}$ and $b_n = \frac{1}{\sqrt{n^3}} = \frac{1}{n^{3/2}}$. $\lim_{n \to \infty} \frac{a_n}{b_n} = \lim_{n \to \infty} \frac{n^{3/2}}{\sqrt{4n^3 - 5n}}$. This limit can be evaluated easily since the highest power of n in both the numerator and denominator are equal. The limit will simply be the ratio of the leading coefficients, which is $\frac{1}{\sqrt{4}} = \frac{1}{2} > 0$. S converges by (11.27) since $\sum b_n$ converges by (11.25) with $p = \frac{3}{2}$.

$\boxed{25}$ The term having the greatest effect in the numerator is n^2 and in the denominator is $e^n n^2$. This ratio is $\frac{n^2}{e^n n^2} = \frac{1}{e^n}$. Thus, let $a_n = \frac{8n^2 - 7}{e^n(n + 1)^2}$ and $b_n = \frac{1}{e^n}$. $\lim_{n \to \infty} \frac{a_n}{b_n} = \lim_{n \to \infty} \frac{8n^2 - 7}{e^n(n + 1)^2} \cdot \frac{e^n}{1} = \lim_{n \to \infty} \frac{8n^2 - 7}{(n + 1)^2} = \frac{8}{1} = 8 > 0$. S converges by (11.27) since $\sum b_n$ is a geometric series with $r = \frac{1}{e} < 1$ and converges by (11.15).

$\boxed{31}$ $\dfrac{1 + 2^n}{1 + 3^n} < \dfrac{1 + 2^n}{3^n} = \dfrac{1}{3^n} + \dfrac{2^n}{3^n} = \left(\dfrac{1}{3}\right)^n + \left(\dfrac{2}{3}\right)^n$ and

$$\sum_{n=1}^{\infty}\left[\left(\tfrac{1}{3}\right)^n + \left(\tfrac{2}{3}\right)^n\right] = \sum_{n=1}^{\infty}\left(\tfrac{1}{3}\right)^n + \sum_{n=1}^{\infty}\left(\tfrac{2}{3}\right)^n \text{ converges by (11.15) and (11.20)(i).}$$

Let $a_n = \dfrac{1 + 2^n}{1 + 3^n}$ and $b_n = \left(\dfrac{1}{3}\right)^n + \left(\dfrac{2}{3}\right)^n$. Then, S converges by (11.26).

$\boxed{33}$ The term having the greatest effect in the numerator is 1 and in the denominator is

$\sqrt[3]{n^2} = n^{2/3}$. Thus, let $a_n = \dfrac{1}{\sqrt[3]{5n^2 + 1}}$ and $b_n = \dfrac{1}{n^{2/3}}$.

$$\lim_{n \to \infty} \dfrac{a_n}{b_n} = \lim_{n \to \infty} \dfrac{1}{\sqrt[3]{5n^2 + 1}} \cdot \dfrac{n^{2/3}}{1} = \lim_{n \to \infty} \dfrac{n^{2/3}}{\sqrt[3]{5n^2 + 1}} = \dfrac{1}{\sqrt[3]{5}} > 0.$$

S diverges by (11.27) since $\sum b_n$ diverges by (11.25) with $p = \tfrac{2}{3}$.

$\boxed{37}$ If we let $f(x) = xe^{-x}$, then $f'(x) = (1 - x)e^{-x} < 0$ for $x > 1$, and f is a positive-valued, continuous, decreasing function on $[1, \infty)$. Since we can readily integrate f using integration by parts or Formula 96, p. A25, we will apply the integral test.

$$\int_1^{\infty} f(x)\, dx = \lim_{t \to \infty} \int_1^t xe^{-x}\, dx = \lim_{t \to \infty}\left[-xe^{-x} - e^{-x}\right]_1^t = 0 - (-2e^{-1}) = \tfrac{2}{e}; \text{ C}$$

$\boxed{39}$ Let $a_n = \sin\dfrac{1}{n^2}$ and $b_n = \dfrac{1}{n^2}$.

$$\lim_{n \to \infty} \dfrac{a_n}{b_n} = \lim_{n \to \infty} \dfrac{\sin(1/n^2)}{1/n^2}\, \{\tfrac{0}{0}\} = \lim_{n \to \infty} \dfrac{(-2/n^3)\cos(1/n^2)}{-2/n^3} = \lim_{n \to \infty} \cos(1/n^2) =$$

$\cos 0 = 1 > 0$. S converges by (11.27) since $\sum b_n$ converges by (11.25) with $p = 2$.

$\boxed{41}$ After deleting terms of least magnitude, we have $\dfrac{(2n)^3}{(n^3)^2} = \dfrac{8n^3}{n^6} = \dfrac{8}{n^3}$.

Let $a_n = \dfrac{(2n + 1)^3}{(n^3 + 1)^2}$ and $b_n = \dfrac{1}{n^3}$. $\lim_{n \to \infty} \dfrac{a_n}{b_n} = \lim_{n \to \infty} \dfrac{(2n + 1)^3 \cdot n^3}{(n^3 + 1)^2} = 8 > 0.$

S converges by (11.27) since $\sum b_n$ converges by (11.25) with $p = 3$.

$\boxed{45}$ When series contain terms with ln and polynomial terms, we often use the fact that

$\ln n < n$ for $n \geq 1$, as illustrated in this solution. Since $\dfrac{\ln n}{n^3} < \dfrac{n}{n^3} = \dfrac{1}{n^2}$ and $\sum_{n=1}^{\infty} \dfrac{1}{n^2}$

converges, S converges by (11.26) with $a_n = \dfrac{\ln n}{n^3}$ and $b_n = \dfrac{1}{n^2}$.

$\boxed{47}$ Since the value of k will affect how we determine the convergence of the given series, we will consider four cases for values of k.

(i) $\quad k > 1$. Let $a_n = \dfrac{1}{n^k \ln n}$ and $b_n = \dfrac{1}{n^k}$. Then since

$$\frac{1}{n^k \ln n} < \frac{1}{n^k} \ \{\ln n > 1 \text{ if } n \geq 3\} \text{ and } \sum_{n=2}^{\infty} \frac{1}{n^k} \text{ converges by (11.25), S converges.}$$

(ii) $\quad k = 1$. Using the integral test, $\displaystyle\int_2^{\infty} \frac{1}{x \ln x}\, dx =$

$$\lim_{t \to \infty} \int_{\ln 2}^{t} \frac{1}{u}\, du \ \{u = \ln x; \ du = \tfrac{1}{x}\, dx\} = \lim_{t \to \infty} \Big[\ln |u|\Big]_{\ln 2}^{t} = \infty; \ D.$$

(iii) $\ 0 \leq k < 1$. $\ u = \ln x, \ x = e^u$, and $dx = e^u\, du \Rightarrow \displaystyle\int_2^{\infty} \frac{1}{x^k \ln x}\, dx =$

$$\int_{\ln 2}^{\infty} \frac{e^u}{e^{ku}\, u}\, du = \int_{\ln 2}^{\infty} \frac{e^{(1-k)u}}{u}\, du. \text{ Since } 1 - k > 0, \text{ the integrand approaches } \infty$$

as $u \to \infty$. Thus, the integral must diverge and S also diverges.

(iv) $\quad k < 0$. Since $\dfrac{1}{n^k \ln n} = \dfrac{n^{-k}}{\ln n} > \dfrac{1}{\ln n} > \dfrac{1}{n}\ (n \geq 2)$ and $\displaystyle\sum_{n=2}^{\infty} \frac{1}{n}$ diverges,

S diverges.

Thus, S converges iff $k > 1$.

$\boxed{49}$ (a) First use the results of the proof of (11.23) with $f(x) = \dfrac{1}{x + 1}$.

(i)

$$\sum_{k=2}^{n} f(k) \leq \int_1^n f(x)\, dx \leq \sum_{k=1}^{n-1} f(k) \Rightarrow$$

$$\sum_{k=2}^{n} \frac{1}{k+1} \leq \int_1^n \frac{1}{x+1}\, dx \leq \sum_{k=1}^{n-1} \frac{1}{k+1} \Rightarrow$$

$$\tfrac{1}{3} + \tfrac{1}{4} + \cdots + \frac{1}{n+1} \leq \Big[\ln |x+1|\Big]_1^n \leq \tfrac{1}{2} + \tfrac{1}{3} + \cdots + \tfrac{1}{n} \Rightarrow$$

$$\tfrac{1}{3} + \tfrac{1}{4} + \cdots + \frac{1}{n+1} \leq \ln(n+1) - \ln 2 \leq \tfrac{1}{2} + \tfrac{1}{3} + \cdots + \tfrac{1}{n} \Rightarrow$$

$$\ln(n+1) \leq \ln 2 + \tfrac{1}{2} + \tfrac{1}{3} + \cdots + \tfrac{1}{n}.$$

$$\{\text{since } \ln 2 < 1\} \ \ln(n+1) < 1 + \tfrac{1}{2} + \tfrac{1}{3} + \cdots + \tfrac{1}{n}.$$

(ii) Now repeat (i) with $f(x) = \tfrac{1}{x}$.

$$\sum_{k=2}^{n} f(k) \leq \int_1^n f(x)\, dx \leq \sum_{k=1}^{n-1} f(k) \Rightarrow$$

$$\sum_{k=2}^{n} \frac{1}{k} \leq \int_1^n \frac{1}{x}\, dx \leq \sum_{k=1}^{n-1} \frac{1}{k} \Rightarrow$$

$$\tfrac{1}{2} + \tfrac{1}{3} + \cdots + \tfrac{1}{n} \leq \ln n \leq 1 + \tfrac{1}{2} + \tfrac{1}{3} + \cdots + \frac{1}{n-1} \Rightarrow$$

$$1 + \tfrac{1}{2} + \tfrac{1}{3} + \cdots + \tfrac{1}{n} \leq 1 + \ln n.$$

By (i) and (ii), $\ln(n+1) < 1 + \tfrac{1}{2} + \tfrac{1}{3} + \cdots + \tfrac{1}{n} < 1 + \ln n \ (n > 1).$

(b) From part (a), $S_n = 1 + \frac{1}{2} + \frac{1}{3} + \cdots + \frac{1}{n} > 100$ if $\ln(n+1) > 100$.

Thus, $n + 1 > e^{100} \Rightarrow n > e^{100} - 1 \approx 2.688 \times 10^{43}$.

51 Since $\lim\limits_{n \to \infty} \frac{a_n}{b_n} = 0$, then by (11.3), there is an $N \geq 1$ such that if $n > N$,

then $\frac{a_n}{b_n} < 1$, or $a_n < b_n$. Since $\sum b_n$ converges and $a_n < b_n$ for all but at most a

finite number of terms, $\sum a_n$ must also converge by (11.26).

53 $\sum\limits_{n=1}^{\infty} a_n = \sum\limits_{n=1}^{N} a_n + \sum\limits_{n=N+1}^{\infty} a_n$. The error in approximating

$\sum\limits_{n=1}^{\infty} a_n$ by $\sum\limits_{n=1}^{N} a_n$ is $\sum\limits_{n=N+1}^{\infty} a_n$ since all the a_n are positive.

Thus, we must show that $\sum\limits_{n=N+1}^{\infty} a_n < \int_{N}^{\infty} f(x)\, dx$.

Since $a_{n+1} < \int_{n}^{n+1} f(x)\, dx$ for every n and f is decreasing, it follows that the error

$E = a_{N+1} + a_{N+2} + a_{N+3} + \cdots$

$= \sum\limits_{n=N+1}^{\infty} a_n < \int_{N}^{N+1} f(x)\, dx + \int_{N+1}^{N+2} f(x)\, dx + \int_{N+2}^{N+3} f(x)\, dx + \cdots = \int_{N}^{\infty} f(x)\, dx.$

(See Figure 11.8.)

55 We must determine an N such that $\int_{N}^{\infty} \frac{1}{x^3}\, dx < 0.01$.

Error $< \int_{N}^{\infty} \frac{1}{x^3}\, dx = \lim\limits_{t \to \infty} \left[-\frac{1}{2x^2} \right]_{N}^{t} = \frac{1}{2N^2} \leq 0.01 \Rightarrow N^2 \geq 50 \Rightarrow N \geq \sqrt{50}$; 8 terms.

57 Since $\sum a_n$ converges, $\lim\limits_{n \to \infty} a_n = 0$ by (11.26) and therefore, $\lim\limits_{n \to \infty} \frac{1}{a_n} = \infty$.

By (11.17), $\sum \frac{1}{a_n}$ diverges.

61 From the graphs, it is apparent that $x \geq \ln(x^k)$ for

$k = 1, 2, 3$ and $x \geq 5$. Thus, $\frac{1}{n} \leq \frac{1}{\ln(n^k)}$ for $k = 1$,

$2, 3$ and $n \geq 5$. By the basic comparison test, since

$\sum\limits_{n=1}^{\infty} \frac{1}{n}$ diverges, $\sum\limits_{n=1}^{\infty} \frac{1}{\ln(n^k)}$ also diverges for $k = 1, 2,$

and 3.

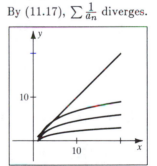

Figure 61

Exercises 11.4

1 $\displaystyle\lim_{n\to\infty}\frac{a_{n+1}}{a_n} = \lim_{n\to\infty}\left(a_{n+1}\cdot\frac{1}{a_n}\right) = \lim_{n\to\infty}\frac{3(n+1)+1}{2^{n+1}}\cdot\frac{2^n}{3n+1} =$

$\displaystyle\lim_{n\to\infty}\frac{2^n(3n+4)}{2^{n+1}(3n+1)} = \lim_{n\to\infty}\frac{3n+4}{2(3n+1)} = \frac{1}{2} < 1.$

Since $\displaystyle\lim_{n\to\infty}\frac{a_{n+1}}{a_n} < 1$, the series is convergent.

Note: For limits involving only powers of n, we will not divide by the highest power of n, but will merely compare leading terms to determine the limit. For example, in this exercise, the highest power of n in the numerator and denominator of $\dfrac{3n+4}{2(3n+1)}$ is 1. The ratio of their coefficients is $\dfrac{3}{2(3)} = \dfrac{1}{2}$ and this ratio is equal to the limit.

3 $\displaystyle\lim_{n\to\infty}\frac{a_{n+1}}{a_n} = \lim_{n\to\infty}\frac{5^{n+1}}{(n+1)3^{n+2}}\cdot\frac{n\,3^{n+1}}{5^n}$

$\displaystyle = \lim_{n\to\infty}\frac{5^{n+1}\cdot 3^{n+1}\cdot n}{5^n\cdot 3^{n+2}\cdot(n+1)} = \lim_{n\to\infty}\frac{5n}{3(n+1)} = \frac{5}{3} > 1;\ \text{D}$

Since $\displaystyle\lim_{n\to\infty}\frac{a_{n+1}}{a_n} > 1$, the series is divergent.

7 $\displaystyle\lim_{n\to\infty}\frac{a_{n+1}}{a_n} = \lim_{n\to\infty}\frac{n+4}{(n+1)^2+2(n+1)+5}\cdot\frac{n^2+2n+5}{n+3} = 1.$ The highest power of n in the numerator and denominator is 3. The ratio of their coefficients is $\frac{1}{1}$. Note that you do *not* need to multiply the expressions out to determine this. It can be done by inspecting only the leading terms. Since $\displaystyle\lim_{n\to\infty}\frac{a_{n+1}}{a_n} = 1$, the ratio test is inconclusive. Using the limit comparison test with $b_n = \frac{1}{n}$, we see that $\displaystyle\lim_{n\to\infty}\frac{a_n}{b_n} =$

$\displaystyle\lim_{n\to\infty}\frac{n+3}{n^2+2n+5}\cdot\frac{n}{1} = 1$ and $\sum a_n$ diverges since $\sum\frac{1}{n}$ diverges. However, the purpose here is to call attention to the condition under which the ratio test fails.

11 $\displaystyle\lim_{n\to\infty}\sqrt[n]{a_n} = \lim_{n\to\infty}\sqrt[n]{\frac{1}{n^n}} = \lim_{n\to\infty}\frac{1}{n} = 0 < 1;\ \text{C}$

13 $\displaystyle\lim_{n\to\infty}\sqrt[n]{a_n} = \lim_{n\to\infty}\sqrt[n]{\frac{2^n}{n^2}} = \lim_{n\to\infty}\frac{2}{n^{2/n}}.$

Since $\displaystyle\lim_{n\to\infty}n^{2/n} = \lim_{n\to\infty}(n^{1/n})^2 = (1)^2$ { see Exercise 11.1.37 },

$\displaystyle\lim_{n\to\infty}\frac{2}{n^{2/n}} = \frac{2}{(1)^2} = 2 > 1$, and $\sum a_n$ diverges.

17 $\displaystyle\lim_{n\to\infty}\sqrt[n]{a_n} = \lim_{n\to\infty}\sqrt[n]{\left(\frac{n}{2n+1}\right)^n} = \lim_{n\to\infty}\frac{n}{2n+1} = \frac{1}{2} < 1;\ \text{C}$

Note: The following may be solved using several methods. For a convenience, we list an abbreviation of the test used at the beginning of the problem.

NTH — nth-term test (11.17)   INT — integral test (11.23)

BCT — basic comparison test (11.26)   LCT — limit comparison test (11.27)

RAT — ratio test (11.28)   ROT — root test (11.29)

$\boxed{19}$ LCT Deleting terms of least magnitude gives $\dfrac{\sqrt{n}}{n^2} = \dfrac{1}{n^{3/2}}$. Let $a_n = \dfrac{\sqrt{n}}{n^2 + 1}$ and

$b_n = \dfrac{1}{n^{3/2}}.$ $\displaystyle\lim_{n \to \infty} \dfrac{a_n}{b_n} = \lim_{n \to \infty} \dfrac{n^{1/2}}{n^2 + 1} \cdot \dfrac{n^{3/2}}{1} = \lim_{n \to \infty} \dfrac{n^2}{n^2 + 1} = 1 > 0.$

S converges since $\sum b_n$ converges.

$\boxed{23}$ BCT Since $\dfrac{2}{n^3 + e^n} < \dfrac{2}{n^3}$ and $2 \displaystyle\sum_{n=1}^{\infty} \dfrac{1}{n^3}$ converges since it is a p-series with

$p = 3 > 1$, S converges.

$\boxed{25}$ RAT $\displaystyle\lim_{n \to \infty} \dfrac{a_{n+1}}{a_n} = \lim_{n \to \infty} \dfrac{2^{n+1}(n+1)!}{(n+1)^{n+1}} \cdot \dfrac{n^n}{2^n\, n!}$

$\qquad = \displaystyle\lim_{n \to \infty} \dfrac{2^{n+1}(n+1)\, n!\, n^n}{2^n\, n!\,(n+1)^{n+1}} = \lim_{n \to \infty} \dfrac{2(n+1)\, n^n}{(n+1)^n (n+1)^1}$

$\qquad = 2 \displaystyle\lim_{n \to \infty} \dfrac{n^n}{(n+1)^n} = 2 \lim_{n \to \infty} \left(\dfrac{n}{n+1}\right)^n = 2 \lim_{n \to \infty} \left(\dfrac{1}{\frac{n+1}{n}}\right)^n$

$\qquad = 2 \displaystyle\lim_{n \to \infty} \dfrac{1^n}{\left(1 + \frac{1}{n}\right)^n} = 2 \cdot \dfrac{1}{e} < 1;$ C. *Note:* Using Example 2,

$\qquad \displaystyle\lim_{n \to \infty} \left(\dfrac{n}{n+1}\right)^n$ is the reciprocal of $\displaystyle\lim_{n \to \infty} \left(\dfrac{n+1}{n}\right)^n$, which is e.

$\boxed{27}$ ROT $\displaystyle\lim_{n \to \infty} \sqrt[n]{a_n} = \lim_{n \to \infty} \sqrt[n]{\dfrac{n^n}{10^{n+1}}} = \lim_{n \to \infty} \dfrac{(n^n)^{1/n}}{(10^{n+1})^{1/n}}$

$\qquad = \displaystyle\lim_{n \to \infty} \dfrac{n}{10^{1+(1/n)}} = \lim_{n \to \infty} \dfrac{n}{10(10^{1/n})} = \infty$ since $\displaystyle\lim_{n \to \infty} n = \infty$ and

$\qquad\qquad \displaystyle\lim_{n \to \infty} 10(10^{1/n}) = 10(10^0) = 10(1) = 10;$ D

$\boxed{31}$ INT $\displaystyle\int_2^{\infty} \dfrac{1}{x\sqrt[3]{\ln x}}\, dx$

$\qquad = \displaystyle\int_2^{\infty} (\ln x)^{-1/3} \cdot \dfrac{1}{x}\, dx \ \{ u = \ln x,\ du = \dfrac{1}{x}\, dx;\ x = 2, \infty \Rightarrow u = \ln 2, \infty \}$

$\qquad = \displaystyle\lim_{t \to \infty} \int_{\ln 2}^t u^{-1/3}\, du = \lim_{t \to \infty} \left[\dfrac{3}{2} u^{2/3}\right]_{\ln 2}^t = \infty;$ D

$\boxed{35}$ NTH $\displaystyle\lim_{n \to \infty} n \tan\dfrac{1}{n} \ \{\infty \cdot 0\} = \lim_{n \to \infty} \dfrac{\tan(1/n)}{1/n} \ \{\tfrac{0}{0}\} = \lim_{n \to \infty} \dfrac{(-1/n^2)\sec^2(1/n)}{-1/n^2} =$

$\qquad\qquad \displaystyle\lim_{n \to \infty} \sec^2(1/n) = \sec^2 0 = 1 \neq 0.$ Thus, S diverges.

39 RAT $\lim\limits_{n \to \infty} \dfrac{a_{n+1}}{a_n} = \lim\limits_{n \to \infty} \dfrac{1 \cdot 3 \cdot 5 \cdot \cdots \cdot (2n-1)(2n+1)}{(n+1)!} \cdot \dfrac{n!}{1 \cdot 3 \cdot 5 \cdot \cdots \cdot (2n-1)} =$

$$\lim\limits_{n \to \infty} \dfrac{2n+1}{n+1} = 2 > 1;\ D$$

Exercises 11.5

1   (a) $a_k = \dfrac{1}{k^2 + 7}.$ $a_{k+1} = \dfrac{1}{(k+1)^2 + 7} = \dfrac{1}{(k^2 + 7) + (2k+1)}.$

Since the denominator of a_{k+1} is larger than the denominator of a_k, $a_k > a_{k+1}.$

Also, $a_{k+1} > 0$ and hence, condition (i) of (11.30) is satisfied.

Condition (ii) is satisfied since $\lim\limits_{n \to \infty} a_n = \lim\limits_{n \to \infty} \dfrac{1}{n^2 + 7} = 0.$

(b) The series converges by (11.30) since conditions (i) and (ii) are satisfied.

3   (a) As in the beginning of Example 1, we will show that $f(x) = (1 + e^{-x})$ is decreasing. $f(x) = 1 + e^{-x} \Rightarrow f'(x) = -e^{-x} < 0$ and f is decreasing. Since $e^{-n} > 0$ for all n, $a_n = 1 + e^{-n} > 0$ for all n. Hence, $a_k > a_{k+1} > 0$ and condition (i) is satisfied. $\lim\limits_{n \to \infty} a_n = \lim\limits_{n \to \infty} (1 + e^{-n}) = 1 + 0 = 1 \neq 0$ and condition (ii) is not satisfied.

(b) The series diverges by the nth-term test.

Note: Let AC denote Absolutely Convergent; CC, Conditionally Convergent; D, Divergent; AST, the Alternating Series Test; and S, the given series.

5   $a_n = \dfrac{1}{\sqrt{2n+1}} > 0.$ To show that $a_n \geq a_{n+1}$, we will show that $\dfrac{a_{n+1}}{a_n} \leq 1.$

$\dfrac{a_{n+1}}{a_n} = \dfrac{1}{\sqrt{2(n+1)+1}} \div \dfrac{1}{\sqrt{2n+1}} = \dfrac{\sqrt{2n+1}}{\sqrt{2n+3}} < 1$ and $\lim\limits_{n \to \infty} a_n = 0 \Rightarrow$

S converges by AST. To determine whether or not this series converges absolutely, we will use the LCT. Let $b_n = \dfrac{1}{\sqrt{n}}.$ Then $\lim\limits_{n \to \infty} \dfrac{a_n}{b_n} = \lim\limits_{n \to \infty} \dfrac{1}{\sqrt{2n+1}} \cdot \dfrac{\sqrt{n}}{1} = \dfrac{1}{\sqrt{2}} > 0$

and $\sum a_n$ diverges since $\sum \dfrac{1}{\sqrt{n}}$ is a p-series with $p = \frac{1}{2}$ and it diverges. Thus, S is not AC, but S is CC.

9   Since $\lim\limits_{n \to \infty} \dfrac{n}{\ln n} \left\{ \frac{\infty}{\infty} \right\} = \lim\limits_{n \to \infty} \dfrac{1}{1/n} = \lim\limits_{n \to \infty} n = \infty$, S is D by the nth-term test.

11   Since $\dfrac{5}{n^3 + 1} < \dfrac{5}{n^3}$ and $5 \sum\limits_{n=1}^{\infty} \dfrac{1}{n^3}$ converges by (11.25), $\sum\limits_{n=1}^{\infty} \dfrac{5}{n^3 + 1}$ converges by a basic comparison test and hence, S is AC.

Note: Absolute convergence implies *both* convergence and conditional convergence.

[13] $\lim\limits_{n\to\infty}\left|\dfrac{a_{n+1}}{a_n}\right| = \lim\limits_{n\to\infty}\dfrac{10^{n+1}}{(n+1)!}\cdot\dfrac{n!}{10^n} = \lim\limits_{n\to\infty}\dfrac{10^{n+1}\,n!}{10^n\,(n+1)\,n!} =$

$$\lim\limits_{n\to\infty}\dfrac{10}{n+1} = 0 \Rightarrow \text{S is AC by (11.35)(i).}$$

[17] $a_n = \dfrac{n^{1/3}}{n+1} > 0.$ $f(x) = \dfrac{x^{1/3}}{x+1} \Rightarrow f'(x) = \dfrac{1-2x}{3x^{2/3}(x+1)^2} < 0$ for $x > \frac{1}{2}$ and

$$a_n \geq a_{n+1}.\ \text{Since}\ \lim\limits_{n\to\infty} a_n = 0,\ \text{it follows that S converges by AST.}$$

To determine the absolute convergence of S, we will use the LCT.

Deleting terms of least magnitude results in $\dfrac{\sqrt[3]{n}}{n} = \dfrac{1}{n^{2/3}}.$ Let $b_n = \dfrac{1}{n^{2/3}}.$

Then $\lim\limits_{n\to\infty}\dfrac{a_n}{b_n} = \lim\limits_{n\to\infty}\dfrac{n^{1/3}}{n+1}\cdot\dfrac{n^{2/3}}{1} = \lim\limits_{n\to\infty}\dfrac{n}{n+1} = 1 > 0$ and

$$\sum a_n\ \text{diverges since}\ \sum 1/n^{2/3}\ \text{diverges by (11.25).\ Thus, S is CC.}$$

[19] Since $\left|\dfrac{\cos\frac{\pi}{6}n}{n^2}\right| \leq \dfrac{1}{n^2}$ and $\sum\limits_{n=1}^{\infty}\dfrac{1}{n^2}$ converges by (11.25), S is AC.

$$\text{Note that}\ \left|\cos\tfrac{\pi}{6}n\right| \leq 1\ \text{for all}\ n.$$

[21] Since $\lim\limits_{n\to\infty} n\sin\dfrac{1}{n}\ \{\infty\cdot 0\} = \lim\limits_{n\to\infty}\dfrac{\sin(1/n)}{1/n}\ \{\tfrac{0}{0}\} = \lim\limits_{n\to\infty}\dfrac{(-1/n^2)\cos(1/n)}{-1/n^2} =$

$$\lim\limits_{n\to\infty}\cos(1/n) = \cos 0 = 1 \neq 0,\ \text{S is D by the}\ n\text{th-term test.}$$

[25] Since both the numerator and denominator are raised to the nth power, we will apply

the root test. $\lim\limits_{n\to\infty}\sqrt[n]{|a_n|} = \lim\limits_{n\to\infty}\sqrt[n]{\left|\dfrac{n^n}{(-5)^n}\right|} = \lim\limits_{n\to\infty}\dfrac{n}{5} = \infty \Rightarrow \text{S is D.}$

[29] Since $\cos\pi n = -1$ when n is odd and $\cos\pi n = 1$ when n is even, it follows that

$$\sum_{n=1}^{\infty}(-1)^n\dfrac{\cos\pi n}{n} = \sum_{n=1}^{\infty}(-1)^n\cdot\dfrac{(-1)^n}{n} = \sum_{n=1}^{\infty}\dfrac{(-1)^{2n}}{n} = \sum_{n=1}^{\infty}\dfrac{\left[(-1)^2\right]^n}{n} =$$

$$\sum_{n=1}^{\infty}\dfrac{(1)^n}{n} = \sum_{n=1}^{\infty}\dfrac{1}{n} = 1 + \tfrac{1}{2} + \tfrac{1}{3} + \cdots.\ \text{This is the harmonic series.\ S is D.}$$

Note: The value of n in Exercises 33–42 was found by trial and error when the

inequality could not be solved using basic algebraic operations.

Note: If the sum S_n is to be approximated to three decimal places, then $a_{n+1} <$

$0.5 \times 10^{-3}.$ Equivalently, we use a_n and S_{n-1} to simplify the computations.

[33] In order to apply (11.31), we must determine a value of n such that $a_n < 0.5 \times 10^{-3}.$

$a_n = \dfrac{1}{n!} < 0.0005 \Rightarrow n! > 2000 \Rightarrow n \geq 7.$ It now follows that

$a_0 + a_1 + a_2 + a_3 + a_4 + a_5 + a_6$ will be an approximation of S to within three

decimal places. Thus, $S \approx S_6 = 1 - \dfrac{1}{1!} + \dfrac{1}{2!} - \dfrac{1}{3!} + \dfrac{1}{4!} - \dfrac{1}{5!} + \dfrac{1}{6!} \approx 0.368.$

$\boxed{37}$ $a_n = \dfrac{n+1}{5^n} < 0.0005 \Rightarrow 2000(n+1) < 5^n \Rightarrow n \ge 6$.

Thus, $S \approx S_5 = \dfrac{2}{5} - \dfrac{3}{25} + \dfrac{4}{125} - \dfrac{5}{625} + \dfrac{6}{3125} \approx 0.306$.

$\boxed{39}$ This exercise is similar to Exercise 33. We must determine an n such that

$a_{n+1} < 0.5 \times 10^{-4}$. However, we do not need to estimate S.

$a_{n+1} = \dfrac{1}{(n+1)^2} < 0.00005 \Rightarrow (n+1)^2 > 20{,}000 \Rightarrow n > \sqrt{20{,}000} - 1 \Rightarrow n \ge 141$.

$\boxed{43}$ (i) $a_n = \dfrac{(\ln n)^k}{n} > 0 \ (n \ge 3)$. $f(x) = \dfrac{(\ln x)^k}{x} \Rightarrow f'(x) = \dfrac{(\ln x)^{k-1}(k - \ln x)}{x^2} < 0$ for

$x > e^k \Rightarrow a_{n+1} < a_n$ for all but a finite number of terms. Remember that a

finite number of terms *does not* affect the convergence or divergence of a series.

(ii) We will show that $\lim\limits_{n \to \infty} a_n = 0$ by repeatedly using L'Hôpital's rule (10.2).

$\lim\limits_{x \to \infty} \dfrac{(\ln x)^k}{x} \ \{\tfrac{\infty}{\infty}\} = \lim\limits_{x \to \infty} \dfrac{k(\ln x)^{k-1} \cdot (1/x)}{1} = \lim\limits_{x \to \infty} \dfrac{k(\ln x)^{k-1}}{x} \ \{\tfrac{\infty}{\infty}\} =$

$\lim\limits_{x \to \infty} \dfrac{k(k-1)(\ln x)^{k-2}}{x} \ \{\tfrac{\infty}{\infty}\} = \cdots = \lim\limits_{x \to \infty} \dfrac{k!}{x} = 0$. Thus, by AST, S converges.

$\boxed{45}$ No. If $a_n = b_n = (-1)^n/\sqrt{n}$, then both $\sum a_n$ and $\sum b_n$ converge by the alternating

series test. However, $\sum a_n b_n = \sum 1/n$, which diverges.

Exercises 11.6

Note: Let u_n denote the nth term of the power series. Let AC denote Absolutely

Convergent; C, Convergent; D, Divergent; and AST, the Alternating Series Test.

$\boxed{1}$ $u_n = \dfrac{x^n}{n+4} \Rightarrow \lim\limits_{n \to \infty} \left| \dfrac{u_{n+1}}{u_n} \right| = \lim\limits_{n \to \infty} \left| \dfrac{x^{n+1}}{n+5} \cdot \dfrac{n+4}{x^n} \right|$

$= \lim\limits_{n \to \infty} \left| \dfrac{n+4}{n+5} \cdot \dfrac{x^{n+1}}{x^n} \right|$

$= \lim\limits_{n \to \infty} \left(\dfrac{n+4}{n+5} \right) |x| = 1 \cdot |x| = |x|$.

The series will converge whenever $|x| < 1$. $|x| < 1 \Leftrightarrow -1 < x < 1$.

The end points of the interval of convergence are $x = \pm 1$. To determine the

convergence at these points, we must substitute $x = 1$ and $x = -1$ into the series.

If $x = 1$, $\sum\limits_{n=0}^{\infty} \dfrac{1}{n+4}(1)^n = \sum\limits_{n=0}^{\infty} \dfrac{1}{n+4}$ is D { use LCT with $b_n = \dfrac{1}{n}$ }.

If $x = -1$, $\sum\limits_{n=0}^{\infty} (-1)^n \dfrac{1}{n+4}$ is C by AST. $\bigstar \ [-1, 1)$

$\boxed{3}$ $u_n = \dfrac{n^2 x^n}{2^n} \Rightarrow \lim\limits_{n \to \infty} \left| \dfrac{u_{n+1}}{u_n} \right| = \lim\limits_{n \to \infty} \left| \dfrac{(n+1)^2 x^{n+1}}{2^{n+1}} \cdot \dfrac{2^n}{n^2 x^n} \right|$

$$= \lim_{n \to \infty} \left| \dfrac{(n+1)^2}{n^2} \cdot \dfrac{2^n}{2^{n+1}} \cdot \dfrac{x^{n+1}}{x^n} \right| = 1 \cdot \tfrac{1}{2} \cdot |x| = \tfrac{1}{2}|x|.$$

The series will converge whenever $\frac{1}{2}|x| < 1$. $\frac{1}{2}|x| < 1 \Leftrightarrow -2 < x < 2$.

The end points of the interval of convergence are $x = \pm 2$.

If $x = 2$, $\sum\limits_{n=0}^{\infty} \dfrac{n^2}{2^n} (2)^n = \sum\limits_{n=0}^{\infty} n^2$ is D { nth-term test }.

If $x = -2$, $\sum\limits_{n=0}^{\infty} \dfrac{n^2}{2^n} (-2)^n = \sum\limits_{n=0}^{\infty} \dfrac{n^2}{2^n}(-1)^n (2)^n = \sum\limits_{n=0}^{\infty} (-1)^n n^2$ is D. $\bigstar$ $(-2, 2)$

$\boxed{9}$ $u_n = \dfrac{(\ln n)\, x^n}{n^3} \Rightarrow \lim\limits_{n \to \infty} \left| \dfrac{u_{n+1}}{u_n} \right| = \lim\limits_{n \to \infty} \left| \dfrac{\left[\ln(n+1)\right] x^{n+1}}{(n+1)^3} \cdot \dfrac{n^3}{(\ln n)\, x^n} \right|$

$$= \lim_{n \to \infty} \left| \dfrac{\ln(n+1)}{\ln n} \cdot \dfrac{n^3}{(n+1)^3} \cdot \dfrac{x^{n+1}}{x^n} \right|$$

$$= 1 \cdot 1 \cdot |x| = |x|. \quad |x| < 1 \Leftrightarrow -1 < x < 1.$$

If $x = 1$, $\sum\limits_{n=2}^{\infty} \dfrac{\ln n}{n^3} \le \sum\limits_{n=2}^{\infty} \dfrac{n}{n^3} = \sum\limits_{n=2}^{\infty} \dfrac{1}{n^2}$ is C by BCT.

If $x = -1$, $\sum\limits_{n=2}^{\infty} (-1)^n \dfrac{\ln n}{n^3}$ is C by AST. $\bigstar$ $[-1, 1]$

$\boxed{11}$ $u_n = \dfrac{(n+1)(x-4)^n}{10^n} \Rightarrow \lim\limits_{n \to \infty} \left| \dfrac{u_{n+1}}{u_n} \right| = \lim\limits_{n \to \infty} \left| \dfrac{(n+2)(x-4)^{n+1}}{10^{n+1}} \cdot \dfrac{10^n}{(n+1)(x-4)^n} \right|$

$$= \lim_{n \to \infty} \left| \dfrac{n+2}{n+1} \cdot \dfrac{10^n}{10^{n+1}} \cdot \dfrac{(x-4)^{n+1}}{(x-4)^n} \right|$$

$$= 1 \cdot \tfrac{1}{10} \cdot |x-4| = \tfrac{1}{10}|x-4|.$$

$\frac{1}{10}|x-4| < 1 \Leftrightarrow |x-4| < 10 \Leftrightarrow -10 < x - 4 < 10 \Leftrightarrow -6 < x < 14.$

If $x = -6$, $\sum\limits_{n=0}^{\infty} \dfrac{n+1}{10^n}(-10)^n = \sum\limits_{n=0}^{\infty} (-1)^n (n+1)$ is D { nth-term test }.

If $x = 14$, $\sum\limits_{n=0}^{\infty} \dfrac{n+1}{10^n} 10^n = \sum\limits_{n=0}^{\infty} (n+1)$ is D { nth-term test }. $\bigstar$ $(-6, 14)$

$\boxed{13}$ $u_n = \dfrac{n!\, x^n}{100^n} \Rightarrow \lim\limits_{n \to \infty} \left| \dfrac{u_{n+1}}{u_n} \right| = \lim\limits_{n \to \infty} \left| \dfrac{(n+1)!\, x^{n+1}}{100^{n+1}} \cdot \dfrac{100^n}{n!\, x^n} \right|$

$$= \lim_{n \to \infty} \left| \dfrac{(n+1)\, n!}{n!} \cdot \dfrac{100^n}{100^{n+1}} \cdot \dfrac{x^{n+1}}{x^n} \right|$$

$$= \lim_{n \to \infty} \left(\dfrac{n+1}{100} \right) |x| = \infty.$$

Remember, the limit variable is n, not x, and hence the last limit increases without bound. Since the limit is always greater than 1, the series converges only for $x = 0$.

$\boxed{15}$ $u_n = \dfrac{x^{2n+1}}{(-4)^n} \Rightarrow \lim\limits_{n \to \infty} \left| \dfrac{u_{n+1}}{u_n} \right| = \lim\limits_{n \to \infty} \left| \dfrac{x^{2n+3}}{(-4)^{n+1}} \cdot \dfrac{(-4)^n}{x^{2n+1}} \right| = \tfrac{1}{4} x^2.$

$\tfrac{1}{4} x^2 < 1 \Leftrightarrow x^2 < 4 \Leftrightarrow |x| < 2 \Leftrightarrow -2 < x < 2.$

If $x = \pm 2$, $|u_n| = \left| \dfrac{(\pm 2)^{2n+1}}{(-4)^n} \right| = \dfrac{2^{2n+1}}{4^n} = \dfrac{2^1 \cdot 2^{2n}}{4^n} = \dfrac{2 \cdot (2^2)^n}{4^n} = \dfrac{2 \cdot 4^n}{4^n} = 2 \; \forall n,$

and both series are D { nth-term test }. $\qquad\qquad\qquad \star \; (-2, 2)$

$\boxed{17}$ $u_n = \dfrac{2^n x^{2n}}{(2n)!} \Rightarrow \lim\limits_{n \to \infty} \left| \dfrac{u_{n+1}}{u_n} \right| = \lim\limits_{n \to \infty} \left| \dfrac{2^{n+1} x^{2n+2}}{(2n+2)!} \cdot \dfrac{(2n)!}{2^n x^{2n}} \right|$

$\qquad\qquad\qquad\qquad = \lim\limits_{n \to \infty} \left| \dfrac{2^{n+1}}{2^n} \cdot \dfrac{(2n)!}{(2n+2)(2n+1)(2n)!} \cdot \dfrac{x^{2n}(x^2)}{x^{2n}} \right|$

$\qquad\qquad\qquad\qquad = \lim\limits_{n \to \infty} \dfrac{2x^2}{(2n+2)(2n+1)} = 0.$

The last limit equals 0 since the degree of the denominator is two and the degree of the numerator (with respect to n) is zero. Since the limit is always less than 1, the series converges $\forall x$. $\qquad\qquad\qquad \star \; (-\infty, \infty)$

$\boxed{23}$ $u_n = (-1)^n \dfrac{n^n (x-3)^n}{n+1} \Rightarrow$

$\lim\limits_{n \to \infty} \left| \dfrac{u_{n+1}}{u_n} \right| = \lim\limits_{n \to \infty} \left| \dfrac{(n+1)^{n+1}(x-3)^{n+1}}{n+2} \cdot \dfrac{n+1}{n^n (x-3)^n} \right|$

$\qquad\qquad = \lim\limits_{n \to \infty} \left| \dfrac{n+1}{n+2} \cdot \dfrac{(n+1)^{n+1}}{n^n} \cdot \dfrac{(x-3)^{n+1}}{(x-3)^n} \right|$

$\qquad\qquad = \left[\lim\limits_{n \to \infty} \dfrac{n+1}{n+2} \right] \cdot \left[\lim\limits_{n \to \infty} \dfrac{(n+1)^n (n+1)}{n^n} \cdot |x-3| \right]$

$\qquad\qquad = 1 \cdot \lim\limits_{n \to \infty} \left(\dfrac{n+1}{n} \right)^n (n+1)|x-3| = \lim\limits_{n \to \infty} e(n+1)|x-3| = \infty.$

$\left\{ \lim\limits_{n \to \infty} \left(\dfrac{n+1}{n} \right)^n = \lim\limits_{n \to \infty} \left(1 + \dfrac{1}{n} \right)^n = e \text{ by } (7.32)(\text{ii}) \right\}$

$\qquad\qquad\qquad\qquad$ Converges only when $|x - 3| = 0$, i.e., for $x = 3$.

$\boxed{27}$ $u_n = (-1)^n \dfrac{(2x-1)^n}{n 6^n} \Rightarrow \lim\limits_{n \to \infty} \left| \dfrac{u_{n+1}}{u_n} \right| = \lim\limits_{n \to \infty} \left| \dfrac{(2x-1)^{n+1}}{(n+1) 6^{n+1}} \cdot \dfrac{n 6^n}{(2x-1)^n} \right|$

$\qquad\qquad\qquad\qquad\qquad = \lim\limits_{n \to \infty} \left| \dfrac{n}{n+1} \cdot \dfrac{6^n}{6^{n+1}} \cdot \dfrac{(2x-1)^{n+1}}{(2x-1)^n} \right|$

$\qquad\qquad\qquad\qquad\qquad = \tfrac{1}{6} |2x - 1|.$

$\tfrac{1}{6} |2x - 1| < 1 \Leftrightarrow |2x - 1| < 6 \Leftrightarrow -6 < 2x - 1 < 6 \Leftrightarrow -\tfrac{5}{2} < x < \tfrac{7}{2}.$

If $x = -\tfrac{5}{2}$, $\sum\limits_{n=1}^{\infty} (-1)^n \dfrac{1}{n 6^n} \cdot (-6)^n = \sum\limits_{n=1}^{\infty} (-1)^n \dfrac{1}{n 6^n} (-1)^n 6^n = \sum\limits_{n=1}^{\infty} \dfrac{1}{n}$ is D.

If $x = \tfrac{7}{2}$, $\sum\limits_{n=1}^{\infty} (-1)^n \dfrac{1}{n 6^n} \cdot 6^n = \sum\limits_{n=1}^{\infty} (-1)^n \dfrac{1}{n}$ is C by AST. $\qquad \star \; (-\tfrac{5}{2}, \tfrac{7}{2}]$

$\boxed{29}$ $u_n = (-1)^n \dfrac{3^n (x - 4)^n}{n!} \Rightarrow \lim\limits_{n \to \infty} \left| \dfrac{u_{n+1}}{u_n} \right| = \lim\limits_{n \to \infty} \left| \dfrac{3^{n+1} (x - 4)^{n+1}}{(n + 1)!} \cdot \dfrac{n!}{3^n (x - 4)^n} \right|$

$$= \lim\limits_{n \to \infty} \left| \dfrac{n!}{(n + 1) n!} \cdot \dfrac{3^{n+1}}{3^n} \cdot \dfrac{(x - 4)^{n+1}}{(x - 4)^n} \right|$$

$$= \lim\limits_{n \to \infty} \dfrac{3}{n + 1} |x - 4| = 0.$$

Since the limit is always less than 1, the series converges $\forall\, x$. $\bigstar$ $(-\infty, \infty)$

$\boxed{31}$ $\lim\limits_{n \to \infty} \left| \dfrac{u_{n+1}}{u_n} \right| =$

$\lim\limits_{n \to \infty} \left| \dfrac{1 \cdot 3 \cdot 5 \cdot \ldots \cdot (2n - 1)(2n + 1)\, x^{n+1}}{3 \cdot 6 \cdot 9 \cdot \ldots \cdot (3n)(3n + 3)} \cdot \dfrac{3 \cdot 6 \cdot 9 \cdot \ldots \cdot (3n)}{1 \cdot 3 \cdot 5 \cdot \ldots \cdot (2n - 1)\, x^n} \right| =$

$\lim\limits_{n \to \infty} \left(\dfrac{2n + 1}{3n + 3} \right) |x| = \tfrac{2}{3} |x|.\ \tfrac{2}{3} |x| < 1 \Leftrightarrow |x| < \tfrac{3}{2}.$ $\bigstar$ $r = \tfrac{3}{2}.$

Note: The convergence/divergence at the end points does not affect the radius of convergence.

$\boxed{33}$ $\lim\limits_{n \to \infty} \left| \dfrac{u_{n+1}}{u_n} \right| = \lim\limits_{n \to \infty} \left| \dfrac{(n + 1)^{n+1}\, x^{n+1}}{(n + 1)!} \cdot \dfrac{n!}{n^n\, x^n} \right|$

$$= \lim\limits_{n \to \infty} \left| \dfrac{(n + 1)(n + 1)^n\, x}{(n + 1) n^n} \right|$$

$$= \lim\limits_{n \to \infty} \left(\dfrac{n + 1}{n} \right)^n |x|$$

$$= \lim\limits_{n \to \infty} \left(1 + \tfrac{1}{n} \right)^n |x| = e|x|.\ e|x| < 1 \Leftrightarrow |x| < \tfrac{1}{e}.$$ $\bigstar$ $r = \tfrac{1}{e}.$

$\boxed{35}$ $\lim\limits_{n \to \infty} \left| \dfrac{u_{n+1}}{u_n} \right| = \lim\limits_{n \to \infty} \left| \dfrac{(n + c + 1)!\, x^{n+1}}{(n + 1)!\, (n + d + 1)!} \cdot \dfrac{n!\, (n + d)!}{(n + c)!\, x^n} \right|$

$$= \lim\limits_{n \to \infty} \left| \dfrac{(n + c + 1)(n + c)!}{(n + c)!} \cdot \dfrac{(n + d)!}{(n + d + 1)(n + d)!} \cdot \dfrac{n!}{(n + 1) n!} \cdot \dfrac{x^{n+1}}{x^n} \right|$$

$$= \lim\limits_{n \to \infty} \dfrac{n + c + 1}{(n + 1)(n + d + 1)} |x| = 0\ \{ \deg(\text{num}) = 1 < 2 = \deg(\text{den}) \}.$$

The limit is always less than 1. $\bigstar$ $r = \infty.$

$\boxed{41}$ $\lim\limits_{n \to \infty} \left| \dfrac{u_{n+1}}{u_n} \right| = \lim\limits_{n \to \infty} \left| \dfrac{a_{n+1}\, x^{n+1}}{a_n\, x^n} \right| = \lim\limits_{n \to \infty} \left| \dfrac{a_{n+1}}{a_n} \right| |x| = k|x|.$

$k|x| < 1 \Leftrightarrow |x| < \tfrac{1}{k}\ (k \neq 0).$ $\bigstar$ $r = \tfrac{1}{k}$

$\boxed{45}$ We will prove this by contradiction. Assume that $\sum a_n x^n$ is absolutely convergent at $x = r$. Then $\sum |a_n r^n|$ is convergent by (11.32). Let $x = -r$. Then $\sum |a_n(-r)^n| = \sum |(-1)^n a_n r^n| = \sum |a_n r^n|$ is absolutely convergent, which implies that $\sum a_n(-r)^n$ is convergent by (11.34). This is a contradiction.

1 (a) Since $|x| < \frac{1}{3}$, it follows that $|3x| < 1$. Moreover, $\dfrac{1}{1 - 3x}$ can be written as

$\dfrac{a}{1 - r}$, where $|r| < 1$. By (11.15)(i) with $a = 1$ and $r = 3x$,

$$f(x) = \frac{1}{1 - 3x} = 1 + (3x) + (3x)^2 + (3x)^3 + \cdots + (3x)^n + \cdots = \sum_{n=0}^{\infty} 3^n x^n.$$

(b) To find the derivative of a power series, we differentiate each term of the series.

By (11.40)(i), $f'(x) = \displaystyle\sum_{n=1}^{\infty} na^n x^{n-1} = \sum_{n=1}^{\infty} n3^n x^{n-1}$. To find

$\displaystyle\int_0^x f(t)\, dt$, where $f(t)$ is a power series, we integrate each term of the series.

By (11.40)(ii), $\displaystyle\int_0^x f(t)\, dt = \sum_{n=0}^{\infty} \frac{a^n}{n+1} x^{n+1} = \sum_{n=0}^{\infty} \frac{3^n}{n+1} x^{n+1}$.

3 (a) Since $|x| < \frac{2}{7}$, it follows that $\left|\frac{7}{2}x\right| < 1$. We must rewrite $\dfrac{1}{2 + 7x}$ in the form of

$\dfrac{a}{1 - r}$. $f(x) = \dfrac{1}{2 + 7x} = \dfrac{1}{2 - (-7x)} = \dfrac{1}{2} \cdot \dfrac{1}{1 - (-\frac{7}{2}x)}$ {the denominator must

be of the form $1 - r$}.

By (11.15)(i), with $a = \frac{1}{2}$ and $r = -\frac{7}{2}x$,

$$f(x) = \frac{1}{2}\left[1 + (-\tfrac{7}{2}x) + (-\tfrac{7}{2}x)^2 + (-\tfrac{7}{2}x)^3 + \cdots + (-1)^n(\tfrac{7}{2}x)^n + \cdots\right]$$

$$= \frac{1}{2}\sum_{n=0}^{\infty} (-1)^n (\tfrac{7}{2})^n x^n.$$

(b) By (11.40)(i), $f'(x) = \dfrac{1}{2}\displaystyle\sum_{n=1}^{\infty} (-1)^n \left(\tfrac{7}{2}\right)^n nx^{n-1} = \dfrac{1}{2}\sum_{n=1}^{\infty} (-1)^n \dfrac{n7^n}{2^n} x^{n-1}$. By

(11.40)(ii), $\displaystyle\int_0^x f(t)\, dt = \dfrac{1}{2}\sum_{n=0}^{\infty} (-1)^n \left(\tfrac{7}{2}\right)^n \dfrac{x^{n+1}}{(n+1)} = \dfrac{1}{2}\sum_{n=0}^{\infty} (-1)^n \dfrac{7^n}{(n+1)\,2^n} x^{n+1}$.

5 By (11.15), $\dfrac{1}{1 - r} = 1 + r + r^2 + r^3 + \cdots$, where $|r| < 1$.

If we let $r = x^2$, then $\dfrac{1}{1 - x^2} = 1 + x^2 + (x^2)^2 + (x^2)^3 + \cdots$. Thus,

$$\frac{x^2}{1 - x^2} = x^2 \cdot \frac{1}{1 - x^2}$$

$$= x^2\left[1 + (x^2) + (x^2)^2 + (x^2)^3 + \cdots\right]$$

$$= x^2 \sum_{n=0}^{\infty} (x^2)^n = \sum_{n=0}^{\infty} x^{2n+2}. \quad |x^2| < 1 \Rightarrow |x| < 1 \Rightarrow r = 1.$$

9 Using long division, $\dfrac{x^2 + 1}{x - 1} = x + 1 + \dfrac{2}{x - 1}$. We will rewrite $\dfrac{2}{x - 1}$ in the form

$\dfrac{a}{1 - r}$ before we represent it as a power series. $\dfrac{2}{x - 1} = -\dfrac{2}{1 - x} = -2\left(\dfrac{1}{1 - x}\right) =$

$-2(1 + x + x^2 + x^3 + \cdots)$ by (11.15) with $a = 1$ and $r = x$. Thus,

$$\dfrac{x^2 + 1}{x - 1} = x + 1 + \dfrac{2}{x - 1}$$

$$= 1 + x - \dfrac{2}{1 - x}$$

$$= 1 + x - 2(1 + x + x^2 + x^3 + \cdots)$$

$$= 1 + x - 2 - 2x - 2(x^2 + x^3 + x^4 + x^5 + \cdots)$$

$$= -1 - x - 2 \sum_{n=2}^{\infty} x^n. \qquad\qquad |x| < 1 \Rightarrow r = 1.$$

11 (a) $f(x) = \ln(1 - x) \Rightarrow f'(x) = -\dfrac{1}{1 - x}$. We do not know the power series for

$\ln(1 - x)$. However, we do know the power series for its derivative. We can

integrate this power series term-by-term to find the power series for $\ln(1 - x)$.

$$f'(x) = -\dfrac{1}{1 - x} = -(1 + x + x^2 + \cdots + x^n + \cdots) = -\sum_{n=0}^{\infty} x^n \Rightarrow$$

$$f(x) = \int_0^x f'(t)\,dt = -\sum_{n=0}^{\infty} \dfrac{x^{n+1}}{n + 1} = -\sum_{n=1}^{\infty} \dfrac{x^n}{n}.$$

(b) To find $\ln(1.2)$ using this series, we must let $x = -0.2$ in the expression

$\ln(1 - x)$. In this case, the series in part (a) becomes an alternating series that

satisfies (11.30). Using (11.31), $\ln(1.2) = \ln\big[1 - (-0.2)\big] \approx$

$-(-0.2) - \dfrac{(-0.2)^2}{2} - \dfrac{(-0.2)^3}{3} \approx 0.183$, with $|\text{error}| < \dfrac{(-0.2)^4}{4} = 0.4 \times 10^{-3}$.

Calculator value ≈ 0.182321557.

15 Using (11.41), we substitute $3x$ for x. Then,

$e^{3x} = 1 + (3x) + \dfrac{(3x)^2}{2!} + \cdots + \dfrac{(3x)^n}{n!} + \cdots$. Multiplying both sides by x gives us

$$f(x) = xe^{3x} = x\left[1 + (3x) + \dfrac{(3x)^2}{2!} + \dfrac{(3x)^3}{3!} + \cdots\right] = x\sum_{n=0}^{\infty} \dfrac{(3x)^n}{n!} = \sum_{n=0}^{\infty} \dfrac{3^n}{n!} x^{n+1}.$$

19 From Example 3, $\ln(1 + x) = x - \dfrac{x^2}{2} + \dfrac{x^3}{3} - \cdots + (-1)^n \dfrac{x^{n+1}}{n + 1} + \cdots$.

Substituting x^2 for x gives us

$$\ln(1 + x^2) = (x^2) - \dfrac{(x^2)^2}{2} + \dfrac{(x^2)^3}{3} - \dfrac{(x^2)^4}{4} + \cdots + (-1)^n \dfrac{(x^2)^{n+1}}{n + 1} + \cdots . \text{ Thus,}$$

$$f(x) = x^2 \ln(1 + x^2)$$

$$= x^2\left[(x^2) - \dfrac{(x^2)^2}{2} + \dfrac{(x^2)^3}{3} - \dfrac{(x^2)^4}{4} + \cdots\right]$$

$$= x^2 \sum_{n=0}^{\infty} (-1)^n \dfrac{(x^2)^{n+1}}{n + 1} = \sum_{n=0}^{\infty} (-1)^n \dfrac{1}{n + 1} x^{2n+4}.$$

21 From Example 5, $\arctan x = x - \dfrac{x^3}{3} + \dfrac{x^5}{5} - \cdots + (-1)^n \dfrac{x^{2n+1}}{2n+1} + \cdots$.

Substituting $\sqrt{x}$ for x gives us

$$f(x) = \arctan \sqrt{x}$$

$$= \left[(x^{1/2}) - \frac{(x^{1/2})^3}{3} + \frac{(x^{1/2})^5}{5} - \frac{(x^{1/2})^7}{7} + \cdots \right]$$

$$= \sum_{n=0}^{\infty} (-1)^n \frac{(x^{1/2})^{2n+1}}{2n+1} = \sum_{n=0}^{\infty} (-1)^n \frac{1}{2n+1} x^{(2n+1)/2}.$$

25 Using the series for $\cosh x$ (just before Example 6), and substituting x^3 for x gives us

$$1 + \frac{(x^3)^2}{2!} + \frac{(x^3)^4}{4!} + \frac{(x^3)^6}{6!} + \cdots . \text{ Thus,}$$

$$f(x) = x^2 \cosh(x^3)$$

$$= x^2 \left[1 + \frac{(x^3)^2}{2!} + \frac{(x^3)^4}{4!} + \frac{(x^3)^6}{6!} + \cdots \right]$$

$$= x^2 \sum_{n=0}^{\infty} \frac{(x^3)^{2n}}{(2n)!} = \sum_{n=0}^{\infty} \frac{1}{(2n)!} x^{6n+2}.$$

27 Since x is integrated over the interval $[0, \frac{1}{3}]$, $|x^6| < 1$.

We will use (11.15) and write $\dfrac{1}{1+x^6}$ as $\dfrac{a}{1-r}$ with $a = 1$ and $r = -x^6$.

Then, $\dfrac{1}{1+x^6} = \dfrac{1}{1-(-x^6)} = 1 + (-x^6) + (-x^6)^2 + (-x^6)^3 + \cdots$.

$$\int_0^{1/3} \frac{1}{1+x^6} \, dx = \int_0^{1/3} (1 - x^6 + x^{12} - x^{18} + \cdots) \, dx$$

$$= \left[x - \tfrac{1}{7}x^7 + \tfrac{1}{13}x^{13} - \tfrac{1}{19}x^{19} + \cdots \right]_0^{1/3}.$$

The series inside the brackets is an alternating series that satisfies (11.30) and hence

(11.31) applies. Using the first two terms, $I \approx \left[x - \tfrac{1}{7}x^7 \right]_0^{1/3} = \tfrac{1}{3} - \tfrac{1}{7}(\tfrac{1}{3})^7 \approx 0.3333,$

with $|\text{error}| < \tfrac{1}{13}(\tfrac{1}{3})^{13} < 0.5 \times 10^{-7}.$

29 Dividing the series for $\arctan x$ (found in Example 5) by x, we have $\dfrac{\arctan x}{x} =$

$$\tfrac{1}{x}\left(x - \tfrac{x^3}{3} + \tfrac{x^5}{5} - \tfrac{x^7}{7} + \cdots \right) = \left(1 - \tfrac{x^2}{3} + \tfrac{x^4}{5} - \tfrac{x^6}{7} + \cdots \right). \text{ Thus,}$$

$$\int_{0.1}^{0.2} \frac{\arctan x}{x} \, dx = \int_{0.1}^{0.2} (1 - \tfrac{1}{3}x^2 + \tfrac{1}{5}x^4 - \tfrac{1}{7}x^6 + \cdots) \, dx$$

$$= \left[x - \tfrac{1}{9}x^3 + \tfrac{1}{25}x^5 - \tfrac{1}{49}x^7 + \cdots \right]_{0.1}^{0.2}. \text{ Using the first two terms,}$$

$I \approx \left[x - \tfrac{1}{9}x^3 \right]_{0.1}^{0.2} \approx 0.0992.$ Since the series satisfies (11.30), (11.31) applies.

Our estimate of $\arctan 0.2$ is $\left[0.2 - \tfrac{1}{9}(0.2)^3 \right]$ and has an error of at most $\tfrac{1}{25}(0.2)^5.$

(cont.)

Our estimate of $\arctan 0.1$ is $\left[0.1 - \frac{1}{9}(0.1)^3\right]$ and has an error of at most $\frac{1}{25}(0.1)^5$.

If we *subtract* these two values, the error will not exceed the *sum* of the two errors.

(If the first estimate is high and the second estimate is low, then the total error

would be the sum of the two individual errors.)

$$\text{Thus, } |\text{error}| < \tfrac{1}{25}(0.2)^5 + \tfrac{1}{25}(0.1)^5 = 0.0000132 < 0.5 \times 10^{-4}.$$

$\boxed{33}$ First, notice that $\dfrac{d}{dx}\left(\dfrac{1}{1-x^2}\right) = \dfrac{2x}{(1-x^2)^2}.$ If we can write the power series for

$\dfrac{1}{1-x^2}$, we can differentiate it term-by-term to obtain the power series for $\dfrac{2x}{(1-x^2)^2}.$

By (11.15) with $a = 1$ and $r = x^2$, $\dfrac{1}{1-x^2} = 1 + x^2 + x^4 + x^6 + \cdots = \displaystyle\sum_{n=0}^{\infty} x^{2n}.$

Differentiating, we have $\dfrac{2x}{(1-x^2)^2} = \dfrac{d}{dx}\left(\dfrac{1}{1-x^2}\right) = \dfrac{d}{dx}\left(\displaystyle\sum_{n=0}^{\infty} x^{2n}\right) = \displaystyle\sum_{n=1}^{\infty} (2n)x^{2n-1}$

$$\text{by (11.40)(i).}$$

$\boxed{35}$ (a) $J_\alpha(x) = \displaystyle\sum_{n=0}^{\infty} \dfrac{(-1)^n}{n!\,(n+\alpha)!}\left(\dfrac{x}{2}\right)^{2n+\alpha} \Rightarrow$

$D_x\left[J_0(x)\right] = D_x \displaystyle\sum_{n=0}^{\infty} \dfrac{(-1)^n}{n!\,n!}\left(\dfrac{x}{2}\right)^{2n}$

$\qquad\qquad = \displaystyle\sum_{n=1}^{\infty} \dfrac{(-1)^n}{n!\,n!}(2n)\left(\dfrac{x}{2}\right)^{2n-1}\left(\dfrac{1}{2}\right)$

$\qquad\qquad = \displaystyle\sum_{n=0}^{\infty} \dfrac{(-1)^{n+1}}{(n+1)!\,(n+1)!}(n+1)\left(\dfrac{x}{2}\right)^{2n+1}$

$\qquad\qquad = \displaystyle\sum_{n=0}^{\infty} \dfrac{(-1)^n(-1)}{n!\,(n+1)!}\left(\dfrac{x}{2}\right)^{2n+1} = -J_1(x).$

(b) $\displaystyle\int x^3 J_2(x)\,dx = \int\left[\displaystyle\sum_{n=0}^{\infty} \dfrac{(-1)^n}{n!\,(n+2)!}\left(\dfrac{x}{2}\right)^{2n+2} x^3\right] dx$

$\qquad\qquad = \displaystyle\int\left[\displaystyle\sum_{n=0}^{\infty} \dfrac{(-1)^n}{n!\,(n+2)!}\left(\dfrac{x}{2}\right)^{2n+2}\cdot\left(\dfrac{x}{2}\right)^3\cdot 8\right] dx$

$\qquad\qquad = \displaystyle\int\left[\displaystyle\sum_{n=0}^{\infty} \dfrac{(-1)^n}{n!\,(n+2)!}(8)\left(\dfrac{x}{2}\right)^{2n+5}\right] dx$

$\qquad\qquad = \left[\displaystyle\sum_{n=0}^{\infty} \dfrac{(-1)^n}{n!\,(n+2)!}\cdot\dfrac{8\cdot 2}{2(n+3)}\cdot\left(\dfrac{x}{2}\right)^{2n+6}\right] + C$

$\qquad\qquad = \left[\displaystyle\sum_{n=0}^{\infty} \dfrac{(-1)^n}{n!\,(n+3)!}\left(\dfrac{x}{2}\right)^{2n+3} x^3\right] + C = x^3 J_3(x) + C.$

37 Using Exercise 11, $\ln(1-x) = -\sum_{n=1}^{\infty} \frac{x^n}{n} = -\left(x + \frac{x^2}{2} + \frac{x^3}{3} + \frac{x^4}{4} + \cdots\right) \Rightarrow$

$$\frac{\ln(1-t)}{t} = -\frac{1}{t}\left(t + \frac{t^2}{2} + \frac{t^3}{3} + \frac{t^4}{4} + \cdots\right)$$

$$= -\left(1 + \frac{t}{2} + \frac{t^2}{3} + \frac{t^3}{4} + \cdots\right) = -\sum_{n=1}^{\infty} \frac{t^{n-1}}{n}.$$

$$f(x) = \int_0^x \left(-\sum_{n=1}^{\infty} \frac{t^{n-1}}{n}\right) dt = \left[-\sum_{n=1}^{\infty} \frac{t^n}{n^2}\right]_0^x = -\sum_{n=1}^{\infty} \frac{1}{n^2} x^n, \ |x| < 1.$$

Exercises 11.8

Note: Let *DERIV* denote the beginning of the sequence:

$f(x), f(c)\ddagger f'(x), f'(c)\ddagger f''(x), f''(c)\ddagger \dots$. For a Maclaurin series, $c = 0$.

We have used the double dagger symbol ($\ddagger$) to separate the terms.

1 In order to determine $a_n = \dfrac{f^{(n)}(0)}{n!}$, we must find a general formula for $f^{(n)}(0)$.

To do this, we will begin by calculating the first three derivatives of f and try to identify a pattern.

$$f(x) = e^{3x}, \quad f(0) = 1$$
$$f'(x) = 3e^{3x}, \quad f'(0) = 3$$
$$f''(x) = 3^2 e^{3x}, \quad f''(0) = 3^2$$
$$f'''(x) = 3^3 e^{3x}, \quad f'''(0) = 3^3$$

As mentioned in the previous note, we will use the following notation to denote the function, its derivatives, and their values at c.

DERIV: $e^{3x}, 1\ddagger 3e^{3x}, 3\ddagger 3^2 e^{3x}, 3^2\ddagger 3^3 e^{3x}, 3^3\ddagger \dots$.

From this pattern, we can see that $f^{(n)}(x) = 3^n e^{3x}$ and $f^{(n)}(0) = 3^n$.

$$\text{Thus, } a_n = \frac{f^{(n)}(0)}{n!} = \frac{3^n}{n!}.$$

3 Let $f(x) = \sin 2x$. *DERIV:* $\sin 2x, 0\ddagger 2\cos 2x, 2\ddagger -2^2 \sin 2x, 0\ddagger -2^3 \cos 2x, -2^3\ddagger \dots$.
From this pattern, we see that the derivatives of even order, evaluated at $x = 0$, are equal to zero. The derivatives of odd order, evaluated at $x = 0$, are equal to 2^n, with alternating signs. To obtain an expression that represents these alternating signs, we introduce another variable, k, and let $k = 0, 1, 2, \dots$. Then $f^{(2k)}(0) = 0$ and $f^{(2k+1)}(0) = (-1)^k 2^{2k+1}$. Thus, if $n = 2k$, then $a_n = 0$, and if $n = 2k + 1$, then

$$a_n = \frac{f^{(n)}(0)}{n!} = (-1)^k \frac{2^{2k+1}}{(2k+1)!}.$$

$\boxed{7}$ In order to show that $\lim_{n \to \infty} R_n(x) = 0$, we will try to find an $M \geq 0$ such that $\left| f^{(n+1)}(z) \right| \leq M$ for all z. Let $f(x) = \cos x$. *DERIV:* $\cos x$, $1\ddagger$ $-\sin x$, $0\ddagger$ $-\cos x$, $-1\ddagger$ $\sin x$, $0\ddagger$ $\cos x$, $1\ddagger$ ... . From this pattern, we see that the higher derivatives are always either $\pm \sin x$ or $\pm \cos x$. Since $|\pm \sin x| \leq 1$ and $|\pm \cos x| \leq 1$, let $M = 1$.

(a) $\left| f^{(n+1)}(z) \right| \leq 1 \Rightarrow |R_n(x)| = \left| \dfrac{f^{(n+1)}(z)}{(n+1)!} x^{n+1} \right| \leq \left| \dfrac{x^{n+1}}{(n+1)!} \right| \to 0$ as $n \to \infty$

by (11.47). Thus, $f(x) = \cos x$ is represented by its Maclaurin series.

(b) By (11.42), $f(x) = f(0) + f'(0)x + \dfrac{f''(0)}{2!}x^2 + \cdots + \dfrac{f^{(n)}(0)}{n!}x^n + \cdots \Rightarrow$

$$\cos x = 1 + 0 \cdot x + \frac{-1}{2!}x^2 + \frac{0}{3!}x^3 + \frac{1}{4!}x^4 + \frac{0}{5!}x^5 + \frac{-1}{6!}x^6 + \cdots$$

$$= 1 - \frac{x^2}{2!} + \frac{x^4}{4!} - \frac{x^6}{6!} + \cdots = \sum_{n=0}^{\infty} (-1)^n \frac{1}{(2n)!} x^{2n}.$$

$\boxed{9}$ Using (11.48)(a), we substitute $3x$ for x. Then,

$$\sin 3x = 3x - \frac{(3x)^3}{3!} + \frac{(3x)^5}{5!} - \frac{(3x)^7}{7!} + \cdots + (-1)^n \frac{(3x)^{2n+1}}{(2n+1)!} + \cdots$$

$$= \sum_{n=0}^{\infty} (-1)^n \frac{(3x)^{2n+1}}{(2n+1)!}. \quad \text{Multiplying by } x \text{ gives us}$$

$$f(x) = x \sin 3x = x \sum_{n=0}^{\infty} (-1)^n \frac{(3x)^{2n+1}}{(2n+1)!} = \sum_{n=0}^{\infty} (-1)^n \frac{3^{2n+1}}{(2n+1)!} x^{2n+2}.$$

$\boxed{13}$ Using a half-angle formula and then (11.48)(b),

$$f(x) = \cos^2 x = \frac{1}{2} + \frac{1}{2}\cos 2x = \frac{1}{2} + \frac{1}{2}\sum_{n=0}^{\infty} (-1)^n \frac{(2x)^{2n}}{(2n)!}$$

$$= \frac{1}{2} + \sum_{n=0}^{\infty} (-1)^n \frac{2^{2n-1}}{(2n)!} x^{2n}$$

$$= \frac{1}{2} + (-1)^0 \frac{2^{-1}}{0!} x^0 + \sum_{n=1}^{\infty} (-1)^n \frac{2^{2n-1}}{(2n)!} x^{2n}$$

$$= 1 + \sum_{n=1}^{\infty} (-1)^n \frac{2^{2n-1}}{(2n)!} x^{2n}.$$

$\boxed{15}$ *DERIV:* 10^x, $1\ddagger$ $10^x \ln 10$, $\ln 10 \ddagger$ $10^x (\ln 10)^2$, $(\ln 10)^2 \ddagger$ $10^x (\ln 10)^3$, $(\ln 10)^3 \ddagger$ ... .

$f^{(n)}(x) = 10^x (\ln 10)^n$ and $f^{(n)}(0) = (\ln 10)^n$. Thus, $10^x =$

$$\sum_{n=0}^{\infty} \frac{f^{(n)}(0)}{n!} x^n = 1 + (\ln 10)x + \frac{(\ln 10)^2}{2!} x^2 + \frac{(\ln 10)^3}{3!} x^3 + \cdots = \sum_{n=0}^{\infty} \frac{(\ln 10)^n}{n!} x^n.$$

$\boxed{17}$ *DERIV*: $\sin x$, $\frac{1}{\sqrt{2}}\ddagger \cos x$, $\frac{1}{\sqrt{2}}\ddagger -\sin x$, $-\frac{1}{\sqrt{2}}\ddagger -\cos x$, $-\frac{1}{\sqrt{2}}\ddagger \ldots$.

$$\sin x = f(\tfrac{\pi}{4}) + f'(\tfrac{\pi}{4})(x - \tfrac{\pi}{4}) + \frac{f''(\tfrac{\pi}{4})}{2!}(x - \tfrac{\pi}{4})^2 + \frac{f'''(\tfrac{\pi}{4})}{3!}(x - \tfrac{\pi}{4})^3 + \cdots$$

$$= \frac{1}{\sqrt{2}} + \frac{1}{\sqrt{2}}(x - \tfrac{\pi}{4}) - \frac{1}{\sqrt{2}\cdot 2!}(x - \tfrac{\pi}{4})^2 - \frac{1}{\sqrt{2}\cdot 3!}(x - \tfrac{\pi}{4})^3 + \cdots$$

$$= \left[\frac{1}{\sqrt{2}}(x - \tfrac{\pi}{4}) - \frac{1}{\sqrt{2}}\frac{(x - \tfrac{\pi}{4})^3}{3!} + \frac{1}{\sqrt{2}}\frac{(x - \tfrac{\pi}{4})^5}{5!} - \cdots\right]\{\text{odd powered terms}\} +$$

$$\left[\frac{1}{\sqrt{2}} - \frac{1}{\sqrt{2}}\frac{(x - \tfrac{\pi}{4})^2}{2!} + \frac{1}{\sqrt{2}}\frac{(x - \tfrac{\pi}{4})^4}{4!} - \cdots\right]\{\text{even powered terms}\}$$

$$= \sum_{n=0}^{\infty} (-1)^n \frac{1}{\sqrt{2}\,(2n+1)!}(x - \tfrac{\pi}{4})^{2n+1} + \sum_{n=0}^{\infty} (-1)^n \frac{1}{\sqrt{2}\,(2n)!}(x - \tfrac{\pi}{4})^{2n}.$$

Note: Because the series is complicated, it's easier to represent it in summation

notation by breaking it into two parts. However, it is not necessary to do this.

$\boxed{19}$ *DERIV*: x^{-1}, $\frac{1}{2}\ddagger -x^{-2}$, $-\frac{1}{2^2}\ddagger 2x^{-3}$, $\frac{2}{2^3}\ddagger -6x^{-4}$, $-\frac{6}{2^4}\ddagger \ldots$.

$f^{(n)}(x) = (-1)^n\, n!\, x^{-n-1}$ and $f^{(n)}(2) = (-1)^n\, n!/2^{n+1}$. Thus,

$$\tfrac{1}{x} = f(2) + f'(2)(x - 2) + \frac{f''(2)}{2!}(x - 2)^2 + \frac{f'''(2)}{3!}(x - 2)^3 + \cdots$$

$$= \tfrac{1}{2} - \tfrac{1}{2^2}(x - 2) + \frac{2!}{2^3\cdot 2!}(x - 2)^2 - \frac{3!}{2^4\cdot 3!}(x - 2)^3 + \cdots$$

$$= \tfrac{1}{2} - \tfrac{1}{2^2}(x - 2) + \tfrac{1}{2^3}(x - 2)^2 - \tfrac{1}{2^4}(x - 2)^3 + \cdots$$

$$= \sum_{n=0}^{\infty} (-1)^n \frac{1}{2^{n+1}}(x - 2)^n.$$

$\boxed{21}$ If we are finding a series representation in powers of $x + 1$, then $x - c =$

$x + 1 = x - (-1) \Rightarrow c = -1$. *DERIV*: e^{2x}, $e^{-2}\ddagger 2e^{2x}$, $2e^{-2}\ddagger 2^2 e^{2x}$, $2^2 e^{-2}\ddagger \ldots$.

$f^{(n)}(x) = 2^n e^{2x}$ and $f^{(n)}(-1) = 2^n e^{-2}$. Thus,

$$e^{2x} = f(-1) + f'(-1)(x + 1) + \frac{f''(-1)}{2!}(x + 1)^2 + \frac{f'''(-1)}{3!}(x + 1)^2 + \cdots$$

$$= e^{-2} + 2e^{-2}(x + 1) + \frac{2^2 e^{-2}}{2!}(x + 1)^2 + \frac{2^3 e^{-2}}{3!}(x + 1)^3 + \cdots$$

$$= \sum_{n=0}^{\infty} \frac{2^n}{e^2\, n!}(x + 1)^n.$$

$\boxed{23}$ *DERIV*: $\sec x$, $2\ddagger \sec x \tan x$, $2\sqrt{3}\ddagger \sec^3 x + \sec x \tan^2 x$, 14.

$$\sec x = f(\tfrac{\pi}{3}) + f'(\tfrac{\pi}{3})(x - \tfrac{\pi}{3}) + \frac{f''(\tfrac{\pi}{3})}{2!}(x - \tfrac{\pi}{3})^2 + \cdots$$

$$= 2 + 2\sqrt{3}(x - \tfrac{\pi}{3}) + 7(x - \tfrac{\pi}{3})^2 + \cdots.$$

The *first three* terms of the Taylor series for $f(x) = \sec x$ at $x = \tfrac{\pi}{3}$ are

$$2 + 2\sqrt{3}(x - \tfrac{\pi}{3}) + 7(x - \tfrac{\pi}{3})^2.$$

$\boxed{25}$ *DERIV*: $\sin^{-1} x$, $\tfrac{\pi}{6}\ddagger \dfrac{1}{(1 - x^2)^{1/2}}$, $\tfrac{2}{\sqrt{3}}\ddagger \dfrac{x}{(1 - x^2)^{3/2}}$, $\dfrac{4}{3\sqrt{3}}$.

$$\sin^{-1} x = f(\tfrac{1}{2}) + f'(\tfrac{1}{2})(x - \tfrac{1}{2}) + \frac{f''(\tfrac{1}{2})}{2!}(x - \tfrac{1}{2})^2 + \cdots$$

$$= \tfrac{\pi}{6} + \frac{2}{\sqrt{3}}(x - \tfrac{1}{2}) + \frac{2}{3\sqrt{3}}(x - \tfrac{1}{2})^2 + \cdots.$$

Note: In Exercises 29–38, all series are alternating, so (11.31) applies. S denotes the sum

of the first two nonzero terms and E the absolute value of the maximum error.

$\boxed{29}$ By (11.48)(iii),

$$e^{-x} = 1 + (-x) + \frac{(-x)^2}{2!} + \frac{(-x)^3}{3!} + \frac{(-x)^4}{4!} + \cdots$$

$$= 1 - x + \frac{x^2}{2!} - \frac{x^3}{3!} + \frac{x^4}{4!} - \cdots$$

$$= \sum_{n=0}^{\infty} (-1)^n \frac{x^n}{n!}.$$

$\dfrac{1}{\sqrt{e}} = e^{-1/2} \approx 1 - (\tfrac{1}{2}) + \dfrac{(\tfrac{1}{2})^2}{2!}$ $(x = \tfrac{1}{2})$. $S = 1 - \tfrac{1}{2} = 0.5$. $E = \dfrac{(\tfrac{1}{2})^2}{2!} = \tfrac{1}{8} = 0.125$.

$\boxed{33}$ By (11.48)(e), $\tan^{-1} x = \displaystyle\sum_{n=0}^{\infty} (-1)^n \frac{x^{2n+1}}{2n + 1}$. $\tan^{-1} 0.1 \approx (0.1) - \dfrac{(0.1)^3}{3} + \dfrac{(0.1)^5}{5}$.

$$S = (0.1) - \frac{(0.1)^3}{3} \approx 0.0997. \quad E = \frac{(0.1)^5}{5} = 2 \times 10^{-6}.$$

$\boxed{35}$ Substituting $-x^2$ into (11.48)(c) gives us $e^{-x^2} = \displaystyle\sum_{n=0}^{\infty} \frac{(-x^2)^n}{n!} = \sum_{n=0}^{\infty} (-1)^n \frac{x^{2n}}{n!} \Rightarrow$

$$\int_0^1 e^{-x^2}\, dx = \int_0^1 \left[\sum_{n=0}^{\infty} (-1)^n \frac{x^{2n}}{n!} \right] dx$$

$$= \left[\sum_{n=0}^{\infty} (-1)^n \frac{x^{2n+1}}{(2n + 1)\, n!} \right]_0^1 \quad \{\text{using (11.40)(ii)}\}$$

$$= \sum_{n=0}^{\infty} (-1)^n \frac{1}{(2n + 1)\, n!}.$$

$I \approx 1 - \tfrac{1}{3} + \tfrac{1}{10}$. $S = 1 - \tfrac{1}{3} = \tfrac{2}{3} \approx 0.6667$. $E = \tfrac{1}{10} = 0.1$.

Note: In Exercises 39–42, all series are alternating and satisfy (11.30), so (11.31) applies.

39 $\dfrac{1 - \cos x}{x^2} = \dfrac{1 - \left(1 - \frac{x^2}{2!} + \frac{x^4}{4!} - \frac{x^6}{6!} + \cdots\right)}{x^2}$

$\qquad = \dfrac{1}{2!} - \dfrac{x^2}{4!} + \dfrac{x^4}{6!} - \cdots = \displaystyle\sum_{n=0}^{\infty} (-1)^n \dfrac{x^{2n}}{(2n+2)!}.$

$\displaystyle\int_0^1 \dfrac{1 - \cos x}{x^2}\,dx = \int_0^1 \left[\sum_{n=0}^{\infty} (-1)^n \dfrac{x^{2n}}{(2n+2)!} \right] dx$

$\qquad = \left[\displaystyle\sum_{n=0}^{\infty} (-1)^n \dfrac{x^{2n+1}}{(2n+1)(2n+2)!} \right]_0^1$

$\qquad = \displaystyle\sum_{n=0}^{\infty} (-1)^n \dfrac{1}{(2n+1)(2n+2)!} = \sum_{n=0}^{\infty} (-1)^n a_n.$

We must determine an n so that $a_n < 0.5 \times 10^{-4}$. By substituting 0, 1, 2, and 3 for

n, we find that $|\text{Error}| < \dfrac{1}{(2n+1)(2n+2)!} < 0.5 \times 10^{-4}$ for $n \geq 3$.

$\qquad\qquad$ Thus, $I \approx a_0 - a_1 + a_2 = \dfrac{1}{(1)(2!)} - \dfrac{1}{(3)(4!)} + \dfrac{1}{(5)(6!)} \approx 0.4864.$

41 By (11.48)(d), $\dfrac{\ln(1+x)}{x} = \dfrac{1}{x} \displaystyle\sum_{n=0}^{\infty} (-1)^n \dfrac{x^{n+1}}{n+1} = \sum_{n=0}^{\infty} (-1)^n \dfrac{x^n}{n+1}.$

$\displaystyle\int_0^{1/2} \dfrac{\ln(1+x)}{x}\,dx = \int_0^{1/2} \left[\sum_{n=0}^{\infty} (-1)^n \dfrac{x^n}{n+1} \right] dx$

$\qquad = \left[\displaystyle\sum_{n=0}^{\infty} (-1)^n \dfrac{x^{n+1}}{(n+1)^2} \right]_0^{1/2}$

$\qquad = \displaystyle\sum_{n=0}^{\infty} (-1)^n \dfrac{\left(\frac{1}{2}\right)^{n+1}}{(n+1)^2} = \sum_{n=0}^{\infty} (-1)^n a_n.$

We must determine an n so that $a_n < 0.5 \times 10^{-4}$. By substituting successive integer

values for n, we find that $|\text{Error}| < \dfrac{\left(\frac{1}{2}\right)^{n+1}}{(n+1)^2} < 0.5 \times 10^{-4}$ for $n \geq 8$. Thus,

$I \approx a_0 - a_1 + a_2 - a_3 + a_4 - a_5 + a_6 - a_7$

$= \dfrac{1}{2} - \dfrac{(1/2)^2}{4} + \dfrac{(1/2)^3}{9} - \dfrac{(1/2)^4}{16} + \dfrac{(1/2)^5}{25} - \dfrac{(1/2)^6}{36} + \dfrac{(1/2)^7}{49} - \dfrac{(1/2)^8}{64} \approx 0.4484.$

43 (a) $\displaystyle\int_0^1 \sin(x^2)\,dx \approx \int_0^1 (x^2 - \tfrac{1}{6}x^6)\,dx = \left[\tfrac{1}{3}x^3 - \tfrac{1}{42}x^7 \right]_0^1 = \tfrac{13}{42} \approx 0.309524.$

$\qquad \displaystyle\int_1^2 \sin(x^2)\,dx \approx \int_1^2 (x^2 - \tfrac{1}{6}x^6)\,dx = \left[\tfrac{1}{3}x^3 - \tfrac{1}{42}x^7 \right]_1^2 = -\tfrac{29}{42} \approx -0.690476.$

(b) From the graph, we can see that $g(x)$ becomes a worse approximation for $f(x)$ the

$\qquad$ further x is from zero. The first approximation is more accurate.

$\boxed{45}$ $\ln\left(\dfrac{1 + x}{1 - x}\right) = \ln\left(1 + x\right) - \ln\left(1 - x\right).$

$\ln\left(1 - x\right) = \ln\left[1 + (-x)\right]$

$$= \sum_{n=0}^{\infty} (-1)^n \frac{(-x)^{n+1}}{n + 1}$$

$$= \sum_{n=0}^{\infty} (-1)^n \frac{(-1)^{n+1} x^{n+1}}{n + 1}$$

$$= \sum_{n=0}^{\infty} (-1)^{2n+1} \frac{x^{n+1}}{n + 1}$$

$$= -\sum_{n=0}^{\infty} \frac{x^{n+1}}{n + 1} \quad \text{since } 2n + 1 \text{ is always an odd integer. Thus,}$$

$$\ln\left(1 + x\right) - \ln\left(1 - x\right) = \sum_{n=0}^{\infty} (-1)^n \frac{x^{n+1}}{n + 1} + \sum_{n=0}^{\infty} \frac{x^{n+1}}{n + 1}$$

$$= \left(x - \frac{x^2}{2} + \frac{x^3}{3} + \cdots\right) + \left(x + \frac{x^2}{2} + \frac{x^3}{3} + \cdots\right)$$

$$= 2(x) + 2\left(\frac{x^3}{3}\right) + 2\left(\frac{x^5}{5}\right) + \cdots = 2 \sum_{n=0}^{\infty} \frac{x^{2n+1}}{2n + 1}.$$

Since the first series is valid for $-1 < x \le 1$ and the second series is valid for

$$-1 < -x \le 1 \Leftrightarrow -1 \le x < 1, \text{ the final series is valid for } |x| < 1.$$

$\boxed{49}$ (a) The central angle θ of a circle of radius R subtended by an arc of length s is

$\theta = \frac{s}{R}$. From the figure in the text, $\cos\theta = \dfrac{R}{R + C} \Rightarrow \sec\theta = \dfrac{R + C}{R} \Rightarrow$

$$R\sec\theta = R + C \Rightarrow C = R(\sec\theta - 1) = R\left[\sec\left(s/R\right) - 1\right].$$

(b) *DERIV:* $\sec x$, $1\ddagger \sec x \tan x$, $0\ddagger \sec^3 x + \sec x \tan^2 x$, $1\ddagger 5\sec^3 x \tan x + \sec x \tan^3 x$,

$0\ddagger 18\sec^3 x \tan^2 x + 5\sec^5 x + \sec x \tan^4 x$, 5.

$f(x) = \sec x = f(0) + f'(0) + \dfrac{f''(0)}{2!}x^2 + \dfrac{f'''(0)}{3!}x^3 + \cdots \approx 1 + \dfrac{1}{2!}x^2 + \dfrac{5}{4!}x^4.$

If we let $x = \frac{s}{R}$, then $C = R\left[\sec\left(s/R\right) - 1\right]$

$$\approx R\left[\left(1 + \frac{s^2}{2R^2} + \frac{5s^4}{24R^4}\right) - 1\right] = \frac{s^2}{2R} + \frac{5s^4}{24R^3}.$$

(c) $R = 3959$, $s = 5 \Rightarrow C \approx \dfrac{5^2}{2(3959)} + \dfrac{5(5)^4}{24(3959)^3} \approx 0.003157 \text{ mi} \approx 16.7 \text{ ft.}$

Exercises 11.9

Note: In the following exercises, $\sin x$ and $\cos x$ have been bound by 1.

In certain intervals, it may be possible to bound them by a smaller value.

1. (a) Use (11.48)(a) to find $P_n(x)$, which consists of
   terms through x^n. Thus, $P_1(x) = P_2(x) = x$, and
   $P_3(x) = x - \frac{1}{6}x^3$.

   (c) $f(x) = \sin x$ and $f(0.05) \approx P_3(0.05) =$
   $0.05 - \frac{1}{6}(0.05)^3 \approx 0.0500$. $f^{(4)}(x) = \sin x$ and
   $\left| f^{(4)}(z) \right| \le 1$ for all z. Error $\le \left| R_3(0.05) \right| =$
   $\left| \frac{f^{(4)}(z)}{4!}(0.05)^4 \right| = \left| \frac{\sin z}{4!}(0.05)^4 \right| \le \frac{(1)}{4!}(0.05)^4 \approx$
   2.6×10^{-7}.

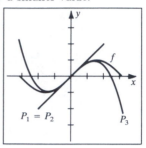

Figure 1

Note: Let *DERIV* denote the beginning of the sequence:

$$f(x), \; f(c)\ddagger \; f'(x), \; f'(c)\ddagger \; \ldots \ddagger \; f^{(n)}(x), \; f^{(n)}(c)\ddagger \; f^{(n+1)}(z).$$

7. *DERIV:* $\sin x$, $1\ddagger \cos x$, $0\ddagger -\sin x$, $-1\ddagger -\cos x$, $0\ddagger \sin z$.

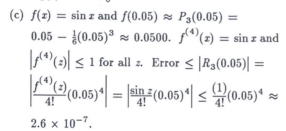

$$\underline{\sin x} = f(\tfrac{\pi}{2}) + f'(\tfrac{\pi}{2})(x - \tfrac{\pi}{2}) + \frac{f''(\tfrac{\pi}{2})}{2!}(x - \tfrac{\pi}{2})^2 + \frac{f'''(\tfrac{\pi}{2})}{3!}(x - \tfrac{\pi}{2})^3 + \frac{f^{(4)}(z)}{4!}(x - \tfrac{\pi}{2})^4$$

$$= 1 + 0 \cdot (x - \tfrac{\pi}{2}) + \frac{-1}{2!}(x - \tfrac{\pi}{2})^2 + \frac{0}{3!}(x - \tfrac{\pi}{2})^3 + \frac{\sin z}{4!}(x - \tfrac{\pi}{2})^4$$

$$= 1 - \tfrac{1}{2}(x - \tfrac{\pi}{2})^2 + \frac{\sin z}{24}(x - \tfrac{\pi}{2})^4, \; z \text{ is between } x \text{ and } \tfrac{\pi}{2}.$$

9. *DERIV:* $x^{1/2}$, $2\ddagger \frac{1}{2}x^{-1/2}$, $\frac{1}{4}\ddagger -\frac{1}{4}x^{-3/2}$, $-\frac{1}{32}\ddagger \frac{3}{8}x^{-5/2}$, $\frac{3}{256}\ddagger -\frac{15}{16}z^{-7/2}$.

$$\sqrt{x} = f(4) + f'(4)(x - 4) + \frac{f''(4)}{2!}(x - 4)^2 + \frac{f'''(4)}{3!}(x - 4)^3 + \frac{f^{(4)}(z)}{4!}(x - 4)^4$$

$$= 2 + \tfrac{1}{4}(x - 4) - \tfrac{1}{64}(x - 4)^2 + \tfrac{1}{512}(x - 4)^3 - \tfrac{5}{128}z^{-7/2}(x - 4)^4,$$

z is between x and 4.

13. *DERIV:* x^{-1}, $-\frac{1}{2}\ddagger -x^{-2}$, $-\frac{1}{4}\ddagger 2x^{-3}$, $-\frac{1}{4}\ddagger -6x^{-4}$, $-\frac{3}{8}\ddagger$

$24x^{-5}$, $-\frac{3}{4}\ddagger -120x^{-6}$, $-\frac{15}{8}\ddagger 720z^{-7}$.

$$\frac{1}{x} = f(-2) + f'(-2)(x + 2) + \frac{f''(-2)}{2!}(x + 2)^2 + \frac{f'''(-2)}{3!}(x + 2)^3 +$$

$$\frac{f^{(4)}(-2)}{4!}(x + 2)^4 + \frac{f^{(5)}(-2)}{5!}(x + 2)^5 + \frac{f^{(6)}(z)}{6!}(x + 2)^6$$

$$= -\tfrac{1}{2} - \tfrac{1}{4}(x + 2) - \tfrac{1}{8}(x + 2)^2 - \tfrac{1}{16}(x + 2)^3 - \tfrac{1}{32}(x + 2)^4 - \tfrac{1}{64}(x + 2)^5 +$$

$$z^{-7}(x + 2)^6, \; z \text{ is between } x \text{ and } -2.$$

15 *DERIV*: $\tan^{-1} x$, $\frac{\pi}{4}$‡ $(1 + x^2)^{-1}$, $\frac{1}{2}$‡ $-2x(1 + x^2)^{-2}$, $-\frac{1}{2}$‡ $(6z^2 - 2)(1 + z^2)^{-3}$.

$$\underline{\tan^{-1} x} = f(1) + f'(1)(x - 1) + \frac{f''(1)}{2!}(x - 1)^2 + \frac{f'''(z)}{3!}(x - 1)^3$$

$$= \frac{\pi}{4} + \frac{1}{2}(x - 1) - \frac{1}{4}(x - 1)^2 + \frac{3z^2 - 1}{3(1 + z^2)^3}(x - 1)^3, \ z \text{ is between } x \text{ and } 1.$$

Note: Exer. 19-30: Since $c = 0$, z is between x and 0.

19 *DERIV*: $\ln(x + 1)$, 0‡ $(x + 1)^{-1}$, 1‡ $-1(x + 1)^{-2}$, -1‡ $2(x + 1)^{-3}$, 2‡

$$-6(x + 1)^{-4}, \ -6‡ \ 24(z + 1)^{-5}.$$

$$\underline{\ln(x + 1)} = f(0) + f'(0)x + \frac{f''(0)}{2!}x^2 + \frac{f'''(0)}{3!}x^3 + \frac{f^{(4)}(0)}{4!}x^4 + \frac{f^{(5)}(z)}{5!}x^5$$

$$= 0 + 1 \cdot x + \frac{-1}{2}x^2 + \frac{2}{6}x^3 + \frac{-6}{24}x^4 + \frac{24}{120(z + 1)^5}x^5$$

$$= x - \frac{1}{2}x^2 + \frac{1}{3}x^3 - \frac{1}{4}x^4 + \frac{x^5}{5(z + 1)^5}$$

21 *DERIV*: $\cos x$, 1‡ $-\sin x$, 0‡ $-\cos x$, -1‡ $\sin x$, 0‡ $\cos x$, 1‡ $-\sin x$, 0‡ $-\cos x$, -1‡

$\sin x$, 0‡ $\cos x$, 1‡ $-\sin z$. {Note that odd-numbered derivatives will equal 0.}

$$\underline{\cos x} = f(0) + f'(0)x + \frac{f''(0)}{2!}x^2 + \frac{f'''(0)}{3!}x^3 + \frac{f^{(4)}(0)}{4!}x^4 + \frac{f^{(5)}(0)}{5!}x^5 +$$

$$\frac{f^{(6)}(0)}{6!}x^6 + \frac{f^{(7)}(0)}{7!}x^7 + \frac{f^{(8)}(0)}{8!}x^8 + \frac{f^{(9)}(z)}{9!}x^9$$

$$= 1 + \frac{-1}{2!}x^2 + \frac{1}{4!}x^4 + \frac{-1}{6!}x^6 + \frac{1}{8!}x^8 + \frac{-\sin z}{9!}x^9$$

$$= 1 - \frac{x^2}{2!} + \frac{x^4}{4!} - \frac{x^6}{6!} + \frac{x^8}{8!} - \frac{\sin z}{9!}x^9$$

23 Since $f'(x) = 2e^{2x}$, $f''(x) = 2^2 e^{2x}$, $f'''(x) = 2^3 e^{2x}$,

we see that $f^{(k)}(x) = 2^k e^{2x}$, $f^{(k)}(0) = 2^k$, and $f^{(6)}(z) = 64e^{2z}$.

$$\underline{e^{2x}} = f(0) + f'(0)x + \frac{f''(0)}{2!}x^2 + \frac{f'''(0)}{3!}x^3 + \frac{f^{(4)}(0)}{4!}x^4 + \frac{f^{(5)}(0)}{5!}x^5 + \frac{f^{(6)}(z)}{6!}x^6$$

$$= 1 + 2x + \frac{4}{2!}x^2 + \frac{8}{3!}x^3 + \frac{16}{4!}x^4 + \frac{32}{5!}x^5 + \frac{64e^{2z}}{6!}x^6$$

$$= 1 + 2x + 2x^2 + \frac{4}{3}x^3 + \frac{2}{3}x^4 + \frac{4}{15}x^5 + \frac{4}{45}e^{2z}x^6$$

25 *DERIV*: $(x - 1)^{-2}$, 1‡ $-2(x - 1)^{-3}$, 2‡ $6(x - 1)^{-4}$, 6‡ $-24(x - 1)^{-5}$, 24‡

$120(x - 1)^{-6}$, 120‡ $-720(x - 1)^{-7}$, 720‡ $5040(z - 1)^{-8}$.

$$\frac{1}{(x - 1)^2} = 1 + 2x + 3x^2 + 4x^3 + 5x^4 + 6x^5 + \frac{7x^6}{(z - 1)^8}$$

29 *DERIV:* $2x^4 - 5x^3$, 0‡ $8x^3 - 15x^2$, 0‡ $24x^2 - 30x$, 0‡ $48x - 30$, -30‡ 48, 48‡ 0.

For both values of n,

$$f(x) = f(0) + f'(0)\,x + \frac{f''(0)}{2!}\,x^2 + \frac{f'''(0)}{3!}\,x^3 + \frac{f^{(4)}(0)}{4!}\,x^4 + \frac{f^{(5)}(z)}{5!}\,x^5$$

$$= 0 + 0 \cdot x + \frac{0}{2!}\,x^2 + \frac{-30}{3!}\,x^3 + \frac{48}{4!}\,x^4 + \frac{0}{5!}\,x^5$$

$$= -5x^3 + 2x^4.$$

Note: If f is a polynomial of degree n, then $P_n(x) = f(x)$ and $R_n(x) = 0$.

31 First, we must write $89°$ in terms of radian measure.

Since $\frac{\pi}{2}$ corresponds to $90°$ and $\frac{\pi}{180}$ to $1°$, let $x = \frac{\pi}{2} - \frac{\pi}{180}$.

Then, $\sin x \approx 1 - \frac{1}{2}(x - \frac{\pi}{2})^2 \Rightarrow \sin 89° \approx 1 - \frac{1}{2}(-\frac{\pi}{180})^2 \approx 0.9998$.

Since $|\sin z| \le 1$, $\left|R_3(x)\right| \le \left|\frac{1}{24}(-\frac{\pi}{180})^4\right| \approx 4 \times 10^{-9}$.

33 Let $x = 4 + 0.03$. Then, $\sqrt{x} \approx 2 + \frac{1}{4}(x - 4) - \frac{1}{64}(x - 4)^2 + \frac{1}{512}(x - 4)^3 \Rightarrow$

$\sqrt{4.03} \approx 2 + \frac{1}{4}(0.03) - \frac{1}{64}(0.03)^2 + \frac{1}{512}(0.03)^3 \approx 2.0075$. We must determine where

$\left|f^{(4)}(z)\right| = \left|\frac{1}{z^{7/2}}\right|$ is maximum when z is between 4 and 4.03. The maximum occurs

when $z = 4$ because that value minimizes the denominator and hence, maximizes the

value of the fraction. Since $\left|\frac{1}{z^{7/2}}\right| \le \left|\frac{1}{4^{7/2}}\right| \approx 0.0078$ on $(4, 4.03)$,

$$\left|R_3(x)\right| \le \left|\frac{5}{128}(0.0078)(0.03)^4\right| < 3 \times 10^{-10}.$$

35 Let $x = -2 - 0.2$ or $x + 2 = -0.2$. Then, $\frac{1}{x} = -\frac{1}{2.2} \approx$

$-\frac{1}{2} - \frac{1}{4}(-0.2) - \frac{1}{8}(-0.2)^2 - \frac{1}{16}(-0.2)^3 - \frac{1}{32}(-0.2)^4 - \frac{1}{64}(-0.2)^5 \approx -0.454545$.

We must determine where $\left|f^{(6)}(z)\right| = \left|\frac{1}{z^7}\right|$ is maximum when z is between -2.2 and

-2. $\left|\frac{1}{z^7}\right|$ is maximized when $z = -2$. Thus, $\left|R_5(x)\right| \le \left|-2^{-7}(-0.2)^6\right| = 5 \times 10^{-7}$.

37 Let $x = 0.25$. Then, $\ln(1 + x) \approx x - \frac{1}{2}x^2 + \frac{1}{3}x^3 - \frac{1}{4}x^4 \Rightarrow$

$\ln 1.25 \approx 0.25 - \frac{1}{2}(0.25)^2 + \frac{1}{3}(0.25)^3 - \frac{1}{4}(0.25)^4 \approx 0.22298$.

We must determine where $\left|f^{(5)}(z)\right| = \left|\frac{24}{(z + 1)^5}\right|$ is maximum when z is between 0 and

0.25. This occurs when $z = 0$. Thus, $\left|R_4(x)\right| \le \left|\frac{(0.25)^5}{5(0 + 1)^5}\right| \approx 2 \times 10^{-4}$.

41 Using Exercise 21 with $n = 3$, $\cos x = 1 - \frac{x^2}{2!} + \frac{\cos z}{4!}\,x^4$.

Since $|\cos z| \le 1$, $|x^4| \le (0.1)^4$ when $-0.1 \le x \le 0.1$, and z is between 0 and x,

$$\left|R_3(x)\right| \le \left|\frac{(1)(0.1)^4}{24}\right| \approx 4.2 \times 10^{-6} < 0.5 \times 10^{-5} \Rightarrow \underline{\text{five decimal places}}.$$

45 Using Exercise 19 with $n = 3$, $\ln(x + 1) = x - \frac{1}{2}x^2 + \frac{1}{3}x^3 - \dfrac{x^4}{4(z + 1)^4}$. Since

$$\left|\frac{1}{4(z + 1)^4}\right| \le \frac{1}{4(0.9)^4}, \quad |x^4| \le (0.1)^4 \text{ when } -0.1 \le x \le 0.1, \text{ and } z \text{ is between } 0 \text{ and } x,$$

$$|R_3(x)| \le \left|\frac{(-0.1)^4}{4(0.9)^4}\right| \approx 0.000038 < 0.5 \times 10^{-4} \Rightarrow \text{four decimal places}.$$

Exercises 11.10

1 (a) Let $k = \frac{1}{2}$ in (11.50). Then,

$$(1 + x)^{1/2} = 1 + \frac{1}{2}x + \frac{(\frac{1}{2})(-\frac{1}{2})}{2!}x^2 + \frac{(\frac{1}{2})(-\frac{1}{2})(-\frac{3}{2})}{3!}x^3 + \cdots$$

$$= 1 + \frac{1}{2}x - \frac{1}{8}x^2 + \sum_{n=3}^{\infty} (-1)^{n-1}\frac{1 \cdot 3 \cdot 5 \cdot \cdots \cdot (2n - 3)}{2^n\, n!}x^n.$$

We started the summation notation with $n = 3$ because this is when the pattern started to become apparent. We could have started the summation notation with another value for n. The "$2n - 3$" was obtained by noticing that the pattern of the product in the numerator, $1 \cdot 3 \cdot 5 \cdot \cdots$, will have a last term of the form $(2n + k)$ for each value of n. In particular, we see that $2n + k = 3$ when $n = 3$, and hence $6 + k = 3$, or $k = -3$, and our term is $2n - 3$. The series is valid when $|x| < 1$. Thus, $r = 1$.

(b) Substituting $-x^3$ for x in part (a) yields

$$(1 - x^3)^{1/2} = 1 - \frac{1}{2}x^3 - \frac{1}{8}x^6 - \sum_{n=3}^{\infty} \frac{1 \cdot 3 \cdot 5 \cdot \cdots \cdot (2n - 3)}{2^n\, n!}x^{3n}.$$

The series is valid when $\left|-x^3\right| < 1 \Leftrightarrow |x| < 1$. Thus, $r = 1$.

3 Let $k = -\frac{2}{3}$ in (11.50). Then,

$$(1 + x)^{-2/3} = 1 - \frac{2}{3}x + \frac{(-\frac{2}{3})(-\frac{5}{3})}{2!}x^2 + \frac{(-\frac{2}{3})(-\frac{5}{3})(-\frac{8}{3})}{3!}x^3 + \cdots$$

$$= 1 - \frac{2}{3}x + \frac{5}{9}x^2 + \sum_{n=3}^{\infty} \frac{(-2)(-5)(-8)\cdots(1 - 3n)}{3^n\, n!}x^n; \ r = 1.$$

To obtain the "$1 - 3n$" term, we notice that each term of the product $(-2)(-5)\cdots$ is 3 *less* than its predecessor, so the general term must be of the form $-3n + k$. In particular, when $n = 3$, $-3n + k = -8 \Rightarrow -9 + k = -8 \Rightarrow k = 1$, and our term is $-3n + 1$, or, equivalently, $1 - 3n$.

5 Let $k = \frac{3}{5}$ and substitute $(-x)$ for x in (11.50). Then,

$$(1 - x)^{3/5} = 1 + \frac{3}{5}(-x) + \frac{(\frac{3}{5})(-\frac{2}{5})}{2!}(-x)^2 + \frac{(\frac{3}{5})(-\frac{2}{5})(-\frac{7}{5})}{3!}(-x)^3 + \cdots$$

$$= 1 - \frac{3}{5}x - \frac{3}{25}x^2 + \sum_{n=3}^{\infty} \frac{(3)(-2)(-7)\cdots(8 - 5n)}{5^n\, n!}(-x)^n; \ r = 1.$$

11 In order to apply (11.50), we must write $\sqrt[3]{8 + x}$ in the form $(1 + x)^k$, where the "1"

is obtained by factoring out 8. $(8 + x)^{1/3} = \left[8(1 + \frac{1}{8}x)\right]^{1/3} = 8^{1/3}(1 + \frac{1}{8}x)^{1/3} =$

$2(1 + \frac{1}{8}x)^{1/3}$. Let $k = \frac{1}{3}$ and substitute $\frac{1}{8}x$ for x in (11.50). Then,

$$2(1 + \tfrac{1}{8}x)^{1/3} = 2\left[1 + \tfrac{1}{3}(\tfrac{1}{8}x) + \frac{(\frac{1}{3})(-\frac{2}{3})}{2!}(\tfrac{1}{8}x)^2 + \frac{(\frac{1}{3})(-\frac{2}{3})(-\frac{5}{3})}{3!}(\tfrac{1}{8}x)^3 + \cdots\right]$$

$$= 2\left[1 + \tfrac{1}{24}x - \tfrac{1}{576}x^2 + \sum_{n=3}^{\infty}\frac{(-2)(-5)\cdots(4 - 3n)}{3^n\,n!\,8^n}x^n\right]$$

$$= 2 + \tfrac{1}{12}x - \tfrac{1}{288}x^2 + 2\sum_{n=3}^{\infty}(-1)^{n-1}\frac{2\cdot 5\cdot\cdots\cdot(3n - 4)}{24^n\,n!}x^n;$$

$$\left|\tfrac{1}{8}x\right| < 1 \Rightarrow r = 8.$$

13 (a) Let $k = -\frac{1}{2}$ and substitute $-t^2$ for x in (11.50). Then, $(1 - t^2)^{-1/2}$

$$= 1 + (-\tfrac{1}{2})(-t^2) + \frac{(-\frac{1}{2})(-\frac{3}{2})}{2!}(-t^2)^2 + \frac{(-\frac{1}{2})(-\frac{3}{2})(-\frac{5}{2})}{3!}(-t^2)^3 + \cdots$$

$$= 1 + \frac{1}{2^1\,1!}t^2 + \frac{1\cdot 3}{2^2\,2!}t^4 + \frac{1\cdot 3\cdot 5}{2^3\,3!}t^6 + \cdots$$

$$= 1 + \sum_{n=1}^{\infty}\frac{1\cdot 3\cdot 5\cdot\cdots\cdot(2n - 1)}{2^n\,n!}t^{2n}.$$

To find a power series representation for $f(x)$, we will use (11.40)(ii).

$$\sin^{-1}x = \int_0^x(1 - t^2)^{-1/2}\,dt = x + \sum_{n=1}^{\infty}\frac{1\cdot 3\cdot 5\cdot\cdots\cdot(2n - 1)}{2^n\,n!\,(2n + 1)}x^{2n+1}.$$

(b) $t^2 < 1 \Rightarrow |t| < 1 \Rightarrow |x| < 1 \Rightarrow r = 1$.

15 Substituting x^3 for x in Exercise 1(a), $(1 + x^3)^{1/2} = 1 + \frac{1}{2}x^3 - \frac{1}{8}x^6 + \cdots \Rightarrow$

$$\int_0^{1/2}(1 + x^3)^{1/2}\,dx \approx \int_0^{1/2}(1 + \tfrac{1}{2}x^3)\,dx = \left[x + \tfrac{1}{8}x^4\right]_0^{1/2} \approx 0.508.$$

Since the series is alternating in sign and satisfies (11.30), (11.31) applies.

$$\text{Hence, } |\text{Error}| \le \int_0^{1/2}\left|-\tfrac{1}{8}x^6\right|\,dx = \left[\tfrac{1}{8(7)}x^7\right]_0^{1/2} = \tfrac{1}{8(7)}(\tfrac{1}{2})^7 < 0.5 \times 10^{-3}.$$

17 Substituting x^2 for x in Exercise 5, $(1 - x^2)^{3/5} = 1 - \frac{3}{5}x^2 - \frac{3}{25}x^4 - \cdots \Rightarrow$

$$\int_0^{0.2}(1 - x^2)^{3/5}\,dx \approx \int_0^{0.2}(1 - \tfrac{3}{5}x^2)\,dx = \left[x - \tfrac{1}{5}x^3\right]_0^{0.2} \approx 0.198.$$

The actual value is approximately 0.198392.

23 (a) Since $(1 - x)^{-1/2} \approx 1 - \frac{1}{2}(-x) = 1 + \frac{1}{2}x$,

let $x = k^2 \sin^2 u$ and $\dfrac{1}{\sqrt{1 - k^2 \sin^2 u}} = (1 - k^2 \sin^2 u)^{-1/2} \approx 1 + \frac{1}{2}k^2 \sin^2 u$. Then,

$$T \approx 4\sqrt{\frac{L}{g}} \int_0^{\pi/2} (1 + \tfrac{1}{2}k^2 \sin^2 u)\, du$$

$$= 4\sqrt{\frac{L}{g}} \int_0^{\pi/2} \left[1 + \tfrac{1}{2}k^2 \left(\frac{1 - \cos 2u}{2}\right)\right] du$$

$$= 4\sqrt{\frac{L}{g}} \int_0^{\pi/2} (1 + \tfrac{1}{4}k^2 - \tfrac{1}{4}k^2 \cos 2u)\, du$$

$$= 4\sqrt{\frac{L}{g}} \left[u + \tfrac{1}{4}k^2 u - \tfrac{1}{8}k^2 \sin 2u\right]_0^{\pi/2}$$

$$= 4\sqrt{\frac{L}{g}} \left[\frac{\pi}{2} + \tfrac{1}{4}k^2 \left(\frac{\pi}{2}\right)\right] = 2\pi \sqrt{\frac{L}{g}} \left(1 + \tfrac{1}{4}k^2\right).$$

(b) $\theta_0 = \frac{\pi}{6} \Rightarrow k = \sin \frac{1}{2}\theta_0 = \sin\left(\frac{1}{2} \cdot \frac{\pi}{6}\right) = \sin \frac{\pi}{12}.$

$$T \approx 2\pi \sqrt{L/g}\, (1 + \tfrac{1}{4}\sin^2 \tfrac{\pi}{12}) \approx 6.39 \sqrt{L/g}.$$

11.11 Review Exercises

Note: Let AC denote Absolutely Convergent; CC, Conditionally Convergent; D,

Divergent; C, Convergent; and AST, the Alternating Series Test.

1 Using (11.5)(i), $\displaystyle\lim_{x \to \infty} \frac{\ln(x^2 + 1)}{x} \left\{\frac{\infty}{\infty}\right\} = \lim_{x \to \infty} \frac{1/(x^2 + 1) \cdot 2x}{1} = \lim_{x \to \infty} \frac{2x}{x^2 + 1} = 0.$

Thus, $L = 0$; C.

5 $\displaystyle\lim_{n \to \infty} \left(\frac{n}{\sqrt{n} + 4} - \frac{n}{\sqrt{n} + 9}\right) = \lim_{n \to \infty} \frac{n(\sqrt{n} + 9) - n(\sqrt{n} + 4)}{(\sqrt{n} + 4)(\sqrt{n} + 9)} =$

$$\lim_{n \to \infty} \frac{5n}{(\sqrt{n} + 4)(\sqrt{n} + 9)} = \lim_{n \to \infty} \frac{5n}{n + 13\sqrt{n} + 36} = 5;\ \text{C}.$$

Note: In Exercises 7–32, let S denote the given series.

7 Let $a_n = \dfrac{1}{\sqrt[3]{n(n + 1)(n + 2)}}$ and $b_n = \dfrac{1}{n}$. $\displaystyle\lim_{n \to \infty} \frac{a_n}{b_n} =$

$$\lim_{n \to \infty} \frac{n}{\sqrt[3]{n(n + 1)(n + 2)}} = 1 > 0. \text{ S diverges by (11.27) since } \sum b_n \text{ diverges.}$$

9 This is a geometric series with $r = -\frac{2}{3}$ and contains negative terms.

Since $|r| = \left|-\frac{2}{3}\right| < 1$, S is AC by (11.15).

13 $\displaystyle\lim_{n \to \infty} \frac{a_{n+1}}{a_n} = \lim_{n \to \infty} \frac{(n + 1)!}{\ln(n + 2)} \cdot \frac{\ln(n + 1)}{n!} = \lim_{n \to \infty} \frac{(n + 1)\ln(n + 1)}{\ln(n + 2)} = \infty \Rightarrow$

S is D by (11.28).

$\boxed{15}$ This series contains negative terms.

$$\lim_{n \to \infty} \left| \frac{a_{n+1}}{a_n} \right| = \lim_{n \to \infty} \left| \frac{\left[(n+1)^2 + 9\right](-2)^{1-(n+1)}}{(n^2+9)(-2)^{1-n}} \right|$$

$$= \lim_{n \to \infty} \left| \frac{(n+1)^2 + 9}{n^2 + 9} \cdot \frac{(-2)^{-n}}{(-2)^{1-n}} \right|$$

$$= \lim_{n \to \infty} \left| \frac{(n+1)^2 + 9}{n^2 + 9} \cdot \frac{(-2)^{n-1}}{(-2)^n} \right| = 1 \cdot \tfrac{1}{2} = \tfrac{1}{2} < 1 \Rightarrow \text{S is AC by (11.35)}.$$

$\boxed{19}$ This series contains negative terms.

$$\lim_{n \to \infty} \left| (-1)^n \frac{1}{\sqrt[n]{n}} \right| = \lim_{n \to \infty} \frac{1}{n^{1/n}} \; \{ \text{see Exer. 11.1.37} \} = 1 \neq 0 \Rightarrow \text{S is D by (11.17)}.$$

$\boxed{23}$ This series contains negative terms.

We will show that this series satisfies (11.30) and, therefore, converges.

$$a_n = \frac{\sqrt{n}}{n+1} > 0. \quad f(x) = \frac{\sqrt{x}}{x+1} \Rightarrow f'(x) = \frac{1-x}{2\sqrt{x}(x+1)^2} < 0 \text{ for } x > 1 \Rightarrow$$

$a_n \geq a_{n+1}$, and $\lim_{n \to \infty} a_n = 0$. S converges by AST. To show that this series does

not converge absolutely, we will apply the LCT. If $b_n = \frac{1}{\sqrt{n}}$, $\lim_{n \to \infty} \frac{a_n}{b_n} = 1 > 0$ and

$\sum a_n$ diverges since $\sum b_n$ diverges by (11.25) $\{ p = \tfrac{1}{2} \}$. Thus, S is CC.

$\boxed{25}$ Note that $1 - \cos n$ is always nonnegative and less than or equal to 2. Using a basic

comparison test, $\frac{1 - \cos n}{n^2} \leq \frac{2}{n^2}$ and $2 \sum_{n=1}^{\infty} \frac{1}{n^2}$ converges by (11.25) $\Rightarrow$ S is C.

$\boxed{27}$ $\lim_{n \to \infty} \sqrt[n]{a_n} = \lim_{n \to \infty} \sqrt[n]{\frac{(2n)^n}{n^{2n}}} = \lim_{n \to \infty} \frac{2n}{n^2} = 0 < 1 \Rightarrow$ S is C by (11.29).

$\boxed{31}$ $a_n = \frac{\sqrt{\ln n}}{n} > 0$ for $n > 1$. $f(x) = \frac{\sqrt{\ln x}}{x} \Rightarrow f'(x) = \frac{1 - 2\ln x}{2x^2\sqrt{\ln x}} < 0$ for

$x > e^{1/2} \; \{ \approx 1.65 \}$ and $\lim_{n \to \infty} a_n = 0 \Rightarrow$ S converges by AST.

However, $\frac{\sqrt{\ln n}}{n} > \frac{1}{n}$ for $n > e$ and $\sum_{n=2}^{\infty} \frac{1}{n}$ diverges. Thus, S is CC.

Note: In Exer. 33–38, each $f(x)$ is positive, continuous, and decreasing on the interval
of integration.

$\boxed{33}$ Let $f(x) = \frac{1}{(3x+2)^3}$.

$$\int_1^\infty f(x)\, dx = \lim_{t \to \infty} \int_1^t (3x+2)^{-3}\, dx = \lim_{t \to \infty} \left[-\frac{1}{6(3x+2)^2} \right]_1^t = 0 - \left(-\frac{1}{150} \right); \; C$$

$\boxed{37}$ Let $f(x) = \frac{10}{\sqrt[3]{x+8}}$.

$$\int_1^\infty f(x)\, dx = \lim_{t \to \infty} \int_1^t 10(x+8)^{-1/3}\, dx = \lim_{t \to \infty} \left[15(x+8)^{2/3} \right]_1^t = \infty; \; D$$

[39] This series satisfies (11.30) and therefore, we will apply (11.31). We need to find an

n such that $a_n < 0.5 \times 10^{-3}$. $a_n = \dfrac{1}{(2n+1)!} < 0.0005 \Rightarrow (2n+1)! > 2000 \Rightarrow$

$2n + 1 \geq 7$ { since $6! = 720$ and $7! = 5040$ } $\Rightarrow n \geq 3$.

$$\text{Thus, } S \approx S_2 = \frac{1}{3!} - \frac{1}{5!} \approx 0.158.$$

Note: In Exer. 41–46, let u_n denote the nth term of the power series.

[41] We will apply (11.35) to the given power series. $u_n = \dfrac{(n+1)x^n}{(-3)^n} \Rightarrow$

$$\lim_{n \to \infty} \left| \frac{u_{n+1}}{u_n} \right| = \lim_{n \to \infty} \left| \frac{(n+2)x^{n+1}}{(-3)^{n+1}} \cdot \frac{(-3)^n}{(n+1)x^n} \right|$$

$$= \lim_{n \to \infty} \left| \frac{n+2}{n+1} \cdot \frac{(-3)^n}{(-3)^{n+1}} \cdot \frac{x^{n+1}}{x^n} \right| = \tfrac{1}{3}|x|. \quad \tfrac{1}{3}|x| < 1 \Leftrightarrow -3 < x < 3.$$

If $x = 3$, $\sum\limits_{n=0}^{\infty} (-1)^n (n+1)$ is D. If $x = -3$, $\sum\limits_{n=0}^{\infty} (n+1)$ is D. $\qquad \bigstar \ (-3, 3)$

[45] $\lim\limits_{n \to \infty} \left| \dfrac{u_{n+1}}{u_n} \right| = \lim\limits_{n \to \infty} \left| \dfrac{(2n+2)! \, x^{n+1}}{(n+1)!(n+1)!} \cdot \dfrac{n! \, n!}{(2n)! \, x^n} \right|$

$$= \lim_{n \to \infty} \left| \frac{(2n+2)(2n+1)(2n)!}{(2n)!} \cdot \frac{n! \, n!}{(n+1) \, n!(n+1) \, n!} \cdot \frac{x^{n+1}}{x^n} \right|$$

$$= \lim_{n \to \infty} \frac{(2n+2)(2n+1)}{(n+1)(n+1)} |x| = 4|x|. \quad 4|x| < 1 \Leftrightarrow |x| < \tfrac{1}{4}. \qquad \bigstar \ r = \tfrac{1}{4}.$$

[47] If $x \neq 0$, using (11.48)(b), we have

$$\frac{1 - \cos x}{x} = \frac{1 - \left(1 - \dfrac{x^2}{2!} + \dfrac{x^4}{4!} - \dfrac{x^6}{6!} + \cdots \right)}{x}$$

$$= \left(\frac{1}{x} \right) \left(\frac{x^2}{2!} - \frac{x^4}{4!} + \frac{x^6}{6!} - \cdots \right)$$

$$= \frac{x}{2!} - \frac{x^3}{4!} + \frac{x^5}{6!} - \cdots$$

$$= \sum_{n=1}^{\infty} (-1)^{n+1} \frac{x^{2n-1}}{(2n)!}. \quad \text{At } x = 0, \text{ the series equals 0.}$$

Since (11.48)(b) converges for every x, it follows that $r = \infty$ for this series too.

[49] $\sin x \cos x = \tfrac{1}{2}(2 \sin x \cos x) = \tfrac{1}{2} \sin 2x.$

By (11.48)(a), $\tfrac{1}{2} \sin 2x = \tfrac{1}{2} \sum\limits_{n=0}^{\infty} (-1)^n \dfrac{(2x)^{2n+1}}{(2n+1)!} = \sum\limits_{n=0}^{\infty} (-1)^n \dfrac{2^{2n} x^{2n+1}}{(2n+1)!}. \quad r = \infty.$

$\boxed{55}$ $\sqrt{x} = \sqrt{4 + (x - 4)} = \sqrt{4\left[1 + \frac{1}{4}(x - 1)\right]} = 2\left[1 + \frac{1}{4}(x - 4)\right]^{1/2}$.

Using the binomial series with $k = \frac{1}{2}$ and substituting $\frac{1}{4}(x - 4)$ for x gives us

$$\sqrt{x} = 2\left[1 + \frac{1}{8}(x - 4) + \frac{(\frac{1}{2})(-\frac{1}{2})}{2!}\left[\frac{1}{4}(x - 4)\right]^2 + \frac{(\frac{1}{2})(-\frac{1}{2})(-\frac{3}{2})}{3!}\left[\frac{1}{4}(x - 4)\right]^3 + \cdots\right]$$

$$= 2 + \frac{1}{4}(x - 4) + 2\sum_{n=2}^{\infty}(-1)^{n-1}\frac{1 \cdot 3 \cdot 5 \cdot \cdots \cdot (2n - 3)}{2^n\, n!\, 4^n}(x - 4)^n$$

$$= 2 + \frac{1}{4}(x - 4) + \sum_{n=2}^{\infty}(-1)^{n-1}\frac{1 \cdot 3 \cdot 5 \cdot \cdots \cdot (2n - 3)}{2^{3n-1}\, n!}(x - 4)^n.$$

$$\{2^n 4^n = 2^n \cdot 2^{2n} = 2^{3n}\}$$

$\boxed{57}$ Substituting $(-x^2)$ for x in (11.41), we find that

$$\int_0^1 x^2\, e^{-x^2}\, dx = \int_0^1 x^2\left[1 - x^2 + \frac{x^4}{2!} - \frac{x^6}{3!} + \frac{x^8}{4!} - \frac{x^{10}}{5!} + \frac{x^{12}}{6!} - \cdots\right] dx$$

$$= \int_0^1\left[x^2 - x^4 + \frac{x^6}{2!} - \frac{x^8}{3!} + \frac{x^{10}}{4!} - \frac{x^{12}}{5!} + \frac{x^{14}}{6!} - \cdots\right] dx$$

$$= \left[\frac{1}{3}x^3 - \frac{1}{5}x^5 + \frac{x^7}{7(2!)} - \frac{x^9}{9(3!)} + \frac{x^{11}}{11(4!)} - \frac{x^{13}}{13(5!)} + \frac{x^{15}}{15(6!)} - \cdots\right]_0^1.$$

Using the first 6 terms, I $\approx \frac{1}{3} - \frac{1}{5} + \frac{1}{14} - \frac{1}{54} + \frac{1}{264} - \frac{1}{1560} \approx 0.189$.

Since the series is alternating in sign and satisfies (11.30), (11.31) applies.

Hence, with $|\text{error}| \leq \dfrac{1}{15(6!)} = \dfrac{1}{10{,}800} \approx 9.26 \times 10^{-5} < 0.5 \times 10^{-3}$.

$\boxed{61}$ *DERIV:* $\ln\cos x$, $\ln(\frac{\sqrt{3}}{2})\ddagger$ $-\tan x$, $-\frac{\sqrt{3}}{3}\ddagger$ $-\sec^2 x$, $-\frac{4}{3}\ddagger$ $-2\sec^2 x\tan x$, $-\frac{8\sqrt{3}}{9}\ddagger$

$$-2\sec^4 z - 4\sec^2 z\tan^2 z.$$

$$\underline{\ln\cos x} = f(\tfrac{\pi}{6}) + f'(\tfrac{\pi}{6})(x - \tfrac{\pi}{6}) + \frac{f''(\tfrac{\pi}{6})}{2!}(x - \tfrac{\pi}{6})^2 + \frac{f'''(\tfrac{\pi}{6})}{3!}(x - \tfrac{\pi}{6})^3 + \frac{f^{(4)}(z)}{4!}(x - \tfrac{\pi}{6})^4$$

$$= \ln(\tfrac{\sqrt{3}}{2}) - \frac{\sqrt{3}}{3}(x - \tfrac{\pi}{6}) - \frac{2}{3}(x - \tfrac{\pi}{6})^2 - \frac{4\sqrt{3}}{27}(x - \tfrac{\pi}{6})^3$$

$$- \frac{1}{12}(\sec^4 z + 2\sec^2 z\tan^2 z)(x - \tfrac{\pi}{6})^4,\ z \text{ is between } x \text{ and } \tfrac{\pi}{6}.$$

$\boxed{63}$ *DERIV:* e^{-x^2}, $1\ddagger$ $-2xe^{-x^2}$, $0\ddagger$ $(4x^2 - 2)e^{-x^2}$, $-2\ddagger$ $(-8x^3 + 12x)e^{-x^2}$, $0\ddagger$

$$(16z^4 - 48z^2 + 12)e^{-z^2}.$$

$$\underline{e^{-x^2}} = f(0) + f'(0)x + \frac{f''(0)}{2!}x^2 + \frac{f'''(0)}{3!}x^3 + \frac{f^{(4)}(z)}{4!}x^4$$

$$= 1 - x^2 + \frac{1}{6}(4z^4 - 12z^2 + 3)e^{-z^2}x^4,\ z \text{ is between } x \text{ and } 0.$$

$\boxed{65}$ Let $f(x) = \cos x$ and $c = \frac{\pi}{4}$. *DERIV*: $\cos x$, $\frac{1}{\sqrt{2}}\ddagger$ $-\sin x$, $-\frac{1}{\sqrt{2}}\ddagger$ $-\cos x$, $-\frac{1}{\sqrt{2}}\ddagger$ $\sin z$.

$\underline{\cos x} = \frac{1}{\sqrt{2}} - \frac{1}{\sqrt{2}}(x - \frac{\pi}{4}) - \frac{1}{2\sqrt{2}}(x - \frac{\pi}{4})^2 + \frac{\sin z}{6}(x - \frac{\pi}{4})^3$, z is between x and $\frac{\pi}{4}$.

Since $43° = 45° - 2° = \frac{\pi}{4} - \frac{\pi}{90}$, let $x = \frac{\pi}{4} - \frac{\pi}{90}$ and then $(x - \frac{\pi}{4}) = -\frac{\pi}{90}$.

Thus, $\cos x = \frac{1}{\sqrt{2}} - \frac{1}{\sqrt{2}}(-\frac{\pi}{90}) - \frac{1}{2\sqrt{2}}(-\frac{\pi}{90})^2 + \frac{\sin z}{6}(-\frac{\pi}{90})^3$.

Using the first three terms, $\cos 43° \approx 0.7314$,

$$\text{with |error|} \leq \left| R_3(x) \right| = \left| \frac{\sin z}{6}(-\frac{\pi}{90})^3 \right| < \frac{\pi^3}{6 \cdot 90^3} \approx 7.09 \times 10^{-6} < 0.5 \times 10^{-4}.$$